Integrating Food Science and Engineering Knowledge Into the Food Chain

Volume 16

The ISEKI-Food book series is a collection where various aspects of food safety and environmental issues are introduced and reviewed by scientists specializing in the field. In all of the books a special emphasis is placed on including case studies applicable to each specific topic. The books are intended for graduate students and senior level undergraduate students as well as professionals and researchers interested in food safety and environmental issues applicable to food safety.

"ISEKI-Food" is an acronym for "Integrating Safety and Environmental Knowledge Into Food Studies". Participants in the ISEKI-Food network, coordinated by Professor Cristina Silva at The Catholic University of Portugal, come from 29 countries in Europe and most of the institutes and universities involved with Food Science education at the university level in Europe are represented. Some international companies and non teaching institutions have also participated in the network. The main objectives of ISEKI-Food are to improve the harmonization of studies in food science and engineering in Europe and to develop and adapt food science curricula emphasizing the inclusion of safety and environmental topics.

Gerhard Schleining • Saverio Mannino
Prisana Suwannaporn
Editors

Statistics in Food and Biotechnology

From Theory to Practical Applications

Springer

ISEKI - Food Association

Editors
Gerhard Schleining
BOKU – Department of Food Science and Technology
Vienna, Austria

Saverio Mannino
Department of Food Science and Technologies
University of Milan
Milan, Italy

Prisana Suwannaporn
Faculty of Medical and Health Food Technology, Chulabhorn Royal Academy
Bangkok, Thailand

Bioscience, Faculty of Science
Kasetsart University
Bangkok, Thailand

ISSN 2512-2223 ISSN 2512-2258 (electronic)
Integrating Food Science and Engineering Knowledge Into the Food Chain
ISBN 978-3-031-51567-5 ISBN 978-3-031-51568-2 (eBook)
https://doi.org/10.1007/978-3-031-51568-2

This Springer imprint is published by the registered company Springer Nature Switzerland AG
The registered company address is: Gewerbestrasse 11, 6330 Cham, Switzerland

Paper in this product is recyclable.

Preface

The proper use of statistical methods is essential in food processing and biotechnology, and very few publications have focused on the practical application of these methods so far. This book utilizes case studies to explain the corresponding basics with a focus on understanding how and when to use the appropriate statistical methods. The text is structured according to practical applications, starting from data collection and data treatment, moving to product development, process control and analytical methods.

The book *Practical Examples of Using Statistics in Food and Biotechnology* is structured in six chapters beginning with a detailed introduction to statistics and their use in food processing.

The first main section focuses on statistics for data collection and treatment featuring scales of measurement and presenting data. Another important section looks at statistics for food product and process development, examining mixture experiments for different food groups from meat and fish to vegetables, juices, rice and olive oils. Further sections focus on statistical process control, statistics for sensory science and a wide range of analytical methods including LC-MS/MS and other cutting-edge methods.

Chapter 2 is completely devoted to sampling, whose importance is often neglected. The chapter covers important aspects of sampling from data collection and treatment to the development of a sampling plan for attribute and variable data.

Chapters have been designated to stand alone, but many cross references can be very useful to solve practical problems. Furthermore, it is assumed that readers will rely upon commercial software to carry out statistical tests.

In providing practical, hands-on information on the use of statistics in food processing, this work will be of great importance for professionals, industry staff and researchers in food and biotechnology.

Vienna, Austria — Gerhard Schleining
Milano, Italy — Saverio Mannino
Bangkok, Thailand — Prisana Suwannaporn

Contents

About the Editors

Gerhard Schleining graduated in Food and Biotechnology and is retired after more than 35 years of experience in research and teaching in food physics and food quality and food safety management at the University of Natural Resources and Life Sciences, Vienna (BOKU). He was responsible for education and coordinated several years an international Master programme "Safety in the food chain". Besides ISEKI, where he was Secretary General for 15 years, he is active in several international organisations like Global Harmonisation Initiative (GHI), chair of regional section Austria and certified trainer of the European Hygienic Equipment and Design Group (EHEDG) and was President of the International Association for Cereal Science and Technology (ICC) from 2022–2023. He has many years experiences in EU-funded projects related to innovative teaching materials, methods and tools, and he coordinated two Erasmus+ projects, FooD-STA a Food Studies and Training Alliance between the food industry and universities in Europe and SEA-ABT, a capacity building project to establish an Academy of Beverage Technology in South East Asia.

Saverio Mannino is retired as Full Professor in Food Quality Control at the Faculty of Agriculture of the University of Milan. Actually, he is an international consultant in food quality control and laboratory accreditation. His research interest and experience were centred mainly in the field of food quality control. In particular, he was interested in the development of methods for food analysis based on different analytical techniques with emphasis in the area of microelectrodes, electrochemical sensors and biosensors. He is the author/co-author of more than 200 publications and has edited three books in Food Analysis and Analytical Techniques applied to food analysis. He is currently member of the board of "Accredia", the Italian Institution for Laboratory Accreditation. He also participated, as expert in international programmes for emerging countries and in EC Commission for Methods in Food Analysis.

Prisana Suwannaporn is Associate Professor in the Department of Food Science and Technology, Kasetsart University, Thailand, and Liaison Professor at Niigata University, Japan. Her research interests are starch and cereal technology (focusing on rice and rice products physico-chemical and rheology properties), carbohydrate health food, baking technology, consumer and new food products studies, and statistical research techniques. Her current research includes physically modified starch and its applications, bioactivities and rheology of young rice grain, food mimicry, plant-based protein and cereal grain digestibility.

Chapter 1
Basic Statistics for Data Collection and Treatment

Victoria A. Jideani, Afam I. O. Jideani, Ivanka Zheleva, Todor Todorov, Hüseyin Bozkurt, Fahrettin Göğüş, Gerhard Schleining, Zsuzsa H. Horváth, Cecília Hodúr, Prisana Suwannaporn, Teresa Brandão, Mafalda Quintas, Cristina Silva, and Edward Muntean

1.1 Scales of Measurement and Types of Data

Victoria A. Jideani and Afam I. O. Jideani

1.1.1 Objectives and Learning Outcomes

After completion of this chapter the reader should be able to classify data according to the types of measurement scale.

V. A. Jideani (✉)
Department of Food Science and Technology, Cape Peninsula University of Technology, Cape Town, South Africa

A. I. O. Jideani · H. Bozkurt · F. Göğüş
Present Address: Department of Food Science and Technology, University of Venda, Thohoyandou, Limpopo Province, South Africa

I. Zheleva · T. Todorov
Rousse University, Rousse, Bulgaria

G. Schleining
BOKU – Department of Food Science and Technology, Vienna, Austria

Z. H. Horváth · C. Hodúr
Faculty of Engineering, University of Szeged, Szeged, Hungary
e-mail: hodur@bibl.szef.u-szeged.hu

G. Schleining et al. (eds.), *Statistics in Food and Biotechnology*, Integrating Food Science and Engineering Knowledge Into the Food Chain 16,
https://doi.org/10.1007/978-3-031-51568-2_1

1.1.2 Introduction

Statistics is the practice or science of collecting and analyzing numerical data to gain information about the population from which it was drawn. In pursuits of this aim, statistics divides the study of data into three parts; (1) Collecting data; (2) Organising, summarizing and presenting of data (Descriptive Statistics); (3) Drawing conclusions from data (Statistical inference).

Collection of data is a fundamental step in the conduct of research. For data to be collected there must be a form of measurement: the process through which observations are translated into numbers. In its broadest sense, measurement is the assignment of numerals to objects or events according to rules. Researchers begin with variables, then use rules to determine how these variables will be expressed in numerical form. The variable texture of bread crumb may be measured according to the numbers indicated by sensory panellists who are asked to select one among (1) gumminess (2) cohesiveness (3) brittleness (4) firmness. The variable weight may be measured as the numbers observed when food materials are weighed with a weighing balance.

Data may be thought of as the results of observation that one uses to reach conclusions or generate interpretations. If data is to be treated statistically, the observations must be expressed in numerical form. That is they must be quantitative. Consequently, statistical data always consists of a collection of numbers. When the phenomena that a researcher observes are expressed as or translated into numerical values these data are called measurements. The nature of the measurement process that produces the numbers determines the interpretation that can be made from them and the statistical procedures that can be meaningfully used with them. What can be done with data depends on how much information the numbers carry. We speak of the kind of information a measurement carries by saying what kind of scale the measurement is made in. A basic understanding about the data types is helpful for choosing statistical procedures. We begin this chapter with a discussion of various scales of measurement. For the information to be useful for decision making, it must

P. Suwannaporn
Faculty of Medical and Health Food Technology, Chulabhorn Royal Academy, Bangkok, Thailand

Bioscience, Faculty of Science, Kasetsart University, Bangkok, Thailand
e-mail: prisana.s@ku.th

T. Brandão · M. Quintas · C. Silva
Universidade Católica Portuguesa – Escola Superior de Biotecnologia, Porto, Portugal
e-mail: tbrandao@ucp.pt

E. Muntean
University of Agricultural Sciences and Veterinary Medicine Cluj Napoca, Cluj-Napoca, Romania

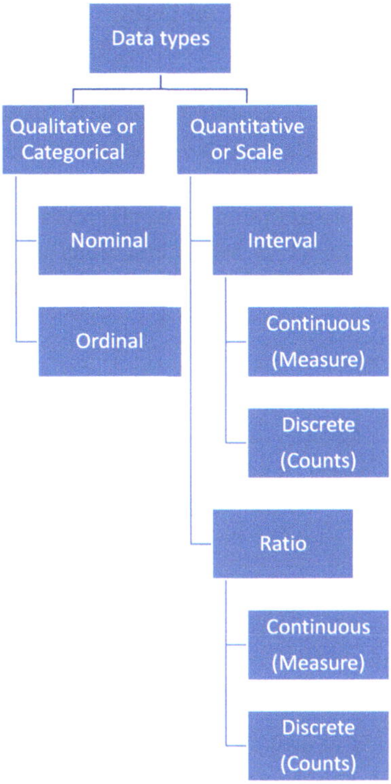

Fig. 1.1 Data types or scales of measurement

be first organised and displayed properly. Hence, we will discuss some methods of data summary and presentation.

Often a researcher will want to obtain or convey information about particular characteristics of data. In this chapter we will introduce several numerical summary measures that describe where a sample or distribution is centred. Another important aspect of a sample or distribution is the extent of spread about the centre. We will develop the most useful measures of variability and how they can be combined to yield concise yet informative data descriptions.

Generally two major types of data or scales of measurements can be identified (Fig. 1.1).

1.1.3 Qualitative or Categorical Data

Qualitative or categorical variables are data with a limited number of distinct values or categories (for example, gender or marital status). Categorical variables can be string (alphanumeric) data or numeric variables that use numeric codes to represent categories (for example, 0 = Unmarried and 1 = Married). There are two basic types of categorical data, Nominal and Ordinal data.

1.1.3.1 Nominal Data

A measurement of a property has a nominal scale if the measurement tells only what class a variable falls in with respect to the property. Measurements in a nominal scale, place variables in categories, nothing more. In other words, nominal data are categorical data where there is no inherent order to the categories. The required empirical operation at this level of measurement involves only recognizing whether an object or individual does or does not belong in a given category. The only relationship between categories is that they are different from each other; there is no suggestion that they represent "more" or "less" of a characteristic.

Example 1.1
Classifying panellists according to gender would constitute nominal measurement. Such properties as race, sex, religion, blood type and employment status are measured in a nominal scale.

Numbers may be used at the nominal level, but only when arbitrarily assigned to serve as labels to identify the categories. The numbers used in nominal measurements has no mathematical relationship.

Example 1.2
The numeral 0 might be used to represents males and the numeral 1 to represent females- but the 1 does not indicate more of something than the 0, and numbers could be interchanged without affecting anything but the labeling scheme used.

Note:

The data values are non-numeric group labels. The numbers at the nominal level do not represent either absolute or relative or relative amounts of any characteristic, but merely to identify members of a given category.

The identifying numbers in a nominal scale cannot, be arithmetically manipulated through addition, subtraction, multiplication, or division. One may use only those statistical procedures based on mere counting, such as reporting the numbers of observations in each category or expressing those numbers as percentages of the total number of subjects (Table 1.41).

1.1.3.2 Ordinal Data

Categorical data where there is a meaningful order of categories, but there isn't a measurable distance between categories. The measurement has an ordinal scale if it also tells when one unit has more of the property than does another unit. In an ordinal scale, the order of numbers is meaningful.

Example 1.3
Ordinal data lists include

1. Sensory panellists rank three bread products from 1 (weakest) to 10 (strongest), the sample ranked 8 is better than the sample ranked 6.
2. Lecturer ratings: Poor, fair, good and superior.
3. Academic ranks: Technician, lecturer, senior lecturer, associate professor and professor. A professor has higher academic rank than a lecturer.
4. Grades of school: First, second, third and so on.

The essential requirement for measurement at this level is that the relationship must be such that if object a is greater than object b, and object b is greater than object c, then object a is greater than object c, i.e. if $(a > b)$ and $(b > c)$, then $(a > c)$. When appropriate, other wording may be substituted for greater than- stronger than, precedes, has more of, and so on.

The empirical operation in ordinal measurement involves direct comparison of objects or individuals solely in terms of the extent to which they possess the attribute in question. Thus when numbers are assigned in ordinal measurement they indicate only the order of the observations and nothing more. Neither the difference between the numbers nor their ratio has meaning.

The statistics appropriate for an ordinal scale are limited. Since the size of the interval between the categories is unknown, one cannot appropriately use any statistical procedure that assumes equal intervals. Statistics that indicate the points below which certain percentages of the cases fall are appropriate with an ordinal scale.

1.1.4 Quantitative or Scale Variables

The data values are counts or numerical measurements. Data measured on an interval or ratio scale, where the data values indicate both the order of values and the distance between values. For example, a salary of $72,195 is higher than a salary of $52,398, and the distance between the two values is $19,797. These data types are also referred to as quantitative or continuous data.

1.1.4.1 Interval Data

A more precise measurement system is the interval scale, which not only orders objects or events according to the amount of the attribute they represent but also has an arbitrary origin and establishes equal intervals between the units of measure. Equal differences in the numbers represent equal differences in the attribute being measured. On an interval scale, both the order and distance relationships among the numbers have meaning. However, the ratio of two interval data is not meaningful.

Example 1.4
Interval data lists include

1. The fahrenheit and centigrade thermometers are examples of interval scales. We may say that the difference between 48 and 49 °C is equal to the difference between 20 and 21 °C. We could not say however, that 40° is twice as hot as 20°. This is because there is no true zero point on an interval scale. A zero point is established by convention, as in the centigrade scale, which assigns the value 0° to the freezing point of water.
2. A student may occasionally receive a score of 0 on a food rheology test, but this does not mean that he has zero knowledge of food rheology. That is on interval scale, data values are ranged in a real interval, which can be as large as from negative infinity to positive infinity.
3. A test score of 90 is five points higher than a test score of 85 (ordinal data). Not only can we say that a test score of 90 is higher than that of 85, we can also say that it is five points higher. But we cannot say that a person with a test score of 100 is twice as smart as a person with a test score of 50.

Numbers on an interval scale may be manipulated by addition and subtraction, but because the zero is arbitrary, multiplication and division of the numbers are not appropriate. As we have seen, ratios between the numbers on an interval scale are meaningless. However, the difference between positions on an interval scale may be reported or the numbers may be added. Any statistical procedures based on addition may be used with this scale level along with the procedures appropriate for the lower level scales. These include most of the common statistical procedures.

1.1.4.2 Ratio Data

The ratio scale is the highest level of measurement scale. It is one that has a true zero point as well as equal intervals. Both difference and ratio of two values are meaningful. The measurement has a ratio scale if it tells us that one unit has so many times as much of the property as does another unit.

Example 1.5
Ratio scales include

1. Scales commonly used to measure units such as kilogram, centimetres, etc. A ruler used to measure the length in units of millimetre or centimetre is a ratio scale, because the origin on the scale is an absolute zero corresponding to not length at all. Thus it is possible to state that a stick 10 cm long is twice as long as a stick 5 cm long.

(continued)

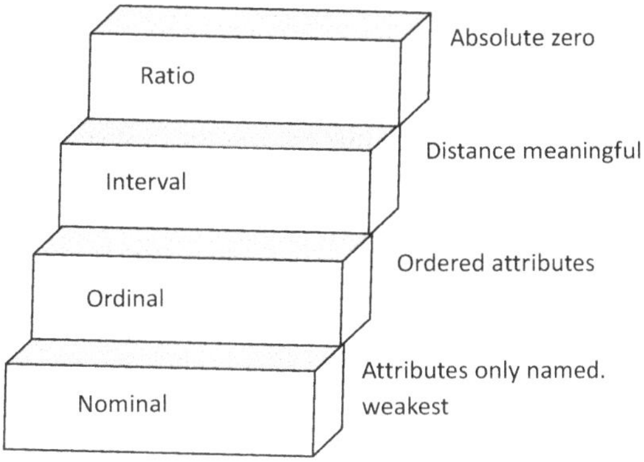

Fig. 1.2 Hierarchy of the measurement scales

Example 1.5 (continued)

2. Results of counting objects. Ten oranges are twice as many as five oranges.
3. A 60-kg person will always weight twice as much as a 30-kg person.
4. Other examples of ratio data are salary, volume, etc.

With a ratio scale, it is possible to multiply or divide each of the values by a certain number without changing the properties of the scale. For example, we can multiply 2 kg by 1000 to change the unit of measurement to gram, or we can multiply 20 metres by 100 to change the unit to centimetres. We can multiply and maintain the same ratio as before the multiplication. All types of statistical procedures are appropriate with ratio scale.

The complexity of the scales of measurements is depicted in Fig. 1.2 The most primitive scale of measurement is the nominal scale which involves placing of objects or individuals into categories that are qualitatively rather than quantitatively different. Next on the hierarchy of the measurement scales is the ordinal scale that ranks objects or individuals with respect to how much or how little of the attribute under consideration they possess. A more precise measurement is the interval scale and the highest level of measurement is the ratio scale.

1.1.4.3 Continuous and Discrete Data

Quantitative data can be classified as discrete or continuous. A variable is continuous if its possible values consist of an entire interval on the number line. Continuous data arise from making measurements. For example if the pH of a food substance is measured the pH could be any number between 0 and 14.

Example 1.6
Continuous data cannot be counted. Weight (in kilograms), height (in centimetres), time (in minutes), distance (in kilometres) are all examples of continuous data.

Data resulting from a process that can be counted are **discrete data**. Discrete variables almost always result from counting, in which case possible values are 0, 1, 2, 3, . . . or some subset of these integers.

Example 1.7
Discrete data include

Number of bacteria or moulds growing on a petri dish; number of students in your class, number of cars in the parking lots, number of people in a checkout line, number of oranges in your refrigerator.

Ratio-, interval- and ordinal-scale data may be either continuous or discrete. Nominal scale data by their nature are discrete.

1.1.5 Limits of Numbers

Most variables are given discrete values because of certain problems in measuring on continuous scales. Account is taken of the mathematical true limit of a number in order to interpret such values. Generally, true limits of a number are said to extend one half unit below and one half unit above that number and any score within those limits is rounded off to the number.

Example 1.8
If after a performance test it was found that ten students scored 75, does this really mean that they are all equal in performance ability? The likelihood of them all being exactly alike in ability is extremely small. It is more likely that some scored just below 90 and some scored just above 75, but the measuring instrument was not sophisticated enough to detect the differences.

To overcome this problem, the number 75 is said to have limits ranging from 74.5 to 75.5 and any score between these limits are given to score 75.

(continued)

Example 1.8 (continued)

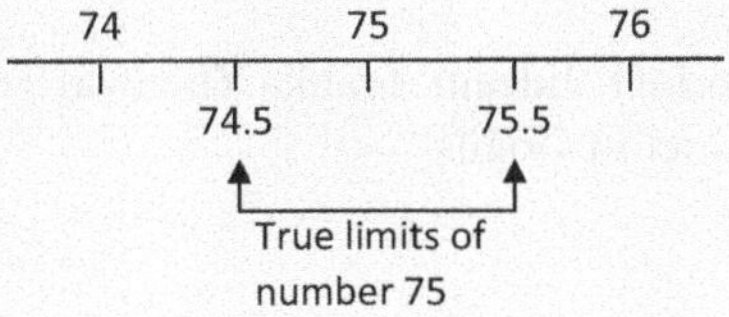

Limits of score 35.5

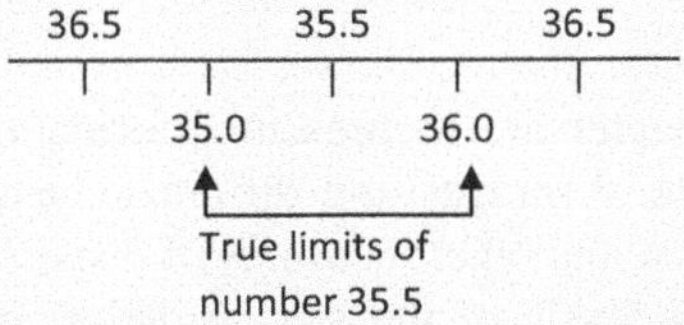

The 'true limit' concept is particularly useful when classifying data into groups or intervals.

Note: It is not just whole numbers that have limits. Any number has limits.

1.1.6 Exercises

1.1.6.1 Exercise 1: Classification of Data

The table below contains the distribution panellists ranking of a food product on a 5-point hedonic scale.

(a) Classify the data in each of the three columns as quantitative or qualitative.
(b) Classify the data in the third column as discrete or continuous.
(c) Classify the data in each of the three columns as nominal, ordinal or ratio.

Panellist	Gender	Overall acceptability
Adamu	Male	5
Roseline	Female	4
John	Male	3
Agnes	Female	5
Peter	Male	3

1.2 Summarizing and Presenting Data

Victoria A. Jideani, Afam I. O. Jideani, Ivanka Zheleva, Todor Todorov, Hüseyin Bozkurt, and Fahrettin Göğüş

1.2.1 Objectives and Learning Outcomes

The objectives of this chapter are to present information for the analysing and representing statistical data. Averages and different measures of central tendency as well as standard deviation and other measures of dispersion are presented in order to obtain reliable initial analysis of the raw statistical data. On the base of this analysis correct conclusions about the processes under consideration or further its investigations could be made.

After completion of this chapter the reader should be able to:

- Organise and summarise a given set of data using tables.
- Organise and display a given set of data using various types of graphs.
- Calculate the measures of central tendency for grouped and ungrouped data.
- Calculate the percentiles for grouped and ungrouped data.
- Calculate the measures of dispersion for grouped and ungrouped data.
- Identify outliers in a set of data.

1.2.2 Introduction

Data collected by researchers are often expressed as or translated into numerical values. These data are called measurements. Weight expressed in kilograms is a measurement of body mass. Height expressed in metres is a measurement of linear body size. When events are expressed as numbers, the term observations are understood to mean numerical observations, i.e., measurements. However, some observations cannot be translated into numbers. Such observations are qualitative or categorical, rather than quantitative. Whether the events one observes are quantitative (expressed as numbers) or qualitative, one can record the number of times each event occurs. Data of this type are called frequencies.

Example 1.9
Examples of frequencies would include the number of students who obtained a particular score in an examination, the number of lecturers who participated in a sensory evaluation session, etc.

A small-scaled research may involve dozens of observations, and studies that yield several hundred observations are uncommon. Before interpretation can be made on observations, the data must be organized in some form so that they are reduced to manageable proportions. The organization of data in such ways also helps to recognize and characterize the pattern inherent in the data. The most common organization schemes are tabular and graphic representations.

1.2.3 *Tabular Representation of Data*

Data that are listed as observed and not organized in some fashion are referred to as raw data. Raw data can be arranged by constructing an ordered array, i.e. arranging in increasing or decreasing order. If the number of data is large, the data array may be difficult to manage or comprehend. Hence, tables are used as a general approach to organizing raw data. The type of data- nominal, ordinal, interval or ratio will determine the type of display.

Data may be listed by individual value or frequency. The frequency of a measurement or category is the number of times the measurement or category occurs in a collection of data. The use of frequencies is suited for qualitative or discrete data. The symbol f is used to denote the frequency of a measurement.

Example 1.10
The ages of the 10 panellists involved in sensory evaluation of a beverage are listed below:

$$23, 20, 21, 21, 30, 25, 23, 35, 24, 21$$

Age 21 occurs with a frequency of $f = 3$, 23 occurs with a frequency of $f = 2$, and 20, 24, 25, 30, 35 each occur with frequency of $f = 1$.

Two general types of tables are employed for reporting data frequencies: ungrouped frequency tables and grouped frequency tables. Both are referred to as frequency tables. Such a table involves a listing of all the observed values of the variable and how many times each value is observed. Tabular representations include frequency tables, relative frequency tables, cumulative frequency tables and cumulative relative frequency tables.

1.2.3.1 Ungrouped Frequency Tables

The ages of the sensory panellists cited above can be summarized as shown in Table 1.1, where x denotes the panellist age and f denotes the frequency of each age. Table 1.1 is an example of an ungrouped frequency table for discrete data.

Table 1.1 Frequency table of panellists age

Panellist age (x)	Frequency (f)
20	1
21	3
23	2
24	1
25	1
30	1
35	1

1.2.3.2 Grouped Frequency Tables

Frequencies tables, such as the one shown in Table 1.1, are correctly termed ungrouped frequency tables because each measurement has its own corresponding frequency. Often the ungrouped frequency data does not give a good idea of how the data is distributed because most observed values are different, so the frequency is low. To improve the efficiency of studying variables, the data may be grouped. Grouped frequency table shows frequency according to groups or classes of measurements. Grouped frequency can also be used when the raw data is large. Grouped frequency tables are typically used for summarizing large amounts of continuous data, containing relatively few repetitions. Such data summaries facilitate certain statistical calculations and graphic displays when the computer is not used. In order to use a grouped frequency table the data must be measured by at least an interval scale. Ungrouped frequency tables may be used for large amounts of data not measured by at least an interval scale.

Example 1.11

Suppose we were to count the number of seeds in 20 oranges and group the oranges according to the number of seeds. We may obtain Table 1.2.

Table 1.2 Grouped frequency table for seeds of 20 oranges

Class	Frequency (f)
3–6	4
7–10	6
11–14	7
15–18	3

The characteristics of classes are class limit, class interval, and class boundary. The classes in grouped frequency have **class limits**. For the class 3–6, 3 is the lower class limit and 6 the upper class limit. There are 4 oranges that have between 3 and 6 seeds, inclusive; 6 oranges have between 7 and 10 seeds, inclusive, and so on. The distance between any two consecutive upper limits or any two consecutive lower limits is called **class width, class size or class strength**. In Table 1.2 the width of each class is 4, i.e., the distance between the upper limit of the first class and the upper limit of the

(continued)

Example 1.11 (continued)
second class is 10–6 = 4. Each class in a frequency table has theoretical class limits (true limits of the numbers as discussed earlier) called **class boundaries.** An upper theoretical class limit is called the upper boundary, and a lower theoretical class limit is called the lower boundary. The upper class boundary for each class is found by adding 0.5 to the upper class limit, and the lower class boundary for each class is found by subtracting 0.5 from the lower class limit. The lower boundary for the first class in this example is 2.5 and the upper boundary for the first class is 6.5.

Every grouped frequency table should possess the following three characteristics:

1. Uniformity: Each class should have the same width.
2. Uniqueness: No two classes should overlap.
3. Completeness: Each piece of data should belong to some class.

Class boundaries and class widths for a grouped frequency table are determined by considering the unit of precision of measurement. The precision of measurement in Table 1.2 is the nearest whole number (since we are counting seeds) so the unit of measurement is 1. The lower class boundary of an interval is located one-half unit below the lower class limit, and the upper class boundary of an interval is located one-half unit above the upper class limit. Therefore, for the first class in Table 1.2, the lower class boundary is [3–0.5(1)] = 2.5 and the upper boundary is [6 + 0.5 (1)] = 6.5. None of the raw data can fall at the boundaries of an interval. Thus, the measurements 2.5 and 6.5 could not fall in the first class, but any measurement between 2.5 and 6.5 could.

The width w for any class in a grouped frequency table can be found by subtracting the lower class boundary from the upper class boundary. Thus for the first class in Table 1.2 the class width w = 6.5–2.5 = 4.

Example 1.12
The following is a grouped frequency table for the weights (in grams) of 18 oranges.

Class	Frequency (f)
13.0–14.4	1
14.5–15.9	1
16.0–17.4	7
17.5–18.9	8
19.0–20.4	1

The precision of measurement for the classes is 0.1 gram. For the class 13.0–14.4, subtracting one-half the unit from the lower class limit yields

(continued)

Example 1.12 (continued)
[13.0 – (0.5)(0.1)] = 12.95. The upper class boundary is obtained by adding one-half the unit to the upper class limit, [14.4 + (0.5)(0.1)] = 14.45.

Note

No weight can correspond to either boundary because the precision of measurement is to the nearest one-tenth of a gram.

1.2.3.2.1 Determination of Class Width

The class width of any grouped frequency table can be obtained by:

- Subtracting two consecutive upper class limits or two consecutive lower class limits, or
- Subtracting two consecutive lower boundaries or two consecutive upper boundaries, or
- Subtracting the lower boundary of a class from the upper boundary of the same class.

For Table 1.2 we can compute the class width as indicated in Table 1.3.

1.2.3.2.2 Determination of Classes for Grouped Frequency Tables

To construct a grouped frequency table for a given collection of data, three questions concerning the classes need to be answered.

- How many classes should be used?
- What should be the width of each class?
- At what value should the first class start?

Table 1.3 Computation of class width for Table 1.2.

Class		Frequency (*f*)
3-6	w = 10 – 6 = 4	4
7-10		6
11-14	w = 15 – 11 = 4	7
15-18		3

Note that the class width, is not found by subtracting the lower class limit from the upper class limit.

If the data are grouped into a small number of classes, characteristics of the original data are hidden and pertinent information may be lost. On the other hand, too many classes provide too much detail and defeat the purpose of grouping, which is to condense the data in a meaningful manner that is easy to interpret. In addition, too many classes may produce too many empty classes and render the grouping of data meaningless.

The number of classes denoted by *c,* depends on the amount of data involved. There is no general agreement among statisticians regarding the number of classes to use and the choice is arbitrary, between 5 and 20 is generally adequate for biological data. A useful suggestion for the number of classes is given by Sturges' rule, which states that the number of classes is approximately.

$$c = 3.3(\log\ n) + 1$$

where n is the number of measurements and log n is the logarithm of n to the base 10. The value of c is usually rounded to the nearest whole number.

Example 1.13
If n = 30 measurements were to be grouped. Sturges' rule would suggest using since

$$\begin{aligned} c &= 3.3(\log\ n) + 1 \\ &= 3.3(\log\ 30) + 1 \\ &= 3.3(1.4771) + 1 \cong 6 \end{aligned}$$

Once it has been agreed on the number of class intervals to use, the width of each class is found by using the range R, the difference between the largest measurement U and the smallest measurement L in the sample.

$$\text{Range}, \text{R} = \text{U} - \text{L}$$

Since the number of classes, c must cover the range, class width, w is obtained by dividing the range by the number of classes.

$$\text{Class width, w} = \frac{R}{c}$$

The value is determined by rounding the quotient R/c to the next highest integer value. Generally, the first class starts with the smallest measurement. However, in practice, the first class begins at a point that permits the classes to be expressed in convenient intervals.

1.2.3.2.3 Class Mark

As mentioned earlier, summarizing data into a grouped frequency table, results in some information loss. Nothing is known about the exact values of the measurements that lie in each class. An approximation of the exact values of the measurements is the class mark. Class mark is the midpoint of each class and is denoted by X. Using class marks, instead of the raw data results in a lack of accuracy. For a given class, the class mark is obtained using the formula

$$X = \frac{l_1 + l_2}{2}$$

where l_1 is the lower class limit and l_2 is the upper class limit.

Exercise 1.1
Problem 2.2

In an experiment the length of the seeds of a legume was measured in millimeter. The following data represent the measurement for 50 seeds. Construct a grouped frequency table for the length of the seeds.

16	15	21	75	19	31	40	60	77	26
14	22	42	30	12	29	47	72	12	42
44	64	77	60	12	76	72	37	13	70
70	67	26	43	17	23	72	42	37	67
17	13	41	75	60	19	47	18	21	15

Solution:
Step 1. We first determine the class interval c using Sturges' rule.

Since there are 50 measurements,
$$\begin{aligned} c &= 3.3(\log\ 50) + 1 \\ &\cong 7 \end{aligned}$$

Step 2. Determine the range R. Since the largest measurement is U = 77 and the smallest is L = 12, the range is

$$\begin{aligned} R &= U–L \\ &= 77–12 = 65 \end{aligned}$$

Step 3. Determine the width of each class, w. Number of classes, $c = 7$ (from step 1)

$$w = \frac{R}{c} = \frac{65}{7} = 9.3$$

We choose the least integer greater than 9.3 to be the value of our width. Hence $w = 10$.

(continued)

Exercise 1.1 (continued)

Step 4. We start at L = 12 and construct the first class with a width of 10. Suppose the first class extends from 12 to x, where x represents the unknown upper class limit typified below.

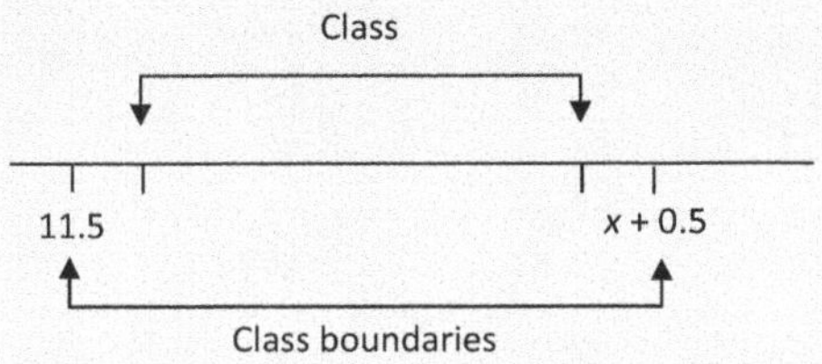

Since the unit of measurements is 1 and 0.5(1) = 0.5, the lower boundary will be 12–0.5 = 11.5, the upper boundary can represented by $x + 0.5$. By subtracting the lower class boundary from the upper class boundary, we obtain the width of the first class. Hence,

$$w = (x + 0.5) - 11.5$$

Since $w = 10$, we obtain

$$10 = x - 11$$

$$x = 21$$

Note:
31 = 21 +10
32 = 22 + 10

Hence the upper class limit is 21,and the first class becomes 12–21.

Step 5. To obtain each class following the first class, we add w = 10 to the lower and upper limits of each preceding class. Thus,

12–21
22–31
32–41
42–51
52–61
62–71
72–81

Step 6. A tally is used to determine the frequency of each class. Our grouped frequency table for the 50 bean measurement is shown in Table 1.4.

(continued)

Exercise 1.1 (continued)

Table 1.4 Group frequency table for bean measurement

Class	Tally	Frequency (f)
12–21	~~////~~ ~~////~~ ~~////~~ /	16
22–31	~~////~~ //	7
32–41	////	4
42–51	~~////~~ //	7
52–61	///	3
62–71	~~////~~	5
72–81	~~////~~ ///	8
		50

Problem 2.3

The following data represents the amounts (in Rand) spent in a particular weekend by 25 undergraduate students. Construct a grouped frequency table containing five classes.

Solution:

Step 1. The number of class interval, $c = 5$ (given)

Step 2. Compute the range R. Since $U = 82.71$ and $L = 17.89$, the range

39.78	28.30	28.31	17.95	44.47
46.65	31.47	33.45	29.17	48.39
82.71	43.63	41.17	47.32	52.16
25.94	50.32	35.25	35.70	17.89
60.20	48.14	22.78	38.22	23.25

$$R = U – L$$
$$= 82.71 – 17.89 = 64.82$$

Step 3. Determine the class width w. Since $c = 5$, then

$$w = \frac{R}{c} = \frac{64.82}{5} = 12.96$$

The least integer greater than 12.96 is 13. Hence, the class width w = 13.

Step 4. Starting at $L = 17.89$ construct a class with width, $w = 13$. Since the unit of measurement is 0.01, then (0.05)(0.01) = 0.005. Let x represent the upper boundary of the first class. The width is obtained by subtracting the lower class boundary (17.89–0.005 = 17.885) from the upper class boundary (x + 0.005).

(continued)

Exercise 1.1 (continued)

$$w = (x + 0.005) - 17.885$$
$$13 = x - 17.88$$
$$x = 30.88$$

Hence, the first class is 17.89–30.88.

Step 5. The class width, $w = 13$ is added to the preceding class limits to obtain the remaining classes (Table 1.5).

Step 6. The frequencies are determined using tallies (Table 1.5).

Table 1.5 Group frequency table for data of Application 2.5

Class	Tally	Frequency (f)	X
17.89–30.88	~~IIII~~ III	8	24.385
30.89–43.88	~~IIII~~ III	8	37.385
43.89–56.88	~~IIII~~ II	7	50.385
56.89–69.88	1	1	63.385
69.89–82.88	1	1	76.385
		25	

Step 7. Class marks are computed for each class following the formula given earlier

$$X = \frac{l_1 + l_2}{2}$$

For the first class, its class mark, $X = \frac{17.89 + 30.88}{2} = 24.385$

1.2.3.3 Relative Frequency Tables

Sometimes it is necessary to express each value or class in a frequency table as a function or percentage of the total number of measurements. The relative frequency helps us achieve this. The relative frequency for a measurement (or class) is the fraction or proportion of time the value occurs:

$$\text{Relative frequency of a value} = \frac{\text{number of times the value occurs}}{\text{number of observations in the data set}}$$

The table is called a relative frequency table (see Table 1.6). Multiplying a relative frequency by 100 gives a percentage. The relative frequencies or percentages, generally has several advantages over a frequency table. One important advantage is that we can make meaningful comparisons between similar sets of data having the same classes but different total frequencies.

Table 1.6 Relative frequency table for Table 1.4

Class	Relative frequency
12–21	0.32
22–31	0.14
32–41	0.08
42–51	0.14
52–61	0.06
62–71	0.10
72–81	0.16
	1.00

Example 1.14
In Table 1.4 the relative frequency for the first class is 16/50 = 0.32. The relative frequency table for Table 1.4 is detailed in Table 1.6.

Note: Theoretically, the sum of the relative frequencies is 1, but in practice the sum may differ slightly from 1 because of rounding.

1.2.3.4 Cumulative Frequency Tables

For any measurement (or class), the total of the frequency for that measurement (or class) and the frequencies of all measurements (or classes) of smaller value is termed Cumulative frequency. Cumulative frequency is very useful in instances where we are interested in the number of observations less than or equal to some value.

Example 1.15
Situations where cumulative frequency is of interest include:

1. A food technologist might want to know how many days a production process produced at most 100 defective items.
2. A lecturer might want to know the number of students that received a score less than or equal to 50% on an examination.
3. A food technologist might want to know how many items during sampling fell below the accepted quality criteria.

Table 1.7 details the cumulative frequency table for the data in Table 1.4.

1.2.3.4.1 Cumulative Relative Frequency Table

Cumulative relative frequency table is obtained when cumulative frequency table is constructed from tables containing relative frequencies or percentages. Table 1.8

Table 1.7 Cumulative frequency table for Table 1.4

Class	f	Cumulative frequency
12–21	16	16
22–31	7	23 = (16 + 7)
32–41	4	27 = (23 + 4)
42–51	7	34 = (27 + 7)
52–61	3	37 = (34 + 3)
62–71	5	42 = (37 + 5)
72–81	8	50 = (42 + 8)
	50	

Table 1.8 Cumulative relative frequency table for Table 1.4

Class	f	Cumulative frequency
12–21	16	16/50 = 0.32
22–31	7	23/50 = 0.46
32–41	4	27/50 = 0.54
42–51	7	34/50 = 0.68
52–61	3	37/50 = 0.74
62–71	5	42/50 = 0.84
72–81	8	50/50 = 1.00
	50	

indicates a cumulative relative frequency of Table 1.4 obtained by computing cumulative relative frequencies for the cumulative frequencies of Table 1.7.

Cumulative relative frequencies are useful in determining medians, percentiles and other quartiles (More of this later). Test scores are often given as percentiles. A percentile score tells what part of the tested population scored less than the give score.

1.2.4 Graphical Representation of Data

"One picture is worth a thousand words" i.e., data are often better understood if represented in graphic form than in tabular form. Graphical procedures are the shortest path to gaining insight into a data set in terms of (1) testing assumptions, (2) model selection, (3) model validation, (4) estimator selection, (5) relationship identification, (6) factor effect determination, and (7) outlier detection. Graphs are of many types. Some of the common types are pie charts, bar graph, stem-and-leaf diagram, box plot, histogram and ogive.

1.2.4.1 Bar and Pie Graphs

The bar and the pie graphs are both used for categorical or nominal data. Pie charts/ graphs are used only to portray parts of a total; they are very useful for displaying budget information. Both types of graphs will be illustrated using the data in

Table 1.9 Colours of bean in the bean-mixture

Colour of beans	Number of beans observed
White	28
Red	30
Brown	23
Black	25

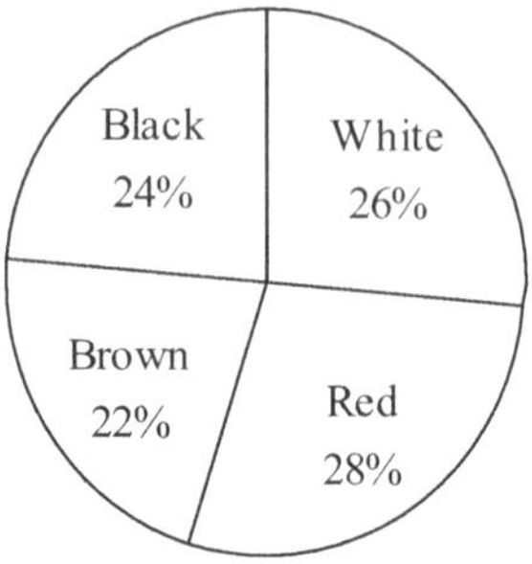

Fig. 1.3 Pie chart for Table 1.10

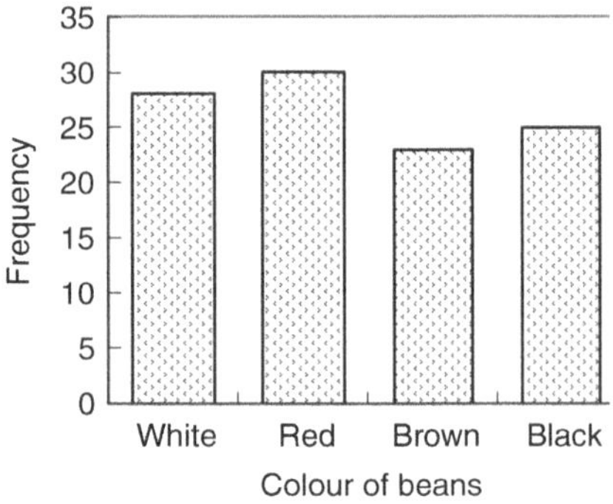

Fig. 1.4 A bar graph of bean colours

Tables 1.9 and 1.10, which shows the frequency distribution of bean colours sorted from a mixture of beans (Fig. 1.4). Certain computations necessary for constructing a pie graph for the bean colours is outlined in Table 1.10.

The pie chart of Table 1.10 is shown in Fig. 1.3.

1.2.4.2 Stem-and-Leaf Diagrams

The use of a grouped frequency table has one disadvantage of original data loss in the grouping process. A stem-and-leaf plot overcomes this by separating each observation into two parts: (1) a stem, consisting of one or more leading digits, and (2) a leaf, consisting of the remaining or trailing digit(s). Stem-and-leaf plots offer a new and novel way for displaying numerical data. For example, the numerals 462 could be split in two ways:

Table 1.10 Computations for constructing pie chart for the data on Table 1.9

Colour of beans	Number of beans observed	%	Degree	
White	28	26.4	95.0	← (0.264 x 360)
Red	30	28.3	101.9	
Brown	23	21.7	78.1	
Black	25	23.6	85.0	← $\left(\frac{25}{106} \times 100\right)$
Total	106	100.0	360.0	

Stem		Leaf		Stem		Leaf
4	|	62		46	|	2

Once stem values have been chosen, they should be listed in a single column. Then the leaf of each observation should be placed on the row of the corresponding stem. A stem-and-leaf display conveys information about the following aspects of the data:

- Identification of a typical or representative value
- Extent of spread about the typical value
- Presence of any gaps in the data
- Extent of symmetry in the distribution of values
- Number and location of peaks
- Presence of any outlying values

Example 1.16

Let us construct a stem-and-leaf diagram for the following collection of scores for food engineering test by 25 students.

67	54	74	56	84
71	46	69	57	94
71	46	34	55	98
66	51	61	46	82
90	22	65	63	79

All the scores fall between 22 and 98, we can use the tens digit in each case to serve as the stem and the units digits to serve as the leaf.

(continued)

Example 1.16 (continued)

Step 1. Place the stems in a vertical array using a vertical line segment, called the trunk, to separate the leaves from the stems.

2	
3	
4	
5	
6	
7	
8	
9	

Step 2. Place each leaf to the right of its stem. Since the first score is 67, we place the leaf, 7, on its stem, 7.

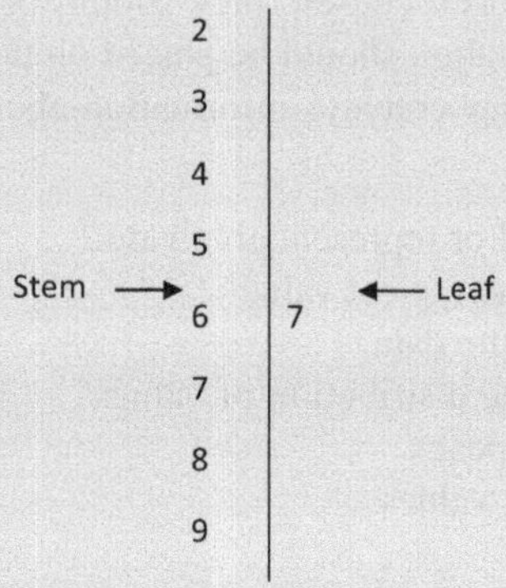

We continue this process until all the data is displayed we obtain the following stem-and-leaf diagram:

2	2					
3	4					
4	6	6	6			
5	4	1	6	7	5	
6	7	6	9	1	5	3
7	1	1	4	9		
8	4	2				
9	0	4	8			

By observing the above stem-and-leaf diagram we can conclude that

- The largest score is 98.
- The smallest score is 22.
- The scores vary from 22 to 98.
- Stem 2 and 3 has the fewest leaves.
- Stems 6 contain the most leaves.
- The total number of leaves represents the size of the data set.

An ordered stem-and-leaf diagram for our example is as follows

Note: The order in which the leaves are placed on a stem does not make any difference. If the leaves on each stem are ordered from smallest to largest, the diagram is called an ordered stem-and-leaf diagram.

2	2					
3	4					
4	6	6	6			
5	1	4	5	6	7	
6	1	3	5	6	7	9
7	1	1	4	9		
8	2	4				
9	0	4	8			

1.2.4.3 Box Plot

The box plot is a graphical display that simultaneously describes several important features of a data set, such as centre, spread, departure from symmetry, and identification of observations that lie unusually far from the bulk of the data called outliers. A box plot displays the three quartiles on a rectangular box, aligned either horizontally or vertically. The box encloses the interquartile range (IQR) with the left (or lower) edge at the first quartile, q_1, and the right (or upper) edge at the third quartile, q_3. A line is drawn through the box at the second quartile (which is the 50th percentile or the median). A line or whisker extends from each end of the box. The lower whisker is a line from the first quartile to the smallest data point within 1.5 interquartile ranges from the first quartile. The upper whisker is a line from the third quartile to the largest data point within 1.5 interquartile ranges from the third quartile. Data farther from the box than the whiskers are plotted as individual points. A point beyond a whisker, but less than three interquartile ranges from the box edge, is called an outlier. A point more than three interquartile ranges from a box edge is called an extreme outlier.

Fig. 1.5 Skewed box plots

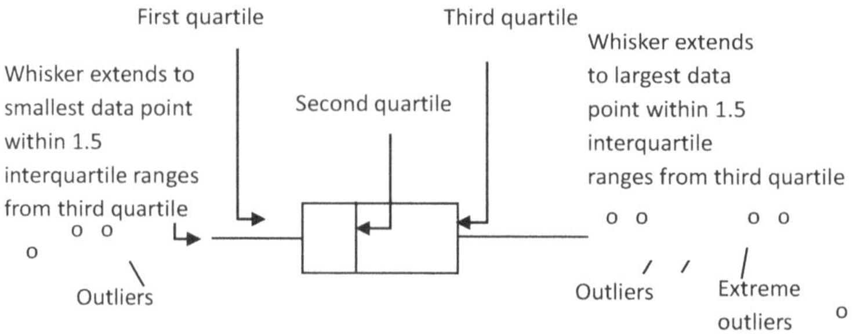

For a box plot that is perfectly symmetrical, the median (second quartile) line is exactly in the middle of the box, and the two whiskers are of the same length. The location of the box is a good indicator of symmetry. If the distribution is skewed to the left, the median is to the right of the centre of the box; and if the distribution is skewed to the right, then the median is to the left of the centre of the box (Fig. 1.5). The width of the box is the interquartile range, and thus provides a measure of dispersion for the data. If one whisker were especially long, this would indicate that the extreme measurement is a possible outlier (Fig. 1.5).

1.2.4.4 Histograms

A histogram is a type of bar graph for a frequency distribution. Histograms can be constructed for grouped and ungrouped frequency distributions.

The idea of constructing a histogram for ungrouped frequency data is to represent each frequency by a bar whose area is proportional to the frequency. Typically, the width of each bar is chosen to be 1 so that the area of the bar is equal to the frequency of the measurement.

Constructing a histogram for a grouped data measured using an interval or ratio scale follows two steps (1) the data are organized into a grouped frequency table as discussed earlier; (2) A bar graph is constructed using the class boundaries to locate the bar and the frequencies to indicate the heights of the bars.

A histogram improves our ability to compare the frequencies of corresponding classes; a class frequency can be easily compared with those of its neighbouring classes.

1.2.4.5 Relative Frequency Histogram

Histograms in addition to representing frequency distributions, can be used to represent relative frequency distributions. The height of the bars on a relative frequency histogram will indicate the proportion of the total represented by each class. The basic shape of a relative frequency histogram is similar to the shape of the corresponding frequency histogram.

A *frequency polygon* is a line graph of class frequency plotted against class mark. It can be obtained by connected midpoints of the tops of rectangles in the histogram.

The *relative frequency* of a class is the frequency of the class divided by the total frequency of all classes and is generally expressed as a percentage. The sum of the relative frequency of all classes is clearly 1 or 100%. If frequencies in frequency distribution are replaced by corresponding relative frequencies, the resulting table is called *relative frequency distribution*. Graphical representation of relative frequency distribution can be obtained from the histogram or frequency polygon by simply changing the vertical scale from frequency to relative frequency, keeping exactly the same diagram. The resulting graphs are called *relative frequency histograms* or *percentage histogram* and *relative frequency polygons* or *percentage polygons*, respectively.

The total frequency of all values less than the upper class boundary of a given class interval is called *the cumulative frequency* up to and including the class interval, A table presenting such cumulative frequencies is called a *cumulative frequency distribution, cumulative frequency table*, or briefly *cumulative distribution*. A graph, showing the cumulative frequency less than any upper class boundary plotted against the upper class boundary is called a *cumulative frequency polygon*.

The *relative cumulative frequency* or *percentage cumulative frequency* is the cumulative frequency divided by the total frequency. If relative cumulative frequencies are used in cumulative tables in place of cumulative frequencies, the results are called *relative cumulative frequency distribution* or *percentage cumulative distribution* and *relative cumulative frequency polygon*.

The barcharts and histograms provide an easily understood illustration of the distribution of the data. As well as showing where most observations lie and how variable the data are.

Frequency curves, arising in the practice take on some characteristic shapes illustrated in the following Figs. 1.6, 1.7, 1.8, 1.9, 1.10, 1.11 and 1.12.

1.2.5 Descriptive Statistics

1.2.5.1 Introduction

Data are results of observations that one uses to reach conclusions or generate interpretations. Conclusions and interpretations, however, are not based on the raw numerical observations themselves, but on the information they contain. Organisation is the first step that a researcher takes when trying to extract information from

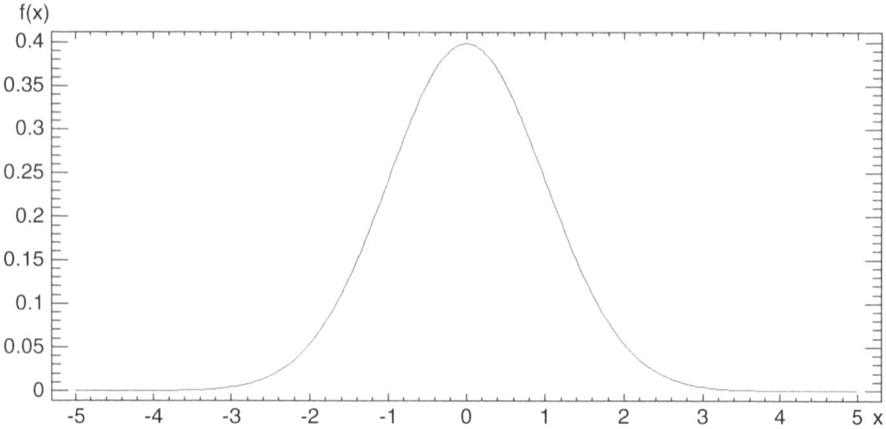

Fig. 1.6 Symmetrical or bell distribution

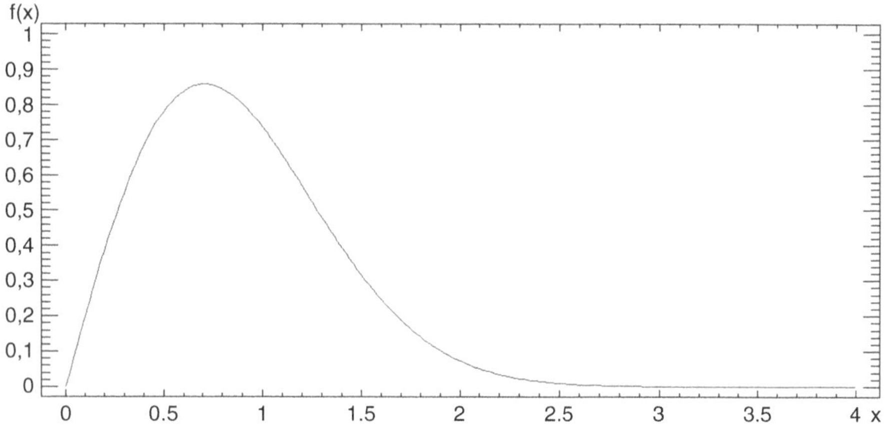

Fig. 1.7 Skewed to the right (positive skewness)

observations. This was the object of the previous section. The second step is to describe or summarise one or more properties for the numerical data that have been collected. To this end statisticians have developed a body of conventional indices called descriptive statistics:

A descriptive statistic is a numerical index that describes or summarises some characteristics of a frequency or relative frequency distribution.

A fundamental concept in describing the numerical data that have been collected is that of a *distribution*. Distributions arise because any measurement process output may not yield the same value every time it is measured. There will be a natural scattering of the measured values about some central tendency value. This scattering

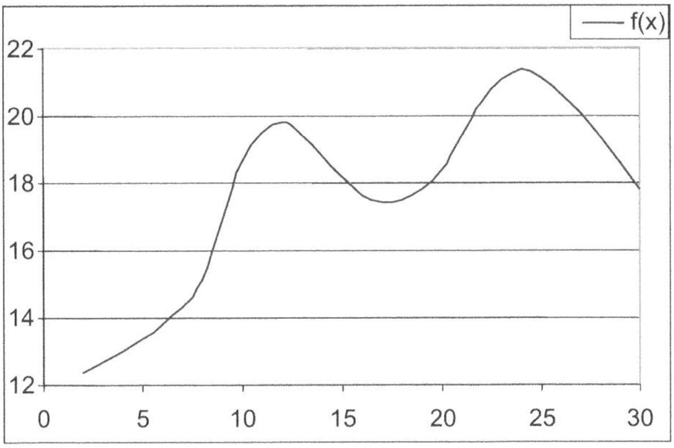

Fig. 1.8 Bimodal distribution

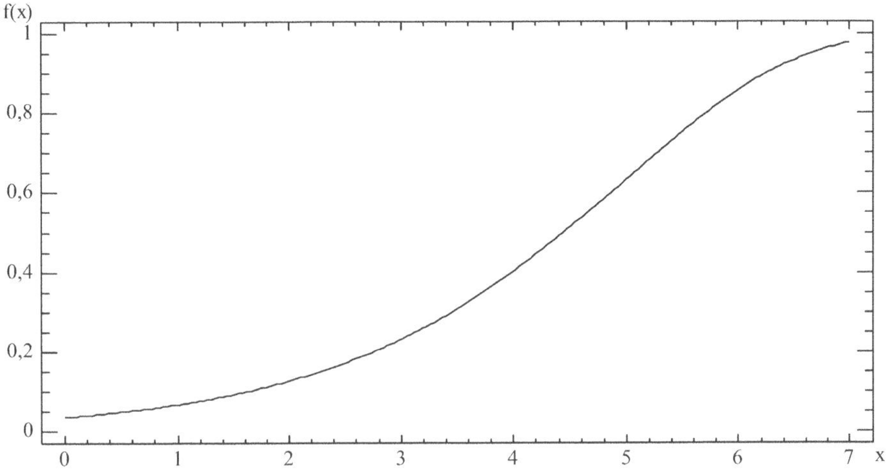

Fig. 1.9 J – shaped distribution

about a central value is known as a distribution. A distribution is characterized by three values:

Location
The location is the expected value of the output being measured. For a stable process, this is the value around which the process has stabilized.

Spread
The spread is the expected amount of variation associated with the output. This tells us the range of possible values that we would expect to see.

Shape
The shape shows how the variation is distributed about the location. This tells us if our variation is symmetric about the mean or if it is skewed or possibly multimodal.

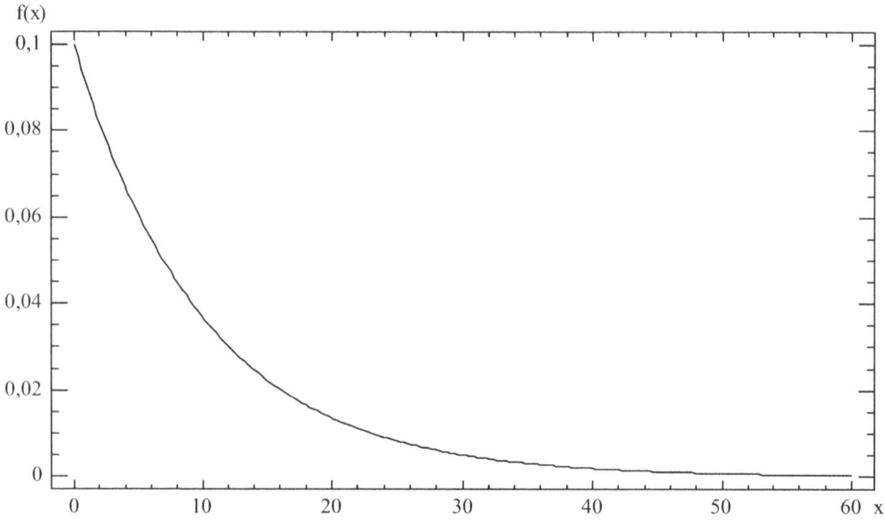

Fig. 1.10 L – shaped distribution

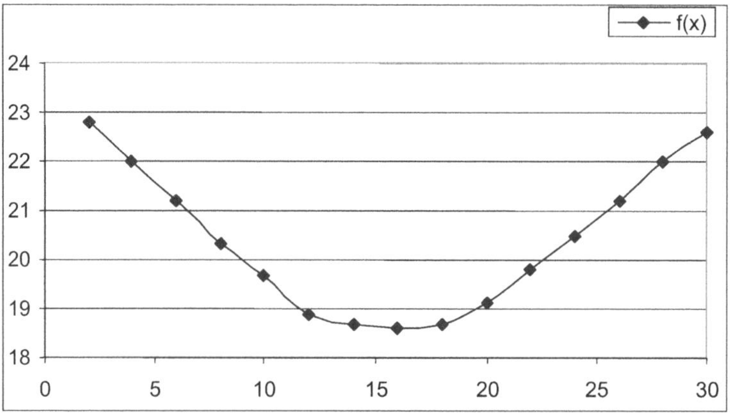

Fig. 1.11 U – shaped distribution

Example 1.17

Suppose, for example, that a lecturer administers an examination with 100 possible points. If the lecturer wants to be sure that the test is pitched at an appropriate level of difficulty for this class, it is important to know whether the majority of students scored toward the low level end to the total possible

(continued)

Fig. 1.12 Mode, median and mean for nonsymmetrical data

Example 1.17 (continued)
range of points, somewhere in the middle or toward the high end. This information is summarized in a statistic that describes the *central tendency or location* of a set of observations. The lecturer might also wish to determine the extent to which test scores are spread across their total possible range of values. This property is captured by statistics that describe dispersion, or variability.

Table 1.11 shows the most common numerical and graphical measures of location, spread and shape.

Measure of Central Tendency or location include mean – arithmetic average; median – midpoint of the observations when arranged in increasing order and the mode – any value having the highest frequency among the observations. Useful as the measures of centre are, they are usually incomplete and often misleading without some accompanying indication of how spread out or dispersed the data are. The simplest adequate summary of a univariate distribution usually requires both a measure of centre and a measure of variability. Statistics that describes the dispersion in a distribution indicates how much variation there is among the observations and include variance or standard deviations and percentiles. Measure of shape is described by skew- its departure from symmetry about the mean and kurtosis- degree to which the frequency polygon is "peaked".

Table 1.11 Common numerical and graphical measures of location, spread and shape

Parameter	Numerical	Graphical
Location	Mean Median	Scatter plot Boxplot Histogram
Spread	Variance Range Inter-quartile range	Boxplot Histogram
Shape	Skewness Kurtosis	Boxplot Histogram Probability plot

1.2.5.2 Measures of Central Tendency

1.2.5.2.1 Arithmetic Mean

Mean of a sample of numerical data is given by

$$\overline{X} = \frac{\sum x}{n} \tag{1.1}$$

where $\overline{X}$ denotes the sample mean, x denotes a sample measurement, Σx denotes the sum of the sample measurements, and n is the size of the sample.

Sample mean for data in a frequency table is given by

$$\overline{X} = \frac{\sum fx}{\sum f} \text{ or } \overline{X} = \frac{f_1x_1 + f_2x_2 + \ldots + f_mx_m}{f_1 + f_2 + \ldots + f_m} = \frac{1}{N}\sum_{i=1}^{m} f_ix_i \tag{1.2}$$

where $\overline{X}$ denotes the sample mean, f denotes frequency of the measurement; x denotes a sample measurement, Σfx denotes the sum of the product of frequency and sample measurement.

Mean deficiency	The mean is affected by extreme measurements on one end of a distribution. Since the mean depends on the value of every measurement, extreme values can lead to the mean misrepresenting the data.

1.2.5.2.2 Weighted Arithmetic Mean

Sometimes we associate with the numbers $x_1, x_2, \ldots, x_m$ certain weighting factor or weights $w_1, w_2, \ldots, w_m$ depending on the significance or importance attached to the numbers. In this case

$$\overline{X} = \frac{w_1x_1 + w_2x_2 + \ldots + w_mx_m}{w_1 + w_2 + \ldots\ldots + w_m} = \frac{\sum_{i=1}^{m} w_ix_i}{\sum_{i=1}^{m} w_i} \tag{1.3}$$

It is called the *weighted arithmetic mean*. It is obvious that (1.2) can be considered also as a weighted mean with weights $f_1, f_2, \ldots, f_m$.

Some properties of the arithmetic mean could be mentioned here:

- The algebraic sum of the deviations of a set of numbers from their arithmetic mean is zero.
- The sum of the squares of the deviations of a set of numbers $x_1, x_2, \ldots, x_n$ from any number a is a minimum if and only if $a = \overline{X}$.

1.2.5.2.3 Harmonic Mean

The *harmonic mean* H of a set of numbers $x_1, x_2, \ldots, x_n$ is the reciprocal of the arithmetic mean of the reciprocals of these numbers:

$$H = \frac{1}{\overline{\overline{X}}} = \frac{N}{\sum_{i=1}^{N} \frac{1}{x_i}} \tag{1.4}$$

If the data are grouped the harmonic mean is calculated by

$$H = \frac{\sum_{i=1}^{m} f_i}{\sum_{i=1}^{m} \frac{1}{x_i} f_i} \tag{1.5}$$

1.2.5.2.4 Geometric Mean

The *geometric mean* G of a set of numbers $x_1, x_2, \ldots, x_n$ is

$$G = \sqrt[n]{\prod_{i=1}^{n} x_i} \tag{1.6}$$

1.2.5.2.5 Cubic Mean

The *root mean square (RMS)* or *quadratic mean* $\overline{X}_q$ of a set of numbers $x_1, x_2, \ldots, x_n$ is

$$(RMS) = \overline{X}_q = \sqrt{\frac{\sum_{i=1}^{N} x_i^2}{N}} \tag{1.7}$$

and

$$(RMS) = \overline{X}_q = \sqrt{\frac{\sum_{i=1}^{m} x_i^2 f_i}{\sum_{i=1}^{m} f_i}} \tag{1.8}$$

for weighted data.

The *cubic mean* $\overline{X}_{cub}$ of a set of numbers $x_1, x_2, \ldots, x_n$ is

$$\overline{X}_{cub} = \sqrt[3]{\frac{\sum_{i=1}^{N} x_i^3}{N}} \tag{1.9}$$

and for the weighted data

$$\overline{X}_{cub} = \sqrt[3]{\frac{\sum_{i=1}^{m} x_i^3 f_i}{\sum_{i=1}^{m} f_i}}. \tag{1.10}$$

If for a set $x_1, x_2, \ldots, x_n$ the different means are calculated, then

$$H \leq G \leq \overline{X} \leq \overline{X}_q \leq \overline{X}_{cub}.$$

The equality sign here holds only if all the numbers $x_1, x_2, \ldots, x_n$ are identical.

1.2.5.2.6 Median

An alternative measure of centre that resists the effects of outliers is the median. The median, denoted by $\widetilde{x}$, is obtained by first ordering the sample observations from smallest to largest. Then

$$\tilde{x} = \begin{pmatrix} \text{single middel value} = \left(\frac{n+1}{2}\right)\text{th value on ordered list } n \text{ odd} \\ \text{average of two middle values} = \text{average of } \frac{n}{2}\text{th and } \left(\frac{n}{2}+1\right)\text{th values } n \text{ even} \end{pmatrix}$$

1.2.5.2.7 Mode

The *mode*, or *modal value*, is the most frequently occurring value. Thus, the *mode* *Mo* of a set of numbers is that value which occurs with the greatest frequency. The mode may not exist, and even if it does exist it may not be unique.

For continuous data, the simplest definition of the mode is the midpoint of the interval with the highest rectangle in the histogram.

There is a more complicated definition involving the frequencies of neighboring intervals.

$$Me = L + \left(\frac{n+1}{2} - C_{me-1}\right)\frac{h}{f} \qquad (1.11)$$

where L is lower boundary of the median group; C_{me-1} is cumulative frequency in the previous group; h is the range of the median group; and f is the frequency of the median group. This definition is only useful if there are a large number of observations.

In the case of grouped data where a frequency curve has been constructed to fit the data, the mode will be the value (or values) of x corresponding to the maximum point (or points) on the curve.

From a frequency distribution or histogram the mode can be obtained from the formula

$$Mo = L + \frac{(f_{mo} - f_{mo-1})h}{(f_{mo} - f_{mo-1}) + (f_{mo} - f_{mo+1})} \qquad (1.12)$$

where L is lower class boundary of the modal class (i.e. class containing the mode); f_{mo} is the frequency in the modal group, f_{mo-1} is the frequency in the previous group; f_{mo+1} is the frequency in the next group; and h is size of the modal group.

1.2.5.3 Measures of Variation

1.2.5.3.1 Range

Reporting a measure of centre gives only partial information about a data set or distribution. Different samples or distributions may have identical measures of centre yet differ from one another in other important ways. The simplest measure of variability in a sample is the range, which is the difference between the largest and smallest sample values. The range depends on only the two most extreme observations and disregards the positions of the remaining values. It is therefore affected by the presence of extreme data values.

The *range R* of a set of numbers $x_1, x_2, \ldots, x_n$ is the difference between the largest and the smallest numbers in the set:

$$R = x_{\max} - x_{\min} \tag{1.13}$$

1.2.5.3.1.1 Standard Deviation and Variance

Our primary measures of variability involve the deviations from the mean which helps us to determine the standard deviation, the positive square root of the variance.

The sample variance, denoted by s^2, is given by

$$s^2 = \frac{\sum (x_i - \overline{x})^2}{n-1} \text{ or } \frac{\text{SS}}{\text{n-1}} \text{ where SS} = \sum x^2 - \frac{(\sum x)^2}{n}$$

where SS denotes sum of squares. Sample standard deviation, denoted by s, is the (positive) square root of the variance:

$$s = \sqrt{s^2}$$

Computational formula for SS using frequencies

$$\text{SS} = \sum fx^2 - \frac{(\sum fx)^2}{n}$$

A rough interpretation of the sample standard deviation is that it is the size of a typical or representative deviation from the sample mean within the given sample. The sample deviation indicates the dispersion of the data about the sample mean. If the data values are clustered closely about the mean, then s is small; if the values are considerably spread about the mean, then s is large. But how shall we determine what values of s are large and what values are small? The Empirical Rule provides some insight as to how the magnitude of the standard deviation of any set of data relates to the concentration of the data about its mean. This rule is:

- Approximately 60–78% of the data are within a distance of one standard deviation from the mean $(\overline{x} - s, \overline{x} + s)$.
- Approximately 90–98% of the data are within a distance of two standard deviations from the mean $(\overline{x} - 2s, \overline{x} + 2s)$.
- More than 99% of the data are within a distance of three standard deviations from the mean $(\overline{x} - 3s, \overline{x} + 3s)$.

The variance of a set of data is defined as the square of the standard deviation and thus is given by s^2 in (1.37), (1.38) and (1.39):

$$\sigma^2 = \frac{\sum_{i=1}^{N} (x_i - \overline{X})^2}{N} \quad (1.14)$$

$$s^2 = \frac{\sum_{i=1}^{N} (x_i - \overline{X})^2}{N - 1} \quad (1.15)$$

$$s^2 = \frac{\sum_{i=1}^{N} (x_i - \overline{X})^2 f_i}{\sum_{i=1}^{N} f_i}$$

Shortcomings of the variance and standard deviation	The variance and standard deviation can be seriously affected by the presence of outliers, since they both depend on the mean, which is affected by extreme measurements. When outliers are present in a data set and a measure of resistant to outliers is desired, the interquartile range should be used.

1.2.5.3.2 Relative Standard Deviation

1.2.5.3.2.1 Interquartile Range

Quartiles are numbers that divide an ordered set of measurements extending from the smallest measurement to the largest measurement into four parts so that each part has approximately 25% of the measurements. There are three quartile points, denoted by Q_1, Q_2 and Q_3. The first quartile Q_1 is the 25th percentile, the second quartile Q_2 is the 50th percentile or the median, and the third quartile Q_3 is the 75th percentile. The interquartile range, IQR is defined as

$$\text{IQR} = Q_3 - Q_1$$

where Q_3 is the third quartile and Q_1 is the first quartile.

1.2.5.4 Standard Scores and Outliers

A measure that allows us to make comparisons from different distributions and takes into account the dispersion of the scores is the standard score. A standard score (z) or z score is defined as

$$z = \frac{x - \overline{x}}{s}$$

where $\overline{x}$ is the mean measurement; x is a measurement and s is the standard deviation. The standard score represents the number of standard deviations a score is from the mean.

1.2.5.4.1 Detection of Outliers

In the evaluation of data, one often encounters unusual observations in the sense that they are extremely below or above expectations. These extreme measurements in a data set are often called outlier, wild or aberrant. If they are truly outliers they should be discarded, and inclusion of these data points will affect the mean and standard deviation of the data set, thus distorting the centre and variability. In some instances, they are known in advance to be an outlying observation, such as the result of instrument malfunction, instrument miscalibration, or solution contamination. In this case one may discard the data with no further statistical test of significance. In another case, the data is suspiciously large or small with no apparent reasons, and it is advisable to conduct a statistical analysis to determine whether to reject or retain the suspicious data.

It is important to point out that subjective data obtained from hedonic, preference, or acceptability measurement do not contain outlying observations because this type of data is an expression of likes and dislikes, and therefore is never wrong. However, subjective data from descriptive evaluation is subject to wild judgment and can be discarded by a statistical test.

Note	Rejection of outlying observation can be applied not only to sensory evaluation results but also to estimated mean values from various experiments. For example, estimates of shelf-life of a product conducted at different times of the year may contain an outlying estimate which is atypical of the true shelf-life of the product under normal condition of use.

A measurement is an outlier of a sample if either rule holds:

Rule 1 The sample size is greater than 10, the frequency distribution is bell shaped, and the z score for the measurement is more than three standard deviations from the mean.

Rule 2 The measurement falls more than three interquartile ranges (IQRs) below the lower quartile or more than three IQR above the upper quartile.

Before abnormal observations can be singled out, it is necessary to characterize normal observations. Two activities are essential for characterizing a set of data:

1. Examination of the overall shape of the graphed data for important features, including symmetry and departures from assumptions.
2. Examination of the data for unusual observations that are far removed from the mass of data. These points are often referred to as outliers. Two graphical techniques for identifying outliers are scatter plots and box plots.

1.2.5.4.1.1 Box Plot Construction

The box plot is a useful graphical display for describing the behaviour of the data in the middle as well as at the ends of the distributions. The box plot uses the median and the lower and upper quartiles (defined as the 25th and 75th percentiles). If the lower quartile is Q1 and the upper quartile is Q2, then the difference (Q2 – Q1) is called the interquartile range or IQ.

A box plot is constructed by drawing a box between the upper and lower quartiles with a solid line drawn across the box to locate the median. The following quantities (called *fences*) are needed for identifying extreme values in the tails of the distribution:

1. Lower inner fence: Q1–1.5*IQ
2. Upper inner fence: Q2 + 1.5*IQ
3. Lower outer fence: Q1–3*IQ
4. Upper outer fence: Q2 + 3*IQ

A point beyond an inner fence on either side is considered a **mild outlier**. A point beyond an outer fence is considered an **extreme outlier**.

The data set of N = 90 ordered observations as shown below is examined for outliers:

30, 171, 184, 201, 212, 250, 265, 270, 272, 289, 305, 306, 322, 322,
336, 346, 351, 370, 390, 404, 409, 411, 436, 437, 439, 441, 444, 448,
451, 453, 470, 480, 482, 487, 494, 495, 499, 503, 514, 521, 522, 527,
548, 550, 559, 560, 570, 572, 574, 578, 585, 592, 592, 607, 616, 618,
621, 629, 637, 638, 640, 656, 668, 707, 709, 719, 737, 739, 752, 758,
766, 792, 792, 794, 802, 818, 830, 832, 843, 858, 860, 869, 918, 925,
953, 991, 1000, 1005, 1068, 1441

The computations are as follows:

- Median = (n + 1)/2 largest data point = the average of the 45th and 46th ordered points = (559 + 560)/2 = 559.5
- Lower quartile = 0.25(N + 1) = 0.25*91 = 22.75th ordered point = 411 + .75 (436–411) = 429.75
- Upper quartile = 0.75(N + 1) = 0.75*91= = 68.25th ordered point = 739 + 0.25 (752–739) = 742.25
- Interquartile range = 742.25–429.75 = 312.5
- Lower inner fence = 429.75–1.5 (312.5) = −39.0
- Upper inner fence = 742.25 + 1.5 (313.5) = 1211.0
- Lower outer fence = 429.75–3.0 (313.5) = −507.75
- Upper outer fence = 742.25 + 3.0 (313.5) = 1679.75

From an examination of the fence points and the data, one point (1441) exceeds the upper inner fence and stands out as a mild outlier; there are no extreme outliers.

1.2.6 Examples

Example 1.18
Moisture content (%) of 20 biscuits is given below. Construct a frequency table for moisture content data and represent graphically.

Solution: Sort the given data in increasing order.

5.5	*5.4*	*5.5*	*6.6*	*5.7*	*5.8*	*5.9*	*5.8*	*5.5*	*6.1*
6.2	*5.8*	*6.3*	*5.7*	*5.8*	*6.3*	*6.6*	*5.7*	*6.4*	*5.7*

Moisture content (%)	Frequency
5.4	*1*
5.5	*3*
5.7	*4*
5.8	*4*
5.9	*1*
6.1	*1*
6.2	*1*
6.3	*2*
6.4	*1*
6.6	*2*

The same information is shown graphically by barchart (Fig. 1.13):

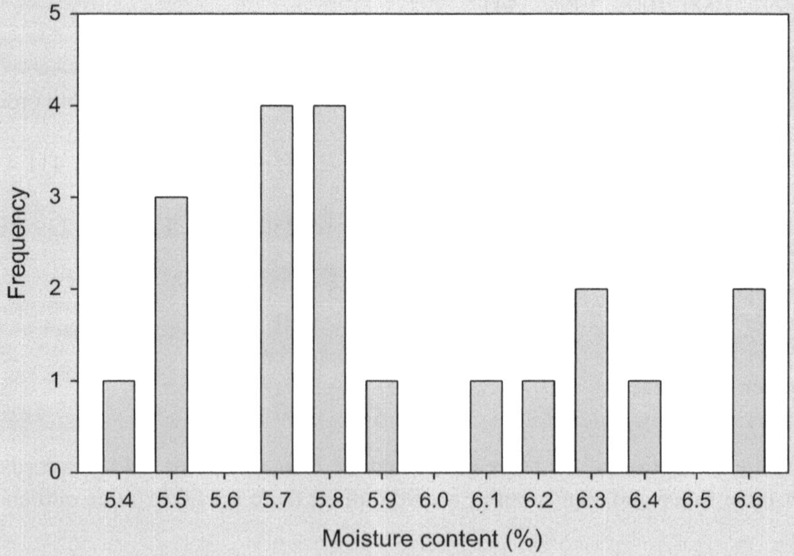

Fig. 1.13 Barchart representing of data

(continued)

Example 1.18 (continued)

It is well known, and it can be easily seen from the plot that it is more understandable if graphical methods are used for data illustration.

Example 1.19 World Wheat production (million tons) is given in the following Table (FAO http://faostat.fao.org/site/567/default.aspx). Plot this information on line, radar and pie chart type.

Solution: Information about the wheat production in world is given in the next three Figures – three different kinds of plots. First of them is a line (Fig. 1.14), second – radar (Fig. 1.15) and third – pie chart type (Fig. 1.16).

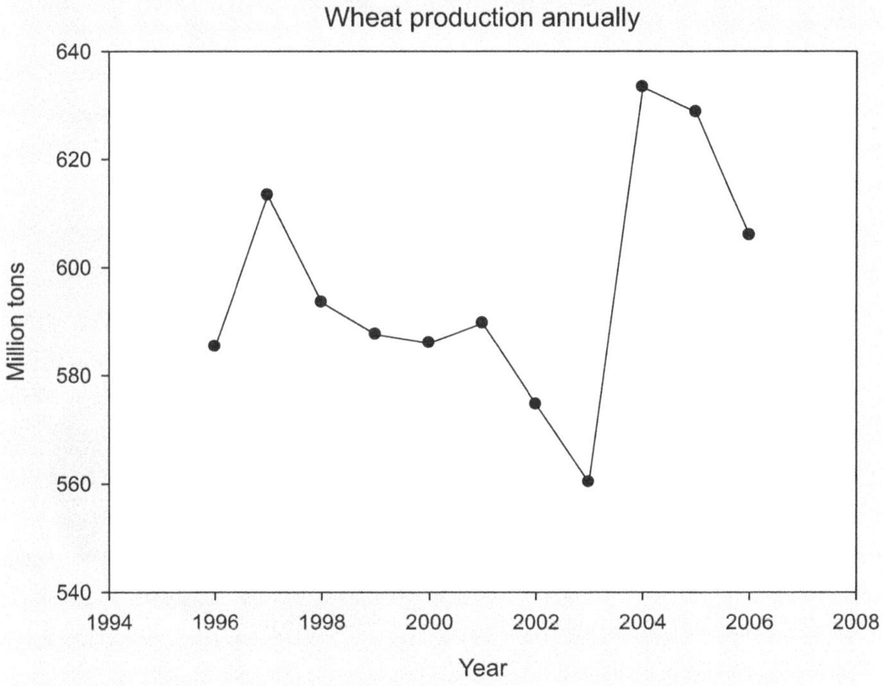

Fig. 1.14 Wheat production annually (Line representing of data)

Year	1996	1997	1998	1999	2000	2001	2002	2003	2004	2005	2006
X	585.38	613.38	593.55	587.67	586.06	589.69	574.68	560.29	633.30	628.70	605.95

(continued)

Example 1.19 (continued)

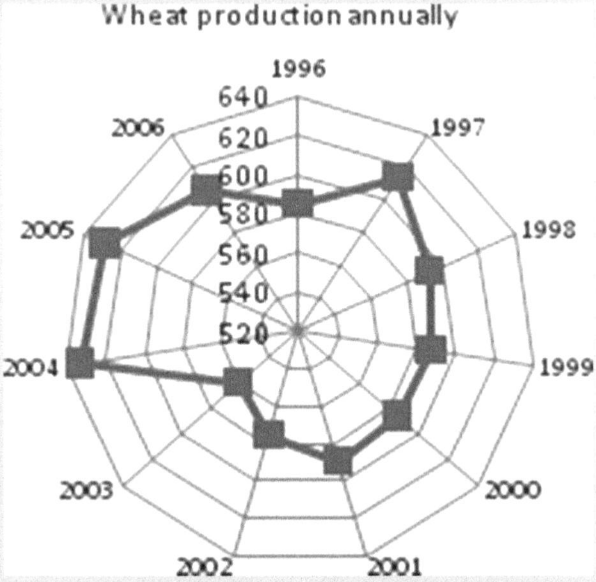

Fig. 1.15 Wheat production annually (Radar representing of data)

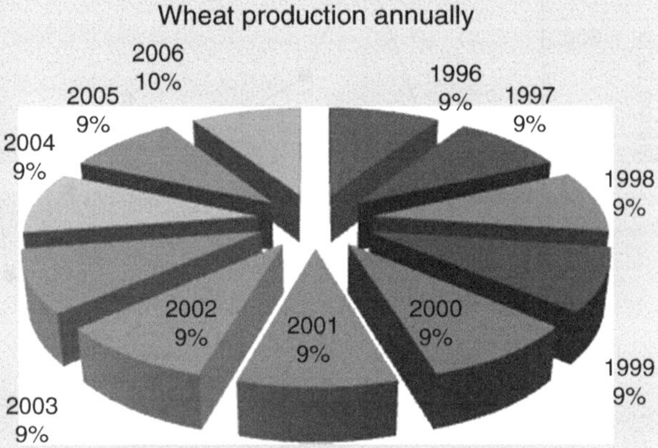

Fig. 1.16 Wheat production annually (Pie Chart representing of data)

Example 1.20 Find the range for data given in Example 1.19.

***Solution*:** First sort the data in decreasing order

633.30	628.70	613.38	605.95	593.55	589.69	587.67	586.06	585.38	574.68	560.29

and then calculate range as;

Range for wheat production = 633.30–560.29 = 73.01 million tons.

Example 1.21 Group the data from Example 1.18 (divided in groups of 4, 5 and 6 classes). Find their relative frequency table, cumulative frequency table and cumulative relative frequency table.

Solution: Range for this data = 6.6–5.4 = 1.2

(a) For 4 groups 1.2/4 = 0.30

5.5	*5.4*	*5.5*	*6.6*	*5.7*	*5.8*	*5.9*	*5.8*	*5.5*	*6.1*
6.2	*5.8*	*6.3*	*5.7*	*5.8*	*6.3*	*6.6*	*5.7*	*6.4*	*5.7*

Class	Lower Limit	Upper Limit	Midpoint	Frequency	Relative Frequency	Cumulative Frequency	Cum. Rel. Frequency
1	5.40	5.70	5.55	8	0.40	8	0.40
2	5.71	6.00	5.85	5	0.25	13	0.65
3	6.01	6.30	6.15	4	0.20	17	0.85
4	6.31	6.60	6.45	3	0.15	20	1.00

Histogram and frequency polygon of the data is given in Fig. 1.17.

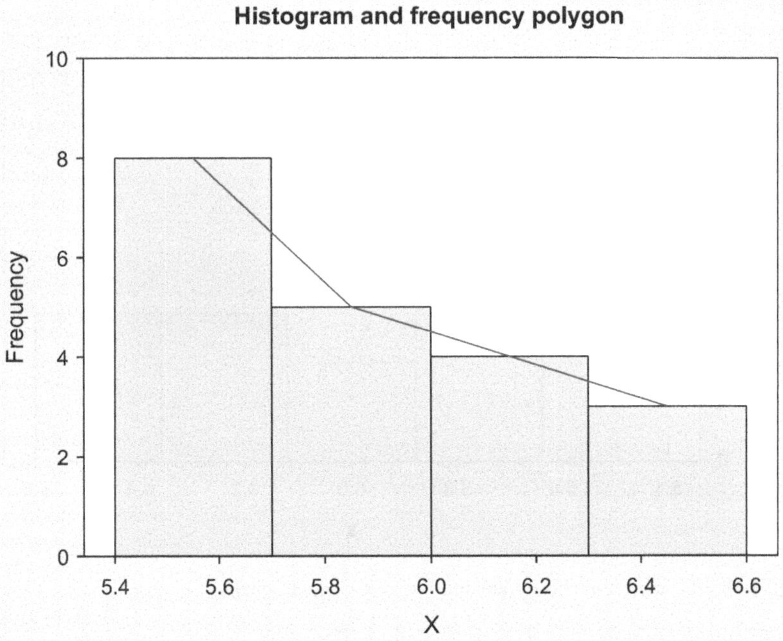

Fig. 1.17 Bar represents histogram and line represents frequency polygon

(continued)

Example 1.21 (continued)

(b) For 5 groups 1.2/5 = 0.24.

Class	Lower Limit	Upper Limit	Midpoint	Frequency	Relative Frequency	Cumulative Frequency	Cum. Rel. Frequency
1	5.40	5.64	5.52	4	0.20	4	0.20
2	5.65	5.88	5.76	8	0.40	12	0.60
3	5.89	6.12	6.00	2	0.10	14	0.70
4	6.13	6.36	6.24	3	0.15	17	0.85
5	6.37	6.60	6.48	3	0.15	20	1.00

Histogram and frequency polygon of the data is given in Fig. 1.18.

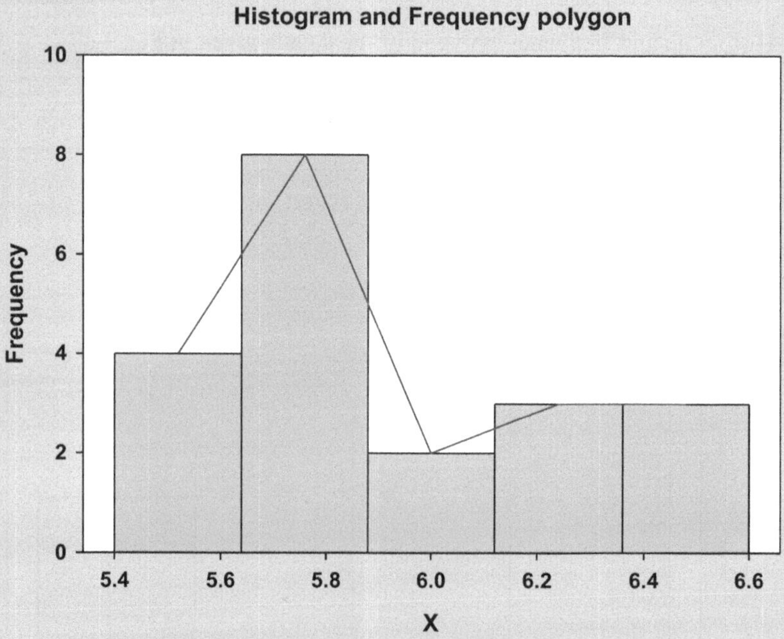

Fig. 1.18 Bar represents histogram and line represents frequency polygon

(continued)

Example 1.21 (continued)

(c) For 6 groups 1.2/6 = 0.20.

Class	Lower Limit	Upper Limit	Midpoint	Frequency	Relative Frequency	Cumulative Frequency	Cum. Rel. Frequency
1	5.40	5.60	5.50	4	0.20	4	0.20
2	5.61	5.80	5.70	8	0.20	8	0.40
3	5.81	6.00	5.90	1	0.25	13	0.65
4	6.01	6.20	6.10	2	0.05	14	0.70
5	6.21	6.40	6.30	3	0.15	17	0.85
6	6.41	6.60	6.50	2	0.15	20	1.00

Histogram and frequency polygon of the data is given in Fig. 1.19.

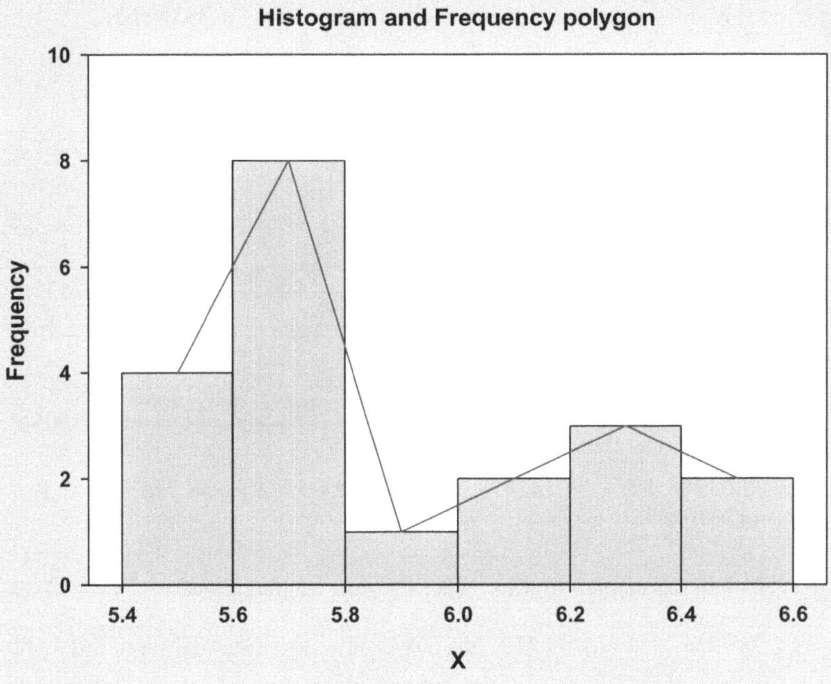

Fig. 1.19 Bar represents histogram and line represents frequency polygon

From this Example one can see that the histograms for the three cases (4, 5 or 6 classes) differ much which means that the number of classes is very important parameter for investigating data.

Example 1.22 Research and development unit of a chocolate factory developed a new form of chocolate. They produced 15 samples only. They measured sugar content and data are given as;

Sample	1	2	3	4	5	6	7	8	9	10	11	12	13	14	15
Sugar (%)	38	37	36	39	41	34	42	36	35	38	42	41	39	38	37

Calculate the arithmetic mean, the harmonic mean, the quadratic mean, the geometric mean, the cubic mean and the median, Q_1 and Q_3.

Solution: Using formula (1.1); (1.4); (1.6); (1.7) and (1.9):

$$\overline{X} = \frac{\sum_{i=1}^{N} x_i}{N} = \frac{38 + 37 + 36 + 39 + \ldots + 39 + 38 + 37}{15} = 38.20000;$$

$$H = \frac{N}{\sum_{i=1}^{N} \frac{1}{x_i}} = \frac{15}{\frac{1}{38} + \frac{1}{37} + \frac{1}{36} + \ldots + \frac{1}{39} + \frac{1}{38} + \frac{1}{37}} = 38.04965;$$

$$G = \sqrt[n]{\prod_{i=1}^{n} x_i} . = \sqrt[15]{38^*37^*36^* \ldots ^*39^*38^*37} = 38.142473.$$

$$\overline{X}_q = \sqrt{\frac{\sum_{i=1}^{N} x_i^2}{N}} = \sqrt{\frac{38^2 + 37^2 + 36^2 + \ldots + 39^2 + 38^2 + 37^2}{15}} = 38.27532;$$

$$\overline{X}_{cub} = \sqrt[3]{\frac{\sum_{i=1}^{N} x_i^3}{N}} = \sqrt[3]{\frac{38^3 + 37^3 + 36^3 + \ldots + 39^3 + 38^3 + 37^3}{15}} = 38.35054$$

$H = 38.04965 < G = 38.142473 < \overline{X} = 38.20000 < \overline{X}_q = 38.27532 < \overline{X}_{cub} = 38.35054$

To calculate the median first we order the data by magnitude:

34	35	36	36	37	37	38	38	38	39	39	41	41	42	42

Here N = 15, odd number. Thus the median is:

$$Me = \widetilde{x}_{\frac{N+1}{2}} = \widetilde{x}_{\frac{15+1}{2}} = \widetilde{x}_8 = 38.$$

(continued)

Example 1.22 (continued)
Using SPSS 16.0 for calculating the same data;

Statistics

Sugar

N	Valid	15
	Missing	0
Mean		38.2
Median		38.0
Mode		38.0

		Frequency	Percent	Valid Percent	Cumulative Percent
Valid	34.00	1	6.7	6.7	6.7
	35.00	1	6.7	6.7	13.3
	36.00	2	13.3	13.3	26.7
	37.00	2	13.3	13.3	40.0
	38.00	3	20.0	20.0	60.0
	39.00	2	13.3	13.3	73.3
	41.00	2	13.3	13.3	86.7
	42.00	2	13.3	13.3	100.0
	Total	15	100.0	100.0	

Box Whisker representation of sugar in chocolate is given in Fig. 1.20.

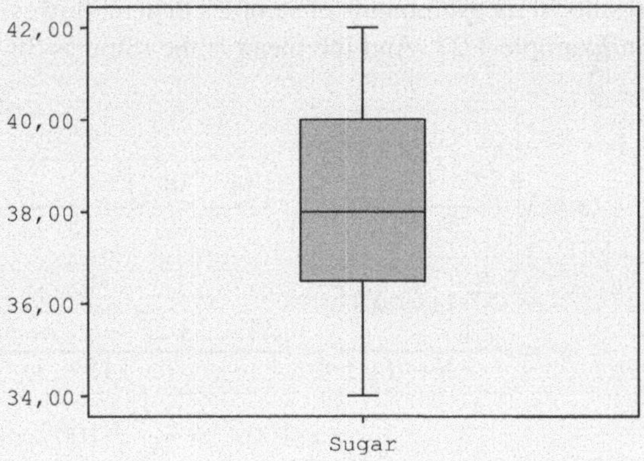

Fig. 1.20 Box-whisker representation of sugar in chocolate

Example 1.23 Find the range, standard deviation, variance and coefficient of variation for the data given in Example 1.22.

Solution: The data given in Example 1.22 represents the whole data (population) and mean of the data is 38.2, so;

$$R = x_{\max} - x_{\min} = 42–34 = 8.0;$$

$$\sigma = \sqrt{\frac{\sum_{i=1}^{N} \left(x_i\text{-}\overline{X}\right)^2}{N}} = \sqrt{\frac{(34 - 38.2)^2 + \ldots + (37 - 38.2)^2}{15}} = 2.40;$$

$$\sigma^2 = \frac{\sum_{i=1}^{N} \left(x_i - \overline{X}\right)^2}{N - 1} = \frac{(34 - 38.2)^2 + \ldots + (37 - 38.2)^2}{15} = 5.76;$$

$$\widehat{V}_S = \frac{\sigma}{\overline{X}} 100\% = \frac{2.4}{38.2} 100\% = 6.28\%..$$

Using SPSS 16.0 for calculating the same data;

Descriptive Statistics.

	Range	Minimum	Maximum	Mean	Std. Deviation	Variance
Sugar	8.000	34.000	42.000	38.200	2.400	5.760

Example 1.24 Find the range, standard deviation, variance and coefficient of variation for the ordinary chocolate. Suppose 15 samples taken from production line randomly. Suppose that the same data were found as given in Example 1.22.

Solution: This situation is different than that in Example 1.23. The data represent the results of measurements of samples in here, however, it was the whole population in Example 1.23. And the mean is the same as 38.2, so;

$$s = \sqrt{\frac{\sum_{i=1}^{N} \left(x_i - \overline{X}\right)^2}{N - 1}} = \sqrt{\frac{(34 - 38.2)^2 + \ldots + (37 - 38.2)^2}{15 - 1}} = 2.484;$$

$$s^2 = \frac{\sum_{i=1}^{N} \left(x_i - \overline{X}\right)^2}{N - 1} = \frac{(34 - 38.2)^2 + \ldots + (37 - 38.2)^2}{15 - 1} = 6.171;$$

$$\widehat{V}_S = \frac{s}{\overline{X}} 100\% = \frac{2.484}{38.2} 100\% = 6.50\%.$$

Using SPSS 16.0 for calculating the same data;

Descriptive Statistics.

	Range	Minimum	Maximum	Mean	Std. Deviation	Variance
Sugar	8.000	34.000	42.000	38.200	2.484	6.171

1.3 Variance Components Analysis: What Are and How Big Are the Sources of the Variability of My Results?

Gerhard Schleining

1.3.1 Objectives and Learning Outcomes

The objective of this chapter is to explain the principle of Variance Components Analysis and to design the experiments accordingly.

After completion of this chapter the reader is able to quantify the contribution of different of sources of variation (e.g. process steps like sampling and analysis) to the total variation of test results.

1.3.2 Introduction

The analysis of variance is a method of separating the total variance of a response into various components, corresponding to the sources of variation which can be identified. There are 2 basic types of classification: Hierarchical Classification (or Nested Design) and Cross Classification.

1.3.2.1 Nested Designs

To assess for example the average moisture content of a bag of flour, the estimate subjected to 2 sources of variation respectively 2 types of errors, those arising in the sampling (inhomogeneity) and those arising in the analysis (moisture content). Similar problems arise in all types of processes – manufacturing processes or testing processes.

The simplest form of hierarchical data involves only 2 independent sources of variation, e.g. the process (or step) of sampling and the process (or step) of testing. In a further step the confidence limits of the sampling variance and the testing variance will be estimated. The pattern can be easily extended to 3 or more sources of variation:

Example: From one batch **n samples** (e.g. cans) are drawn at random and **k analysis** of each sample are performed at randomised sequence (Fig. 1.21).

The Variance Component Analysis divides the variance of a measured variable Y into variance components, one for each process (or step). Each process (or step) is nested in the one above. The goal of such an analysis is usually to estimate the

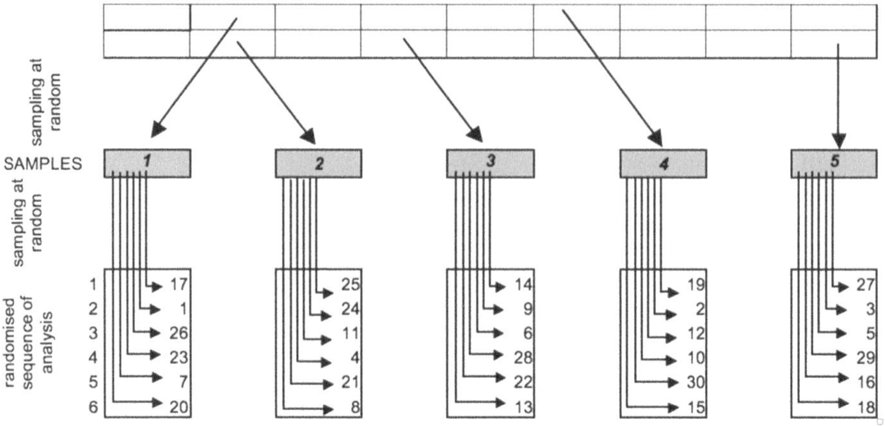

Fig. 1.21 Design of experiments

Table 1.12 Calculation of Variance Components

Source	SQ	Df	MQ	EMQ
Within samples	SQ_0	$Df_0 = a*n*(k-1)$	MQ_0	σ_0^2
Between samples	SQ_1	$Df_1 = a*(n-1)$	MQ_1	$\sigma_0^2 + k^*\sigma_1^2$
Between batches	SQ_2	$Df_2 = a-1$	MQ_2	$\sigma_0^2 + k^*\sigma_1^2 + n^*k^*\sigma_2^2$
etc.				

Table 1.13 Calculation of the Confidence Limits of the standard deviations s_i (for L_1, L_2 see Table 1.25)

Lower limit		Upper limit	
s_0*L_1	$< s_0 <$	s_0*L_2	$L_1, L_2\ (\alpha, Df_0, \infty)$
$\sqrt{\frac{MQ_1*L_1^2-MQ_0}{k}}$	$< s_1 <$	$\sqrt{\frac{MQ_1*L_2^2-MQ_0}{k}}$	$L_1, L_2\ (\alpha, Df_1, Df_0)$
$\sqrt{\frac{MQ_2*L_1^2-MQ_1}{k}}$	$< s_2 <$	$\sqrt{\frac{MQ_2*L_2^2-MQ_1}{k}}$	$L_1, L_2\ (\alpha, Df_2, Df_1)$

If $(MQ_{i+1}*L_1^2 - MQ_i) < 0 => $ lower limit $= 0$

amount of variability contributed by each process (or step), called the variance components (Table 1.12).

The Mean Squares $MQ = SQ/Df$ of the Variance Component Analysis are used to calculate the variances σ_i^2 and the standard deviations $si = \sqrt{\sigma_i^2}$ contributed by each process (or step) i according to the EMQ (Expected Mean Square).

$$\sigma_0^2 = MQ_0$$

$$\sigma_1^2 = (MQ_1 - MQ_0)/k$$

$$\sigma_2^2 = (MQ_2 - MQ_1)/k^*n \ldots \text{etc.}$$

If $\sigma_i^2 < 0 \Rightarrow$ the contribution of process (or step) i cannot be estimated (Table 1.13).

1.3.3 Problem to Solve

Estimate the variance of sampling and the variance of analysis and the corresponding confidence limits for the moisture content of a bag of flour.

1.3.4 Procedure

1.3.4.1 Design of Measurements

n samples will be drawn at random and **k analysis** of the **response variable Y** from each sample should be performed at randomised sequence. An appropriate overview of the design of experiments must be given.

Like in Fig. 1.21 **n = 5** samples will be drawn at random from a bag of flour. Each sample was homogenized and the moisture content has been analysed **k = 6** times from each sample.

1.3.4.2 Visualisation of Data

Some summary statistics (Table 1.14) should be calculated, which are used for the analysis (e.g. average, standard deviation, range of each sample, average and standard deviation of the sample averages). The data should be plotted in an

Table 1.14 Summary statistics for the measured data (%dm..%dry matter)

Sample		1	2	3	4	5
		%dm	%dm	%dm	%dm	%dm
Replications	1	90.92	91.29	89.77	90.57	90.24
	2	90.60	90.86	89.87	90.82	90.35
	3	90.40	90.82	89.88	90.63	90.15
	4	90.52	91.09	89.66	90.98	90.08
	5	90.77	91.01	89.99	90.44	90.46
	6	90.66	91.02	89.95	90.65	90.29
Summary statistics						
av(i)		90.645	91.015	89.853	90.682	90.262
s(i)		0.184	0.169	0.121	0.191	0.137
av(n)				90.491		
s(n)				0.445		
s^2				0.198		
min(i)		90.400	90.820	89.660	90.440	90.080
max(i)		90.920	91.290	89.990	90.980	90.460
range(i)		0.520	0.470	0.330	0.540	0.380
range(i)/s(i)		2.827	2.776	2.725	2.825	2.773

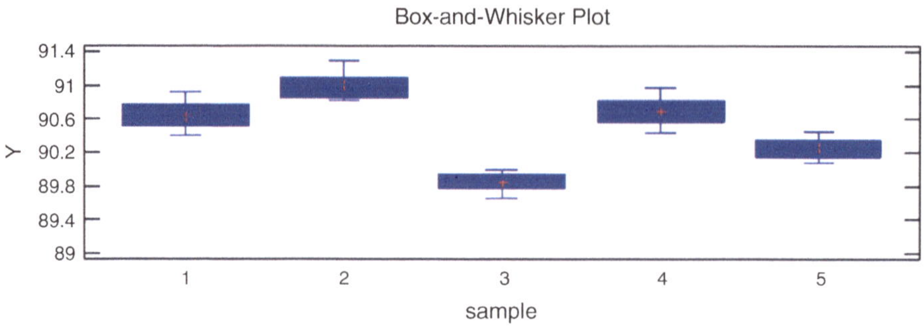

Fig. 1.22 Distribution of the data

Table 1.15 Table of calculated P-values for the Grubbs-Test

Sample				
1	2	3	4	5
1.4951	1.6241	0.6881	0.5842	0.1581
0.2447	0.9154	0.1376	0.7237	0.6446
1.3320	1.1517	0.2202	0.2703	0.8149
0.6796	0.4429	1.5964	1.5608	1.3258
0.6796	0.0295	1.1285	1.2643	1.4474
0.0816	0.0295	0.7982	0.1657	0.2068

appropriate manner (e.g. Box-Whisker-Plot, Fig. 1.22) to give an idea about the distribution of the data and about possible outliers.

The data of the samples seem to be more or less normal distributed, with no outliers and with equal variance. The variability of the sample means seems to be higher than the variability within the samples.

1.3.4.3 Test for Outliers Within Samples

Before applying any statistics, each sample should be tested for outliers. Statistically significant outliers should only be removed, if there is a plausible reason (e.g. something went wrong during the analysis). If an outlier is removed, the outlier may be replaced by the average of the remaining data and the test must be repeated with the new data set. (Table 1.15)

e.g. Grubbs-Test: significant outlier if: $P(y_i) = |y_i - av(y_i)|/s_i > T(k;\alpha)$ (Table 1.22)

Since all P-values are smaller than $T(k;\alpha) = T(6;0.05) = 1.82$ (Table 1.22), there are **no statistically significant outliers**.

1.3.4.4 Check for Normal Distribution Within Samples

The data of each sample must be normal distributed to perform the Variance Component Analysis.

e.g. David-Test: normal distributed if: $T_L(k; \alpha) < P = \text{range}(i)/s(i) < T_U(k; \alpha)$

Since all P-values (Table 1.16) are between $T_L(6;0.05) = 2.280$ and $T_U(6;0.05) = 3.012$ (Table 1.23), the data of **all samples are normal distributed**.

Table 1.16 Calculated P-values for the David-Test

Sample	P
1	2.827
2	2.776
3	2.725
4	2.825
5	2.773

1.3.4.5 Variance Check

The standard deviations of the data within each sample must be the same to perform the Variance Component Analysis.

e.g. Hartley's test: standard deviations within each of the samples are the same if:

$$P = \max\left(s^2\right)/\min\left(s^2\right) < T(\alpha; n; k-1)$$

Since $P = 0.191^2/0.121^2 = 2.49$ is smaller than $T(0.05;5;5) = 16.30$ (Table 1.24), there is **not a statistically significant difference amongst the standard deviations** at the 95.0% confidence level.

1.3.4.6 Variance Component Analysis

The analysis can be done with the use of any statistical software (Table 1.17).

The factor contributing the most variance is **sample**. Its contribution represents 87.99% of the total variation in Y.

Table 1.17 ANOVA Table for Y by sample

Source	Sum of Squares	Df	Mean Square	Var. Comp	%
TOTAL (corrected)	5.42555	29			
sample	4.76315	4	1.19079	0.194048	87.99
ERROR	0.6624	25	0.026496	0.026496	12.01

Table 1.18 Errors and Confidence Limits for $\alpha = 0.05$, *) L_1 and L_2 see Table 1.11

Source of error	Lower limit	Error	Upper limit
Analysis	$s_0{}^*L_1$ $0.163*0.81 = \mathbf{0.132}$	$<s_0<$ < 0.163 $<$	$s_0{}^*L_2$ $0.163*1.32 = \mathbf{0.215}$
	$L_1(\alpha, Df_0 = 25, \infty) = 0.81$*)		$L_2(\alpha, Df_0 = 25, \infty) = 1.32$*)
sampling	$\sqrt{\frac{MQ_1 * L_1{}^2 - MQ_0}{k}}$ $\sqrt{\frac{1.19079 * 0.6^2 - 0.026496}{6}}$ $= 0.259$	$<s_1<$ < 0.441 $<$	$\sqrt{\frac{MQ_1 * L_2{}^2 - MQ_0}{k}}$ $\sqrt{\frac{1.19079 * 2.4^2 - 0.026496}{6}}$ $= 1.067$
	$L_1(\alpha, Df_1 = 4, Df_0 = 25) = 0.60$*)		$L_2(\alpha, Df_1 = 4, Df_0 = 25) = 2.40$*)

1.3.4.7 Calculation of Variance Components and Confidence Limits

$$\sigma_0{}^2 = MQ_0 = 0.026496 \qquad s_0 = \sqrt{\sigma_0{}^2} = 0.163$$

$$\sigma_1{}^2 = (MQ_1 - MQ_0)/k = (1.19079 - 0.026496)/6 = 0.19405 \quad s_1 = \sqrt{\sigma_1{}^2} = 0.441$$

The variability of sampling ($s_1 = 0.441$) is more than 2 times higher than the variability of analysis ($s_0 = 0.163$) (Table 1.18).

1.4 Testing of Hypothesis

Zsuzsa H. Horváth and Cecília Hodúr

1.4.1 Objectives and Learning Outcomes

The objective of this chapter is to review hypothesis tests (Bhattacharyya and Johnson 1997; Byrkit 1987; Rosner 1989) and then to learn about the parametric test for population means and variances. First of all, we write the procedure of the tests. Applying the tests to a specific problem is described step by step, making the method easy to use. The determination of ratios, their transformation into coded factor levels, the expression of component ratios with ratios, the design matrix presenting coded factor levels, and the calculation and interpretation of results are presented by the case studies. This was meant to encourage mindfulness and deeper processing and, therefore, enhance learning outcomes.

1.4.2 Introduction

Decisions are generally based on a belief regarding the true state of things. We often have to decide such questions, that

- the population mean is equal to 50
- the population standard deviation is less, than 0.5
- the means from two populations are equal
- the variance from 6 populations are equal.

The statistical inference provides an excellent systematic method for assessing whether or not the belief likely to be true. This chapter presents methods for testing hypotheses and drawing conclusion about the probability that these hypotheses are true. First we will discuss the general method of formulating and testing hypotheses and then present how to apply these methods to hypotheses about population variances and means.

The hypothesis tests can be divided into two parts: parametric and non-parametric tests. The parametric methods suppose that the examined populations fit to a normal distribution. The non-parametric methods can be used without this assumption.

In this part we will review the parametric tests about means and variances. First, the tests about variance of populations will be described. The χ^2 –test for one population variance, the F-test for comparison of two variances and the tests about homogeneity of variances will be presented. Then we present tests about means, tests for one mean, for comparison of two means and after we consider tests about more than two population means.

1.4.3 Problem to Solve

We can often use the different tests about the mean and the variance of populations in the food industrial practise and in course of the food science research. In food industry the tests have important role in quality control, for example, when we have to check the weight, protein content, fat content or moisture content of the products. These methods are well applicable, if we have to: compare results of two measuring procedures; analyse, how the different circumstances influence the change of one or more parameters of food materials, determine the optimal value of the temperature, pressure for a one food industrial process, and so on.

1.4.4 Procedure

1.4.4.1 Basis of the Hypothesis Tests

A statistical hypothesis is a statement regarding the value of one or more population parameters. A hypothesis supporting the belief is called the research hypothesis. First we formulate two hypotheses, the **null hypothesis** and the **alternative hypothesis**. The **null hypothesis** (H_0) is a statement that the population parameter is equal to a specific value. The **alternative hypothesis** (H_1) is a hypothesis opposing the null hypothesis. Consider a production line where the foreman believes that machine produces too many items with unacceptable weight. One statement about mean of the population supporting this belief is the hypothesis $m = m_0$, where m_0 is the requirement weight of the product. So null hypothesis H_0 is the statement: $m = m_0$; and the alternative hypothesis, H_1 is the statement $m \neq m_0$.

Then we must determine an appropriate sampling distribution of a sample statistic based on the assumption that the null hypothesis about the population parameter is true. If the hypothesis concerns a population mean, the appropriate sampling distribution is the sampling distribution of the mean, if it concerns a population variance, we should use the sampling distribution of the sample variance, and so on. Then we collect data to see whether the sample result is consistent or inconsistent with null hypothesis. **If the sample result supports the alternate hypothesis, we reject the null hypothesis in favour of the alternate hypothesis.**

The possibility of error is implicit in every statistical test. When we use a sample result to test the validity of a null hypothesis, we know that sample statistics vary. Therefore it is possible to reject a true hypothesis as well as to accept a false one. The error is called a **type I error**, if we reject a true null hypothesis, and the error is called a **type II error**, if we accept a false null hypothesis. If we assume that H_0 is actually true, we can usually determine the precise sampling distribution of the statistic used for testing the hypothesis, so we can control the value of the probability of making a type I error (α). The value of $\boldsymbol{\alpha}$ is usually set before the sample is obtained, and $\boldsymbol{\alpha}$ is equal to the probability of making an error in rejecting the null hypothesis. This probability, called the **level of significance** for the test, designates the willingness of the experimenter to make a type I error. The probability of correctly accepting the null hypothesis ($1\text{-}\alpha$) is called the **level of confidence**. The probability of making a type II error is designated β. The ($1\text{-}\beta$) is the **power** of the test. The power of the test is the probability of correctly rejecting a false null hypothesis. It can only be computed for a specific alternative hypothesis.

In most cases an acceptable value of α is decided upon and the possible outcomes are divided into two groups before the sample is selected. One group makes up the **rejection (critical) region**, if the sample result falls into this region we reject the null hypothesis, in favour of the alternative hypothesis. The other group comprises the **acceptance region**, if the sample result falls into this region we fail to reject the null hypothesis, concluding either that the null hypothesis is probably true, the level of confidence is ($1\text{-}\alpha$).

If we want to determine the critical and acceptance region we have to determine the **test statistic**. The test statistic to be used is determined by the parameter appearing in the null hypothesis, the shape of the distribution of the sample statistic of the parameter. For example, suppose, that the H_0: $m = m_0$, the sample size is n, and the sampling distribution is normal. If the H_0 is true the $t_c = \frac{\bar{x} - m_0}{\frac{s}{\sqrt{n}}}$ fits to t distribution with $(n\text{-}1)$ degrees of freedom (t_{n-1}), where $\bar{x}$ is the mean of the sample, s is the standard deviation of the sample.

If we have identified the test statistic then we can determine the critical (rejection) region. We relate the sample result to the probability distribution of the test statistic and divide the area under the graph of this distribution density function into two regions. The area above the critical region is α. The critical region depends on the alternative hypothesis. If the H_0: $m = m_0$ and alternative hypothesis is H_1: $m \neq m_0$ we say, that the alternative hypothesis is two-sided. If the alternative hypothesis is H_1: $m < m_0$ or $m > m_0$, we say, that the alternative hypothesis is one- sided. Supposed, that test statistic fits to a standard normal distribution. The two- sided critical region we can see on Fig. 1.42, and critical regions in case of one- sided alternative hypothesis are presented on Fig. 1.43. In the figure z_p is the value for which $P(X < z_p) = p$, where X is a random variable with standardized normal distribution. Value of z_p is called the **p-quantile** of standardized normal distribution.

So, the hypothesis-testing procedure consists of the following steps:

- Formulating a research hypothesis, usually a statement about a population parameter that we will verify the belief.
- Determining statement of H_0 and H_1.
- The choice of the probability of making a type I error α**.** The choice depends on a concrete problem; practice values of 0.1, 0.05, and 0.01 are commonly used.
- Determining the critical region.
- Calculating of the value of test statistic for sample.
- Decision:

  If the sample result falls to the critical region we reject H_0.

  If the sample result doesn't fall to the critical region we fail to reject H_0.

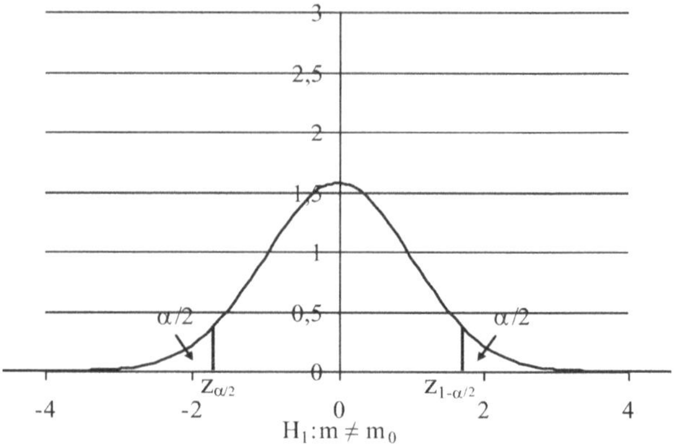

Fig. 1.42 Two-sided critical region for standardized normal distribution

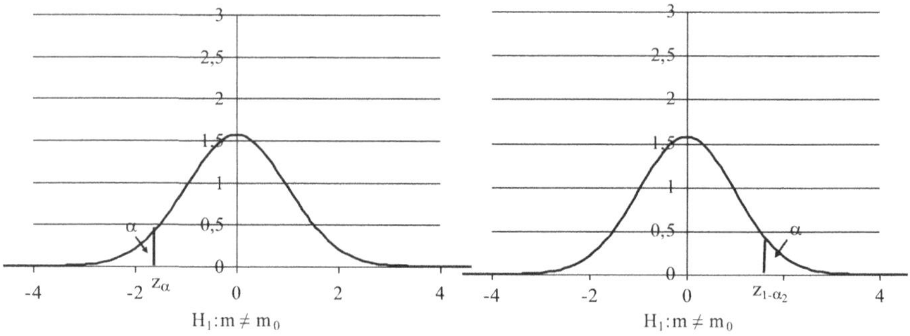

Fig. 1.43 One-sided critical regions for standardized normal distribution

It is clear, that the conclusion of the hypothesis test is based on the choice of the probability of making a type I error α. It is possible, that if $\alpha = 0.05$ was chosen, then H_0 was rejected; and if $\alpha = 0.01$ was chosen, then H_0 was accepted. Therefore is used the experiment observed significance level (P). The observed **significance level** is the smallest α for which the observed sample results in rejecting H_0. The statistical computer programs, such as Statistica, Statgraphics and so on, show the P value as part of the output. So we can make an independent judgment based on the following procedure: if $P < \alpha$ (chosen by us), then H_0 will be rejected, and if $P \geq \alpha$, then H_0 will not be rejected.

In the following parts we will see how the hypothesis-testing procedure can be used to test hypothesis about variance and mean of the populations.

1.4.4.2 Comparison of Variances

We can use the hypothesis tests about the population variance, for example, in the industrial practise, because the variations in the quality of products must be kept to a minimum. Usually the maximum of the variance is given and we have to check weather the variance is less, than the given limit. Often the hypothesis tests for population mean use test concerning the value of the population variance.

1.4.4.2.1 Hypothesis Test for the Normal Distributed Population Variance (χ2-Test)

Suppose, we obtain a random sample of n observation of normal random variable X for which the population variance is hypothesized to be σ_0^2. We test the H_0: $\sigma^2 = \sigma_0^2$. We used the following fact to determine the test statistic. If we have a normal distributed population where σ^2 is the population variance, and s^2 is the variance of a sample of n observations, the random variable $\chi^2 = \frac{(n-1)s^2}{\sigma^2}$ has a χ^2 (chi-square) distribution with $(n$-$1)$ degrees of freedom ($\chi^2{}_{n-1}$). So, if the hypothesis is true, the random variable $\chi_c^2 = \frac{(n-1)s^2}{\sigma_0^2}$ has χ^2 distribution with $(n$-$1)$ degrees of freedom.

So the test for population variance is the following:

- Condition of application: normally distributed population.
- Null hypothesis: H_0: $\sigma^2 = \sigma_0^2$
- Test statistic: $\chi_c^2 = \frac{(n-1)s^2}{\sigma_0^2}$
- Choice of the probability of making a type I error α
- Critical region depends on the alternative hypothesis:

Alternative hypothesis:	Critical region:
$H_1 : \sigma^2 \neq \sigma_0^2$	$C = \left\{\chi_c^2 < \chi_{\frac{\alpha}{2};n-1}^2 \cup \chi_c^2 > \chi_{1-\frac{\alpha}{2};n-1}^2\right\}$
$H_1 : \sigma^2 < \sigma_0^2$	$C = \left\{\chi_c^2 < \chi_{\alpha;n-1}^2\right\}$
H_1: $\sigma^2 > \sigma_0^2$	$C = \left\{\chi_c^2 > \chi_{1-\alpha;n-1}^2\right\}$

The $\chi_{p;n-1}^2$ is the p-quantile of the χ_{n-1}^2 distribution. The values can be found in Table 1.26.

Example 1.25

We have to check the standard deviation of weight of the coffee pack. The allowable standard deviation is $\sigma_0 = 0.15$. We take a sample of size $n = 20$. The standard deviation of sample is $s = 0.16$ and we can assume, that the population is normally distributed. Use $\alpha = 0.05$ to test the hypothesis.

(continued)

Example 1.25 (continued)
Solution:
Hypothesis: H_0: $\sigma^2 = 0.15^2$
H_1: $\sigma^2 > 0.15^2$
Value of test statistic for sample: $\chi_c^2 = \frac{(n-1)s^2}{\sigma_0^2} = \frac{19 \cdot 0.16^2}{0.15^2} = 21.62$
Critical region: $C = \{\chi_c^2 > \chi_{0,95;\,19}^2\} = \{\chi_c^2 > 30.14\}$, because degrees of freedom, $df = 19$.
Conclusion
Value of χ_c^2 doesn't fall into critical region, so we fail to reject H_0. The standard deviation of weight of the pack is acceptable. The confidence level is 0.95.

Hereunder we will not detail the conceptual basis of the test, only the steps of the statistical tests will be described.

1.4.4.2.2 Hypothesis Test to Compare Two Normal Distributed Population Variances (F-Test)

Suppose, that two independent random samples are given, a sample of size n_1 from a normal distributed population with variance σ_1^2, and a sample of size n_2 from a normal distributed population with variance σ_2^2. The variance of the samples is s_1^2 and s_2^2, respectively. We test the hypothesis that the two population variances are equal.
The test to compare two population variance is the following:

- Condition of application: two normally distributed populations.
- Null hypothesis: H_0: $\sigma_1^2 = \sigma_2^2$
- Test statistic: $F_c = \max\left\{\frac{s_1^2}{s_2^2}; \frac{s_2^2}{s_1^2}\right\}$, where F_c is a random variable with F distribution, (n_1-1) degrees of freedom for the numerator and (n_2-1) degrees of freedom for the denominator ($F_{n_1-1;\,n_2-1}$), or (n_2-1) degrees of freedom for the numerator (n_1-1) degrees of freedom for the denominator ($F_{n_2-1;\,n_1-1}$), accordingly.
- Choice of the probability of making a type I error α.
- Critical region depends on the alternative hypothesis:
  Alternative hypothesis: Critical region:
- $H_1 : \sigma_1^2 \neq \sigma_2^2$ if $s_1^2 \geq s_2^2$ $C = \left\{F_c > F_{1-\frac{\alpha}{2};\,n_1-1;\,n_2-1}\right\}$
  - if $s_1^2 < s_2^2$ $C = \left\{F_c > F_{1-\frac{\alpha}{2};\,n_2-1;\,n_1-1}\right\}$
- if $s_1^2 \geq s_2^2$
- $H_1 : \sigma_1^2 > \sigma_2^2$ $C = \{F_c > F_{1-\alpha;\,n_1-1;\,n_2-1}\}$
- if $s_2^2 \geq s_1^2$
- $H_1 : \sigma_2^2 > \sigma_1^2$ $C = \{F_c > F_{1-\alpha;\,n_2-1;\,n_1-1}\}$

$F_{1-\alpha;n_1-1;n_2-1}$ is the p-quantile of the $F_{n_1-1;n_2-1}$ distribution. The values can be found in Table 1.27 for $p = 0.95$, and in Table 1.28 for $p = 0.975$.

Example 1.26

We compare variances of two normally distributed samples. We calculated, that the standard deviation of sample of size $n_1 = 15$ is $s_1 = 2.45$ and the standard deviation of sample of size $n_2 = 18$ is $s_2 = 4.13$. We test if the variances are equal. Use $\alpha = 0.1$ to test the hypothesis.

Solution:

Hypothesis: H_0: $\sigma_1^2 = \sigma_2^2$

H1: $\sigma_1^2 \neq \sigma_2^2$

Value of test statistic for sample: $F_c = \max\left\{\frac{s_1^2}{s_2^2}; \frac{s_2^2}{s_1^2}\right\} = \frac{4.13^2}{2.45^2} = 2.84$

Critical region: $C = \left\{F_c > F_{1-\frac{\alpha}{2};n_2-1;n_1-1}\right\} = \{F_c > F_{0.95;17;14}\} = \{F_c > 2.42\}$

Conclusion

Value of F_c falls into the critical region, so we reject H_0. So the variances are not equal, they differ significantly, $P < 0.1$.

1.4.4.2.3 Test for Homogeneity of Populations Variance

Techniques for comparing means of populations generally assume the populations have the same variance. Before using these tests, it is advisable to test whether this assumption of homogeneity of variance is reasonable. Bartlett's test is a commonly used test for equal variances, but is very sensitive to departures from normality. So we can use if only for normal distributed samples. When non-normality is suspected, the Levene's procedure is usable. These procedures are described as follows.

- **Bartlett's Test for Homogeneity of Variances**

We test variance of samples of size n_i from the i-th population, $i = 1, 2, \ldots, k$, and the usual variance estimates from each sample: $s_1^2, s_2^2, \ldots, s_k^2$.

- Condition of application: normal distributed populations.
- Null hypothesis: H_0: $\sigma_1^2 = \sigma_2^2 = \ldots = \sigma_k^2$
- Alternative hypothesis: H_1: $\sigma_i^2 \neq \sigma_j^2$ for at least one pair $(i;j)$
- Test statistic: $B_c = \dfrac{\nu \ln s^2 - \sum_{i=1}^{k} \nu_i \ln s_i^2}{1 + \frac{1}{3(k-1)}\left(\sum_{i=1}^{k} \frac{1}{\nu_i} - \frac{1}{\nu}\right)}$, where

$$\nu_i = n_i - 1, \quad \nu = \sum_{i=1}^{k} \nu_i, \quad s^2 = \frac{1}{\nu} \sum_{i=1}^{k} \nu_i s_i^2.$$

The test statistic follows a Chi-square distribution with $\nu\text{-}1$ degrees of freedom ($\chi^2_{\nu-1}$).

- Critical region: $C = \left\{B_c > \chi^2_{1-\alpha;\nu-1}\right\}$
- **The Levene's Test for Homogeneity of Variances**

We test variance of samples of size n_i from the i-th population, $i = 1, 2, \ldots, k$, and y_{ij} is the j-th element of sample for i-th population. It is usable in all cases.

- Null hypothesis: H_0: $\sigma_1^2 = \sigma_2^2 = \ldots = \sigma_k^2$
- Alternative hypothesis: H_1: $\sigma_i^2 \neq \sigma_j^2$ for at least one pair $(i;j)$
- Test statistic: $F_c = \dfrac{(k-1)\sum_{i=1}^{k} n_i(\overline{z}_i - \overline{z})^2}{(N-k)\sum_{i=1}^{k}\sum_{j=1}^{n_i}\left(z_{ij} - \overline{z}_i\right)^2}$, where

$N = \sum_{i=1}^{k} n_i$, $z_{ij} = |y_{ij} - y_i|$, y_i mean or median of i-th sample,

$$\overline{z}_i = \frac{1}{n_i}\sum_{j=1}^{n_i}\left|y_{ij} - y_i\right|, \overline{z} = \frac{1}{N}\sum_{i=1}^{k}\sum_{j=1}^{n_i}\left|y_{ij} - y_i\right|$$

F_c follows F-distribution, degree of freedom of numerator is k-1 and degree of freedom of denominator is N-k.

- Critical region: $C = \{F_c > F_{1-\alpha;\ k-1;\ N-k}\}$

Example 1.27
We have to check whether the moisture contents of paprika powders, produced in different days, are equal or not. We can see the measured moisture content values in the Table 1.32.We use Bartlett's and Levene's tests for the samples. Use α = 0.05 to test the hypothesis.

Solution:
Bartlett's Test:

The samples follow normal distribution.

- Value of test statistic for sample: $B_c = 1175$
- Critical region: $C = \left\{B_c > \chi^2_{0.95;20}\right\} = \{B_c > 31.41\}$
- Conclusion: Value of B_c doesn't fall into the critical region, so we fail to reject H_0. The variances don't differ significantly ($\alpha = 0.05$).

  - Levene's Test:

- Value of test statistic for sample: $F_c = \frac{0.0206}{0.0275} = 0.749$
- Critical region: $C = \{F_c > F_{0.95;\ 3;\ 21}\} = \{F_c > 3.072\}$
- Conclusion: Value of F_c doesn't fall into the critical region, so we fail to reject H_0. The variances don't differ significantly ($\alpha = 0.05$).

Table 1.32 Moisture content of paprika powders made in different day

1th day	2th day	3th day	4th day
8.12	8.34	9.01	8.56
8.15	8.52	8.77	8.76
8.92	8.98	8.99	8.32
9.01	9.12	8.95	9.06
8.56	9.06	8.45	9.01
9.07	8.77		8.07
8.67			8.88
$n_1 = 7$	$n_2 = 6$	$n_3 = 5$	$n_4 = 7$

1.4.4.2.4 Hypothesis Tests for Population Mean

Suppose we obtain a random sample of n observations of random variable X with mean m and standard deviation is σ. There are the following types of questions regarding the true mean of the population that can be addressed with the sample data. Does the true mean agree with a known standard or assumed mean? Is the true mean of the population less than a given standard? Is the true mean of the population at least as large as a given standard?

We used the following facts to determine the test statistic.

The mean of the sampling distribution of the mean is the population mean ($m_{\overline{x}} = m$) and the standard deviation of the sampling distribution of the mean is $\sigma_{\overline{x}} = \frac{\sigma}{\sqrt{n}}$. In addition important to know whether or not a sample has come from a normal population and we know or don't know the standard deviation of X.

1.4.4.2.5 Test for Population Mean, If the variance Is Known (z-Test for One Sample)

- Condition of application: – the population variance is known
  - normal distributed population or $n > 30$
- Null hypothesis: H_0: $m = m_0$
- Test statistic: $z_c = \frac{\overline{x} - m_0}{\frac{\sigma}{\sqrt{n}}}$, where z_c following standardized normal distribution.
- Critical region depend on the alternative hypothesis:

Alternative hypothesis:	Critical region:
H_1:$m \neq m_0$	$C = \left\{z_c < z_{\frac{\alpha}{2}} \cup z_c > z_{1-\frac{\alpha}{2}}\right\} = \left\{\lvert z_c \rvert > z_{1-\frac{\alpha}{2}}\right\}$
H_1:$m < m_0$	$C = \{z_c < z_\alpha\} = \{z_c < -z_{1-\alpha}\}$
H_1: $m > m_0$	$C = \{z_c > z_{1-\alpha}\}$

The z_p is p-quantile of the standardized normal distribution. The values of distribution function of standardized normal distribution can be found in Table 1.29. The density function of standardized normal distribution is symmetric, therefore $z_\alpha = -z_{1-\alpha}$.

Example 1.28
We have to check the weight of the coffee pack. The standard deviation is known. $\sigma = 0.15$. The required weight is $m_0 = 25$. We take a sample of size $n = 20$. Mean of sample is $\overline{x} = 24.87$ and we can assume that the population is normal distributed. Use $\alpha = 0.04$ to test the hypothesis.

Solution: The standard deviation of population is known, so we can use z-test for one sample.

$$\text{Hypothesis} : H_0 : m = 25$$
$$H_1 : m \neq 25$$

- Value of test statistic for sample: $z_c = \frac{\overline{x} - m_0}{\frac{\sigma}{\sqrt{n}}} = \frac{24.87 - 25}{\frac{0.15}{\sqrt{20}}} = -3.87$
- Critical region: $C = \left\{ |z_c| > z_{1-\frac{\alpha}{2}} \right\} = \{ |z_c| > z_{0.98} \} = \{ |z_c| > 2.05 \}$, $z_{0.98} = 2.05$, because $\Phi(2.05) = 0.98$
- Conclusion: Value of z_c falls into the critical region, so we reject H_0. Based on samples we can say, that weight of packs are not equal to 25, the mean of the weight of pack differ from 25 significantly. ($P < 0.04$)

1.4.4.2.5.1 *Test for Population Mean, If the Variance Is Unknown* (t-*Test for One Sample*)

- Condition of application: normal distributed population or $n > 30$
- Null hypothesis: H_0: $m = m_0$
- Test statistic: $t_c = \frac{\overline{x} - m_0}{\frac{s}{\sqrt{n}}}$, where t_c follows t distribution with $n-1$ degrees of freedom (t_{n-1}).
- Critical region depends on the alternative hypothesis:

Alternative hypothesis:	Critical region:
$H_1: m \neq m_0$	$C = \left\{ t_c < t_{\frac{\alpha}{2}; n-1} \cup t_c > t_{1-\frac{\alpha}{2}; n-1} \right\} = \left\{ \lvert t_c \rvert > t_{1-\frac{\alpha}{2}; n-1} \right\}$
$H_1: m < m_0$	$C = \{ t_c < t_{\alpha; n-1} \} = \{ t_c < -t_{1-\alpha; n-1} \}$
$H_1: m > m_0$	$C = \{ t_c > t_{1-\alpha; n-1} \}$

The $t_{p; n-1}$ is p-quantile of t_{n-1} distribution. The values can be found in Table 1.30. The density function of t distribution is symmetric, so $t_{\alpha; n} = -t_{1-\alpha; n}$.

Example 1.29
We have to control, whether the fat content of clotted cream is at least 20%. The random sample of size 18 was taken. The sample mean is 19.91 and the sample standard deviation is 0.224. The sample fits to a normal distribution. Use the $\alpha = 0.05$ to test the hypothesis.

(continued)

Example 1.29 (continued)
Solution: The standard deviation of population is unknown, so we have to use *t*-test for one sample.
Hypothesis: $H_0 : m = 20$ $H_1 : m < 20$
Value of test statistic for sample: $t = \frac{\overline{x} - m_0}{\frac{s}{\sqrt{n}}} = \frac{19,91 - 20}{\frac{0.224}{\sqrt{18}}} = -1.70$
Critical region: $C = \{t_c < t_{0.05;\ 17}\} = \{t_c < -t_{0.95;\ 17}\} = \{t_c < -1.740\}$
Conclusion
Value of t_c doesn't fall into the critical region, so we accept H_0. Based on samples we can say that the fat content of clotted cream was at least *20%*. The confidence level is *0.95*.

Henceforth we look the tests about two means. When we compare two population means it is important to know whether the samples are independent or not.

1.4.4.3 Hypothesis Tests to Compare Two Normal Distributed Population Means in Case of Independent Samples

Suppose, that two independent random samples are given, a sample of size n_1 from a normal distributed population with variance σ_1^2 and a sample of size n_2 from a normal distributed population with variance σ_2^2. The means of the samples are $\overline{x}_1$ and $\overline{x}_2$, and variance of the samples are s_1^2 and s_2^2, respectively. We test the hypothesis that the two population means are equal.

We have to distinguish three cases:

- The variances of population are known.
- The variances of population are unknown but supposedly are equal.
- The variances of population are unknown and supposedly are not equal.

1.4.4.3.1 Test for Two Population Means, If the Variances Are Known (z-Test for Two Samples)

- Condition of application: – the variance of populations are known
  - normal distributed populations or $n_1, n_2 > 30$
- Null hypothesis: H_0: $m_1 = m_2$
- Test statistic: $z_c = \frac{\overline{x}_1 - \overline{x}_2}{\sqrt{\frac{\sigma_1^2}{n_1} + \frac{\sigma_2^2}{n_2}}}$, where z_c follows standardized normal distribution.
- Critical region depends on the alternative hypothesis:

Alternative hypothesis:	Critical region:
H_1:$m_1 \neq m_2$	$C = \left\{z_c < z_{\frac{\alpha}{2}} \cup z_c > z_{1-\frac{\alpha}{2}}\right\} = \left\{\lvert z_c \rvert > z_{1-\frac{\alpha}{2}}\right\}$
H_1:$m_1 < m_2$	$C = \{z_c < z_\alpha\} = \{z_c < -z_{1-\alpha}\}$
H_1:$m_1 > m_2$	$C = \{z_c > z_{1-\alpha}\}$

The z_p is p-quantile of the standardized normal distribution. The values of distribution function of standardized normal distribution can be found in Table 1.29.

If the variances of populations are unknown, first we have to check whether the equality of variances is true. We can use F-test that has been described above.

1.4.4.3.2 Test for Two Population Means, If the Variances Are Unknown and Supposedly Equal (t-Test for Two Samples)

- Condition of application: – The variance of populations are unknown and

$$\sigma_1^2 = \sigma_2^2, \text{ supposedly}$$

- Normal distributed populations.
- Null hypothesis: H_0: $m_1 = m_2$
- Test statistic: $t_c = \frac{\bar{x}_1 - \bar{x}_2}{s\sqrt{\frac{1}{n_1}+\frac{1}{n_2}}}$, where $s^2 = \frac{(n_1-1)s_1^2+(n_2-1)s_2^2}{n_1+n_1-2}$.

t_c follows t-distribution with $n_1 + n_1 - 2$ degrees of freedom.

- Critical region depends on the alternative hypothesis:

Alternative hypothesis:	Critical region:
H_1:$m_1 \neq m_2$	$C = \left\{t_c < t_{\frac{\alpha}{2};n_1+n_2-2} \cup t_c > t_{1-\frac{\alpha}{2};n_1+n_2-2}\right\} = \left\{\lvert t_c\rvert > t_{1-\frac{\alpha}{2};n_1+n_2-2}\right\}$
H_1:$m_1 < m_2$	$C = \{t_c < t_{\alpha;n_1-n_2-2}\} = \{t_c < -t_{1-\alpha;n_1-n_2-2}\}$
H_1:$m_1 > m_2$	$C = \{t_c > t_{1-\alpha;n_1-n_2-2}\}$

The $t_{p;n_1+n_2-2}$ is p-quantile of $t_{n_1+n_2-2}$ distribution. The values can be found in Table 1.30. The density function of t distribution is symmetric, so $t_{\alpha;n_1+n_2-2} = -t_{1-\alpha;n_1+n_2-2}$.

1.4.4.3.3 Test for Two Population Means, If the Variances Are Unknown and Supposedly Unequal (Welch-Test)

- Condition of application: – The variance of populations are unknown and

$$\sigma_1^2 \neq \sigma_2^2, \text{ supposedly.}$$

Normal distributed populations.

Null hypothesis: H_0: $m_1 = m_2$

Test statistic: $t_c = \frac{\bar{x}_1 - \bar{x}_2}{\sqrt{\frac{s_1^2}{n_1}+\frac{s_2^2}{n_2}}}$, t_c follows t-distribution, with df degrees of freedom,

where $df = \frac{\left(\frac{s_1^2}{n_1}+\frac{s_2^2}{n_2}\right)^2}{\frac{s_1^4}{n_1^2(n_1-1)}+\frac{s_2^4}{n_2^2(n_2-1)}}$.

To give df the Welch-Satterthwaite approximation was used, because the degrees of freedom are not known exactly.

Critical region depends on the alternative hypothesis:

Alternative hypothesis:	Critical region:
$H_1{:}m_1 \neq m_2$	$C=\left\{t_c < t_{\frac{\alpha}{2};df} \cup t_c > t_{1-\frac{\alpha}{2};df}\right\} = \left\{\lvert t_c\rvert > t_{1-\frac{\alpha}{2};df}\right\}$
$H_1{:}m_1 < m_2$	$C = \{t_c < t_{\alpha;\,df}\} = \{t_c < -t_{1-\alpha;\,df}\}$
$H_1{:}m_1 > m_2$	$C = \{t_c > t_{1-\alpha;\,df}\}$

The $t_{p;\ df}$ is p-quantile of $t_{n_1+n_2-2}$ distribution. The values can be found in Table 1.30. The density function of t distribution is symmetric, so $t_{\alpha;\ df} = -t_{1-\alpha;\ df}$.

1.4.4.4 Hypothesis Tests to Compare Two Normal Distributed Population Means in Case of Dependent Samples (Paired t-test)

In many cases we can pair the members of two samples according to certain criteria. In other cases there may be only one sample but in that case two measurements are made on the sample. Since the subjects are paired the samples are not independent. In this case we can use a paired difference test.

Suppose, we randomly obtain n pairs of observations of random variables X_1 and X_2 with means m_1 and m_2, respectively. We test the hypothesis $m_1 - m_2 = 0$ or $m_1 - m_2 = d_0$, where d_0 is the hypothesized difference between two means.

Let $d_i = x_{1i} - x_{2i}$ difference for each sample pair.

- Condition of application: – Population of d is normal distributed or number of pairs $n \geq 50$.
- Null hypothesis: H_0: $m_1 - m_2 = d_0$
- Test statistic: $t_c = \frac{\overline{d}}{\frac{s_d}{\sqrt{n}}}$, where $\overline{d} = \frac{1}{n}\sum_{i=1}^{n} d_i$ is the mean of differences and

  $s_d^2 = \frac{1}{n-1}\sum_{i=1}^{n}\left(d_i - \overline{d}\right)^2$, s_d is the standard deviation of differences.

t_c follows t-distribution, with n-1 degrees of freedom.

- Critical region depends on the alternative hypothesis:

Alternative hypothesis:	Critical region:
$H_1{:}m_1 - m_2 \neq d_0$	$C=\left\{t_c < t_{\frac{\alpha}{2};n-1} \cup t_c > t_{1-\frac{\alpha}{2};n-1}\right\} = \left\{\lvert t_c\rvert > t_{1-\frac{\alpha}{2};n-1}\right\}$
$H_1{:}m_1 - m_2 < d_0$	$C = \{t_c < t_{\alpha;\,n-1}\} = \{t_c < -t_{1-\alpha;\,n-1}\}$
$H_1{:}m_1 - m_2 > d_0$	$C = \{t_c > t_{1-\alpha;\,n-1}\}$

The t_{n-1} is p-quantile of t_{n-1} distribution. The values can be found in Table 1.30.

Example 1.30
We compare means of two normal distributed samples. We calculated, that the standard deviation of sample of size $n_1 = 15$ is $s_1 = 2.45$, its mean is $\overline{x}_1 = 45.6$ and the standard deviation of sample of size $n_2 = 18$ is $s_2 = 4.13$, its mean is $\overline{x}_2 = 44.5$. We test if the means are equal. Use $\alpha = 0.1$ to test the hypothesis.

Solution

The variances of populations are unknown, so first we have to test, whether they are equal or not, using *F*-test. In Example 1.26 we established, that the variances are not equal, so we have to use Welch-test.

- Hypothesis: H_0: $m_1 = m_2$

$$H_1 : m_1 \neq m_2$$

- Value of test statistic for sample: $t_c = \frac{\overline{x}_1 - \overline{x}_2}{\sqrt{\frac{s_1^2}{n_1} + \frac{s_2^2}{n_2}}} = 0.8613$

$$df = \frac{\left(\frac{s_1^2}{n_1} + \frac{s_2^2}{n_2}\right)^2}{\frac{s_1^4}{n_1^2(n_1 - 1)} + \frac{s_2^4}{n_2^2(n_2 - 1)}} = 28.26$$

- Critical region: $C = \left\{|t_c| > t_{1-\frac{\alpha}{2};df}\right\} = \{|t_c| > t_{0.95;29}\} = \{|t_c| > 1.70\}$
- Conclusion: Value of t_c doesn't fall into the critical region, so H_0 is accepted. The means do not differ significantly ($P = 0.1$).

Example 1.31
We compare two methods for analysis of moisture content. The measurements were made for 12 times with each method on the same sample. In the Table 1.33 we can see the measured values. We test, that the measured values are equal or not. Use $\alpha = 0.5$ to test the hypothesis.

Solution

The two samples are not independent, so we have to use the paired *t*-test.

First we check the normality of population *d* differences, it follows normal disribution.

- Hypothesis: H_0: $m_1 - m_2 = 0$

(continued)

Example 1.31 (continued)

$$H_1: m_1 - m_2 \neq 0$$

- Value of test statistic for sample: $t_c = \frac{\bar{d}}{\frac{s_d}{\sqrt{n}}} = \frac{0.475}{\frac{0.176}{\sqrt{12}}} = 9.35$
- Critical region: $C = \left\{ |t_c| > t_{1-\frac{\alpha}{2}; n-1} \right\} = \{ |t_c| > t_{0.975; 11} \} = \{ |t_c| > 2.20 \}$
- Conclusion: Value of t_c falls into the critical region, so H_0 is rejected. The values, that measured different method differ significantly ($P < 0.05$).

Table 1.33 The moisture content values were measured with two different methods

M1	20.4	19.8	20.2	20.1	20.5	19.9	19.9	19.5	19.7	20.3	19.7	20.2
M2	19.9	19.5	19.7	19.9	20.1	19.4	19.4	19.2	19.2	19.7	19.2	19.3

1.4.4.5 Hypothesis Test for Comparing Means of More Than Two Populations

We often need to compare means of more than two populations. We may wish to determine which of several temperature values of dehydration the best is, or which treatment is the more effective in case of vegetables, and so on. The procedure that has been developed to answer such questions is called **analysis of variance**.

1.4.4.5.1 Analysis of Variance (ANOVA)

The procedure of the analysis of variance analyze the total variability of all data and then shows how much variability is caused by actual differences between population means and how much is due to the variability of sample means within the separate populations. Analysis of variance uses these two sources of variation to obtain estimates for the common population variance σ^2. One estimate is from the sample variances. A second estimate is based on the sample means. If, in fact, the population means are equal, we use the variance of sample means to obtain an estimate for $\sigma_{\bar{x}}^2$, which we can then use to estimate σ^2. The extent to which these two estimates differ from each other is the basis for decision whether or not we can reject the hypothesis that all population means are equal.

Suppose we obtain independent random samples of observations from k population with means $m_1, m_2, \ldots, m_k$, each population is normal and each population has the same variance, σ^2. The samples have $n_1, n_2, \ldots, n_k$ observations, respectively. The mean of ith sample is $\bar{x}_i$, variance is s_i^2. The elements of ith sample are $x_{i1}, x_{i2}, \ldots, x_{in_i}$.

We assumed that each population has the same variance. One estimate for σ^2 is a weighted average of the variances of samples. That is:

$$MSE = \frac{\sum_{i=1}^{k} (n_i - 1)s_i^2}{\sum_{i=1}^{k} (n_i - 1)} = \frac{\sum_{i=1}^{k} \sum_{j=1}^{n_i} (x_{ij} - \overline{x}_i)^2}{N - k},$$

where $N = \sum_{i=1}^{k} n_i$,

MSE is called the **mean square for error**, the numerator is called **sum of squares for error** (*SSE*). The degree of freedom for numerator is *N-k*. The mean square for error measures variability **within** the samples.

A second estimate using the variance of sample means is called **mean square for treatments** (*MSTR*):

$$MSTR = \frac{\sum_{i=1}^{k} n_i (\overline{x}_i - \overline{\overline{x}})^2}{k - 1},$$

where $\overline{\overline{x}} = \frac{\sum_{i=1}^{k} \sum_{j=1}^{n_i} x_{ij}}{N}$.

The numerator of *MSTR* is called **sum of squares for treatments** (*SSTR*). The degree of freedom for numerator is *k-1*.

The mean square for treatments measures variability **between** the samples.

It can be proved, that.

$SSTR + SSE = SSTO = \frac{\sum_{i=1}^{k} \sum_{j=1}^{n_i} (x_{ij} - \overline{\overline{x}})^2}{N - 1}$, which is the **total sum of squares**.

So the **total sum of squares** is partitioned into the various sources of variability. These sources are the sum of square due to treatments (*SSTR*) and the sum of square due to error (*SSE*).

If all population means are equal, then two estimates of σ^2 are equal. We can test using *F*-test, was described above.

Now we can determine hypothesis test for comparing means.

- Condition of application: – *k* independent random samples from

normal distributed populations.

- Each population has the same variance.

- Null hypothesis: H_0: $m_1 = m_2 = \ldots = m_k$
- Alternate hypothesis: H_1: not all m_i are equal
- Test statistic: $F_c = \frac{MSTR}{MSE} = \frac{(N-k) \sum_{i=1}^{k} n_i (\overline{x}_i - \overline{\overline{x}})^2}{(k-1) \sum_{i=1}^{k} \sum_{j=1}^{n_i} (x_{ij} - \overline{x}_i)^2}$
- Critical region: $C = \{F_c > F_{1-\alpha;\, k-1;\, N-k}\}$

We can summarize the results of analysis of variance into an **analysis of variance (ANOVA) table**. (Table 1.34).

Table 1.34 Analysis of variance (ANOVA) table

Source of variation	DF (Degrees of freedom)	SS	MS	F value
Treatments(between samples)	*k-1*	*SSTR*	*MSTR*	F_c
Error (within samples)	*N-k*	*SSE*	*MSE*	
Total	*N-1*	*SSTO*		

Example 1.32

We studied how the lightness color coordinate of Hungarian paprika powders change following decrease of their particle size. The measured values can be seen in Table 1.35. Use $\alpha = 0.5$ to test the hypothesis. (Zs.H.Horváth et al., 2006).

Table 1.35 Values of lightness coordinate for samples with different particle size

Particle size	125–250μm	250–315μm	315–400μm	400–500μm
	35.05	33.78	33.23	34.08
	35.28	34.33	32.46	33.05
	38.32	37.62	35.44	35.27
	34.77	33.74	32.14	31.00
	35.05	34.66	33.65	33.59
	35.14	34.28	32.88	32.47
	36.99	36.47	34.53	33.59
	37.15	36.21	35.04	34.48
	36.83	36.59	35.79	35.33
	37.75	36.69	35.15	34.04
Means	*36.23*	*35.44*	*34.03*	*33.69*
Variances	*1.72*	*2.01*	*1.75*	*1.69*

Solution:

First we have to check the homogeneity of variances and that the samples fit to a normal distribution.

The normality of samples was established.

We compared the variances using Bartlett's test:

- Value of test statistic for sample: $B_c = 0{,}084$
- Critical region: $C = \{B_c > \chi^2_{0.95;35}\} = \{B_c > 49.80\}$
- Conclusion: Value of B_c doesn't fall into critical region, so each populations have the same variance.

Therefore we can use ANOVA.

We compared $k = 4$ populations, $n_1 = n_2 = n_3 = n_4 = 10$, so $N = 40$.

(continued)

Example 1.32 (continued)

$$MSE = = \frac{\sum_{i=1}^{k}\sum_{j=1}^{n_i}(x_{ij}-\bar{x}_i)^2}{N-k} = \frac{64.58}{36} = 1.79, MSTR = \frac{\sum_{i=1}^{k} n_i(\bar{x}_i-\bar{\bar{x}})^2}{k-1}$$

$$= \frac{42.73}{3} = 14.24$$

$$F_c = \frac{MSTR}{MSE} = \frac{14.24}{1.79} = 7.94$$

The ANOVA table is presented by Table 1.36.

Table 1.36 Analysis of variance (ANOVA) table

Source of variation	DF	SS	MS	F value
Treatments(between samples)	*3*	*42.73*	*14.24*	*7.94*
Error (within samples)	*36*	*64.58*	*1.79*	
Total	*39*			

Critical region: $C = \{F_c > F_{0.95;\,3;\,36}\} = \{F_c > 2.87\}$.

Value of F_c falls into critical region, therefore not all means are equal. It is means, that the particle size of powder influences the value of lightness coordinate significantly, $P < 0.05$.

If we established, using analysis of variance, that all mean population means are equal then the evaluation is finished. If we conclude that not all treatment means are equal, the next step is to determine what differences exist between means. We can answer these questions using multiple-comparison tests. Now we will review these methods.

1.4.4.5.2 Multiple Comparison Tests

These methods examine or compare more than one pair of means at the same time. When we compare the means of populations we can take several questions. Which mean(s) differ from a standard or from each other? Does the mean of treatment 1 differ from that of treatment 2? Does the average of treatments 1 and 2 differ from the average of treatments 3 and 4? We choose method for comparison of means depending on the object of our investigation.

1.4.4.5.2.1 Fisher's LSD Method

This method is a multiple t test for mean pairs. MSE is used to estimate s^2.

So. if we compare mean of ith population and mean of jth population, the

test statistic is the following: $t_c = \frac{\bar{x}_i - \bar{x}_j}{\sqrt{MSE\left(\frac{1}{n_i}+\frac{1}{n_j}\right)}}$

The critical region: $C = \left\{|t_c| > t_{1-\frac{\alpha}{2};\,N-k}\right\}$.

We can use this method in the following way too:

We calculate the least significant difference (*LSD*) between samples, and if the difference between $\overline{x}_i$ and $\overline{x}_j$ exceeds *LSD* we conclude, that m_i and m_j differ significantly.

Formula of the least significant difference:

$$LSD = t_{1-\frac{\alpha}{2};N-k}\sqrt{MSE\left(\frac{1}{n_1}+\frac{1}{n_2}\right)}, \text{ if } n_1 = n_2 \; LSD = t_{1-\frac{\alpha}{2};N-k}\sqrt{\frac{2MSE}{n}}.$$

Example 1.33
We see the Example 1.30. It was established, that not all means are equal. We will use *LSD*-test to answer the question, which means differ from each other?

Solution:

In Table 1.37 we present the differences of means and calculated t_c values.

The critical region: $C = \{|t_c| > t_{0.975;\,36}\} = \{|t_c| > 2.03\}$. We check, which t_c values fall into the critical region. We can established, that the following means differ significantly ($P < 0.05$) from each other: m_1 and m_3; m_1 and m_4; m_2 and m_3; m_2 and m_4. So we can say, that the lightness coordinate of paprika with 125–250μm particle size significantly differ from the lightness coordinate of paprika with 315–400μm particle size and from the lightness coordinate of paprika with 400–500μm particle size. The lightness coordinate of paprika with 250–315μm particle size significantly differ from the lightness coordinate of paprika with 315–400μm particle size and from the lightness coordinate of paprika with 400–500μm particle size.

We will get similar results, if the differences of means are calculated and compared with *LSD*.

$$LSD = 2.03\sqrt{\frac{2 \cdot 1.79}{10}} = 1.22$$

Table 1.37 Differences of means and t_c values

Compared means	Differences of means	t_c
$\overline{x}_1 - \overline{x}_2$	0.80	1.33
$\overline{x}_1 - \overline{x}_3$	2.20	3.68
$\overline{x}_1 - \overline{x}_4$	2.54	4.25
$\overline{x}_2 - \overline{x}_3$	1.41	2.35
$\overline{x}_2 - \overline{x}_4$	1.75	2.92
$\overline{x}_3 - \overline{x}_4$	0.34	0.57

1.4.4.5.2.2 Bonferroni's Method

Using *LSD* method leaves us with the following problems: if we make many such *t*-tests, the probability that (at least) one of the tests rejects the null hypothesis, even though it is actually true, can be very high. That is, if we conduct several *t*-tests, each at significance level α, the probability of falsely rejecting at least one H_0 is much larger than α. The more tests we conduct, the higher the probability for falsely rejecting at least one H_0. A traditional way to avoid this problem, is to apply the method of **Bonferroni.**

The Bonferroni's method reduced the significance level (probability of making a type I error). Using the Bonferroni's method we conduct multiple *t* tests with a reduced level of significance for each test. If we conduct *l* tests and desired overall level of significance of α. each test is conducted at an α/l level of significance. For example, if five tests are to be conducted at a family significance level of *0.05*, we will conduct each test at a *0.01* level of significance.

So, the critical region will be the following: $C = \{|t_c| > t_{1-0.005;\, N-k}\}$.

This method is usable if we compared only a few means. If we are comparing many populations, the significance level for the individual tests can end up being so small, that it becomes very difficult to reject any of the individual hypotheses.

Example 1.34
We apply the Bonferroni's method for Example 1.31. Since there are *4* different populations to be tested pairwise, we need to perform *(4x3)/2* = *6* different *t*-tests. If we wish to use the overall significance level α = *0.05*, each of the *t*-tests should be tested at the *0.05/6* = *0.008* significance level. So the critical region will be the following:

$$C = \left\{|t_c| > t_{1-\frac{0.008}{2};\,36}\right\} = \{|t_c| > t_{0.996;\,36}\} = \{|t_c| > 2.81\}.$$

If we apply this critical region, the m_2 and m_3 don't differ from each other.

1.4.4.5.2.3 Tukey's Test

The second method we shall discuss here for comparing means of several populations is *Tukey's method*. In general, this method gives much better (more precise) results than the least significant difference method, described in the previous subsection. Procedure of this test determines the **honestly significant difference** (*HSD*) between samples, and if the difference between $\bar{x}_i$ and $\bar{x}_j$ exceeds *HSD* we conclude, that m_i and m_j differ significantly. This test was developed to perform all pairwise comparisons, and it is exact if all samples are equal size.

Formula of the honestly significant difference:

$$HSD = \frac{q_{1-\alpha;k;N-k}}{\sqrt{2}}\sqrt{MSE\left(\frac{1}{n_1}+\frac{1}{n_2}\right)}\ \text{. if } n_1 = n_2\ HSD = q_{1-\alpha;k;N-k}\sqrt{\frac{MSE}{n}},\ \text{where}$$

$q_{p;\ k;\ N\ -\ k}$ is the p-quantile of a studentized range statistic. The values of q are determined by two parameters, the number of means being compared (k) and the degrees of freedom of MSE $(N\text{-}K)$. The values can be found in Table 1.37.

The method is applicable using test statistic.

The test statistic is the following:

$$q_c = \frac{\left|\overline{x}_i - \overline{x}_j\right|}{\sqrt{MSE\left(\frac{1}{n_i} + \frac{1}{n_j}\right)}} \text{.if } n_1 = n_2\ q_c = \frac{\left|\overline{x}_i - \overline{x}_j\right|}{\sqrt{\frac{MSE}{n}}}$$

The critical region: $C = \{q_c > q_{\alpha;\ k;\ N\ -\ k}\}$.

Note that Tukey's method is generally better than the LSD method. But it does share one of the disadvantages of the LSD test, in that it might not detect any significant pairwise differences, even if the ANOVA F-test has rejected the null hypothesis of equal means in the k populations.

Example 1.35

We apply the Tukey's test for Example 1.30. to compare mean pairs.

Solution:

In Table 1.38 we present the differences of means and calculated q_c values.

The critical region: $C = \{|q_c| > q_{0.95;\ 4;\ 36}\} = \{|q_c| > 3.809\}$

$$HSD = 3.809\sqrt{\frac{1.79}{10}} = 1.61$$

We check, which q_c values fall into the critical region or which differences of means major, than HSD. We give same result, when we used the LSD-test with Bonferroni's method: the following means differ significantly ($P < 0.05$) from each other: m_1 and m_3; m_1 and m_4; m_2 and m_4.

Table 1.38 Differences of means and q_c values

Compared means	Differences of means	q_c
$\overline{x}_1 - \overline{x}_2$	0.80	1.87
$\overline{x}_1 - \overline{x}_3$	2.20	5.19
$\overline{x}_1 - \overline{x}_4$	2.54	6.00
$\overline{x}_2 - \overline{x}_3$	1.41	3.31
$\overline{x}_2 - \overline{x}_4$	1.75	4.12
$\overline{x}_3 - \overline{x}_4$	0.34	0.81

1.4.4.5.2.4 Scheffe's Method

Scheffe's method can be applied in more general situations than both the *LSD* and Tukey's method. The two latter methods compare the means pairwise, while Scheffe's method considers general *contrasts*. Since each difference of pairs of population means is a special case of a contrast. Sheffe's method can be used to compare the means pairwise. But Tukey's method is preferable as it gives more precise results in situations where we are interested in comparing means pairwise, than Scheffe's method. However, if we are interested in contrasts other than pairwise differences of population means, Scheffe's method should be used. A contrast is described by any linear function of the population means.

Let an arbitrary contrast be defined by $C = \sum_{i=1}^{k} c_i \bar{x}_i$. where $\sum_{i=1}^{k} c_i = 0$. The procedure to test this contrast is the following:

The test statistic:

$$F_c = \frac{\left(\sum_{i=1}^{k} c_i \bar{x}_i\right)^2}{(k-1)MSE \sum_{i=1}^{k} \frac{c_i^2}{n_i}}, \text{ for means pairwise } F_c = \frac{\left(\bar{x}_i - \bar{x}_j\right)^2}{(k-1)MSE\left(\frac{1}{n_i}+\frac{1}{n_j}\right)}.$$

Critical region: $C = \{F_c > F_{1-\alpha;\, k-1;\, N-k}\}$.

Example 1.36

We apply the Scheffe's test for Example 1.30.

First we compare the means pairs. The calculated F_c values are seen in Table 1.39.

Critical region: $C = \{F_c > F_{0.95;\, 3;\, 36}\} = \{F_c > 2.87\}$.

The value of F_c falls into the critical region in two case, the following hypotheses were rejected: $H_0 : m_1 \neq m_3$ and $H_0 : m_1 \neq m_4$. We can establish that, using Scheffe's method, significant difference was found only for two mean pairs.

It appears that the paprika powders split into two distinct groups: paprika powder with 125–250μm particle size and powder with 250–315μm particle size in one group, and paprika powder with 315–400μm particle size and powder with 400–500μm particle size in another. Scheffe's method can be used to investigate if the two groups differ or not. That is, we are interested in testing the following contrast:

$$\frac{m_1+m_2}{2} - \frac{m_3+m_4}{2}, \text{ so } c_1 = c_2 = \frac{1}{2},\ c_3 = c_4 = -\frac{1}{2}, \text{ therefore } F_c = \frac{3.880}{0.537} = 7.23.$$

Value of F_c falls into the critical region, so the two groups differ significantly (*P < 0.05*).

(continued)

Example 1.36 (continued)

Table 1.39 Differences of means and F_c values

Compared means	Differences of means	F_c
$\bar{x}_1 - \bar{x}_2$	0.80	0.59
$\bar{x}_1 - \bar{x}_3$	2.20	4.15
$\bar{x}_1 - \bar{x}_4$	2.54	6.02
$\bar{x}_2 - \bar{x}_3$	1.41	1.84
$\bar{x}_2 - \bar{x}_4$	1.75	2.84
$\bar{x}_3 - \bar{x}_4$	0.34	0.11

1.4.4.5.2.5 Dunnet's Test

The Dunnet's test is a t-test, when we compare to a control means the other means. For example, if the control is m_1, then the statistical test:

$$t_c = \frac{\bar{x}_1 - \bar{x}_i}{\sqrt{MSE\left(\frac{1}{n_1} + \frac{1}{n_i}\right)}} .\mathrm{i} = 2, \ldots, \mathrm{k}.$$

These tests are not independent, so we must consider when we determine the significance level of tests. Similarly to Bonferroni's test, if we conduct $(k\text{-}1)$ tests and desired overall level of significance of α, each test is conducted at an $\frac{\alpha}{k-1}$ level of significance.

Example 1.37
We apply the Dunnet's test for Example 1.30. Let be the control m_1. In Table 1.40 the differences of means and calculated t_c values are presented.

We need to perform *3* different t-tests. If we wish to use the overall significance level $\alpha = 0.05$, each of the t-tests should be tested at the $0.05/3 = 0.017$ significance level. So the critical region will be the following:

$$\text{The critical region} : C = \left\{|t_c| > t_{1-\frac{0.017}{2};36}\right\} = \{|t_c| > t_{0.992;36}\} = \{|t_c| > 2.53\}.$$

The two t_c values fall into the critical region, m_1 differ significantly from m_3 and from m_4.

Table 1.40 Differences of means and t_c values

Compared means	Differences of means	t_c
$\bar{x}_1 - \bar{x}_2$	0.80	1.33
$\bar{x}_1 - \bar{x}_3$	2.20	3.68
$\bar{x}_1 - \bar{x}_4$	2.54	4.25

1.5 Finding Correlations, Relationships and Underlying Structures of Multivariate Data

Prisana Suwannaporn

1.5.1 Objective

Study of association helps justify whether are there any relationship and in which directions (negative or positive). If there are any associations, it can tell the strength of the relationship. Strength of the relationship can help decide which variable has more influence on given dependent variables. There are various methods to determine degree of association according to the difference in type of scale as mentioned before. Association analysis methods frequently applied in food science research are mentioned as followings;

1.5.2 Introduction

The analysis of association is applied when researcher want to establish the relationship of two or more variables. Researcher may need to know how are things related and want to prove their hypothesis or statement. For example, they may want to proof statement "there is a strong relationship between smoking and lung cancer" or "there is a strong relationship between eating high saturated fat food and coronary heart disease" etc. Analysis of association is relatively easy to use and understand. However mistakes due to ignorance type of scale are usually happened among scientific researchers who familiar with interval or ratio data. Researchers are facing different types of scales nowadays, especially those work with consumer study or sensory evaluation. Different types of scale require different analysis methods. However, software program can give a warning if scale is not appropriated. But only if it is correctly classified. Association analysis methods for different scale of measurement are summarized in Table 1.41 (Glass and Stanley 1995).

Table 1.41 Appropriate method of association given combination of measurement

	Measurement of Variable X_2		
Measurement of variableX_1	Normianal (Dichotomous)	Ordinal	Interval or Ratio
Norminal (dichotomous)	Crossbreaks Phi coefficient Contingency correlation[a]	Rank-biserial correlation	Point-biserial correlation
Ordinal	Rank-biserial correlation[a]	Spearman Rank Kendall Tau	Convert X_2 variable to Ranks, and use Spearman or Kendall Tau
Interval or Ratio	Point-biserial correlation	Convert X_1 Variable to Ranks, and use Spearman or Kendall Tau	Pearson product moment correlation

[a]Speacial case when there are more than two categories of each variable
Source: Glass, G.V. and Stanley, J.C., 1970. Statistical Methods in Education and Psychology. pp. 155–191

1.5.3 Crosstabulation

Crosstabulation is the easiest way to look at the association of two (or more) nominal data. Both dependent and independent variables should be categorical data. It is used for examine categorical data since the crosstabulation table is interpreted by calculate the cell frequency and the marginal frequency. Generally, row is assigned for independent variable and column as dependent variable.

1.5.3.1 Problem to Solve

Consumer preferences and buying criteria for Jasmine rice. (Suwannaporn and Linnemann 2008).

Researcher wanted to know whether there were any relationship between consumer who preferred rice from Thailand (prefer, not prefer; 2 groups) and consumer from different grain type preference countries (long grain, short grain; 2 groups) (Suwannaporn and Linnemann 2008). The objective of this study was to investigate consumer preferences towards Thai rice in different rice eating habit countries. Result could be useful for Thailand rice export. All scales used were categorical data (nominal) so crosstabulation method is appropriated to investigate relationship in this case. First, we have to calculate expected cell frequencies as in the formula below and example in Table 1.42.

Table 1.42 Relationship between preference for rice from Thailand and grain type preference countries

	Preference for rice from Thailand		Total
	Prefer	Not Prefer	
Long Grain	33	22	55
Preference	60 %	40 %	
Short Grain	17	28	45
Preference	38 %	62 %	
Total	50	50	100

Ref : Adapt from Suwannaporn and Linnemann (2008)

1.5.3.2 Procedure

Step 1 Calculated expected cell frequencies.

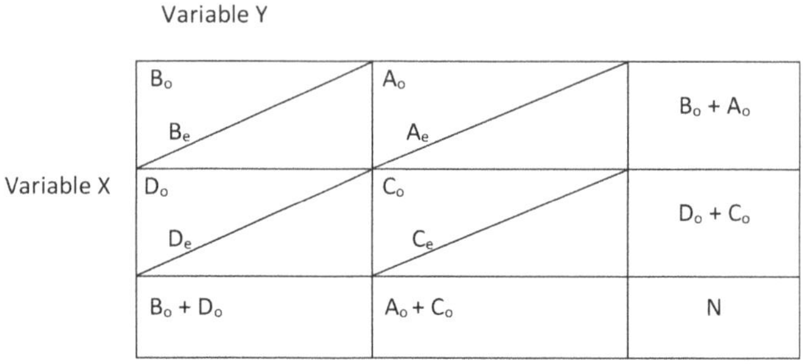

$$B_e = \frac{(B_o + A_o)(B_o + D_o)}{N} \qquad A_e = \frac{(B_o + A_o)(A_o + C_o)}{N}$$

$$D_e = \frac{(D_o + C_o)(B_o + D_o)}{N} \qquad C_e = \frac{(D_o + C_o)(A_o + C_o)}{N}$$

When subcribe o = observed frequency
subcribe e = expected frequency

According to the data in Table 1.42, given the accumulated % count row as 100 (or column is dependent variable), 60% of consumers from long grain preference countries such as Southeast Asian prefer rice from Thailand. 62% of consumers from short grain preference countries such as Japanese, Korean and Northern Chinese did not prefer rice from Thailand.

Step 2 Chi Square test.

Chi-square test is used for prove of any relationship or association in the data set. The significance of Chi Square test in cross tabulation analysis indicates that the relationship is existed. Chi-square test in this study is significance at $P < 0.05$

Table 1.43 Calculation of Chi-square

Preference of rice from Thailand (rows: Grain type preference country; each cell gives observed / expected)

	prefer	not prefer	Total
Long grain preference	22 / 27.5	33 / 27.5	55
Short grain preference	28 / 22.5	17 / 22.5	45
Total	50	50	100

indicated that there are association in these data set. However it does not provide the strength of the relationship. Strength of relationship can be calculated using contingency correlation coefficient. Chi-square test can be calculated as formula below and example in Table 1.43:

$$X^2 = \sum_{i=1}^{r}\sum_{j=1}^{c}\frac{\left(O_{ij}-E_{ij}\right)^2}{E_{ij}} \tag{1.16}$$

$$\mathrm{df} = (\mathrm{r}-1)(\mathrm{c}-1)$$

$$\chi^2 = \sum\left[\frac{(o-e)^2}{e}\right] = \frac{(22-27.5)^2}{27.5}+\frac{(33-27.5)^2}{27.5}+\frac{(28-22.5)^2}{22.5}+\frac{(17-22.5)^2}{22.5}$$

$$= \frac{30.25}{27.5}+\frac{30.25}{27.5}+\frac{30.25}{22.5}+\frac{30.25}{22.5} = 1.1+1.1+1.35+1.35$$

$$\chi^2 = 4.9$$

Chi Square table show that at degree of freedom 1 $\chi^2 = 4.9$ is significant at $p < 0.05$ so there is relationship in this data set. Next calculate strength of association using Contingency correlation coefficient as followings.

Step 3 Calculated Contingency Correlation.

Contingency correlation coefficient (C) indicated the strength of association between categorical variables which its value ranging from 0 to 1. The higher value that approach 1 shows the stronger relationship. Contingency correlation coefficient is calculated as followings. Calculated result of C equals to 0.22 means that consumers who preferred rice from Thailand have moderate relation with type of grain they prefer.

$$C^2 = \frac{X^2}{X^2 + N}; 0 \leq CC \leq 1 \tag{1.17}$$

$$C = \sqrt{\frac{\chi^2}{\chi^2 + N}}$$

$$C = \sqrt{\frac{4.9^2}{4.9^2 + 100}} = 0.22$$

1.5.4 Spearman Rank Correlation

Spearman rank correlation is applied in case researcher is dealing with ordinal data such as ranking scale which highly applied in consumer study or sensory evaluation. It can help investigate whether are there any correlation between two ordinal variables (or rating scale) and how strong the relationship.

1.5.4.1 Problem to Solve

Beverage Consumption of mother-toddler dyads in family with limited incomes (Hoerr et al. 2006).

The objective of this research was to investigate relationship between beverage consumption behaviours of mother and her child. Research question was whether there was any influence of mother's consumption behaviour in beverage to her child's. Beverage intake and diet quality of toddlers from families with limited incomes were described and compared to their mother's beverage intake. Consumption quantity of different types of beverage (milk, juice, soft drinks, sweetened beverages, caffeine drink and water) were rating by order (ranking). Data was collected for 36 months and analyzed using Spearman Rank Correlation Coefficient.

1.5.4.2 Procedure

The appropriate method is Spearman Rank Coefficient (Rho) or Kendall Tau. Since data used is ordinal data which is the ranking scale of consumed quantity of different beverages comparing between mother and child. Data is ranking in similar order direction (low to high or high to low) of mother and child; the different of match pair data (D) is calculated by subtract each pair. Different between match pair data (D) is use for calculate Spearman rank coefficient as following formula;

1.5.4.2.1 Spearman Rank Coefficient (Rho)

$$\rho = 1 - \frac{6\sum D^2}{N\left(N^2 - 1\right)} \tag{1.18}$$

T-test

$$t = \frac{\rho\sqrt{N-2}}{\sqrt{1-\rho^2}} \tag{1.19}$$

N = number of data.
D = Different between match pair data.

T-test in Table 1.44 indicated that there are significant correlation in variables with ** ($p < 0.01$) and * ($p < 0.05$). Value in Table 1.44 showed Spearman rank coefficient. It can be interpreted that mother's soft drink consumption have a negative correlation with child's juice consumption but has positive correlation with child's soft drink consumption. Or the more mother drink soft drink, the less juice and more soft drink her child trend to drink. (at $p < 0.01$). According to Hoerr et al. (2006) mothers who consumed more than 12 fl oz. of soft drinks per day were nearly four times more likely to have a child with poor diet quality. In a contrarily, mother's milk consumption has a positive correlation with child's juice consumption. Or the more mother drink milk, the more juice her child trend to consume, however at lower significant level (at $p < 0.05$).

Table 1.44 Relationship between mother drinking consumption and child consumption for 36 months periods using spearman rank order correlation coefficient

Child	Mother milk	Juice	Soft drinks	Sweet Beverage	Caffein Drink	Water
Milk	–0.03	0.04	–0.01	0.14	0.05	–0.21*
Juice	0.25*	0.03	–0.36**	–0.19	0.11	0.10
Soft drinks	–0.15	–0.14	0.30**	–0.06	0.10	0.01
Sweetened beverages	–0.18	0.15	–0.01	0.17	–0.13	0.05
Water	–0.09	0.12	–0.16	–0.01	0.17	0.09

* $p < 0.05$ ** $p < 0.01$
Ref : Hoerr et al. (2006)

1.5.5 Pearson Product Moment Correlation

Pearson product moment correlation coefficient is the indicator of association between 2 interval variables. It is common use in scientific research because most data are interval or ratio. However the mistake that can violate analysis result was (1) outlier (2) the assumption of linearity. Pearson product moment correlation is a linear relationship (plot of data should be linear). Curvilinear pattern should use other model. Pearson product moment correlation coefficient is used for prediction in simple linear regression. The variance changes in the dependent variable reflects by changes in the independent variables. R^2 or the coefficient of determination at 0.05 means 5% of the variance is explained in the system. Pearson product moment correlation coefficient tells (1) strength of the relationship (2) direction (positive or negative relationship) (3) percentage of variation explained in the relationship (Davis and Cosenza 1993).

1.5.5.1 Problem to Solve 1

To find the correlation between age and wages of 10 employees (Table 1.45). Pearson product moment correlation coefficient is applied in this case since all are ratio data (age and wages).

1.5.5.2 Procedure

Preliminary examine the data set using scatter plot is recommended in order to determine data that may violate the result which are (1) outlier (2) non linearity.

Scatter plot showed that data point 10 (Table 1.45) is possibly an outlier which may violate the result. So it depends on researcher background knowledge about this issue. If researcher assumes that the older age the higher wages in professional or

Table 1.45 Pearson's correlation coefficients between employee's age and wages

Number	Age (X)	Wages (Y)	
1	26	120	
2	22	15	
3	30	15	
4	30	20	
5	19	20	
6	22	14	
7	22	22	
8	20	24	
9	18	15	
10	**55**	**20**	**Outlier**

skill workers, then data point 10 (Table 1.45) may be an outlier. Researcher then has 2 choices of eliminating this data or collecting more data. But in real life collecting new data within different period may not be consistent. However small number of data as in the example is not recommended anyway because of its error prone behaviour. In a contrarily, data point 10 (Table 1.45) may not be an outlier if researcher assumes that the older the employee, the lower wages they will gain in non-professional or non skill workers. So Pearson correlation may not appropriate in this case since scatter plot shows curvilinear or non linear relationship if we include data point 10 (Table 1.45).

It should be cautious here that even same set of data, result may not interpret the same. Computer software calculated according to input. It can not verify as situation in real life is more complicate. Need of human brain and background knowledge are important in data analysis. Most researchers believe all parameters retrieving from statistic software without putting any effort of human consideration. Example below shows the importance of outlier especially when the number of data is small. Calculation of Pearson product moment correlation coefficient (r) is demonstrated both with and without outlier to see its effect on analyzing result. Pearson correlation coefficient and its t-test was calculated as followings;

1.5.5.2.1 Pearson Product Moment Correlation Coefficient (r)

$$r = \frac{N\sum XY - (\sum X)(\sum Y)}{\sqrt{\left\{N\sum X^2 - (\sum X)^2\right\}\left\{N\sum Y^2 - (\sum Y)^2\right\}}} \tag{1.20}$$

X= independent variable.
Y= dependent variable.
N= total number of data set.

T-test

$$t = \frac{r\sqrt{N-2}}{\sqrt{1-r^2}} \tag{1.21}$$

$$r = \frac{10(3614) - 209(157)}{\sqrt{\{10(5013) - (209^2)\}\{10(2875) - (157^2)\}}}$$

r= -0.216 (without outlier data point 10).
r= 0.106 (with outlier data point 10).

According to Table 1.45, include outlier data point, the whole data set trend to have positive correlation (0.106). But exclude this data point, the negative correlation is prevailed (−0.216). So small sample size with outlier can change the direction of relationship, hence creates misinterpretation.

1.5.5.3 Problem to Solve 2

Classification of Rice Amylose Group Using Discriminant Analysis. (Ref: Suwannaporn et al. 2007).

Cooked rice's texture was influenced by many factors such as protein content, amylose content and pasting properties. Researcher tried to investigate relationship between sensory texture attributes and pasting properties of rice grain. So that its sensory texture attributes which was a subjective measurement can be predicted by a more rapid and objective measurement such as pasting measurement. (Suwannaporn et al. 2007). Pearson correlation was applied in this case since all data was ratio data. Result in Table 1.46 showed that only peak viscosity and trough were correlated with sensory texture attributes. Peak viscosity correlated with cohesiveness of mass, tooth-pack and adhesion to lips at 0.76, 0.73 and 0.72 ($p < 0.05$) respectively. Trough correlated with adhesive to lips at 0.69 ($p < 0.05$). As a consequence, peak viscosity and trough were the best predictors which can be use in further analysis.

Table 1.46 Pearson's correlation coefficients between sensory texture attributes and pasting parameters

	Sensory texture attributes				
Pasting property	Adhesion to lips	Hardness	Cohesion of mass	Roughness of mass	Tooth-pack
Peak Visco.	0.72*	–0.66	0.76*	–0.64	0.73*
Trough	0.69*	–0.26	0.31	–0.17	0.65
Breakdown	–0.02	–0.54	0.59	–0.63	0.06
Final Visco.	0.32	0.01	0.04	–0.10	0.65
Setback	0.13	0.13	–0.07	–0.08	0.62

Note : * = significant at p< 0.05.
Ref : Suwannaporn et al. (2007)

1.6 Inferential Statistics

Ivanka Zheleva, Todor Todorov, Hüseyin Bozkurt, and Fahrettin Göğüş

1.6.1 Objectives and Learning Outcomes

The main objectives of this chapter are to present sufficient information about how to make correct decision about population on the base of sample information. According to this; the sampling theory, statistical inference and fundamentals of statistical estimation theory are presented.

The fundamentals of sampling theory – sampling distribution of means, proportions, of differences, sums and standard errors are presented. Estimation of population parameters, unbiased, biased, efficient, point and interval estimates as well as confidence interval estimates, t-distribution and chi-distribution are also discussed.

On the base of this information and sample analysis, correct conclusions about the population or processes under consideration or further their investigations could be properly made.

1.6.2 Introduction

Sampling theory is a study of relationships existing between a population and samples drawn from the population. It is of great value in many connections. It is useful in *estimation* of unknown population quantities (such as population mean, variance, etc.), often called *population parameters* or briefly *parameters*, from the knowledge of corresponding sample quantities (such as sample mean, variance, etc.), often called *sample statistics* or briefly *statistics*.

From a practical point of view, it is often more important to be able to infer information about a population by use of samples drawn from it. Such problems are dealt with in *statistical inference*, which uses principles of sampling theory. Very often in practice we have to make decisions about populations on the basis of sample information. Such decisions are called *statistical decisions*.

In attempting to reach decisions, it is useful to make assumptions or guesses about the populations involved.

1.6.3 Elementary Sampling Theory

In general, a study of inferences made concerning a population by use of samples drawn from it, together with indications of the accuracy of such inferences using probability theory, is called *statistical inference.*

In order that conclusions of sampling theory and statistical inference to be valid, samples must be chosen so as to be *representative* of population. A study of methods of sampling and the related problems which arise is called *the design of the experiment.*

Sampling theory is also useful in determining whether observed differences between two samples are actually due to chance variation or whether they are really significant. Such questions arise, for example, in testing a canned food for the absence of ***Clostridium botulinum*** toxins or in deciding whether one production process is better than another. Their answers involve use of *so-called test of significance and hypotheses* which are important in the theory of decisions.

One way in which a representative sample may be obtained is by a process called *random sampling*, according to which each member of a population has an equal chance of being included in the sample. One technique for obtaining a random sample is to assign number to each member of the population, write these numbers of small pieces of paper, place them in an urn and then draw numbers from the urn, being careful to mix thoroughly before each drawing. This can be replaced by using a table of *random numbers* specially constructed for such purposes.

If we draw a number from an urn, we have the choice of replacing or not replacing the number into the urn before the second drawing. In the first case the number can come up again and again, whereas in the second it can only come up once. Sampling where each member of population may be chosen more than once is called *sampling with replacement,* while if each member cannot be chosen more than once is called *sampling without replacement.*

Populations are either finite or infinite. If for example we draw 10 balls successively without replacement from an urn containing 100 balls, we are sampling from a finite population: while if we toss a coin 50 times and count the number of heads, we are sampling of an infinite population.

A finite population in which sampling is with replacement can theoretically be considered infinite, since any numbers of samples can be drawn without exhausting the population. For many practical purposes, sampling from a finite population which is very large can be considered as sampling from an infinite population.

Consider all possible samples of size N which can be drawn from a given population (either with or without replacement). For each sample we can compute a statistic, such as the mean, standard deviation, etc., which will vary from sample to sample. In this manner we obtain a distribution called *its sampling distribution.* If for example, the particular statistics used is the sample mean, the distribution is called *the sampling distribution of means* or the *sampling distribution of the mean.* Similarly we could have sampling distributions of standard deviations, variances, medians, proportions, etc.

For each sampling distribution we can compute the mean, standard deviation, etc. Thus we can speak of the mean and standard deviation of the sampling distribution of means, etc.

Suppose that all possible samples of size *N* are drawn without replacement from a finite population of size $N_p > N$. If we denote the mean and standard deviation of the sampling distribution of means by μ and σ respectively then

$$\mu_x = \mu \text{ and } \sigma_x = \frac{\sigma}{\sqrt{N}}\sqrt{\frac{N_p - N}{N_p - 1}} \tag{1.22}$$

If the population is infinite or if sampling is with replacement, than above results reduced to.

$$\mu_x = \mu \text{ and } \sigma_x = \frac{\sigma}{\sqrt{N}} \tag{1.23}$$

For large values of N $(N \geq 30)$ the sampling distribution of means is approximately a normal distribution with mean μ_x and standard deviation σ_x irrespective of the population (so long as the population mean and variance are finite and population size is as least twice the sample size). This result for an infinite population is a special case of *the central limit theorem of* advanced probability theory which shows that the accuracy of the approximations improves as *N* gets larger. This is sometimes indicated by saying that the sampling distribution is asymptotically normal.

In case the population is normally distributed, the sampling distribution of means is also normally distributed even for small values of N *(i.e.* $N < 30$*)*.

Suppose that population is infinite and that the probability of occurrence of an event (called its success) is p while the probability of non-occurrence of the event is $q = 1 - p$.

For example, the population may be all possible samples tosses of a fair coin in which the probability of the event is "heads" is $p = \frac{1}{2}$. In a similar way, probability of getting failure and success for total microbial count in milk; probability of success (number of total count within the limits) event is p= $\frac{1}{2}$ and probability of failure (number of total count above the limits) event is p=$\frac{1}{2}$.

Consider all possible samples of size *N* drawn from this population, and for each sample determine the proportion *P* of successes. In the case of the coin *P* would be the proportion of heads turning up in *N* tosses. Then we obtain a sampling distribution whose mean μ_p and standard deviation σ_p is given by:

$$\mu_p = \text{p and } \sigma_p = \sqrt{\frac{pq}{N}} = \sqrt{\frac{p(1-p)}{N}} \tag{1.24}$$

which can be obtained from Eq. 1.2 by placing $\mu = p$ and $\sigma = \sqrt{pq}$.

For large values of N $(N \geq 30)$ the sampling distribution is very closely normally distributed. Note that the population is binomially distributed.

The Eq. 1.3 is also valid for a finite population in which sampling is with replacement. For finite populations in which sampling is without replacement, Eq. 1.3 is replaced by equations with $\mu = p$ and $\sigma = \sqrt{pq}$.

Note that Eq. 1.3 is obtained most easily by dividing the mean and standard deviation (N_p and $\sqrt{N_{pq}}$) of the binomial distribution by N.

Suppose that we are given two populations. For each sample of size N_1 drawn from the first population let us compute a statistic S_1. This yields a sampling distribution and standard deviation we denote by μ_{S1} and σ_{S1}, respectively.

Similarly, for each sample for each size N_2 drawn from the second population let us compute a statistic as two. This yields a sampling distribution for the statistics S_2, whose mean and standard deviation are denoted by μ_{S2} and σ_{S2}, respectively. From all possible combinations of these samples from the two populations we can obtain a distribution of the differences, $S_1 - S_2$ which is called the *sampling distribution of difference of the statistics*. The mean and standard deviation of this sampling distribution denoted respectively by

$$\mu_{S1\text{-}S2} = \mu_{S1}\text{-}\mu_{S2} \ and \ \sigma_{S1\text{-}S2} = \sigma_{S1} - \sigma_{S2} = \sqrt{\sigma_{S1}^2 + \sigma_{S2}^2} \tag{1.25}$$

provided that the samples chosen do not in the anyway depend on each other, i. e. the samples are independent.

If N_1 and N_2 are large $(N_1, N_2 \geq 30)$ the sampling distributions of differences of means or proportions are very closely normally distributed.

It is sometimes useful to speak of the *sampling distribution of the sum of statistics*. The mean and standard deviation of this distribution are given by

$$\mu_{S1+S2} = \mu_{S1} + \mu_{S2} \ \ and \ \ \sigma_{S1+S2} = \sigma_{S1} + \sigma_{S2} = \sqrt{\sigma_{S1}^2 + \sigma_{S2}^2} \tag{1.26}$$

if the samples are independent.

The standard deviation of sampling distribution of a statistic is often called *standard error*.

It is noted that if the sample size N is large enough, the sampling distributions are normal or nearly normal. For this reason, the methods are known as *large sampling methods*. When $N < 30$ samples are called *small*. When population parameters are unknown they may be estimated closely by their corresponding sample statistics if samples are large enough.

1.6.4 Statistical Estimation Theory

One important problem of statistical inference is the estimation of population parameters or briefly parameters (such as population mean, variance, etc.). If the mean of sampling distribution of a statistic equals the corresponding population parameter, the statistic is called *unbiased estimator* of the parameter, otherwise it is called a *biased estimator*. The corresponding values of such statistic are called *unbiased* or *biased estimates* respectively. The sample mean $-X$ is an unbiased estimate of population mean μ. The sample variance s^2 is a biased estimate of the population variance σ^2. By use of the modified variance $s^2 = \frac{N}{N-1}s^2$, we find that s^2 is an unbiased estimate of σ^2. However, s is a biased estimate of σ.

We could say that a statistic is *unbiased* if its expectation equals the corresponding population parameter. Thus $-X$ and s^2 are unbiased since $E\{\overline{X}\} = \mu$ and $E\{s^2\} = \sigma^2$.

If the sampling distributions of two statistics have the same mean (or expectation), the statistic with the smaller variance is called an *efficient estimator* of the mean while the other statistic is called *inefficient estimator*. The corresponding values of the statistics are called *efficient* or *inefficient estimates* respectively.

If we consider all possible statistics whose sampling distributions have same mean, the one with the smallest variance is sometimes called *the most efficient* or *best estimator* of this mean.

Example: The sampling distributions of the mean and median both have the same mean namely the population mean (Eq. 1.3.1.1 Example 1.22, 1.23 and 1.24).

However, the variance of the sampling distribution of means is smaller than that for the sampling distribution of medians. Hence the sample mean gives efficient estimate of the population mean, while the sample median gives inefficient estimate of it. Of all statistics estimating the population mean, the sample mean provides the best or most efficient estimate.

In practice inefficient estimates are often used because of the relative ease which some of them can be obtained.

An estimate of a population parameter given by a single number is called *point estimate* of parameter. An estimate of a population parameter given by two numbers between which the parameter may be considered to lie is called *interval estimate* of parameter.

Interval estimates indicate the precision or accuracy of an estimate and are therefore preferable to point estimates.

A statement of error or precision of an estimate is often called is *reliability*.

Let μ_s and σ_s be the mean and standard deviation (standard error) of the sampling distribution of a statistic *S*. Then if the sampling distribution of *S* is approximately normal (which as we have seen is true for many statistics if the sample size $N \geq 30$) we can expect to find an actual sample statistic S lying in the intervals $\mu_S - \sigma_S$ to $\mu_S + \sigma_S$, $\mu_S - 2\sigma_S$ to $\mu_S + 2\sigma_S$ or $\mu_S - 3\sigma_S$ to $\mu_S + 3\sigma_S$, about 68.27%, 95.45% and 99.73% of the time respectively.

Equivalently we can expect to find, or we can be confident of finding, μ_s in the intervals $S - \sigma_s$ to $S + \sigma_s$, $S - 2\sigma_s$ to $S + 2\sigma_s$ or $S - 3\sigma_s$ to $S + 3\sigma_s$, about 68.27%, 95.45%, 99.73% confidence interval for estimate μ_s. The end numbers of these intervals ($S \pm \sigma_s$, $S \pm 2\sigma_s$, $S \pm 3\sigma_s$) are then called 68.27%, 95.45%, 99.73% confidence limits or as they sometimes called *fiducial limits*.

Similarly, $S \pm 1.96\sigma_s$ and $S \pm 2.58\sigma_s$ are 95%, 99% (or 0.95 and 0.99) confidence limits for *S*. The percentage confidence is often called the *confidence level*. The

Table 1.19 z_c values for different confidence intervals

Confidence level	99.73%	99%	98%	96%	95.45%	95%	90%	80%	68.27%	50%
$\mathbf{z_c}$	3.00	2.58	2.33	2.05	2.00	1.96	1.645	1.28	1.00	0.6745

numbers 1.96, 2.58, etc., in the confidence limits are called *confidence coefficients* or *critical values* and are denoted by z_c. From confidence levels we can find confidence coefficients, and conversely.

Z_c values corresponding to various confidence levels used in practice are given in Table 1.19. For confidence levels not presented in the table, the values of z_c can be found from the normal curve area tables.

If the statistics S is the sample mean $-X$, then 95% and 99% confidence limits for estimation of the population mean μ are given by $-X \pm 1.96\sigma_x$ and $-X \pm 2.58\sigma_x$ respectively. More generally, the confidence limits are given by $-X \pm z_c\mu_x$ where z_c, which depends on the particular level of confidence desired, can be read from Table 1.19. Using the values of σ_x obtained in the previous chapter, we see that the confidence limits for the population mean are given by

$$\overline{X} = z_c \frac{\sigma}{\sqrt{N}} \tag{1.27}$$

in case sampling is from an infinite population or if sampling is with replacement from a finite population, and by

$$\overline{X} = z_c \frac{\sigma}{\sqrt{N}} \sqrt{\frac{N_p - N}{N_p - 1}} \tag{1.28}$$

if sampling is without replacement from a population of finite size N_p.

In general, the population standard deviation σ is unknown, so that to obtain the above confidence limits we use the sample estimate s' or s. This will prove satisfactory if $N \geq 30$. For $N < 30$, the approximation is poor and small sampling theory must be employed.

If this statistic S is the proportion of "successes" in a sample of size N drawn from binomial population in which p is the proportion of successes (i.e. the probability of success), the confidence limits for p are given by $P \pm z_c\sigma_p$, where p is the proportion of successes in the sample of size N. Using the values *of* σ_p obtained earlier, we see that the confidence limits for the population proportion are given by:

$$P \pm z_c \sqrt{\frac{pq}{N}} = P \pm z_c \sqrt{\frac{p(1\text{-}p)}{N}} \tag{1.29}$$

in case sampling is a form an infinite population or if sampling is with replacement from a finite population, and by:

$$P \pm z_c \sqrt{\frac{pq}{N}} \sqrt{\frac{N_p - N}{N_p - 1}} \tag{1.30}$$

if sampling is without replacement from s population of finite size N_p.

To compute these confidence limits we can use the sample estimate P for p, which will generally prove satisfactory if $N \geq 30$.

If S_1 and S_2 are two sample statistics with approximately normal sampling distributions, confidence limit for the differences of the population parameters corresponding to S_1 and S_2 are given by:

$$S_1 - S_2 \pm z_c \sigma_{S1\text{-}S2} = S_1 - S_2 \pm z_c \sqrt{\sigma^2{}_{S_1} + \sigma^2{}_{S_2}} \tag{1.31}$$

while confidence limits for the sum of the population parameters are:

$$S_1 + S_2 \pm z_c \sigma_{S1+S2} = S_1 + S_2 \pm z_c \sqrt{\sigma^2{}_{S1} + \sigma^2{}_{S2}} \tag{1.32}$$

provided that the samples are independent.

For example, confidence limits for the difference of two population means, in the case where the populations are infinite, are given by:

$$X_1 - X_2 \pm z_c \sigma_{X1\text{-}X2} = X_1 - X_2 \pm z_c \sqrt{\frac{\sigma_1^2}{N_1} + \frac{\sigma_2^2}{N_2}} \tag{1.33}$$

where X_1, σ_1, N_1 and X_2, σ_2, N_2 are the respective means, standard deviations and sizes of the two samples drawn from the populations.

Similarly, confidence limits for the difference of two population proportions, where the population are infinite are given by:

$$P_1 - P_2 \pm z_c \sigma_{P1\text{-}P2} = P_1 - P_2 \pm z_c \sqrt{\frac{p_1(1\text{-}P_1)}{N_1} + \frac{p_2(1\text{-}p_2)}{N_2}} \tag{1.34}$$

where P_1 and P_2 are the two sample proportions, N_1 and N_2 are sizes of the two samples drawn from the populations, and p_1 and p_2 are the proportions in the two population (estimated by P_1 and P_2).

The confidence limits for the standard deviation σ of a normally distributed population as estimated from a sample with standard deviation s, are given by:

$$s \pm z_c \sigma_s = s \pm z_c \frac{\sigma}{\sqrt{2N}} \tag{1.35}$$

using Table 1.19. In computing these confidence limits we use s or ŝ to estimate σ.

The 50% confidence limits of the population parameters corresponding to a statistic S are given by $S \pm 0.6745\ \sigma_s$. The quantity $0.6745\sigma_s$ is known as the probable error of the estimate.

"STUDENT'S" T –distribution

Let us define the statistic

$$t = \frac{\_X - \mu}{s}\sqrt{N-1} \tag{1.36}$$

If we consider samples of size N drawn from a normal (or approximately normal) population with mean μ and for each sample we compute t, using the sample mean _X and sample standard deviation s, the sampling distribution for t can be obtained. This distribution is given by

$$Y = \frac{Y_0}{\left(1 + \frac{t^2}{N-1}\right)^{N/2}} \tag{1.37}$$

Here Y_0 is a constant depending on N such that the total area under the curve is one and where the constant υ = (N-1) is called the number of degree of freedom. This distribution is called "Student's" distribution after its discoverer Gosset, who published his work under the pseudonym "Student" during the early part of the twentieth century.

We can define 95%, 99% or other confidence intervals by using the Table of the t-distribution in the Table 1.47 (Appendix).

For example, if $-t_{0.075}$ and $t_{0.075}$ are the values of t for which 2.5% area lies in each tail of the t – distribution, then a 95% confidence interval for t is

$$-t_{0.075} < \frac{\_X - \mu}{s}\sqrt{N-1} < t_{0.075} \tag{1.38}$$

From which we see that μ is estimated to lie in the interval

$$\_X - t_{0.075}\frac{s}{\sqrt{N-1}} < \mu < \_X + t_{0.075}\frac{s}{\sqrt{N-1}} \tag{1.39}$$

with 95% confidence (i.e. probability 0.95). Note that $t_{0.075}$ represents the 97.5 percentile value, while $t_{0.025} = -t_{0.075}$ represents 2.5 percentile value.

The Chi – Square Distribution

Let us define the statistic

$$\chi^2 = \frac{Ns^2}{\sigma^2} = \frac{\left(X_1 - \overline{X}\right)^2 + \left(X_2 - \overline{X}\right) + \ldots\ldots + \left(X_n - \overline{X}\right)^2}{\sigma^2} \tag{1.40}$$

If consider samples of size N drawn from a normal population with standard deviation σ, and if for each sample we compute χ^2, a sampling distribution for χ^2 can be obtained. This distribution, called the chi-square distribution, is given by

$$Y = Y_0\left(\chi^2\right)^{\frac{1}{2}(v-2)e^{-\frac{1}{2}\chi^2}} = Y_0\chi^{v-2}e^{-\frac{1}{2}\chi^2} \tag{1.41}$$

where $\upsilon = N - 1$ is the number of degrees of freedom, and Y_0 is a constant depending on υ such that the total area under the curve is one. The maximum value of Y occurs at $\chi^2 = \upsilon - 2$ *for* $\upsilon \geq 2$.

We can define 95%, 99% or other confidence limits and intervals for χ^2 by use of the table of the χ^2 distribution in the Appendices. In this manner we can estimate within specified limits of confidence the population standard deviation σ in terms of a sample standard deviation s.

For example, if $\chi^2_{0.025}$ and $\chi^2_{0.975}$ are the values of χ^2 (called critical values) or which 2.5% of the area lies in each "tail" of the distribution, then the 95% confidence interval is

$$\chi^2_{0.025} < \frac{Ns^2}{\sigma^2} < \chi^2_{0.975} \tag{1.42}$$

from which we see that σ is estimated to lie in the interval

$$\frac{s\sqrt{N}}{\chi_{0.975}} < \sigma < \frac{s\sqrt{N}}{\chi_{0.025}} \tag{1.43}$$

with 95% confidence. Similarly other confidence interval can be found. The values $\chi_{0.025}$ and $\chi_{0.975}$ represent respectively the 2.5 and 97.5 percentile values.

The Table 1.48 in the Appendix, gives percentile values corresponding to a number degrees of freedom υ. For large values of v υ ($\upsilon \geq 30$) we can use the fact that $\left(\sqrt{2\chi^2} - \sqrt{2v-1}\right)$ is very nearly normally distributed with mean zero and standard deviation one, so that normal distribution tables can be used if $\upsilon \geq 30$. Then if χ^2_p and z_p, are the pth percentiles of the chi-square normal distribution respectively, we have

$$\chi^2_p = \frac{1}{2}\left(z_p + \sqrt{2v-1}\right)^2 \tag{1.44}$$

In order to compute a statistic such as (1.36) to (1.44), it is necessary to use observations obtained from a sample as well as certain population parameters. If these parameters are unknown they must be estimated from the sample.

The number of degrees of freedom of a statistic generally denoted by υ is defined as the number N of independent observation in the sample (i.e. the sample size) minus the number k of population parameters which must be estimated from sample observations. In symbols, $\upsilon = N - k$.

In the case of the statistic (1.36) the number of independent observation in the sample is N, from which we can compute $\overline{X}$ and s. However, since we must estimate μ, $k = 1$ and so $\upsilon = N - 1$.

In the case of statistic (1.14), the number of independent observation in the sample is *N*, from which we can compute s. However, since we must estimate σ, $k = 1$ and so $\upsilon = N - 1$.

1.6.5 Problems to Solve

Example 1.38 Research and Development department of a factory developed a new cracker with 20 pieces. Moisture content of these new crackers investigated and following data were found. Construct 80% confidence limit for moisture content.

Solution: Whether standard deviation of population given or not and sample size should be checked before the solution. When we look at the example, the data belong to whole population although sample size lower than 30. For this case, we have to apply Z test (normal distribution) whatever sample size is.

5.5	5.3	5.4	5.9	6.3	5.4	5.3	6.3	5.8	5.2
5.5	5.9	5.8	6.1	6.0	5.3	5.4	5.4	5.5	6.2

$$\overline{X} = \frac{x_1 + x_2 + \ldots + x_n}{n} = \frac{1}{n}\sum_{i=1}^{n} x_i = 5.675$$

σ is not given in the question, thus we can calculate from data by using Eq. 1.37.:

$$\sigma = \sqrt{\frac{\sum_{i=1}^{N}(x_i - \overline{X})^2}{N}} = 0.355$$

and then for 80% confidence limit α equals to 0.2 and from Table 1.19

$$z_c = 1.28$$

(continued)

Example 1.38 (continued)

$$-X \pm z_c \frac{\sigma}{\sqrt{n}} = 5.675 \pm 1.28 \frac{0.355}{\sqrt{20}}$$

$$5.573 < \mu < 5.777$$

Example 1.39 Histamine levels (ppm) of 40 sausages taken from an ordinary market were investigated and following results were found. Calculate 90%, 95% and 99% confidence limits. Assume that these sausages represent population.

Solution: For solution of this type of problems, first of all we have to check whether standard deviation of population is given or not. If it is given we have to apply Z test (normal distribution). If not, then we check whether sample size (N) is greater than 30 or not. If N is greater than 30, we have to apply normal distribution. If not, we have to apply sampling distribution (t test) for calculation of confidence interval.

279.32	189.15	38.62	165.42	26.26	3.56	7.37	24.18	0.00	11.69
154.12	65.24	32.27	118.86	150.64	6.88	0.85	7.88	103.84	45.91
108.13	58.65	18.88	62.85	108.60	3.92	0.00	0.00	105.15	58.34
81.20	25.21	15.63	36.67	62.14	71.68	28.88	16.14	23.95	9.53

When we look at the data standard deviation for population is not given, however, sample size is greater than 30. Therefore, we apply Z test.

$$\overline{X} = \frac{x_1 + x_2 + \ldots + x_n}{n} = \frac{1}{n} \sum_{i=1}^{n} x_i = 58.190$$

σ is not given in the question, thus we can only calculate "s" as:

$$s = \sqrt{\frac{\sum_{i=1}^{N} (x_i - \overline{X})^2}{n-1}} = 62.602$$

and then for 90% confidence limit α equals to 0.1 and from Table 1.19

(continued)

Example 1.39 (continued)

$$z_c = 1.645$$

$$\overline{X} \pm z_c \frac{s}{\sqrt{n}} = 58.190 \pm 1.645 \frac{62.602}{\sqrt{40}}$$

$$41.907 < \mu < 74.473$$

For 95% confidence limit z_c = 1.96 (Table 1.19)

$$-X \pm z_c \frac{s}{\sqrt{n}} = 58.190 \pm 1.96 \frac{62.602}{\sqrt{40}}.$$

$$38.789\ \mu < 77.591$$

For 99% confidence limit z_c = 2.58 (Table 1.19)

$$\overline{X} \pm z_c \frac{s}{\sqrt{n}} = 58.190 \pm 1.96 \frac{62.602}{\sqrt{40}}$$

$$38.789 < \mu < 77.591$$

Example 1.40 Construct 95% confidence limit of cracker moisture content. The information is taken from production line randomly. Suppose that the same data were found as given in Example 1.38.

Solution: Whether standard deviation of population given or not and sample size should be checked before the solution. When we look at the example, standard deviation for population is not given and sample size is lower than 30. For this case, we have to apply t test (sampling distribution).

5.5	5.3	5.4	5.9	6.3	5.4	5.3	6.3	5.8	5.2
5.5	5.9	5.8	6.1	6.0	5.3	5.4	5.4	5.5	6.2

$$\overline{X} = \frac{x_1 + x_2 + \ldots + x_n}{n} = \frac{1}{n}\sum_{i=1}^{n} x_i = 5.675$$

we can calculate s from data by using Eq. 3.1.17:

(continued)

Example 1.40 (continued)

$$s = \sqrt{\frac{\sum_{i=1}^{N} (x_i - \overline{X})^2}{n-1}} = 0.364$$

$$\overline{X} \pm t_{\alpha/2} \frac{s}{\sqrt{n}}$$

$$\upsilon = n\text{–}1 = 20\text{–}1 = 19$$

and then for 95% confidence limit α equals to 0.05 and α/2 = 0.025 from Appendices Table 1.47.

$$t_{\alpha/2} = 2.093$$

$$\overline{X} \pm t_c \ \frac{s}{\sqrt{n}} = 5.675 \pm 2.093 \ \frac{0.364}{\sqrt{20}}$$

$$5.505 < \mu < 5.845$$

Example 1.41 An examination of presence of ***Clostridium botulinum*** in canned mushroom for a random sample of 80 can in a storage showed 5 of them contain such bacteria. Find a 90% confidence limit for p.

Solution: When problem is examined, it can be seen that this problem related with proportion.

Where p is proportion of success.

$p = 5/80 = 0.0625$.
$q = 1\text{-}\ p = 1\text{–}0.0625 = 0.9375$.

Confidence limit for proportion Eq. 1.8 is;

$$p \pm z_c \sqrt{\frac{pq}{N}}$$

then for 90% confidence limit α equals to 0.1 and from Table 1.19

$$z_c = 1.645$$

$$p \pm z_c \sqrt{\frac{pq}{N}} = 0.0625 \pm 1.645 \ \sqrt{\frac{0.0625 \times 0.9375}{80}}$$

$$0.018\ p < 0.107$$

Example 1.42 Find 99% confidence limit for variance and standard deviation for data given in Example 1.40.

Solution:

$$\overline{X} = \frac{x_1 + x_2 + \ldots + x_n}{n} = \frac{1}{n}\sum_{i=1}^{n} x_i = 5.675$$

$$s = \sqrt{\frac{\sum_{i=1}^{N}(x_i - \overline{X})^2}{n-1}} = 0.364$$

$$s^2 = (0.364)^2 = 0.1325$$

Confidence limit for variance is

$$\frac{(n-1)\,s^2}{\chi^2_{right}} < \sigma^2 < \frac{(n-1)\,s^2}{\chi^2_{left}}$$

Confidence limit for standard deviation

$$\sqrt{\frac{(n-1)\,s^2}{\chi^2_{right}}} < \sigma < \sqrt{\frac{(n-1)\,s^2}{\chi^2_{left}}}$$

From χ^2 table (Appendices Table 1.48) for 99% $\upsilon = 19$.

$$\chi^2_{right} = 36.191\ \chi^2_{left} = 7.633$$

Thus;

$$\frac{(20-1)\times 0.1325}{36.191} < \sigma^2 < \frac{(20-1)\times 0.1325}{7.633}$$

Confidence limit for variance:

$$0.00366 < \sigma^2 < 0.32982$$

Confidence limit for standard deviation:

$$0.0605 < \sigma < 0.5743$$

1.6.6 Procedure

For solution of this type of problems, first of all we have to check whether standard deviation of population is given or not.

If the data belong to whole population, apply Z test (normal distribution) whatever sample size is.

If not, then we check whether sample size (N) is greater than 30 or not. If N is greater than 30, we have to apply normal distribution. If not, we have to apply sampling distribution (t test) for calculation of confidence interval.

Statistical analysis is conducted using the software SPSS 16.0. Data are given as matrix. Descriptive statistics is carried out in SPSS by following the "Analyze" item.

1.7 Multiresponse Modelling

Teresa R. S. Brandão, Mafalda Quintas, and Cristina L. M. Silva

1.7.1 *Objectives and Learning Outcomes*

Multiresponse modelling is a powerful tool for studying complex kinetics of reactions occurring in food products. This modelling technique uses information of reactants and products involved, allowing insightful kinetic parameters estimation and helping in clarifying reaction mechanisms. One example of a complex reaction that occurs in food processing is the caramelisation reaction. Caramelisation is the common name for a group of reactions observed when carbohydrates are exposed to high temperatures.

The objective of this work was to apply the principles of multiresponse modelling in developing a mathematical mechanistic model that describes sucrose thermal degradation and caramelisation products formation in highly concentrated sucrose solutions, with different water contents and under various temperatures, as case study.

It is expected as learning outcomes, the following:

(i) Understand of the basic principles of multiresponse modelling;
(ii) Apply the multiresponse modelling to real experimental data, conveniently (practice on the case study presented);
(iii) Acquirements on regression analysis procedures, parameter estimation and model adequacy assessment;
(iv) Comprehend the multiresponse modelling as a potential tool for understanding reaction mechanisms in food matrices.

1.7.2 *Introduction*

Foods often suffer processing, which has three major aims: (i) to make food safer, (ii) to provide products of the highest quality attributes, and (iii) to transform food into forms that are more convenient or more appellative. To achieve that, studies in food science and technology are mainly concerned with process design, assessment and optimization. Early food science approach was essentially concerned with applying a specific treatment to a food product and reporting end-point data (trial and error). In the early eighties through middle nineties, with the general availability of powerful computer aided tools, an increased interest on the use of modelling and regression analysis tools to describe and predict phenomena in food systems was observed.

In the area of quality/safety loss kinetics during storage and processing, some remarkable examples of the effort made at the time to draw food scientists' attention to the benefits of accurate modelling and reporting of experimental data are available (e.g. are Saguy and Karel 1980; Labuza and Kamman 1983; Lund 1983; Arabshahi and Lund 1985). Later works (Cohen et al. 1994; van Boekel 1996) also played a critical role on alerting to the importance of estimates precision and quality of the regression.

Most of the phenomena observed in food systems are mainly described by non-linear models. Thus, food scientists aiming at modelling their results, need a sound understanding of non-linear regression analysis principles. Assessment of model adequacy to describe the observed phenomena and adequate interpretation of regression inference results, are critical skills for developing useful models.

Changes in food systems often occur within a chain of complex reactions, all dependent on several different environmental conditions. Assessing such complex reaction mechanisms is not an easy task. However, measuring several responses of a food system at the same time and using all the information available when modelling experimental data, can be a powerful tool for solving and clarifying these issues.

This can be done using multiresponse modelling (Box and Draper 1965; Stewart et al. 1992). This modelling approach accounts for kinetic information of both reactants and products involved in the reaction, resulting in insightful parameter estimation and, consequently, in more accurate model predictions.

In food science, multiresponse modelling has been successfully applied to enzymatic reactions catalysed by lipase, such as hydrolysis of acylglycerols (Malcata et al. 1991; Rice et al. 1999) and esterification of conjugated linoleic acid with sorbitol (Torres et al. 2003a), acidolysis (Torres and Hill 2004) and ethanolysis (Torres et al. 2003b) of oils. Other enzymatic reactions, like whey protein hydrolysis by proteases cardosin A and B (Barros and Malcata 2002, 2004) and lactose hydrolysis in skim milk by beta-galactosidase (Bakken et al. 1992), have also been described with multiresponse regression techniques. This modelling approach has also been fruitfully applied in other studies, like chlorophyll degradation in foods (van Boekel 1999) and lactic fermentation during pickled carrot manufacture (Nabais and Malcata 1997). Multiresponse modelling has also been widely used in understanding the Maillard reaction (van Boekel 1998; Martins et al. 2000; Brands and van Boekel 2002; Martins and Van Boekel 2003; Mundt and Wedzicha 2005) and to elucidate the acrylamide formation in cooked foods (Knol et al. 2005).

1.7.3 Regression Analysis

Regression analysis is a statistical tool, which can be used for assessing models adequacy to describe experimental data points. Such tool has been intensely reviewed and discussed (good examples are Johnson and Frasier 1985; Bates and Watts 1988; Seber and Wild 1989; Neter et al. 1996).

A model is a mathematical function that is able to describe the dependence of an X independent random variable with a Y dependent random variable and can be generally described as $f(\theta_j,X)$, where θ_j represents model parameters ($j = 1,\ldots,p$). Each observation of the dependent variable (y_i) ($i = 1,\ldots,n$) can then be expressed as a function of the mathematical model with optimum set of parameter values θ_j, at the corresponding value of the dependent variable ($f(\theta_j, x_i)$):

$$y_i = f(\theta_j, x_i) + \varepsilon_i \tag{1.45}$$

where ε_i is the experimental error of the measurement (or residual).

Such mathematical model can be linear or non-linear in respect to the parameters. In the later, the first and second derivatives of the model function in order to the parameters, depends on the parameters themselves.

In order to estimate the most probable estimates of θ_j, the first step is to collect experimental data. The food scientist then postulates the model which he believes describes the observed phenomenon. Using a regression procedure, parameters are estimated. The final step is assessing parameters and prediction precision and the "quality" of the regression. With the exception of some mathematical treatment, regression analysis basic principles are common to both linear and non-linear models. In the following sections such principles are briefly reviewed.

1.7.3.1 Parameter Estimation

Least-squares, Maximum likelihood, and Bayesian estimation are well-known examples of different statistical approaches to parameter estimation. Under certain specific conditions, these methods provide similar regression results (Bates and Watts 1988). The next step is the calculation of the parameters that minimise or maximise the selected objective function. However, these methods only provide accurate models if two important assumptions are observed: (1) the mathematical function is correct, i.e. the model describes the observed physical phenomenon, and (2) all of the experimental error can be attributed to the measurement of y_i, i.e. the observed response (y_i) is the sum of the mathematical model with the disturbance (ε_i) (Johnson and Frasier 1985; Bates and Watts 1988; van Boekel 1996).

In this section, the basic principals of each parameter estimation method are briefly discussed.

1.7.3.1.1 Least Squares Estimation

Least-squares method is a deterministic approach to parameters estimation and is undoubtedly the oldest and most used regression scheme. To this popularity contributes the simplicity of the methodology mathematical principles, which basically consists on finding values for the parameters that minimise the sum of squares of deviations between experimental values (y_i) and the ones predicted by the mathematical model ($f(\theta_j, x_i)$), i.e. the residuals sum of squares (SSR):

$$SSR = \sum_{i=1}^{n} \left(y_i - f\left(\theta_j, x_i\right)\right)^2 \tag{1.46}$$

However, this estimation procedure is only valid under certain assumptions: i) the disturbance is independent of the mathematical model, ii) each ε_i has a normal distribution with zero mean, and iii) all the disturbances have constant variance and are independently distributed.

The least-squares regression becomes complex if the model assumed is non-linear (in the parameters). In those situations, the parameter estimation is not straightforward, since the minimisation of the residuals, which is mathematically attained by calculation of the first and second derivatives of the model function in order to the parameters, depends on the parameters themselves. Under such circumstances, iterative methods for function minimisation are required. The advantages and disadvantages of the numerical schemes commonly used (e.g. Nelder-Mead simplex concept; Gauss-Newton and Marquardt-Levenberg) have been discussed (Johnson and Frasier 1985).

1.7.3.1.2 Maximum Likelihood Estimation

This method is based on finding the parameter estimates (θ_j) that maximize the probability of finding the observed responses (y_i). One of the main advantages of this method is that is based on the errors (ε_i) distribution, which can be non-normal (Bates and Watts 1988; Seber and Wild 1989). Herein, a brief discussion on this method for normally distributed errors is presented.

If the experimental errors of each measurement are i) due to random factors and several causes can contribute to them, and ii) are uncorrelated, it is possible (from the central limit theorem) to assume a normal distribution for ε_i and, consequently, for each possible values of y_i. The density probability of making the observed measurement at the particular data point $g(y_i)$ can then be written:

$$g(y_i) = \frac{1}{\sigma_i\sqrt{2\pi}} \exp\left[-\frac{1}{2}\left(\frac{y_i - f\left(\theta_j, x_i\right)}{\sigma_i}\right)^2\right] \tag{1.47}$$

where σ_i represents the standard deviation of the random experimental error.

The probability of making a series of measurements at n distinct x_i points is then given by the product of $g(y_i)$ at each point and the best estimates for θ_j can be obtained using the maximum likelihood estimation method (Neter et al. 1996), i.e. by maximizing the likelihood function:

$$P(Y|\theta) = \left\{\prod_{i=1}^{n} \frac{1}{\sigma_i\sqrt{2\pi}}\right\} \left\{\sum_{i=1}^{n} \exp\left[-\frac{1}{2}\left(\frac{y_i - f(\theta_j, x_i)}{\sigma_i}\right)^2\right]\right\} \quad (1.48)$$

(Herein, Y and θ are the vector form representation of y_i and θ_j)

This procedure reduces to minimizing the second term of the equation. Furthermore, if σ_i is constant, this estimation method reduces to the familiar least-squares procedure for estimating parameters.

1.7.3.1.3 Bayesian Estimation

Bayesian estimation is based on Bayesian statistics, which argues that the mathematical theory of probability is applicable to the degree to which a person believes a proposition. Such inference uses the Bayes' theorem as the basis for a rule for updating beliefs in the light of new information.

Contrary to least-squares and maximum likelihood estimators, Bayesian estimation does not estimate parameter values, but their "most probable" probability distribution (Bates and Watts 1988). The conditional (or posterior) density function of the parameters is then determined (Seber and Wild 1989):

$$P(\theta|Y) = \frac{P(\theta, Y)}{P(Y)} = \frac{P(Y|\theta)\ P(\theta)}{P(Y)} \propto P(Y|\theta)\ P(\theta) \quad (1.49)$$

For independent and normally distributed errors (with zero mean and constant variances) and assuming non-informative prior distribution for the parameters (P (θ)), this approach leads to a p-variate student's T distribution. The marginal posterior density for each parameter is a univariate student's T distribution (Bates and Watts 1988).

1.7.3.2 Assessing "Quality" of Regression

Once the parameters are estimated, it is imperative to ensure that the model adequately describes the experimental data and that the assumptions postulated throughout the estimation procedure are observed.

In most regression studies the model fit is assessed using the coefficient of determination (R^2), which measures the proportion of the total variation in the response values "explained" by the model (Neter et al. 1996):

$$R^2 = 1 - \frac{SSR}{SST} \ ; 0 \le R^2 \le 1 \quad (1.50)$$

where SST (Total Sum of Squares) is given by:

$$SST = \sum_{i=1}^{n} (y_i - \bar{y})^2, \tag{1.51}$$

being $\bar{y}$ the response's average value. However, this criterion alone is not always appropriate for analysing both linear (Neter et al. 1996) and non-linear (Bates and Watts 1988) regression models and due caution is necessary when analysing R^2 results.

Another commonly used "measurement" of model fit is the mean square error (MSE), i.e. the variance estimate based on n-p degrees of freedom (Eq. 1.52). However, this is only truly useful for comparing the fit of two different models.

$$MSE = \frac{SSR}{n - p} \tag{1.52}$$

The most important, and sometimes forgotten, regression "quality check" is the residuals analysis. This consists on scrutinizing if the obtained residuals actually reflect the properties assumed for the disturbance (ε_i) (Bates and Watts 1988; Seber and Wild 1989; Neter et al. 1996; van Boekel 1996). The simplest way of undertaking a residual analysis is to assess randomness, normality and constant variance of residuals through simple graphical representations, like plotting the residuals in time sequence or against the model estimated response (Box et al. 1978). Approximate analysis can be made by checking the dispersion of the experimental points along the estimated curve.

1.7.3.3 Assessing "Quality" of Parameters' Estimates

The prime objective of regression analysis is to obtain useful predictions. Even the most complex and sophisticated model, with "good quality regression indicators", is totally useless if such predictions are not accurate. In non-linear regression analysis, and due to possible finding local minima during minimization of SSR, the first question the food scientist must place is: does this parameter have physical meaning? And if so, does the obtained value make sense? Once these questions are surpassed, a "statistical" analysis of quality of parameters' estimates can be done: For comprehensive parameter estimation, collinearity and precision of the estimates must be taken into consideration.

Collinearity (interdependency or correlation) may affect the optimisation procedure inherent to the regression analysis (Bard 1974). It can be reduced by model reparameterisation, thus improving the estimation procedure. The Arrhenius equation expressed in terms of a finite reference temperature is the most well-known reparameterised model.

Precision of the estimates can be evaluated by the confidence intervals (the intervals within the true parameters values are expected to lie, for a chosen significance level). In the case of linear models, the 1-α confidence interval for a parameter is then calculated (Bates and Watts 1988):

$$\theta_j \pm MSE\sqrt{\left[(X^TX)^{-1}\right]_{jj}} t_{\left(n-p,\alpha/2\right)} \tag{1.53}$$

where $[(X^TX)^{-1}]_{jj}$ is the jth diagonal term of the parameters' variance-covariance matrix $(X^TX)^{-1}$ (X is the matrix with elements $\frac{\partial f(\theta_j, x_i)}{\partial \theta_j}$) and $t_{\left(n-p,\alpha/2\right)}$ is the upper $\alpha/2$ quantile for student's T distribution with n-p degrees of freedom.

In the case of non-linear regression analysis, confidence intervals may be approximated by expression 1.53 (Bates and Watts 1988). However, the food scientist must be aware of this approximation when evaluating parameters confidence intervals (van Boekel 1996).

As already discussed, one of the main advantages of modelling any behaviour in food systems is to be able to predict the final properties of a processed product. This is usually done by determining such property using the estimated parameters of the proposed model. Although extensively used in food science and technology literature, this approach does not take into account the fact that estimated parameters are not a specific value, but an interval where the parameters' true value probably lies. To evaluate the interval where the measured response probably lies, the linear approximation of the 95% inference band of a non-linear model can be calculated (Bates and Watts 1988):

$$f(\theta, x) \pm \sqrt{MSE}\sqrt{x^T(X^TX)^{-1}v}\sqrt{pF_{(p,n-p,0.05)}} \tag{1.54}$$

where x is the parameters derivative vector and $F_{(p,n-p,0.05)}$ is the upper 5% of the quantile for Fisher distribution with p and n-p degrees of freedom.

Using a Bayesian approach, the precision of the estimates can be assessed by the contour of the marginal highest posterior density (HPD), i.e. the 1-α HPD interval can be determined by the shortest interval in parameter space that contains 1-α of the posterior probability (Seber and Wild 1989).

1.7.4 Multiresponse Nonlinear Regression

Food scientists often face kinetic problems where more than one response is measured at the same time (e.g. concentrations of different reactants and products). Under these circumstances, the different responses can be considered as random variables with a multinormal distribution (Priestley 1992). In this case, the mathematical model can be naturally extended to a system of equations describing each measured response (Box and Draper 1965; Bates and Watts 1988; Seber and Wild 1989; van Boekel 1996):

$$y_{iu} = f_u(\theta_j, x_{iu}) + \varepsilon_{iu} \quad (1.55)$$

where u denotes the measured response (u = 1,. . .,d).

The assumptions in multiresponse regression are similar to the ones in single-response analysis. The system formed by the mathematical functions should be correctly formulated and all the experimental errors must be due to the measurement of y_{iu}. Furthermore, the random errors (ε_{iu}) are assumed to be normally distributed with zero mean and constant variance for each particular response (u). Errors of the dependent variable measured at different experimental runs are assumed to be independent.

Box and Draper (1965) determined the marginal posterior probability distribution of the θ parameters, given the observed d responses (y_i), by applying a Bayesian approach and assuming non-informative prior distributions for θ and the error structure. These authors concluded that this marginal posterior probability was maximized by simply minimizing with respect to the parameters the determinant of the matrix of cross products of the d responses, with the following elements:

$$v_{uz} = \sum_{i=1}^{n} \left(y_u - f_u(\theta_j, x_i)\right)\left(y_z - f_z(\theta_j, x_i)\right) \quad (1.56)$$

where z is any ≠u.

This came to be known as the "determinant criterion" (Bates and Watts 1988; van Boekel 1996).

From the above, it is easily concluded that the computational effort involved in multiresponse estimation is of great significance. However, several numerical methods and software are presently available (www.athenavisual.com; Bates and Watts 1988; Stewart et al. 1992), thus making this tool accessible to food scientists with less experience in statistical and numerical methods, but with a sound understanding of multiresponse estimation basic principles.

1.7.5 Problem to Solve

Sucrose thermal degradation is an important reaction in the food industry, since it is responsible for either important characteristics of the final food products or can influence the yield of obtainable white sugar in sugar manufacture. This degradation may occur by two different major reaction pathways: the Maillard reaction, which takes place in the presence of amino acids, and caramelisation, that ocurs when sucrose is heated at high temperatures (BeMiller and Whistler 1996), which is the case of traditional sugar syrups production. This is an interesting case to apply multiresponse modelling.

Caramelisation, which is the common name for a group of complex reactions, has been studied by several authors. In the past decades, great insight was gained on the reactions mechanism (Mauch 1971; Richards 1986; Clarke et al. 1997). Due to the large number of reaction pathways and even larger number of products formed during the caramelisation reaction, kinetic studies based on the caramelisation reaction mechanism are of increased complexity. The complexity of the reaction makes the use of multiresponse modelling very attractive, since it also allows clarifying the main reaction pathways that directly depend on the process conditions.

The caramelisation reaction occurs in the presence or absence of catalysts (acids, basis or salts and impurities). It can be observed in diluted or concentrated solutions (amorphous state), and in the crystal form, leading to many different pathways. Each possible pathway may be preponderant under given environmental conditions. In this work we focus on the case where sucrose in very concentrated solutions (above saturation at room temperature) is heated at high temperatures in the absence of impurities and at neutral pH. Under these conditions, the effect of heat is similar to the effect of using lower temperatures in acidic conditions and the reaction mechanisms can be considered to be identical (Mauch 1971). This is due to the formation of weak acids during sucrose hydrolysis.

The first step in the caramelisation reaction is sucrose hydrolysis (please see Fig. 1.23). This takes place by protonation of the glycosidic linkage (Mauch 1971; Richards 1986; Eggleston et al. 1996; Clarke et al. 1997). The H^+ used in this reaction step can be derived from water (or even sucrose) dissociation at high temperatures and also from acidic reaction products (Richards 1986; Lowary and Richards 1988; Eggleston et al. 1996).

Under neutral conditions, sucrose hydrolysis yields D-glucose and a fructose oxocarbonium ion (Richards 1986; Lowary and Richards 1988; Eggleston et al. 1996; Clarke et al. 1997). This fructose carbocation can react with: (i) a water molecule to form fructose and regenerate the catalyst H^+, or (ii) another carbohydrate molecule present in solution (sucrose, glucose, fructose or even another fructose carbocation) to form oligosaccharides. Moreover, this fructose carbocation may undergo non-specific reactions to form a wide range of products, including: anhydrosaccharides, 5-hydroxymethyl-furfural (HMF) and organic acids (Richards 1986; Eggleston et al. 1996; Clarke et al. 1997).

The oligosaccharide and anhydro-sacharide components of caramel have been extensively studied (Manley-Harris and Richards 1991, 1996, 1997; Defaye and Garcia Fernandez 1994, 1995; Ratsimba et al. 1999). However, the extent of production of these substances is dependent on reaction conditions and is typically favoured by extreme acidic/high temperatures reaction conditions (Antal Jr. et al. 1990; Manley-Harris and Richards 1996, 1997), which is not the present case.

Further degradation of fructose and glucose is responsible for the formation of other compounds, with special preponderance to HMF (Sugisawa 1966; Mauch 1971; Lowary and Richards 1988; Antal Jr. et al. 1990; Kroh 1994; Eggleston et al. 1996; Clarke et al. 1997). The mechanism of such reaction is generally accepted to be an open chain β-elimination mechanism, which proceeds via an enediol intermediate to 3-deoxyhexulose (Kroh 1994; Clarke et al. 1997). With the further development of the caramelisation reaction, HMF can be degraded to other products, including: (i) furfural, via fragmentation reaction (with indirect subsequent

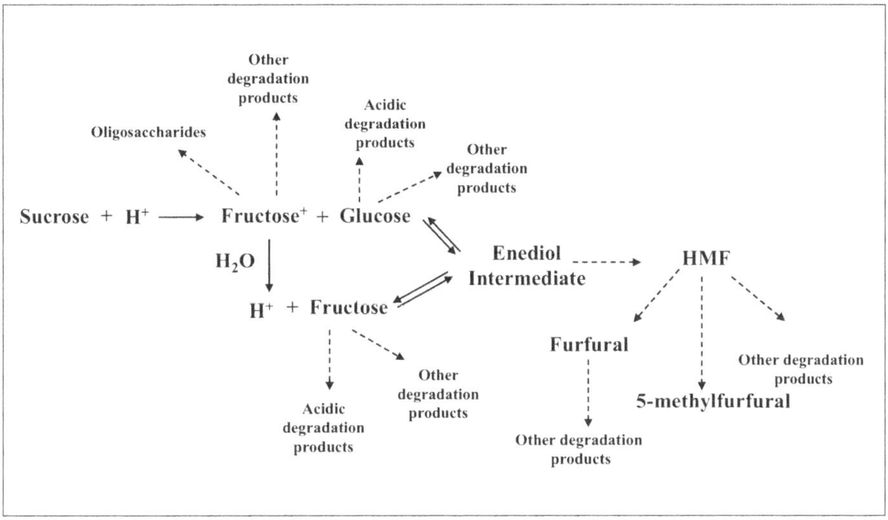

Fig. 1.23 Schematic representation of sucrose caramelisation mechanism based on literature review

H^+ release) and (ii) 5-methylfurfural, via reduction reaction (Mauch 1971; Antal Jr. et al. 1990; Kroh 1994; Clarke et al. 1997).

The global sucrose caramelisation reaction at high temperatures and neutral pH can then be generally described by the scheme in Fig. 1.23. However, one must always bear in mind that this is a simplification of the reaction steps, where intermediate products and reactions are not considered (dashed arrows in Fig. 1.23).

Temperature is known to have a significant effect on food reactions kinetics. This effect is often translated into a dependence of the kinetic parameters on temperature, which has been widely studied and can be generally expressed with an Arrhenius type equation, using a finite reference temperature (Saguy and Karel 1980; Haralampu et al. 1985; Peleg 1992; Cohen et al. 1994; van Boekel 1996):

$$k = k_{ref} \exp\left(-\frac{E_a}{R}\left(\frac{1}{T} - \frac{1}{T_{ref}}\right)\right) \tag{1.57}$$

where k is the reaction rate, E_a the activation energy, k_{ref} the reaction rate at finite reference temperature, T_{ref}, and R the universal gas constant. Several studies indicated that if T_{ref} corresponds to the middle temperature of the experimental range, the correlation between parameters and the confidence intervals will be minimised (Cohen et al. 1994; van Boekel 1996; Brandão 2004), thus improving the mathematical modelling quality.

Reactant concentration and water content may also affect reactions kinetics. This effect can be included in the Arrhenius parameters (Eq. 1.57; k_{ref} and E_a), on the basis of experimental data and fitting procedures. A mixed-model, that includes the influence of one or more main effects and significant interactions, may account empirically for a mathematical description of food reaction kinetics (Saguy and Karel 1980).

The problem to solve it basically the development of a mathematical model that describe conveniently the sucrose thermal degradation and caramelisation products formation in highly concentrated sucrose solutions, including the water and temperature effects.

1.7.6 *Procedure*

1.7.6.1 Experimental Procedures

Sample preparation

Concentrated sucrose solutions of 30.03, 25.30, 16.32, 12.20 and 3.58% (w/w) water content were prepared, subjected to heat treatment (at 100, 120, 140, 160 and 180 °C). The total acidity of the prepared samples was determined by acid-base titration, in order to confirm that solutions were indeed neutral (pH = 7).

Analytical determinations

Sucrose, fructose, glucose, HMF, furfural and 5-methylfurfural content of the sugar syrups were determined using High Pressure Liquid Cromatography (HPLC).

Monitoring of pH changes during the reaction

After heat processing, the samples acidity was assessed using a pHmeter at room temperature.

1.7.6.2 Modelling Procedures

The development of the most appropriate model for describing the caramelisation mechanism was based on proposing possible reaction pathways (please see Fig. 1.23). For each assumed mechanism, the rate of reaction of all components involved was defined. The obtained system of differential equations composed the model of the caramelisation reaction.

The models were fitted to experimental data points by multiresponse regression schemes, provided by Athena Visual Studio software (www.athenavisual.com), which allows estimating the reaction rate constants using the determinant criterion of the Bayesian approach (Box and Draper 1965; Bates and Watts 1988; Seber and Wild 1989; Stewart et al. 1992; van Boekel 1996).

Mathematically, the model that best described the experimental data was selected based on fulfilment of regression analysis: i.e., minimization of the sum of squares of the residuals (SSR) and independent random residuals, normally distributed with zero mean and constant variance (Bates and Watts 1988).

After selecting the best model, the temperature dependence of rate constants was described by an Arrhenius equation and a one-step multiresponse regression analysis was carried out. Water content effects were also included in the Arrhenius parameters (i.e. pre-exponential factors and activation energies), by performing a global fit of all experimental data.

Quality of the regressions was assessed by the sum of squares of the residuals (SSR) and visual inspection of residuals' randomness. Parameters' precision was evaluated by the 95% highest posterior density (HPD) intervals (Box and Draper 1965; Bates and Watts 1988; Seber and Wild 1989; Stewart et al. 1992; van Boekel 1996).

1.7.6.3 Experimental Results

Experimental results showed that after an initial phase of delay, sucrose content rapidly decreases and glucose and fructose start to form (Fig. 1.24). Short time after, HMF is detected, followed by furfural. Later in the reaction, 5-methylfurfural can also be identified. However, the later compound could not be quantified in all of the experiments. This is due to reactions time, which in those cases was not sufficient to allow 5-methylfurfural formation. This observation led to the decision of not including this compound in modelling procedures.

1.7.6.4 Modelling Sucrose Degradation into Caramel

The mathematical description of the sucrose caramelisation reaction was based on mechanisms referred in literature. As already stressed, the scheme presented in Fig. 1.23 is a general representation of the possible reaction pathways. Among all pathways presented, some were selected as preferred to occur. Such selections, and subsequent mechanisms proposed, were based on: (i) experimental measurements of the reactants and products and (ii) mathematically, by the degrees of freedom of the system of differential equations and number of parameters to estimate. A considerable number of mechanisms were tested (results not shown). The pathway proposed in Fig. 1.25 fulfilled multiresponse regression analysis of the experimental data.

The correspondent differential equations, that express the rate of reactions taken into consideration, are enumerated in Eqs. 1.58 to 1.65.

$$\frac{d[Sucrose]}{dt} = -\mathrm{k_1}[\mathrm{Sucrose}][\mathrm{H^+}] \tag{1.58}$$

$$\frac{d[H^+]}{dt} = \mathrm{k_2}[\mathrm{Fructose}] + \mathrm{k_4}[\mathrm{HMF}] \tag{1.59}$$

$$\frac{d[Fructose]}{dt} = \mathrm{k_1}[\mathrm{Sucrose}][\mathrm{H^+}] - \mathrm{k_3}[\mathrm{Fructose}] - \mathrm{k_2}[\mathrm{Fructose}] \tag{1.60}$$

$$\frac{d[Glucose]}{dt} = \mathrm{k_1}[\mathrm{Sucrose}][\mathrm{H^+}] - \mathrm{k_6}[\mathrm{Glucose}] \tag{1.61}$$

$$\frac{d[HMF]}{dt} = \mathrm{k_3}[\mathrm{Fructose}] - \mathrm{k_4}[\mathrm{HMF}] \tag{1.62}$$

$$\frac{d[Furfural]}{dt} = \mathrm{k_4}[\mathrm{HMF}] - \mathrm{k_5}[\mathrm{Furfural}] \tag{1.63}$$

$$\frac{d[Other\ Products_I]}{dt} = \mathrm{k_5}[\mathrm{Furfural}] \tag{1.64}$$

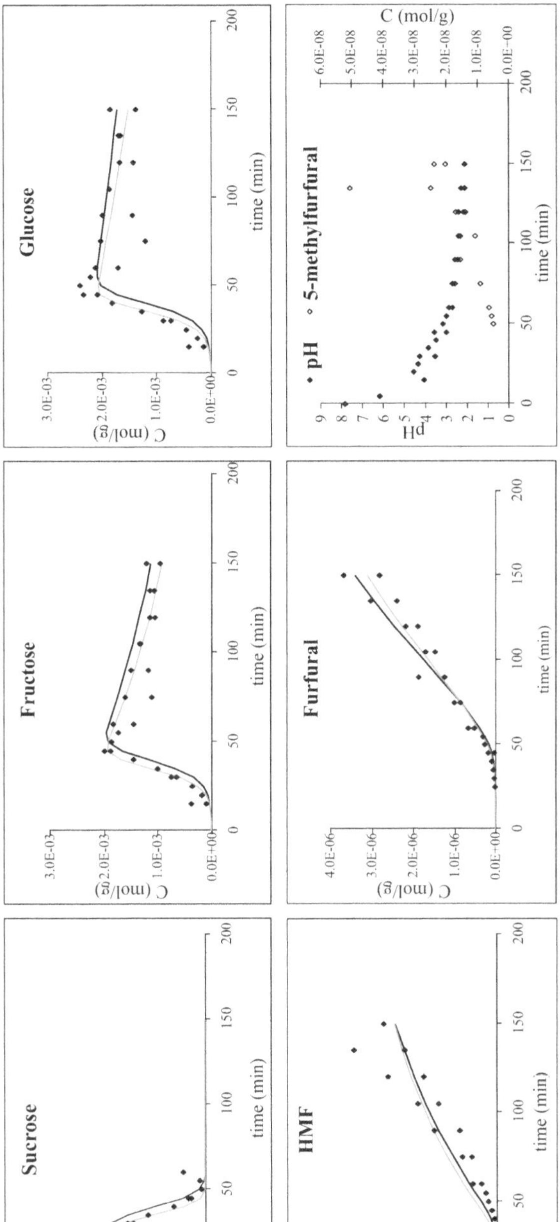

Fig. 1.24 Typical experimental data of sucrose, fructose, glucose, HMF, furfural, 5-methylfurfural and pH of heat processed solutions. Example for a 25.30% (w/w) water content sucrose solution at 140 °C. Predictions based on the mechanism proposed (grey lines) and on the model including temperature and water content effect, model *Global 2* (black lines)

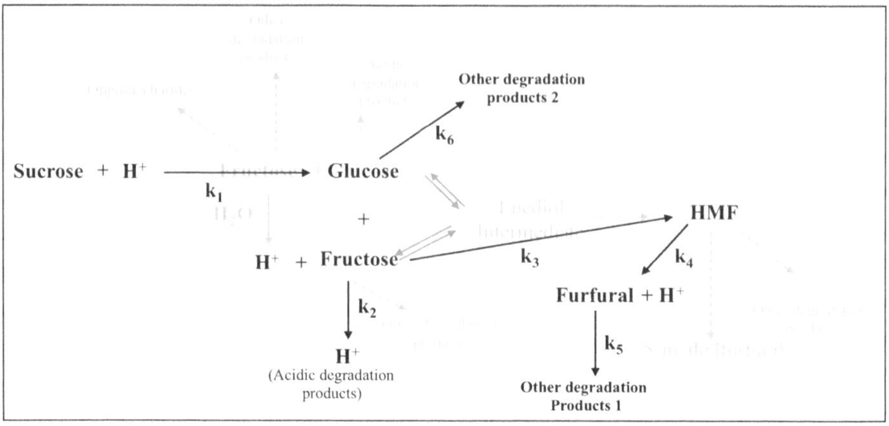

Fig. 1.25 Schematic representation of caramelisation reaction mechanism – proposed reaction pathways based on experimental data and multiresponse modelling procedures

$$\frac{d[Other\ \ Products_2]}{dt} = \mathrm{k}_6[\mathrm{Glucose}] \tag{1.65}$$

The mechanism proposed (Fig. 1.25) only accounts for the main reaction pathways: i.e., the fact that this model does not consider the step of glucose degrading into HMF, does not mean that this step is not occurring, it only indicates that, kinetically and under the experimental conditions imposed, production of HMF from fructose is favoured. This preferential pathway has been described by several authors when discussing the higher reactivity of fructose over glucose (Mauch 1971; Richards 1986; Lee and Nagy 1988; Lowary and Richards 1988; Antal Jr. et al. 1990; Arena et al. 2001). However, other pathways may emerge, if different processing conditions are used.

The model formed by the system of Eqs. 1.58 to 1.65 was fitted to experimental data. Constant rates (k_1 to k_6) were estimated for fixed values of processing temperature and sucrose solutions water content. The proposed model successfully described the experimental data, as can be observed in the example presented in Fig. 1.24. Nevertheless, it is convenient to bear in mind that the mathematical model can only be "tentatively entertained" (Box and Draper 1965). As an example, in the present study, acidity changes were monitored using a pHmeter, which is not able to give an exact H^+ concentration. Measurements of the organic acids during this reaction may have revealed a different mathematical model, since accurate H^+ concentration determination would allow using $[H^+]$ has a response in the regression procedure.

Drawbacks in modelling procedures occurred for sucrose solutions with 3.58% (w/w) water content. This may be explained by the fact that these are amorphous glassy systems at room temperature, with extremely low water content – in sucrose hydrolysis, water can also act as a solvent or reactant (Labuza et al. 1970). Under these conditions a different reaction mechanism may have occurred. In face of such observations, experimental results obtained for this concentration level were excluded from further modelling procedures.

1.7.6.5 Temperature and Water Content Effects

Arrhenius behaviour was studied for describing the effect of temperature on reaction kinetics. The water content effect on reaction kinetics was included in the Arrhenius model parameters, on the basis of experimental data and fitting procedures.

Kinetic parameters, $k_{ref\text{-}s}$ and $E_{a\text{-}s}$, of each reaction step ($s = 1,\ldots 6$ is the pathway identification) were estimated for fixed water content levels, with the purpose of studying the influence of this factor on the estimates.

Results indicated that water content has no significant effect on activation energies of all reaction steps (results not shown). In relation to $k_{ref\text{-}s}$, water content affected significantly all parameters, excepting $k_{ref\text{-}2}$ (i.e. fructose degradation into acidic degradation products) (Fig. 1.26).

This can be an indication that the mechanism of fructose degradation into acidic products is significantly affected by temperature, but not by the amount of water molecules present in solution. This hypothesis is supported by the observed effects of temperature and water content on pHmeter readings drop during the reaction (Fig. 1.27).

The relationships between $k_{ref\text{-}s}$ (s = 1,3,4 and 5) and water content were exponential (Fig. 1.26). Regarding $k_{ref\text{-}s}$ (s = 4,5 and 6), the relationship with water content was less evident, probably due to mathematical drawbacks in parameter estimation, justified by the assumption of reaction pathways of components that were not experimentally measured (i.e., $[H^+]$, for $k_{ref\text{-}4}$, other degradation products 1, for $k_{ref\text{-}5}$ and other degradation products 2, for $k_{ref\text{-}6}$). However, such relationships could be approximately described by exponential relationships for $k_{ref\text{-}s}$ (s = 4 and 5) (Fig. 1.26). For reaction pathway number 6, the water effect on $k_{ref\text{-}6}$ was successfully described by both linear and exponential functions.

Based on these results, two global models, including both temperature and water content effects, were considered. Model *Global 1* was formed by the system of differential Eqs. 1.58 to 1.65 combined with the Arrhenius relation, where $k_{ref\text{-}2}$ = constant, $k_{ref\text{-}6} = a_6 + b_6W$, remaining $k_{ref\text{-}s} = a_s \exp(b_sW)$ (s = 1,3,4 and 5), and all activation energies are constant (W is water content and a_s and b_s are coefficients to be estimated). For model *Global* 2, $k_{ref\text{-}2}$ = constant, remaining $k_{ref\text{-}s} = a_s \exp(b_sW)$ (s = 1,3,4,5 and 6), and all activation energies are constant. Results of multiresponse regression analysis of both models are included in Table 1.20. The sums of squares of the residuals were identical ($SSR_{Global\ 1} = 2.093 \times 10^{-4}$ and $SSR_{Global\ 2} = 2.094 \times 10^{-4}$). Both models were also evaluated in terms of adequacy for describing experimental data and in parameters' precision. The later can be clearly evaluated by the errors listed in Table 1.20, which were calculated dividing the 95% HPD intervals by the estimate. Excepting a_5 and b_6, all other parameters were estimated with satisfactory precision, with errors varying between 2 to 71%

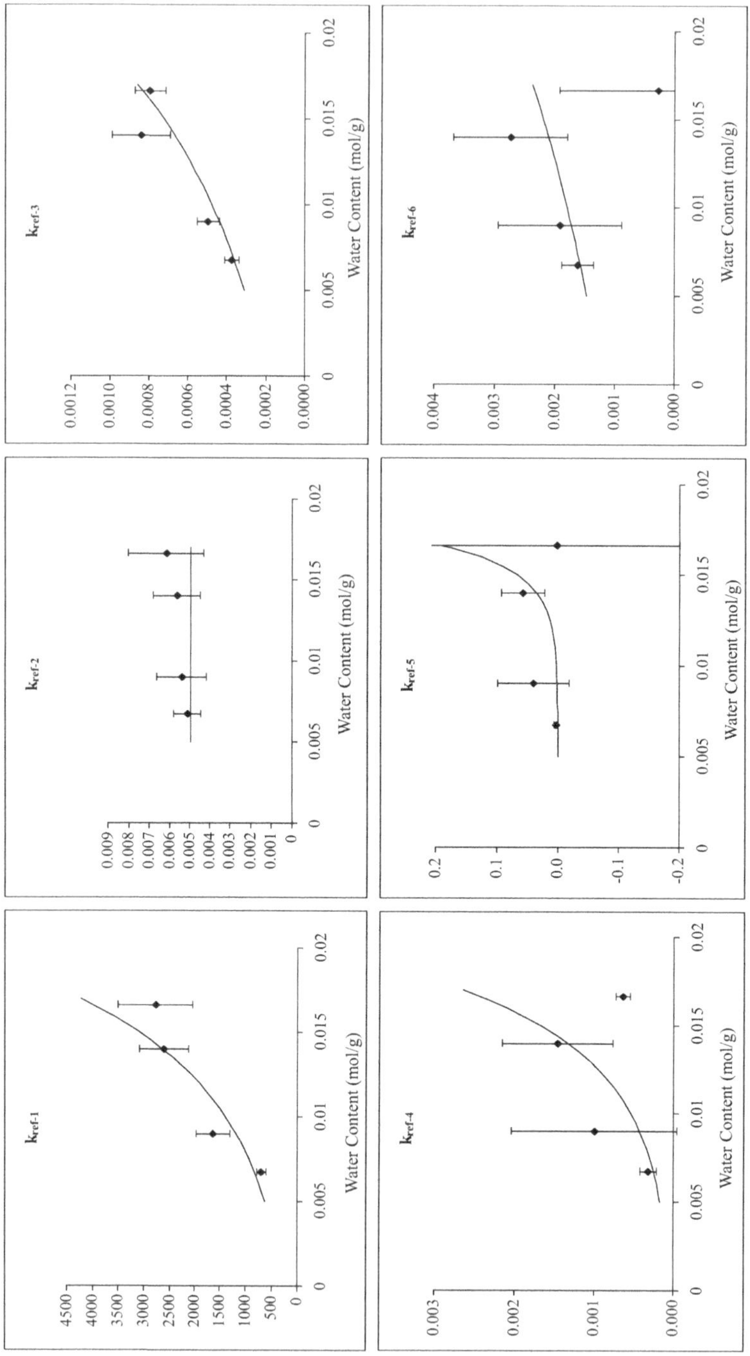

Fig. 1.26 Dependence of $k_{ref\text{-}s}$ on water content (T_{ref} = 140 °C). Points are one-step estimates and bars indicate 95% HPD intervals. Lines show predicted values by the *global model 2*

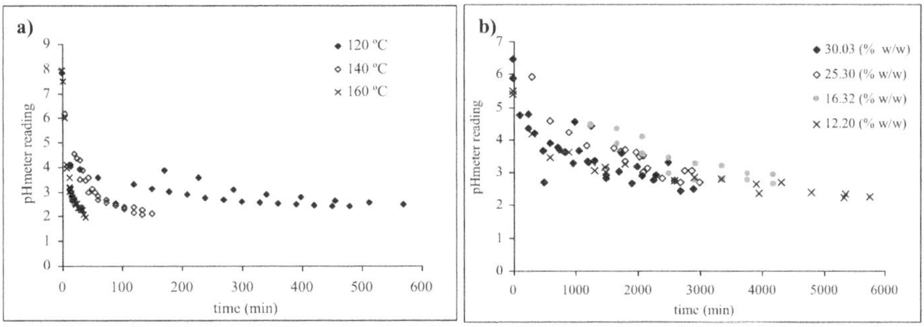

Fig. 1.27 Effect of (**a**) temperature and (**b**) water content on on pHmeter reading of sucrose solutions during heat treatment. Results are for (**a**) 25.30% (w/w) water content solutions processed at different temperatures and for (**b**) different water content solutions processed at 100 °C

(model *Global 1*) and 2 to 45% (model *Global 2*). The inferior precision obtained for a_5 and b_6 may be, once again, justified by the assumption of reaction pathways of components that were not experimentally measured (i.e. other degradation products 1 and 2), hence contributing to mathematical drawbacks in parameter estimation.

Overall, it can be concluded that the two models successfully described the experimental data (sucrose solutions from 12.2 to 30.0% of water content, processed within the temperature range 100–180 °C), with similar regression analysis results and identical parameters' precision. An example of model *Global 2* prediction can be observed in Fig. 1.24.

1.7.6.6 Model Inadequacy for Reactions Occurring at Extremely Low Water Content

As previously justified, experimental data of thermally processed sucrose solutions with 3.58% (w/w) water content were not included in the development of the proposed models. The ability of model *Global 2* to predict sucrose thermal degradation and caramelisation products under such extremely low water content solutions, was tested. Experimental results and model predictions can be compared in Fig. 1.28.

As it can be observed, the model does not express the behaviour of caramelisation kinetics. The inadequacy in fructose concentration description may indicate that, at such low water content levels, the reaction step of the fructose carbocation with water to form fructose described in Fig. 1.28 is not occurring due to the absence of H_2O molecules. The fructose carbocation is probably transformed to anhydrofructose under these conditions, as found by other authors (Simkovic et al. 2003). Observed glucose behaviour may indicate that its degradation is favoured under such conditions. Observed HMF behaviour deviation from model prediction can be an indication of the emergence of two possible reaction pathways: i) HMF production from glucose is favoured under such conditions and/or ii) HMF is a product of non-specific degradation of the fructose carbocation.

Table 1.20 Estimated parameters of the two proposed models, that include temperature and water content effects on the rates of the caramelisation reaction

Parameter	Model			
	Global 1		*Global 2*	
	k_{ref-2} = const. $k_{ref-6} = a_6 + b_6W$ $k_{ref-s} = a_s \exp(b_sW)$ $_{s\ =\ 1,3,4\ \text{and}\ 5}$		k_{ref-2} = const. $k_{ref-s} = a_s \exp(b_sW)$ $_{s\ =\ 1,3,4,\ 5\ \text{and}\ 6}$	
	Estimate ± 95% HPD	Error (%)	Estimate ± 95% HPD	Error (%)
k_{ref-2} (min^{-1})	$4.95 \times 10^{-3} \pm 4.69 \times 10^{-4}$	9	$4.94 \times 10^{-3} \pm 4.69 \times 10^{-4}$	10
E_{a-1} (J mol^{-1})	$9.68 \times 10^{4} \pm 4.16 \times 10^{3}$	4	$9.68 \times 10^{4} \pm 4.15 \times 10^{3}$	4
E_{a-2} (J mol^{-1})	$1.01 \times 10^{5} \pm 4.49 \times 10^{3}$	4	$1.01 \times 10^{5} \pm 4.48 \times 10^{3}$	4
E_{a-3} (J mol^{-1})	$1.28 \times 10^{5} \pm 2.95 \times 10^{3}$	2	$1.28 \times 10^{5} \pm 2.95 \times 10^{3}$	2
E_{a-4} (J mol^{-1})	$1.39 \times 10^{5} \pm 4.37 \times 10^{3}$	3	$1.39 \times 10^{5} \pm 4.37 \times 10^{3}$	3
E_{a-5} (J mol^{-1})	$1.33 \times 10^{5} \pm 2.39 \times 10^{4}$	18	$1.33 \times 10^{5} \pm 2.41 \times 10^{4}$	18
E_{a-6} (J mol^{-1})	$1.25 \times 10^{5} \pm 7.47 \times 10^{3}$	6	$1.25 \times 10^{5} \pm 7.47 \times 10^{3}$	6
a_1 (min^{-1} g mol^{-1})	$2.82 \times 10^{2} \pm 3.46 \times 10^{1}$	12	$2.84 \times 10^{2} \pm 3.47 \times 10^{1}$	12
a_3 (min^{-1})	$2.02 \times 10^{-4} \pm 2.74 \times 10^{-5}$	14	$2.02 \times 10^{-4} \pm 2.74 \times 10^{-5}$	14
a_4 (min^{-1})	$5.40 \times 10^{-5} \pm 2.46 \times 10^{-5}$	46	$5.45 \times 10^{-5} \pm 2.48 \times 10^{-5}$	45
a_5 (min^{-1})	$3.95 \times 10^{-6} \pm 9.62 \times 10^{-6}$	243	$3.91 \times 10^{-6} \pm 9.48 \times 10^{-6}$	242
a_6 (min^{-1})	$1.03 \times 10^{-3} \pm 7.31 \times 10^{-4}$	71	$1.20 \times 10^{-3} \pm 4.61 \times 10^{-4}$	38
b_1 (g mol^{-1})	$1.59 \times 10^{2} \pm 6.73 \times 10^{0}$	4	$1.59 \times 10^{2} \pm 6.73 \times 10^{0}$	4
b_3 (g mol^{-1})	$8.51 \times 10^{1} \pm 1.20 \times 10^{1}$	14	$8.51 \times 10^{1} \pm 1.19 \times 10^{1}$	14
b_4 (g mol^{-1})	$2.29 \times 10^{2} \pm 6.70 \times 10^{1}$	29	$2.27 \times 10^{2} \pm 6.68 \times 10^{1}$	29
b_5 (g mol^{-1})	$6.49 \times 10^{2} \pm 1.56 \times 10^{2}$	24	$6.48 \times 10^{2} \pm 1.56 \times 10^{2}$	24
b_6 [a]	$7.92 \times 10^{-2} \pm 8.95 \times 10^{-2}$	113	$4.01 \times 10^{1} \pm 4.24 \times 10^{1}$	106

[a](g mol^{-1}) and (min^{-1} g mol^{-1}) for *Global 1* and *Global 2*, respectively

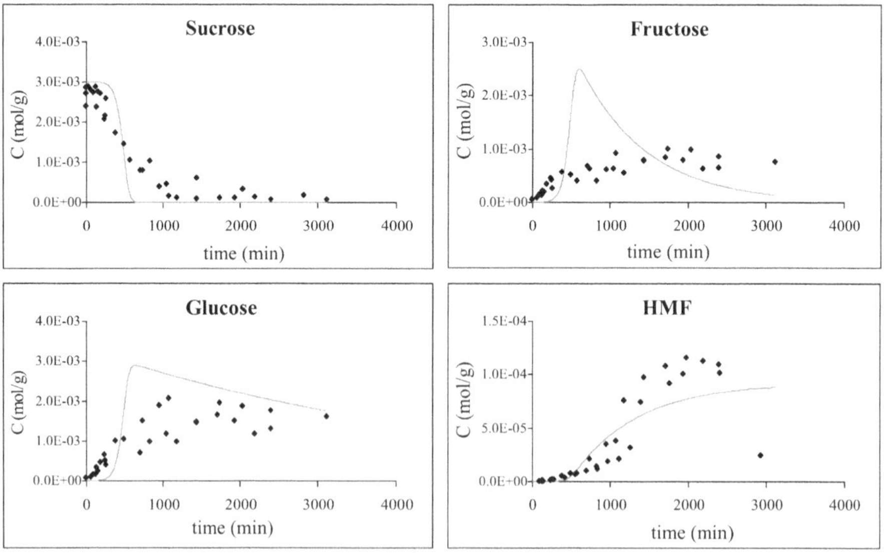

Fig. 1.28 Typical experimental data and predicted behaviour (using model Global 2 – grey lines) of sucrose, fructose, glucose and HMF for a 3.58% (w/w) water content solution. Example is for a process at 120 °C

1.8 Statistics in MathCAD

Edward Muntean

1.8.1 Objectives and Learning Outcomes

This chapter aims to give a basic knowledge in statistical methods for use in chemistry and food science-related areas, while developing the ability to handle modern worksheet software for calculations and data presentation: MathCad.

The objectives of this course are:

- to provide an introduction to the MathCAD environment, in order to make the best use of its problem-solving features;
- to introduce basic principles of working with worksheets in Mathcad;
- to supply a series of examples that enable specialists to use this software for statistical and graphing activities in data analysis.

On completion of this chapter, the readers will be able to:

- use the MathCAD interface to enter data and text, use built in constants and relevant statistical functions, define variables, use units, enter and edit expressions, and perform 2D plots;
- understand basic statistics for data evaluation within a specific context and be capable of calculating the results of descriptive statistics;
- perform linear regression calculations in order to evaluate correlation and construct calibration curves;
- conduct basic statistical analysis using MathCad worksheets, based on several provided templates;
- solve a range of problems using the techniques covered.

1.8.2 Introduction

MathCAD[1] combines a user-friendly WYSIWYG interface of a word processor with worksheet and graphing tools. It solves equations, graphs functions and performs symbolic manipulations, addressing just about any math problem. Much simpler than a programming language, in MathCAD equations are written as they would appear in a mathematical reference book. By integrating text, formulas and graphs in to a single worksheet in natural math notation, users can easily visualize, illustrate,

[1] MathCAD is a trademark of MathSoft Inc., which was acquired in 2006 by Parametric Technology Corp. (140 Kendrick Street, Needham, MA 02494, USA, http://www.ptc.com)

annotate and verify calculations. This simplifies applications development, and enhances design verification, validation and audit processes.

The biggest advantage of the MathCAD is its graphical style, which allows mathematical formulas to be entered as they would be on paper. This feature significantly distinguishes MathCAD and other similar commercial software, which usually requires specialized syntax. The graphical style along with the ability to combine graphics and text, has made from this software a reference tool, not only for numerical and symbolical calculations, but also as an environment for transmitting the engineering and scientific information.

Another big advantage of MathCAD is how easy it is to change the value of a parameter and immediately see the impact on all dependent variables and graphs. After creating a MathCAD worksheet, it can be shared by printing, emailing it or posting it on the World Wide Web.

MathCAD includes a robust built-in help system, and valuable on-line references are available in order to assist users. Expert documentation, sample applications (quick sheet examples of the most common tasks) and context-sensitive help all contribute to completing projects quickly. To gain a more detailed understanding of the possibilities offered by this software it is highly advisable to study the materials available on-line and the recommended references from the engineering domain (Brent 2006; Chapra and Canale 2002) or mathematics (Benker 2000; Fausett 2002), as well as classical handbooks (Kyrianov 2002; Larsen 2001). However, despite the vast amount of available information, the average user needs a quick-start method for using this product, to achieve the desired results in real-time. The purpose of this chapter is to be a helpful instrument for performing basic statistic tasks using MathCAD as quickly as possible.

1.8.3 The Interface

Beginning of a MathCAD session is accomplished by clicking on the MathCAD icon; MathCAD opens a classical Windows interface (Fig. 1.29), with a standard main menu bar at the top and a blank area that allows typing and evaluating mathematical expression, graphing, symbolic computing, entering and running scripting codes and entering text.

Although data, operators and text can be entered in MathCAD using a standard computer keyboard, the software assists users with tools called "palettes" to simplify entering math operations. The View menu allows bringing up palettes in working area (Fig. 1.30); these help users to enter math formulas, mathematical symbols, graphing, programming, etc. By clicking on a palette's icons, one can use each individual item as needed.

To begin, it is advisable to use the Math toolbar located in the upper-left part of the screen:  . If it is not visible, this can be

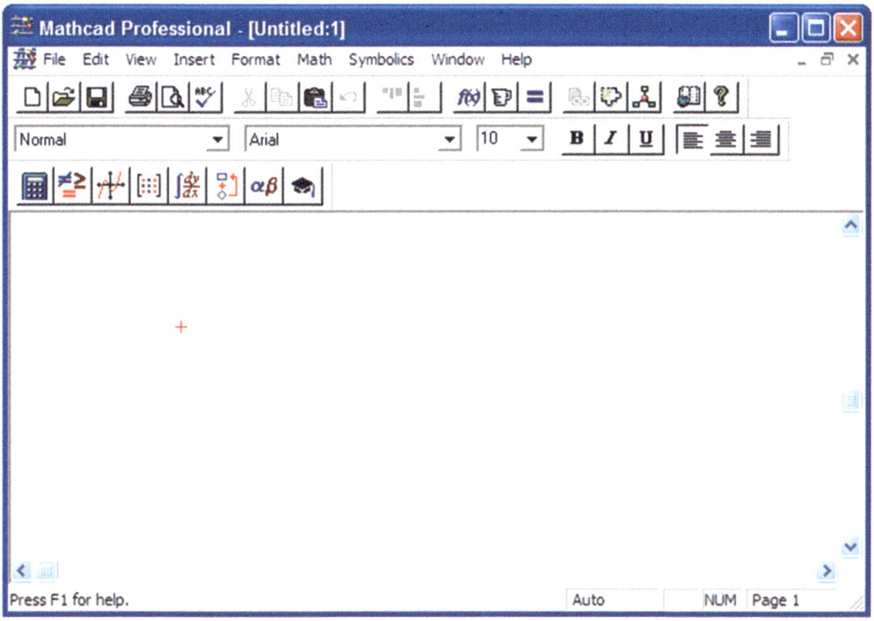

Fig. 1.29 The MathCAD interface

opened from the <View> pull-down menu. The palettes available from here are, in the specified order: Arithmetic palette, Evaluation and Boolean palette, Graph palette, Vector and Matrix palette, Calculus palette, Programming palette, Greek symbol palette and Symbolic keyword palette.

Entering simple equations is very simple: a click in the desired position of the screen, followed by typing the numerical values, together with the operators and finally the "equals" symbol, which evaluates the mathematical sequence and returns the numerical result (Fig. 1.31).

The Arithmetic palette is very convenient and simple to use, similar to a pocket calculator, but using the computer keyboard is much faster. When using the computer keyboard, the operating keys are: "+" for addition, "-" for subtraction, "*" for multiplication, "/" for division and "^" for power. To further assist users, MathCAD automatically formats equations in an easy to read fashion. Besides using menus or palettes to insert items into the working area one can use shortcut keystrokes to accomplish the same goals, which is a much quicker way to invoke menu selection; once a user remembers the shortcut keystrokes for each operation, the palettes became less necessary.

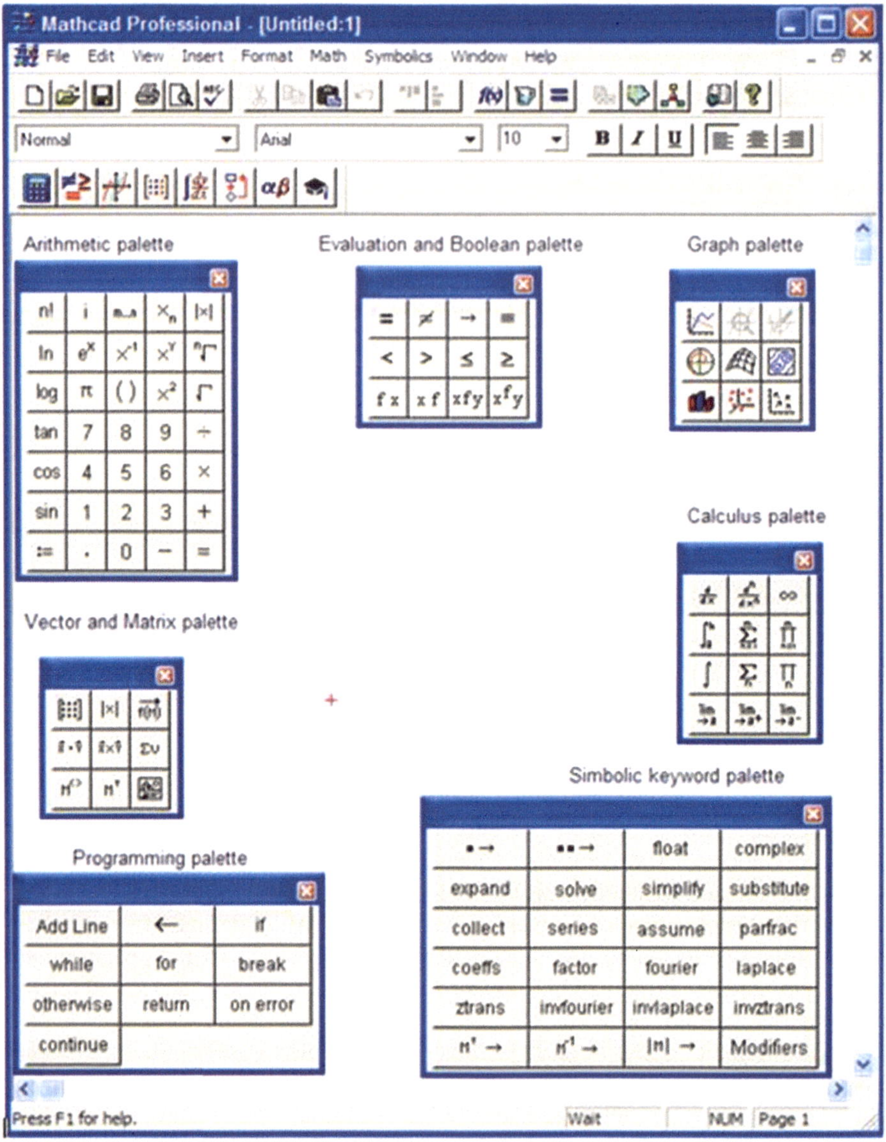

Fig. 1.30 Available palettes in MathCAD

1.8.4 Editing

In a worksheet, any equation, text or plot is a separate object called a region; MathCAD evaluates all regions from top to bottom and from left to right of the screen. Sequential entries of equations and data must follow this order to produce the correct result; entries with an incorrect syntax will result in error messages (Table 1.21).

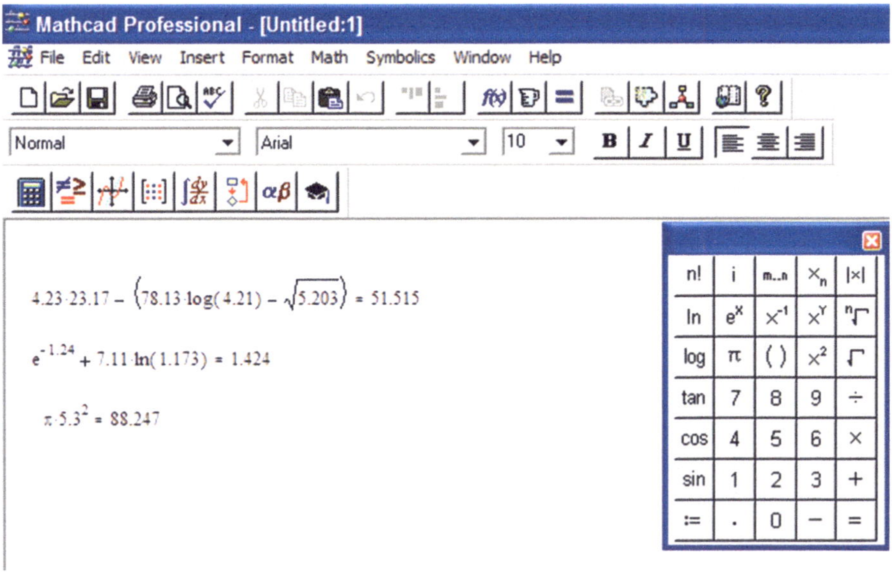

Fig. 1.31 A screenshot of some simple calculations in MathCAD

Table 1.21 Some important keys for editing purposes

Shortcut key	Effect
:	defines a variable
=	evaluates the equation/variable
.	creates subscripts
[	calls out functional subscripts
;	makes range variable
[Ctrl] + g	changes to Greek letter (pressing [Ctrl] together with g
‘	inserts a text region

Starting a text region is possible typing “or by selecting <Text region> from the <Insert> pull-down menu then entering the text. Inside the text region, users can enter any text as in a normal text editor with any format setting that can be specified through the <Format>/ <Style> menu. The text in a text region can be edited much like the text in a word processor; to change the font, font size or font color, the text in the text region will be selected by clicking and dragging over it, then using the buttons from formatting bar or the appropriate commands in the <Format> pull-down menu to change the selected text.

For correcting a typing error, one will place the mouse to the right of the wrong character or symbol, then press the delete key and insert the correct symbol.Whole expressions or text regions can be removed by selecting the desired region and pressing the [Delete] or [Backspace] keys. To select a region, either click and drag into that region, or click once in the region and press [Up arrow] repeatedly until whole region is outlined.

To select texts in a text region, click at any desired point inside the region and either hold the right button down while dragging or hold the [Shift] down and using the arrow keys to make the selection as in most text editors.

Variables are defined using the colon operator, the effect on the screen being the appearance of the symbol "=". As in all programming languages, when a computation involves a variable of function, the variable or function must be defined before its use unless they are predefined in MathCAD. There are two types of variables in MathCAD: ***normal variables***, which can hold only one value and ***range variable*** which, like vectors in math, can hold multiple values. The variable name can be any sequence of letters and digits that begins with a letter. To define a normal variable one just click any clear region and type the name of the variable followed by ":" which brings up the assignment operator ":=" and then the numeral value. To define a range variable, usually an index variable, one will use ';'. Any variables used in the expression to the left of the equals sign must be defined beforehand; if not, any undefined variable will be marked in red. Variable names should not re-define common units, e.g.: "m" for meters, "s" for seconds, "g" for gravity, "e" for exponential function or "c" for the speed of light.

Expressions can be edited for variables and constants or for operators. To remove/ replace an operator, click on the expression and use the up-arrow or down-arrow to select the part of the expression that contains the wrong operator; press the back-arrow to remove it and press the desired operator key to replace it.

Regions can be cut and pasted too: select a region, then pull down the <Edit> menu and choose <Cut>; using the scissor icon on the standard toolbar has the same effect. Then click on the new location where the region is to be pasted – pull down the <Edit> menu and choose <Paste>; the paste icon is also available on the standard toolbar. Faster operations, the shortcuts [Ctrl]/C (for copy) cut [Ctrl]/ X (for cut) and paste [Ctrl]/ V (for paste) can be used, as in most text editors.

Regions can be moved in a worksheet; first, it is necessary to select the desired region by clicking on it. A light bounding box will appear, indicating that the region is selected. To select one or more regions on the screen, the mouse pointer will be placed to the starting point, then holding down the right button while moving the mouse. A selecting box appears, with any item inside the selecting box being surrounded by a dotted line; after releasing the right mouse button, all selected items are bounded by dash box and the mouse cursor changes to a small hand. Clicking and holding the mouse on any of selected region allows moving the selected regions to the desired location. To deselect, simply click on any blank region outside the selecting box. When changing regions positions, users have to keep in mind that MathCAD evaluates the worksheet from top to bottom and from left to right.

Defining a function in MathCAD is the same as writing it down on a piece of paper; typing in the function name, followed by open parenthesis and the argument (s) that are separated by comma and then close parenthesis. Next, type the assignment operator – colon(:) followed by typing in the expression(formula) for the given

Fig. 1.32 A screenshot with simple example of calculating the mean value

Defining variables

$a := 17 \qquad b := 19 \qquad c := 20$

Defining the equation

$$mean := \frac{a + b + c}{3}$$

Requesting the answer

$mean = 18.667$

Defining variables

$a := 17 \qquad b := 19 \qquad c := 20$

Requesting the answer

$mean =$

Defining the equation

$$mean := \frac{a + b + c}{3}$$

Defining the equation

$$mean := \frac{a + b + c}{3}$$

Defining variables

$a := 17 \qquad b := 19 \qquad c := 20$

Requesting the answer

$mean =$

Fig. 1.33 Possible errors in the previous example

function. To find value of the defined function one can type the name of it, followed by "=".

To introduce built-in functions, from the <Insert> pull-down menu, the <Function> option allows users a broad variety of mathematical functions to choose from.

For common calculations, the necessary steps are:

- defining the variables using a colon,
- defining relationships (equations) again using a colon and finally
- requesting the answer using equals.

This protocol is illustrated below, in the screenshot form Fig. 1.32, in a simple example of calculating a mean value. Changing the correct order will result in error messages shown in red symbols, as in Fig. 1.33.

For saving the worksheet, from the <File> pull-down menu one will select <Save as>; after choosing a proper location and file name a click on <Save> will finalize saving the worksheet.

1.8.5 Plots

There are several ways to insert a plot region:

- from the <Insert> pull-down menu, selecting <Graph> then the desired plot type (Cartesian, polar, surface, contour, bar, scatter and vector-field plots), or
- from the Math Palette, with a click on the Graph icon to reveal the Graph palette from which one can select the appropriate graph type, or
- much simpler and rapidly using the @ symbol.

In each case, an empty template appears, with two placeholders on it; the placeholder from the bottom is used to specify x values, while the one from the left axis is for y values. After filling the placeholders with the variable names, MathCAD will generate automatically the graph scale. To quickly plot a vector y vs. x, one can type y@x, which produces a plot over a default range.

It is possible to adjust the graph formatting by double clicking on the graph plot; this action reveals the formatting menu, which enables a lot of options to change the graph appearance, add title, label the axes, etc. The graph can be resized and moved in any region using the mouse. Resizing a graph is possible after selecting it by dragging the pointer into the graph region (a dotted line appears around it), then position the mouse arrow at a border of the graph so that a double arrow appears, and finally hold down the mouse button and drag.

To graph a function, first it is necessary to define the function, then to define the range value for the variable of the function; next, to press @ and to enter the function name and variable name in proper placeholders, to adjust the thickness of the trace and to adjust the y ranges by changing the numbers on the far left. For plotting two or more functions in one graphing box, a comma will be placed after each function name (e.g. f(x), g(x)). If two or more variable names are involved, it is necessary to specify them in the place holder for variable, again separated by comma.

1.8.6 Using Units

Units are one of the most useful components of MathCAD; the complete list of MathCAD units is available from the <Insert> pull-down menu, choosing then <Units>. Pressing simultaneously [Ctrl/U] also opens a window in which the current system of units is identified along with a scroll of units types. The default units system can be set under the <Math> pull down menu by choosing <Options> and then <Unit System>. The mole unit is not supported in early MathCAD versions.

Units are added to values by simple multiplication, typing "*" and then using the <Insert> followed by <Units>. They are important for checking complex equations as MathCAD will automatically check for proper units and do conversions. MathCAD will give error messages, alerting the user when units are incompatible (Fig. 1.34).

Fig. 1.34 Using units

$$\text{mass} := (12.3456\cdot \text{gm}) \qquad \text{volume} := 10.12\cdot \text{mL}$$

$$\rho := \frac{\text{mass}}{\text{volume}}$$

$$\rho = 1.22{+}10^3\ {+}\text{kg}{+}\text{m}^{-3}$$

$$n1 := 1.11 \quad n2 := 1.23 \quad n3 := 1.21 \quad n4 := 1.35 \quad n5 := 1.22$$

$$\text{nmean} := \frac{(n1 + n2 + n3 + n4 + n5)}{5}$$

$$\text{nmean} = 1.224$$

Fig. 1.35 A simple way to calculate a mean in MathCAD

1.8.7 Descriptive Statistics

The following example was built on a dataset obtained from a repeated nitrate determination in a food matrix, with the following values: 1.11 ppm, 1.23 ppm, 1.10 ppm, 1.35 ppm and 1.22 ppm. Calculating the mean value can be accomplished in different ways. First, these values should be entered in a worksheet; for this purpose, different methods are available. The simplest way is to input them individually (Fig. 1.35), then to write the equation for calculating the mean, exactly in the same manner as it would be in a notebook.

Another possibility is to input the values as a vector, then to process them using the well-known formula for calculating means (Fig. 1.36). For this approach it is necessary to use first the Vector and Matrix palette (to input the vector containing nitrate data by clicking on the icon on the upper left corner on the palette, then entering the number of rows and columns and finally introducing each value individually in to the vector), then the Calculus palette (to select the summation). Using this approach, it is possible to further calculate further statistical parameters, such as the variance, the standard deviation, the coefficient of variation, etc.

However, there is no need to write quite complicated mathematical equations, since MathCAD has built-in function and everything can be accomplished faster using these (Fig. 1.37). The data range can be calculated simply using the functions max(X) and min(X) which returns the largest and the smallest data elements. For retrieving sample variance, one can use either the classical formula or the built-in function Var(X). A useful list of the commonly used statistical functions implemented in MathCAD can be accessed in the Appendix.

$$n := \begin{bmatrix} 1.11 \\ 1.23 \\ 1.21 \\ 1.35 \\ 1.22 \end{bmatrix} \qquad N := \text{length}(n) \qquad N = 5$$

Arithmetic mean
$$\text{nmean} := \frac{\sum_{i=1}^{N} n_i}{N} \qquad \text{nmean} = 1.224$$

Variance
$$v := \frac{\sum_{i=1}^{N} \left(n_i - \text{nmean}\right)^2}{N-1} \qquad v = 7.28+10^{-3}$$

Standard deviation
$$\text{std} := \sqrt{\frac{\sum_{i=1}^{N} \left(n_i - \text{nmean}\right)^2}{N-1}} \qquad \text{std} = 0.085$$

Coefficient of variation
$$CV := \frac{\text{Stdev}(n)}{\text{mean}(n)} \qquad CV = 0.07$$

Fig. 1.36 Calculations with user-defined functions

Arithmetic mean	mean(n) = 1.224			
Variance	$\text{Var}(n) = 7.28+10^{-3}$			
Standard deviation	Stdev(n) = 0.085	or	$\text{std2} := \sqrt{\text{Var}(n)}$	std2 = 0.085
Range	ran := max(n) – min(n)		ran = 0.24	
Median	median(n) = 1.22			

Fig. 1.37 Calculations using built-in functions for the same dataset as in Fig. 1.36

1.8.8 Regression

For obtaining the results reported earlier, a calibration was first performed using a set of eight known nitrate concentrations (from 0.05 ppm to 1.2 ppm); the measured absorbances are listed in Fig. 1.38, in which the vector x stores the concentrations values, while the vector y stores the measured absorbances.

Before attempting a fit, a visual inspection of the data point is a must; for this purpose, typing y@x will generate a plot of the measured absorbances vs. concentrations. From this plot it is obvious that experimental points are arranged linearly. The next step will be the calculation of the intercept and slope; for this, one can use either the quite complicated equations which involve a lot of editing effort (Fig. 1.38), or much simpler and faster with the built-in functions (Fig. 1.39). For checking the quality of the fit, one can plot on the same graph the experimental values and the regression function; a numerical measure of the quality of the fit is the correlation coefficient, which in this case is quite high (0.999 – as calculated in Fig. 1.39).

$$x := \begin{bmatrix} 0.0369 \\ 0.0695 \\ 0.1483 \\ 0.3044 \\ 0.4621 \\ 0.6297 \\ 0.7743 \\ 0.8984 \end{bmatrix} \qquad y := \begin{bmatrix} 0.05 \\ 0.1 \\ 0.2 \\ 0.4 \\ 0.6 \\ 0.8 \\ 1 \\ 1.2 \end{bmatrix}$$

Absorbances Concentrations

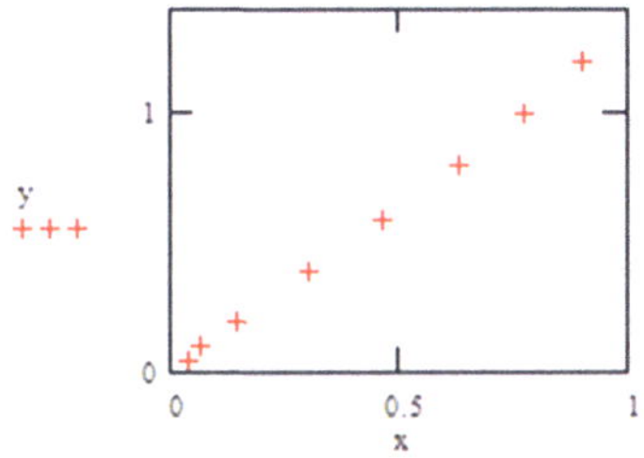

$n := \mathrm{rows}(x) \qquad i := 1..n$

$$a := \frac{\left[\sum_i (x_i)^2\right]\cdot\sum_i y_i - \left(\sum_i x_i\right)\cdot\sum_i (x_i\cdot y_i)}{n\cdot\sum_i (x_i)^2 - \left(\sum_i x_i\right)^2} \qquad b := \frac{n\cdot\sum_i (x_i\cdot y_i) - \left(\sum_i x_i\right)\cdot\sum_i y_i}{n\cdot\sum_i (x_i)^2 - \left(\sum_i x_i\right)^2}$$

$a = 2.238\cdot 10^{-3} \qquad b = 1.303 \qquad fit := a + b\cdot x$

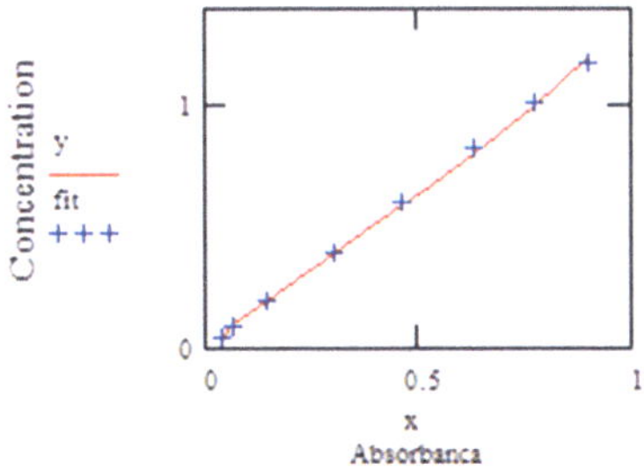

Fig. 1.38 The screenshot of the worksheet used for regression

Fig. 1.39 Regression performed with built-in functions intercept and slope

a := intercept(x, y) a = 2.238+10^{-3}

b := slope(x, y) b = 1.303

R := corr(x, y) R = 0.999

Rsquared := corr(x, y)^2 Rsquard = 0.999

resid := y − fit

$$\text{fit} = \begin{bmatrix} 0.05 \\ 0.093 \\ 0.196 \\ 0.399 \\ 0.605 \\ 0.823 \\ 1.011 \\ 1.173 \end{bmatrix} \qquad \text{resid} = \begin{bmatrix} -3.351\cdot 10^{-4} \\ 7.173\cdot 10^{-3} \\ 4.462\cdot 10^{-3} \\ 9.963\cdot 10^{-4} \\ -4.555\cdot 10^{-3} \\ -0.023 \\ -0.011 \\ 0.027 \end{bmatrix}$$

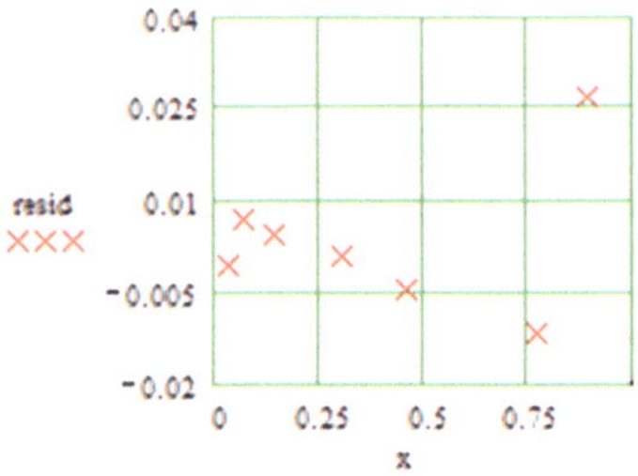

$$\text{stderr} := \sqrt{\frac{\sum_{i=1}^{n} (\text{fit}_i - y_i)^2}{n-2}} \qquad \text{stderr} = 0.016$$

Uncertainty in the slope $\quad \text{unslope} := \text{stderr}\cdot\sqrt{\frac{n}{\left[n\cdot\sum_i (x_i)^2 - \left(\sum_i x_i\right)^2\right]}} \quad \text{unslope} = 0.018$

Uncertainty in the intercept $\quad \text{uninter} := \text{stderr}\cdot\sqrt{\frac{\sum_i (x_i)^2}{\left[n\cdot\sum_i (x_i)^2 - \left(\sum_i x_i\right)^2\right]}} \quad \text{uninter} = 9.303\cdot 1($

Fig. 1.40 Worksheet for checking the quality of the fit

The quality of the fit can be assessed also calculating the residuals (the differences between the data and the fit) and analyzing their graph; a systematic error or unexpected trend in data will be visible in this way. Such an approach is illustrated in the worksheet from Fig. 1.40, in which the standard errors, the uncertainty in the slope ant the uncertainty in the intercept are also calculated.

Measured absorbances	Regression ecuation	Calculated concentrations
$abs := \begin{bmatrix} 0.8502 \\ 0.9417 \\ 0.9265 \\ 1.0341 \\ 0.9343 \end{bmatrix}$	$concalc := a + b \cdot abs$	$concalc = \begin{bmatrix} 1.11 \\ 1.23 \\ 1.21 \\ 1.35 \\ 1.22 \end{bmatrix}$

Fig. 1.41 Worksheet for calculation of the concentrations

Having the regression established, it is now easy to determine the desired concentrations; in Fig. 1.41 the measured absorbances are recorded in the vector abs, with which, using the intercept and the slope, the corresponding values for the concentrations can be easily calculated and displayed.

MathCAD includes a number of functions for performing regression, which generate a curve or surface of a specified type that, in some sense, minimizes the error between itself and the supplied data. The functions differ primarily in the type of curve or surface they use to fit the data.

The in-built regress and loess functions are useful when you have a set of measured y values corresponding to x values and it is necessary to fit a polynomial through those values. One uses regress when fitting a single polynomial to all data values, while the loess function fits a polynomial of any order, as it performs a more localized regression. The coefficients returned by regress and loess are designed to be passed to MathCAD's interp function; this returns a single interpolated y value for a given x value, but as a practical matter you'll probably be evaluating interp for many different points. MathCAD also allows multivariate polynomial regression with regress or loess to fit values corresponding to two or more independent variables.

Two other useful MathCAD functions are linfit and genfit: linfit is designed to model experimental data by a linear combination of arbitrary functions, while genfit is designed to model data by some arbitrary function whose parameters must be chosen. Anything one can do with linfit, you can also do with genfit, albeit less conveniently; the difference between these two functions is in the way MathCAD solves a system of linear equations.

1.8.9 Concluding remarks

MathCAD proved to be a valuable tool for statistical analysis, offering a robust set of features that simplify complex mathematical computations and enhance accuracy. Its intuitive interface allows for seamless integration of units, data visualization, and regression analysis, making it particularly advantageous for handling real-world data where precision is crucial. The ability to use built-in functions significantly reduces the time and effort required to perform detailed statistical analysis, while also minimizing the risk of human error that can arise from manually solving equations.

Moreover, MathCAD's flexibility in handling both simple and complex datasets, coupled with its powerful visualization capabilities, enables users to gain deeper insights into their data. Whether calculating basic descriptive statistics, performing linear or nonlinear regression, or conducting multivariate analyses, MathCAD streamlines these processes, making it accessible to both novice and experienced users alike.

The automated error checking, unit conversions, and the ability to visually inspect data trends before proceeding with analysis further underscore MathCAD's superiority in statistical applications. By leveraging MathCAD, users can focus more on interpreting results and drawing meaningful conclusions, rather than being bogged down by the technical intricacies of calculations. This makes MathCAD an essential tool for anyone engaged in data analysis, research, or any field requiring precise statistical computations.

Appendix

See Tables 1.22, 1.23, 1.24, 1.25, 1.26, 1.27, 1.28, 1.29, 1.30, 1.31, 1.32, 1.33, 1.34, 1.35, 1.36, 1.37, 1.38, 1.39, 1.40, 1.41, 1.42, 1.43, 1.44, 1.45, 1.46, 1.47, 1.48; Figs. 1.42, 1.43, 1.44.

Table 1.22 Critical values for T, when standard deviation is calculated from the same sample at probability level $\alpha = 0.05$ (single sided) (Hartung et al. 2005)

k	T
3	1.15
4	1.46
5	1.67
6	1.82
7	1.94
8	2.03
9	2.11
10	2.18
12	2.26
15	2.41
20	2.71
30	2.75
40	2.87
50	2.96
100	3.21

Table 1.23 Critical limits for the ratio range/s (Sachs 2003)

	Lower limit T_L						Upper limit T_U					
k	α											
k	0,000	0,005	0,01	0,025	0,05	0,10	0,10	0,05	0,025	0,01	0,005	0,000
3	1732	1735	1737	1745	1758	1782	1997	1999	2000	2000	2000	2000
4	1732	1,83	1,87	1,93	1,98	2,04	2409	2429	2439	2445	2447	2449
5	1826	1,98	2,02	2,09	2,15	2,22	2712	2753	2782	2803	2813	2828
6	1826	2,11	2,15	2,22	2,28	2,37	2949	3012	3056	3095	3115	3162
7	1871	2,22	2,26	2,33	2,40	2,49	3143	3222	3282	3338	3369	3464
8	1871	2,31	2,35	2,43	2,50	2,59	3308	3399	3471	3543	3585	3742
9	1897	2,39	2,44	2,51	2,59	2,68	3449	3552	3634	3720	3772	4000
10	1897	2,46	2,51	2,59	2,67	2,76	3,57	3685	3777	3875	3935	4243
11	1915	2,53	2,58	2,66	2,74	2,84	3,68	3,80	3903	4012	4079	4472
12	1915	2,59	2,64	2,72	2,80	2,90	3,78	3,91	4,02	4134	4208	4690
13	1927	2,64	2,70	2,78	2,86	2,96	3,87	4,00	4,12	4244	4325	4899
14	1927	2,70	2,75	2,83	2,92	3,02	3,95	4,09	4,21	4,34	4431	5099
15	1936	2,74	2,80	2,88	2,97	3,07	4,02	4,17	4,29	4,44	4,53	5292
16	1936	2,79	2,84	2,93	3,01	3,12	4,09	4,24	4,37	4,52	4,62	5477
17	1944	2,83	2,88	2,97	3,06	3,17	4,15	4,31	4,44	4,60	4,70	5657
18	1944	2,87	2,92	3,01	3,10	3,21	4,21	4,37	4,51	4,67	4,78	5831
19	1949	2,90	2,96	3,05	3,14	3,25	4,27	4,43	4,57	4,74	4,85	6000
20	1949	2,94	2,99	3,09	3,18	3,29	4,32	4,49	4,63	4,80	4,91	6164
25	1961	3,09	3,15	3,24	3,34	3,45	4,53	4,71	4,87	5,06	5,19	6,93
30	1966	3,21	3,27	3,37	3,47	3,59	4,70	4,89	5,06	5,26	5,40	7,62
35	1972	3,32	3,38	3,48	3.58	3,70	4,84	5,04	5,21	5,42	5,57	8,25
40	1975	3,41	3,47	3,57	3,67	3,79	4,96	5,16	5,34	5,56	5,71	8,83
45	1978	3,49	3,55	3,66	3,75	3,88	5,06	5,26	5,45	5,67	5,83	9,38
50	1980	3,56	3,62	3,73	3,83	3,95	5,14	5,35	5,54	5,77	5,93	9,90

55	1982	3,62	3,69	3,80	3,90	4,02	5,22	5,43	5,63	5,86	6,02	10,39
60	1983	3,68	3,75	3,86	3,96	4,08	5,29	5,51	5,70	5,94	6,10	10,86
65	1985	3,74	3,80	3,91	4,01	4,14	5,35	5,57	5,77	6,01	6,17	11,31
70	1986	3,79	3,85	3,96	4,06	4,19	5,41	5,63	5,83	6,07	6,24	11,75
75	1987	3,83	3,90	4,01	4,11	4,24	5,46	5,68	5,88	6,13	6,30	12,17
80	1987	3,88	3,94	4,05	4,16	4,26	5,51	5,73	5,93	6,18	6,35	12,57
85	1988	3,92	3,99	4,09	4,20	4,33	5,56	5,78	5,98	6,23	6,40	12,96
90	1989	3,96	4,02	4,13	4,24	4,36	5,60	5,82	6,03	6,27	6,45	13,34
95	1990	3,99	4,06	4,17	4,27	4,40	5,64	5,86	6,07	6,32	6,49	13,71
100	1990	4,03	4,10	4,21	4,31	4,44	5,68	5,90	6,11	6,36	6,53	14,07
150	1993	4,32	4,38	4,48	4,59	4,72	5,96	6,18	6,39	6,64	6,82	17,26
200	1995	4,53	4,59	4,68	4,78	4,90	6,15	6,39	6,60	6,84	7,01	19,95
500	1998	5,06	5,13	5,25	5,37	5,49	6,72	6,94	7,15	7,42	7,60	31,59
1000	1999	5,50	5,57	5,68	5,79	5,92	7,11	7,33	7,54	7,80	7,99	44,70

Table 1.24 Distribution of F_{max} to test the homogeneity of variances according to Hartley (Sachs 2003)

k	n										
	2	3	4	5	6	7	8	9	10	11	12
2	39,0	87,5	142	202	266	333	403	475	550	626	704
3	15,4	27,8	39,2	50,7	62,0	72,9	83,5	93,9	104	114	124
4	9,60	15,5	20,6	25,2	29,5	33,6	37,5	41,1	44,6	48,0	51,4
5	7,15	10,8	13,7	16,3	18,7	20,8	22,9	24,7	26,5	28,2	29,2
6	5,82	8,38	10,4	12,1	13,7	15,0	16,3	17,5	18,6	19,7	20,7
7	4,99	6,94	8,44	9,70	10,8	11,8	12,7	13,5	14,3	15,1	15,8
8	4,43	6,00	7,18	8,12	9,03	9,78	10,5	11,1	11,7	12,2	12,7
9	4,03	5,34	6,31	7,11	7,80	8,41	8,95	9,45	9,91	10,3	10,7
10	3,72	4,85	5,67	6,34	6,92	7,42	7,87	8,28	8,66	9,01	9,34
12	3,28	4,16	4,79	5,30	5,72	6,09	6,42	6,72	7,00	7,25	7,48
15	2,86	3,54	4,01	4,37	4,68	4,95	5,19	5,40	5,59	5,77	5,93
20	2,46	2,95	3,29	3,54	3,76	3,94	4,10	4,24	4,37	4,49	4,59
30	2,07	2,40	2,61	2,78	2,91	3,02	3,12	3,21	3,29	3,36	3,39
60	1,67	1,85	1,96	2,04	2,11	2,17	2,22	2,26	2,30	2,33	2,36
-	1,00	1,00	1,00	1,00	1,00	1,00	1,00	1,00	1,00		

Table 1.25 Multipliers L_1 and L_2 for the confidence limits of the ratio of 2 standard deviations at probability level $\alpha = 0.05$ (single sided)

ϕ_o	ϕ_N																		
	1	2	3	4	5	6	7	8	9	10	12	15	20	24	30	40	60	120	∞
1	.079	.071	.063	.067	.066	.065	.065	.065	.064	.064	.064	.064	.063	.063	.063	.063	.063	.063	.063
	12.7	4.30	3.18	2.78	2.57	2.45	2.36	2.31	2.26	2.23	2.18	2.13	2.09	2.06	2.04	2.00	2.02	1.98	1.96
2	0.23	0.23	0.23	0.23	0.23	0.23	0.23	0.23	0.23	0.23	0.23	0.23	0.23	0.23	0.23	0.23	0.23	0.23	0.23
	14.1	4.36	3.09	2.64	2.41	2.27	2.18	2.11	2.06	2.03	1.97	1.92	1.87	1.84	1.82	1.80	1.77	1.75	1.73
3	0.31	0.32	0.33	0.33	0.33	0.33	0.34	0.34	0.34	0.34	0.34	0.34	0.34	0.34	0.34	0.34	0.34	0.34	0.34
	14.7	4.38	3.05	2.57	2.33	2.18	2.08	2.02	1.97	1.93	1.87	1.81	1.76	1.73	1.71	1.68	1.66	1.64	1.61
4	0.36	0.38	0.39	0.40	0.40	0.40	0.41	0.41	0.41	0.41	0.41	0.41	0.42	0.42	0.42	0.42	0.42	0.42	0.42
	15.0	4.39	3.02	2.53	2.28	2.13	2.03	1.96	1.91	1.86	1.81	1.75	1.69	1.67	1.64	1.61	1.59	1.56	1.54
5	0.39	0.42	0.43	0.44	0.45	0.45	0.45	0.46	0.46	0.46	0.46	0.47	0.47	0.47	0.47	0.47	0.48	0.48	0.48
	15.2	4.39	3.00	2.50	2.25	2.09	1.99	1.92	1.87	1.82	1.76	1.70	1.65	1.62	1.59	1.57	1.54	1.51	1.49
6	0.41	0.44	0.46	0.47	0.48	0.48	0.48	0.49	0.49	0.49	0.50	0.50	0.51	0.51	0.51	0.51	0.52	0.52	0.52
	15.3	4.40	2.99	2.48	2.22	2.07	1.97	1.89	1.84	1.79	1.73	1.67	1.61	1.58	1.56	1.53	1.50	1.47	1.42
7	0.42	0.46	0.49	0.50	0.51	0.51	0.52	0.52	0.52	0.52	0.53	0.53	0.54	0.54	0.54	0.55	0.55	0.55	0.55
	15.4	4.40	2.98	2.47	2.21	2.05	1.95	1.87	1.81	1.77	1.71	1.65	1.59	1.56	1.53	1.50	1.47	1.44	1.42
8	0.43	0.47	0.50	0.51	0.52	0.53	0.53	0.54	0.54	0.55	0.55	0.56	0.56	0.57	0.57	0.57	0.58	0.58	0.58
	15.5	4.40	2.97	2.46	2.20	2.04	1.93	1.85	1.80	1.75	1.69	1.63	1.56	1.54	1.51	1.48	1.45	1.42	1.39
9	0.44	0.48	0.51	0.52	0.54	0.54	0.55	0.56	0.56	0.56	0.57	0.58	0.58	0.59	0.59	0.59	0.60	0.60	0.61
	15.5	4.40	2.97	2.45	2.18	2.02	1.92	1.84	1.78	1.74	1.67	1.61	1.55	1.52	1.49	1.46	1.43	1.40	1.37
10	0.45	0.49	0.52	0.54	0.55	0.56	0.56	0.57	0.58	0.58	0.59	0.59	0.60	0.60	0.61	0.61	0.62	0.62	0.63
	15.6	4.04	2.96	2.44	2.18	2.01	1.91	1.83	1.77	1.73	1.66	1.59	1.53	1.50	1.47	1.44	1.41	1.38	1.35
12	0.46	0.51	0.54	0.55	0.57	0.58	0.59	0.59	0.60	0.60	0.61	0.62	0.63	0.63	0.64	0.64	0.65	0.65	0.66
	15.6	4.41	2.96	2.43	2.16	2.00	1.89	1.81	1.75	1.71	1.64	1.57	1.51	1.48	1.45	1.42	1.38	1.35	1.32
15	0.47	0.52	0.55	0.57	0.59	0.60	0.61	0.62	0.62	0.63	0.64	0.65	0.65	0.66	0.67	0.67	0.68	0.69	0.70
	15.7	4.41	2.95	2.42	2.15	1.98	1.87	1.79	1.73	1.69	1.62	1.55	1.48	1.45	1.42	1.39	1.36	1.32	1.29

(continued)

Table 1.25 (continued)

20	0.48	0.54	0.57	0.59	0.61	0.62	0.63	0.64	0.65	0.65	0.66	0.67	0.69	0.69	0.70	0.71	0.72	0.73	0.74
	15.7	4.41	2.94	2.41	2.14	1.97	1.86	1.77	1.71	1.67	1.59	1.53	1.46	1.42	1.39	1.36	1.32	1.29	1.25
24	0.48	0.54	0.58	0.60	0.62	0.63	0.64	0.65	0.66	0.67	0.68	0.69	0.70	0.71	0.72	0.73	0.74	0.75	0.76
	15.8	4.41	2.94	2.40	2.13	1.96	1.85	1.77	1.70	1.65	1.58	1.51	1.44	1.41	1.37	1.34	1.30	1.27	1.23
30	0.49	0.56	0.58	0.61	0.63	0.64	0.65	0.66	0.67	0.68	0.69	0.70	0.72	0.73	0.74	0.75	0.76	0.77	0.79
	15.8	4.41	2.94	2.40	2.12	1.95	1.84	1.75	1.69	1.64	1.57	1.50	1.43	1.39	1.36	1.32	1.28	1.25	1.21
40	0.49	0.56	0.59	0.62	0.64	0.65	0.67	0.68	0.69	0.69	0.71	0.72	0.74	0.75	0.76	0.77	0.78	0.80	0.81
	15.8	4.41	2.93	2.39	2.11	1.94	1.83	1.74	1.68	1.63	1.56	1.48	1.41	1.38	1.34	1.30	1.26	1.22	1.18
60	0.50	0.56	0.60	0.63	0.65	0.67	0.68	0.69	0.70	0.71	0.72	0.74	0.76	0.77	0.78	0.79	0.81	0.83	0.85
	15.9	4.41	2.93	2.38	2.11	1.93	1.82	1.73	1.67	1.62	1.54	1.47	1.40	1.36	1.32	1.28	1.24	1.20	1.15
120	0.51	0.57	0.61	0.64	0.66	0.68	0.69	0.70	0.71	0.72	0.74	0.76	0.78	0.79	0.80	0.82	0.84	0.86	0.89
	15.9	4.41	2.92	2.38	2.10	1.92	1.81	1.72	1.66	1.61	1.53	1.45	1.38	1.34	1.30	1.26	1.21	1.16	1.11
∞	0.51	0.58	0.62	0.65	0.67	0.69	0.71	0.72	0.73	0.74	0.76	0.77	0.80	0.81	0.83	0.85	0.87	0.90	1.00
	15.9	4.42	2.92	2.37	2.09	1.92	1.80	1.71	1.65	1.59	1.52	1.44	1.36	1.32	1.27	1.23	1.18	1.12	1.00

$\mathbf{L_1}$ is given by the upper entry in each cell and $\mathbf{L_2}$ by the lower. $\boldsymbol{\phi_N}$ represents the degrees of freedom of the standard deviation in the numerator and $\boldsymbol{\phi_D}$ represents the degrees of freedom of the standard deviation in the denominator (Davies 1984)

Table 1.26 Table of p-quantile values of χ^2 distribution

	P									
DF	0.01	0.02	0.025	0.05	0.1	0.9	0.95	0.975	0.98	0.99
2	0.02	0.04	0.05	0.10	0.21	4.61	5.99	7.38	7.82	9.21
3	0.11	0.18	0.22	0.35	0.58	6.25	7.81	9.35	9.84	11.34
4	0.30	0.43	0.48	0.71	1.06	7.78	9.49	11.14	11.67	13.28
5	0.55	0.75	0.83	1.15	1.61	9.24	11.07	12.83	13.39	15.09
6	0.87	1.13	1.24	1.64	2.20	10.64	12.59	14.45	15.03	16.81
7	1.24	1.56	1.69	2.17	2.83	12.02	14.07	16.01	16.62	18.48
8	1.65	2.03	2.18	2.73	3.49	13.36	15.51	17.53	18.17	20.09
9	2.09	2.53	2.70	3.33	4.17	14.68	16.92	19.02	19.68	21.67
10	2.56	3.06	3.25	3.94	4.87	15.99	18.31	20.48	21.16	23.21
11	3.05	3.61	3.82	4.57	5.58	17.28	19.68	21.92	22.62	24.72
12	3.57	4.18	4.40	5.23	6.30	18.55	21.03	23.34	24.05	26.22
13	4.11	4.77	5.01	5.89	7.04	19.81	22.36	24.74	25.47	27.69
14	4.66	5.37	5.63	6.57	7.79	21.06	23.68	26.12	26.87	29.14
15	5.23	5.98	6.26	7.26	8.55	22.31	25.00	27.49	28.26	30.58
16	5.81	6.61	6.91	7.96	9.31	23.54	26.30	28.85	29.63	32.00
17	6.41	7.26	7.56	8.67	10.09	24.77	27.59	30.19	31.00	33.41
18	7.01	7.91	8.23	9.39	10.86	25.99	28.87	31.53	32.35	34.81
19	7.63	8.57	8.91	10.12	11.65	27.20	30.14	32.85	33.69	36.19
20	8.26	9.24	9.59	10.85	12.44	28.41	31.41	34.17	35.02	37.57
21	8.90	9.91	10.28	11.59	13.24	29.62	32.67	35.48	36.34	38.93
22	9.54	10.60	10.98	12.34	14.04	30.81	33.92	36.78	37.66	40.29
23	10.20	11.29	11.69	13.09	14.85	32.01	35.17	38.08	38.97	41.64
24	10.86	11.99	12.40	13.85	15.66	33.20	36.42	39.36	40.27	42.98
25	11.52	12.70	13.12	14.61	16.47	34.38	37.65	40.65	41.57	44.31
26	12.20	13.41	13.84	15.38	17.29	35.56	38.89	41.92	42.86	45.64
27	12.88	14.13	14.57	16.15	18.11	36.74	40.11	43.19	44.14	46.96
28	13.56	14.85	15.31	16.93	18.94	37.92	41.34	44.46	45.42	48.28
29	14.26	15.57	16.05	17.71	19.77	39.09	42.56	45.72	46.69	49.59
30	14.95	16.31	16.79	18.49	20.60	40.26	43.77	46.98	47.96	50.89
35	18.51	20.03	20.57	22.47	24.80	46.06	49.80	53.20	54.24	57.34
40	22.16	23.84	24.43	26.51	29.05	51.81	55.76	59.34	60.44	63.69
45	25.90	27.72	28.37	30.61	33.35	57.51	61.66	65.41	66.56	69.96
50	29.71	31.66	32.36	34.76	37.69	63.17	67.50	71.42	72.61	76.15

Table 1.27 Table of *0.95* quantile values of *-F* distribution

		DF of numerator							
DF of denominator 2		3	4	5	6	7	8	9	10
2	19.00	19.16	19.25	19.30	19.33	19.35	19.37	19.38	19.40
3	9.55	*9.28*	9.12	9.01	8.94	8.89	8.85	8.81	8.79
4	6.94	*6.59*	6.39	6.26	6.16	6.09	6.04	6.00	5.96
5	5.79	*5.41*	5.19	5.05	4.95	4.88	4.82	4.77	4.74
6	5.14	*4.76*	4.53	4.39	4.28	4.21	4.15	4.10	4.06
7	4.74	*4.35*	4.12	3.97	3.87	3.79	3.73	3.68	3.64
8	4.46	*4.07*	3.84	3.69	3.58	3.50	3.44	3.39	3.35
9	4.26	*3.86*	3.63	3.48	3.37	3.29	3.23	3.18	3.14
10	4.10	*3.71*	3.48	3.33	3.22	3.14	3.07	3.02	2.98
11	3.98	*3.59*	3.36	3.20	3.09	3.01	2.95	2.90	2.85
12	3.89	*3.49*	3.26	3.11	3.00	2.91	2.85	2.80	2.75
13	3.81	*3.41*	3.18	3.03	2.92	2.83	2.77	2.71	2.67
14	3.74	*3.34*	3.11	2.96	2.85	2.76	2.70	2.65	2.60
15	3.68	*3.29*	3.06	2.90	2.79	2.71	2.64	2.59	2.54
16	3.63	*3.24*	3.01	2.85	2.74	2.66	2.59	2.54	2.49
17	3.59	*3.20*	2.96	2.81	2.70	2.61	2.55	2.49	2.45
18	3.55	*3.16*	2.93	2.77	2.66	2.58	2.51	2.46	2.41
19	3.52	*3.13*	2.90	2.74	2.63	2.54	2.48	2.42	2.38
20	3.49	*3.10*	2.87	2.71	2.60	2.51	2.45	2.39	2.35
21	3.47	*3.07*	2.84	2.68	2.57	2.49	2.42	2.37	2.32
22	3.44	*3.05*	2.82	2.66	2.55	2.46	2.40	2.34	2.30
23	3.42	*3.03*	2.80	2.64	2.53	2.44	2.37	2.32	2.27
24	3.40	*3.01*	2.78	2.62	2.51	2.42	2.36	2.30	2.25
25	3.39	*2.99*	2.76	2.60	2.49	2.40	2.34	2.28	2.24
26	3.37	*2.98*	2.74	2.59	2.47	2.39	2.32	2.27	2.22
27	3.35	*2.96*	2.73	2.57	2.46	2.37	2.31	2.25	2.20
28	3.34	*2.95*	2.71	2.56	2.45	2.36	2.29	2.24	2.19
29	3.33	*2.93*	2.70	2.55	2.43	2.35	2.28	2.22	2.18
30	3.32	*2.92*	2.69	2.53	2.42	2.33	2.27	2.21	2.16
32	3.29	*2.90*	2.67	2.51	2.40	2.31	2.24	2.19	2.14
34	3.28	*2.88*	2.65	2.49	2.38	2.29	2.23	2.17	2.12
36	3.26	*2.87*	2.63	2.48	2.36	2.28	2.21	2.15	2.11
38	3.24	*2.85*	2.62	2.46	2.35	2.26	2.19	2.14	2.09
40	3.23	*2.84*	2.61	2.45	2.34	2.25	2.18	2.12	2.08
42	3.22	*2.83*	2.59	2.44	2.32	2.24	2.17	2.11	2.06
44	3.21	*2.82*	2.58	2.43	2.31	2.23	2.16	2.10	2.05
46	3.20	*2.81*	2.57	2.42	2.30	2.22	2.15	2.09	2.04
48	3.19	*2.80*	2.57	2.41	2.29	2.21	2.14	2.08	2.03
50	3.18	*2.79*	2.56	2.40	2.29	2.20	2.13	2.07	2.03
52	3.18	*2.78*	2.55	2.39	2.28	2.19	2.12	2.07	2.02
54	3.17	*2.78*	2.54	2.39	2.27	2.18	2.12	2.06	2.01
56	3.16	*2.77*	2.54	2.38	2.27	2.18	2.11	2.05	2.00

Table 1.28 Table of *0.975* quantile values of *-F* distribution

		DF of numerator							
DF of denominator 2		3	4	5	6	7	8	9	10
2	39.00	39.17	39.25	39.30	39.33	39.36	39.37	39.39	39.40
3	16.04	15.44	15.10	14.88	14.73	14.62	14.54	14.47	14.42
4	10.65	9.98	9.60	9.36	9.20	9.07	8.98	8.90	8.84
5	8.43	7.76	7.39	7.15	6.98	6.85	6.76	6.68	6.62
6	7.26	6.60	6.23	5.99	5.82	5.70	5.60	5.52	5.46
7	6.54	5.89	5.52	5.29	5.12	4.99	4.90	4.82	4.76
8	6.06	5.42	5.05	4.82	4.65	4.53	4.43	4.36	4.30
9	5.71	5.08	4.72	4.48	4.32	4.20	4.10	4.03	3.96
10	5.46	4.83	4.47	4.24	4.07	3.95	3.85	3.78	3.72
11	5.26	4.63	4.28	4.04	3.88	3.76	3.66	3.59	3.53
12	5.10	4.47	4.12	3.89	3.73	3.61	3.51	3.44	3.37
13	4.97	4.35	4.00	3.77	3.60	3.48	3.39	3.31	3.25
14	4.86	4.24	3.89	3.66	3.50	3.38	3.29	3.21	3.15
15	4.77	4.15	3.80	3.58	3.41	3.29	3.20	3.12	3.06
16	4.69	4.08	3.73	3.50	3.34	3.22	3.12	3.05	2.99
17	4.62	4.01	3.66	3.44	3.28	3.16	3.06	2.98	2.92
18	4.56	3.95	3.61	3.38	3.22	3.10	3.01	2.93	2.87
19	4.51	3.90	3.56	3.33	3.17	3.05	2.96	2.88	2.82
20	4.46	3.86	3.51	3.29	3.13	3.01	2.91	2.84	2.77
21	4.42	3.82	3.48	3.25	3.09	2.97	2.87	2.80	2.73
22	4.38	3.78	3.44	3.22	3.05	2.93	2.84	2.76	2.70
23	4.35	3.75	3.41	3.18	3.02	2.90	2.81	2.73	2.67
24	4.32	3.72	3.38	3.15	2.99	2.87	2.78	2.70	2.64
25	4.29	3.69	3.35	3.13	2.97	2.85	2.75	2.68	2.61
26	4.27	3.67	3.33	3.10	2.94	2.82	2.73	2.65	2.59
27	4.24	3.65	3.31	3.08	2.92	2.80	2.71	2.63	2.57
28	4.22	3.63	3.29	3.06	2.90	2.78	2.69	2.61	2.55
29	4.20	3.61	3.27	3.04	2.88	2.76	2.67	2.59	2.53
30	4.18	3.59	3.25	3.03	2.87	2.75	2.65	2.57	2.51
32	4.15	3.56	3.22	3.00	2.84	2.71	2.62	2.54	2.48
34	4.12	3.53	3.19	2.97	2.81	2.69	2.59	2.52	2.45
36	4.09	3.50	3.17	2.94	2.78	2.66	2.57	2.49	2.43
38	4.07	3.48	3.15	2.92	2.76	2.64	2.55	2.47	2.41
40	4.05	3.46	3.13	2.90	2.74	2.62	2.53	2.45	2.39
42	4.03	3.45	3.11	2.89	2.73	2.61	2.51	2.43	2.37
44	4.02	3.43	3.09	2.87	2.71	2.59	2.50	2.42	2.36
46	4.00	3.42	3.08	2.86	2.70	2.58	2.48	2.41	2.34
48	3.99	3.40	3.07	2.84	2.69	2.56	2.47	2.39	2.33
50	3.97	3.39	3.05	2.83	2.67	2.55	2.46	2.38	2.32
52	3.96	3.38	3.04	2.82	2.66	2.54	2.45	2.37	2.31
54	3.95	3.37	3.03	2.81	2.65	2.53	2.44	2.36	2.30

Table 1.29 Table of values of the standardized normal distribution

Function (**Φ(x)**)									
x	**Φ(x)**	x	**Φ(x)**	x	**Φ(x)**	x	**Φ(x)**	x	**Φ(x)**
0.51	**0.695**	0.91	**0.819**	1.31	**0.905**	1.71	**0.956**	2.11	**0.983**
0.52	**0.698**	0.92	**0.821**	1.32	**0.907**	1.72	**0.957**	2.13	**0.983**
0.53	**0.702**	0.93	**0.824**	1.33	**0.908**	1.73	**0.958**	2.15	**0.984**
0.54	**0.705**	0.94	**0.826**	1.34	**0.910**	1.74	**0.959**	2.17	**0.985**
0.55	**0.709**	0.95	**0.829**	1.35	**0.911**	1.75	**0.960**	2.19	**0.986**
0.56	**0.712**	0.96	**0.831**	1.36	**0.913**	1.76	**0.961**	2.21	**0.986**
0.57	**0.716**	0.97	**0.834**	1.37	**0.915**	1.77	**0.962**	2.23	**0.987**
0.58	**0.719**	0.98	**0.836**	1.38	**0.916**	1.78	**0.962**	2.25	**0.988**
0.59	**0.722**	0.99	**0.839**	1.39	**0.918**	1.79	**0.963**	2.27	**0.988**
0.60	**0.726**	1.00	**0.841**	1.40	**0.919**	1.80	**0.964**	2.29	**0.989**
0.61	**0.729**	1.01	**0.844**	1.41	**0.921**	1.81	**0.965**	2.31	**0.990**
0.62	**0.732**	1.02	**0.846**	1.42	**0.922**	1.82	**0.966**	2.33	**0.990**
0.63	**0.736**	1.03	**0.848**	1.43	**0.924**	1.83	**0.966**	2.35	**0.991**
0.64	**0.739**	1.04	**0.851**	1.44	**0.925**	1.84	**0.967**	2.37	**0.991**
0.65	**0.742**	1.05	**0.853**	1.45	**0.926**	1.85	**0.968**	2.39	**0.992**
0.66	**0.745**	1.06	**0.855**	1.46	**0.928**	1.86	**0.969**	2.41	**0.992**
0.67	**0.749**	1.07	**0.858**	1.47	**0.929**	1.87	**0.969**	2.43	**0.992**
0.68	**0.752**	1.08	**0.860**	1.48	**0.931**	1.88	**0.970**	2.45	**0.993**
0.69	**0.755**	1.09	**0.862**	1.49	**0.932**	1.89	**0.971**	2.47	**0.993**
0.70	**0.758**	1.10	**0.864**	1.50	**0.933**	1.90	**0.971**	2.49	**0.994**
0.71	**0.761**	1.11	**0.867**	1.51	**0.934**	1.91	**0.972**	2.51	**0.994**
0.72	**0.764**	1.12	**0.869**	1.52	**0.936**	1.92	**0.973**	2.53	**0.994**
0.73	**0.767**	1.13	**0.871**	1.53	**0.937**	1.93	**0.973**	2.55	**0.995**
0.74	**0.770**	1.14	**0.873**	1.54	**0.938**	1.94	**0.974**	2.57	**0.995**
0.75	**0.773**	1.15	**0.875**	1.55	**0.939**	1.95	**0.974**	2.60	**0.995**
0.76	**0.776**	1.16	**0.877**	1.56	**0.941**	1.96	**0.975**	2.63	**0.996**
0.77	**0.779**	1.17	**0.879**	1.57	**0.942**	1.97	**0.976**	2.66	**0.996**
0.78	**0.782**	1.18	**0.881**	1.58	**0.943**	1.98	**0.976**	2.69	**0.996**
0.79	**0.785**	1.19	**0.883**	1.59	**0.944**	1.99	**0.977**	2.72	**0.997**
0.80	**0.788**	1.20	**0.885**	1.60	**0.945**	2.00	**0.977**	2.75	**0.997**
0.81	**0.791**	1.21	**0.887**	1.61	**0.946**	2.01	**0.978**	2.78	**0.997**
0.82	**0.794**	1.22	**0.889**	1.62	**0.947**	2.02	**0.978**	2.81	**0.998**
0.83	**0.797**	1.23	**0.891**	1.63	**0.948**	2.03	**0.979**	2.84	**0.998**
0.84	**0.800**	1.24	**0.893**	1.64	**0.949**	2.04	**0.979**	2.87	**0.998**
0.85	**0.802**	1.25	**0.894**	1.65	**0.951**	2.05	**0.980**	2.90	**0.998**
0.86	**0.805**	1.26	**0.896**	1.66	**0.952**	2.06	**0.980**	2.93	**0.998**
0.87	**0.808**	1.27	**0.898**	1.67	**0.953**	2.07	**0.981**	2.96	**0.998**
0.88	**0.811**	1.28	**0.900**	1.68	**0.954**	2.08	**0.981**	2.99	**0.999**
0.89	**0.813**	1.29	**0.901**	1.69	**0.954**	2.09	**0.982**	3.02	**0.999**
0.90	**0.816**	1.30	**0.903**	1.70	**0.955**	2.10	**0.982**	3.05	**0.999**

Table 1.30 Table of p-quantile values of t distribution

	p					
DF	0.9	0.95	0.975	0.98	0.99	0.999
1	3.08	6.31	12.71	15.89	31.82	318.31
2	1.89	2.92	4.30	4.85	6.96	22.33
3	1.64	2.35	3.18	3.48	4.54	10.21
4	1.53	2.13	2.78	3.00	3.75	7.17
5	1.48	2.02	2.57	2.76	3.36	5.89
6	1.44	1.94	2.45	2.61	3.14	5.21
7	1.41	1.89	2.36	2.52	3.00	4.79
8	1.40	1.86	2.31	2.45	2.90	4.50
9	1.38	1.83	2.26	2.40	2.82	4.30
10	1.37	1.81	2.23	2.36	2.76	4.14
11	1.36	1.80	2.20	2.33	2.72	4.02
12	1.36	1.78	2.18	2.30	2.68	3.93
13	1.35	1.77	2.16	2.28	2.65	3.85
14	1.35	1.76	2.14	2.26	2.62	3.79
15	1.34	1.75	2.13	2.25	2.60	3.73
16	1.34	1.75	2.12	2.24	2.58	3.69
17	1.33	1.74	2.11	2.22	2.57	3.65
18	1.33	1.73	2.10	2.21	2.55	3.61
19	1.33	1.73	2.09	2.20	2.54	3.58
20	1.33	1.72	2.09	2.20	2.53	3.55
21	1.32	1.72	2.08	2.19	2.52	3.53
22	1.32	1.72	2.07	2.18	2.51	3.50
23	1.32	1.71	2.07	2.18	2.50	3.48
24	1.32	1.71	2.06	2.17	2.49	3.47
25	1.32	1.71	2.06	2.17	2.49	3.45
26	1.31	1.71	2.06	2.16	2.48	3.43
27	1.31	1.70	2.05	2.16	2.47	3.42
28	1.31	1.70	2.05	2.15	2.47	3.41
29	1.31	1.70	2.05	2.15	2.46	3.40
30	1.31	1.70	2.04	2.15	2.46	3.39
35	1.31	1.69	2.03	2.13	2.44	3.34
40	1.30	1.68	2.02	2.12	2.42	3.31
45	1.30	1.68	2.01	2.12	2.41	3.28
50	1.30	1.68	2.01	2.11	2.40	3.26
55	1.30	1.67	2.00	2.10	2.40	3.25
60	1.30	1.67	2.00	2.10	2.39	3.23

Table 1.31 Table of $p = 0.95$-quantile values of q distribution

	k→2	**3**	**4**	**5**	**6**	**7**	**8**	**9**	**10**	**11**	**12**
DF											
1	18.00	27.00	32.80	37.10	40.40	43.10	45.40	47.40	49.10	50.60	52.00
2	6.09	8.30	9.80	10.90	11.70	12.40	13.00	13.50	14.00	14.40	14.70
3	4.50	5.91	6.82	7.50	8.04	8.48	8.85	9.18	9.46	9.72	9.95
4	3.93	5.04	5.76	6.29	6.71	7.05	7.35	7.60	7.83	8.03	8.21
5	3.64	4.60	5.22	5.67	6.03	6.33	6.58	6.80	6.99	7.17	7.32
6	3.46	4.34	4.90	5.31	5.63	4.89	6.12	6.32	6.49	6.65	6.79
7	3.34	4.16	4.69	5.06	5.36	5.61	5.82	6.00	6.16	6.30	6.43
8	3.26	4.04	4.53	4.89	5.17	5.40	5.60	5.77	5.92	6.05	6.18
9	3.20	3.95	4.42	4.76	5.02	5.24	5.43	5.60	5.74	5.87	5.98
10	3.15	3.88	4.33	4.65	4.91	5.12	5.30	5.46	5.60	5.72	5.83
11	3.11	3.82	4.26	4.57	4.82	5.03	5.20	5.35	5.49	5.61	5.71
12	3.08	3.77	4.20	4.51	4.75	4.95	5.12	5.27	5.40	5.51	5.62
13	3.06	3.73	4.15	4.45	4.69	4.88	5.05	5.19	5.32	5.43	5.53
14	3.03	3.70	4.11	4.41	4.64	4.83	4.99	5.13	5.25	5.36	5.46
16	3.00	3.65	4.05	4.33	4.56	4.74	4.90	5.03	5.15	5.26	5.35
18	2.97	3.61	4.00	4.28	4.49	4.67	4.82	4.96	5.07	5.17	5.27
20	2.95	3.85	3.96	4.23	4.45	4.62	4.77	4.90	5.01	5.11	5.20
24	2.92	3.53	3.90	4.17	4.37	4.54	4.68	4.81	4.92	5.01	5.10
30	2.89	3.49	3.84	4.10	4.30	4.46	4.60	4.72	4.83	4.92	5.00
40	2.86	3.44	3.79	4.04	4.23	4.39	4.52	4.63	4.74	4.82	4.91
60	2.83	3.40	3.74	3.98	4.16	4.31	4.44	4.55	4.65	4.73	4.81
120	2.80	3.36	3.69	3.92	4.10	4.24	4.36	4.48	4.56	4.64	4.72

Table 1.47 The upper quantiles $t(\gamma)$ of t – distributions $t(\gamma, \upsilon)$ (υ-*degree of freedom*)

	γ								
υ	0.75	0.90	0.95	0.975	0.99	0.995	0.9975	0.999	0.9995
1	1.000	3.078	6.314	12.710	31.82	63.66	17.3	318.3	636.6
2	0.816	1.886	2.920	4.303	6.965	9.925	14.09	22.33	31.60
3	0.765	1.638	2.358	3.182	4.541	5.841	7.453	10.21	12.92
4	0.741	1.533	2.132	2.776	3.747	4.604	5.598	7.173	8.610
5	0.727	1.476	2.015	2.571	3.365	4.032	4.778	5.893	6.869
6	0.718	1.440	1.943	2.447	3.143	3.707	4.317	5.208	5.959
7	0.711	1.415	1.895	2.365	2.998	3.499	4.029	4.785	5.408
8	0.706	1.397	1.860	2.306	2.896	3.355	3.833	4.501	5.041
9	0.703	1.383	1.833	2.262	2.821	3.250	3.690	4.297	4.781
10	0.700	1.372	1.812	2.228	2.764	3.169	3.581	4.144	4.587
11	0.697	1.363	1.796	2.201	2.718	3.106	3.497	4.025	4.437
12	0.695	1.356	1.782	2.179	2.681	3.055	3.428	3.930	4.318
13	0.694	1.350	1.771	2.160	2.650	3.012	3.372	3.852	4.221
14	0.692	1.345	1.761	2.145	2.624	2.977	3.326	3.787	4.140
15	0.691	1.341	1.753	2.131	2.602	2.947	3.286	3.733	4.073
16	0.690	1.337	1.746	2.120	2.583	2.921	3.252	3.686	4.015
17	0.689	1.333	1.740	2.110	2.567	2.898	3.222	3.646	3.965
18	0.688	1.330	1.734	2.101	2.552	2.878	3.197	3.610	3.922
19	0.688	1.328	1.729	2.093	2.539	2.861	3.174	3.579	3.883
20	0.687	1.325	1.725	2.086	2.528	2.845	3.153	3.552	3.850
21	0.686	1.323	1.721	2.080	2.518	2.831	3.135	3.527	3.819
22	0.686	1.321	1.717	2.074	2.508	2.819	3.119	3.505	3.792
23	0.685	1.319	1.714	2.069	2.500	2.807	3.104	3.485	3.767
24	0.685	1.318	1.711	2.064	2.492	2.797	3.091	3.467	3.745
25	0.684	1.316	1.708	2.060	2.485	2.787	3.078	3.450	3.725
26	0.684	1.315	1.706	2.056	2.479	2.779	3.067	3.435	3.707
27	0.684	1.314	1.703	2.052	2.473	2.771	3.057	3.421	3.690
28	0.683	1.313	1.701	2.048	2.467	2.763	3.047	3.408	3.674
29	0.683	1.311	1.699	2.045	2.462	2.756	3.038	3.396	3.659
30	0.683	1.310	1.697	2.042	2.457	2.750	3.030	3.385	3.646
40	0.681	1.303	1.684	2.021	2.423	2.704	2.971	3.307	3.551
60	0.679	1.296	1.671	2.000	2.395	2.660	2.915	3.232	3.460
120	0.677	1.289	1.658	1.980	2.358	2.617	2.860	3.160	3.373
∞	0.674	1.282	1.645	1.960	2.326	2.576	2.807	3.090	3.291

Table 1.48 Chi-square distribution (Percentile values χ^2 for chi-square distribution with υ degree of freedom standard area = P)

	P										
υ	0.99	0.975	0.95	0.90	0.75	0.50	0.25	0.10	0.05	0.025	0.01
1	2×10^{-4}	1×10^{-3}	4×10^{-3}	0.02	0.10	0.46	1.32	2.71	3.84	5.02	6.63
2	0.02	0.05	0.10	0.21	0.58	1.39	2.77	4.61	5.99	7.38	9.21
3	0.12	0.22	0.35	0.58	1.21	2.37	4.11	6.25	7.81	9.35	11.34
4	0.30	0.48	0.71	1.06	1.92	3.36	5.39	7.78	9.49	11.14	13.28
5	0.55	0.83	1.15	1.61	2.68	4.35	6.63	9.24	11.07	12.83	15.09
6	0.87	1.24	1.64	2.20	3.46	5.35	7.84	10.64	12.59	14.45	16.81
7	1.24	1.69	2.17	2.83	4.26	6.35	9.04	12.02	14.07	16.01	18.48
8	1.65	2.18	2.73	3.49	5.07	7.34	10.22	13.36	15.51	17.53	20.09
9	2.09	2.70	3.33	4.17	5.90	8.34	11.39	14.68	16.92	19.02	21.67
10	2.56	3.25	3.94	4.87	6.74	9.34	12.55	15.99	18.31	20.48	23.21
11	3.05	3.82	4.57	5.58	7.58	10.34	13.70	17.28	19.68	21.92	24.73
12	3.57	4.40	5.23	6.30	8.44	11.34	14.85	18.55	21.03	23.34	26.22
13	4.11	5.01	5.89	7.04	9.30	12.34	15.98	19.81	22.36	24.74	27.69
14	4.66	5.63	6.57	7.79	10.17	13.34	17.12	21.06	23.68	26.12	29.14
15	5.23	6.26	7.26	8.55	11.04	14.34	18.25	22.31	25.00	27.49	30.58
16	5.81	6.91	7.96	9.31	11.91	15.34	19.37	23.54	26.30	28.85	32.00
17	6.41	7.56	8.67	10.09	12.79	16.34	20.49	24.77	27.59	30.19	33.41
18	7.01	8.23	9.39	10.86	13.68	17.34	21.61	25.99	28.87	31.53	34.81
19	7.63	8.91	10.12	11.65	14.56	18.34	22.72	27.20	30.14	32.85	36.19
20	8.26	9.59	10.85	12.44	15.45	19.34	23.83	28.41	31.41	34.17	37.57
21	8.90	10.28	11.59	13.24	16.34	20.34	24.94	29.62	32.67	35.48	38.93
22	9.54	10.98	12.34	14.04	17.24	21.34	26.04	30.81	33.92	36.78	40.29
23	10.20	11.69	13.09	14.85	18.14	22.34	27.14	32.01	35.17	38.08	41.64
24	10.86	12.40	13.85	15.66	19.04	23.34	28.24	33.20	36.42	39.36	42.98
25	11.52	13.12	14.61	16.47	19.94	24.34	29.34	34.38	37.65	40.65	44.31

26	12.20	13.84	15.38	17.29	20.84	25.34	30.43	35.56	38.89	41.92	45.64
27	12.88	14.57	16.15	18.11	21.75	26.34	31.53	36.74	40.11	43.19	46.96
28	13.56	15.31	16.93	18.94	22.66	27.34	32.62	37.92	41.34	44.46	48.28
29	14.26	16.05	17.71	19.77	23.57	28.34	33.71	39.09	42.56	45.72	49.59
30	14.95	16.79	18.49	20.60	24.48	29.34	34.80	40.26	43.77	46.98	50.89

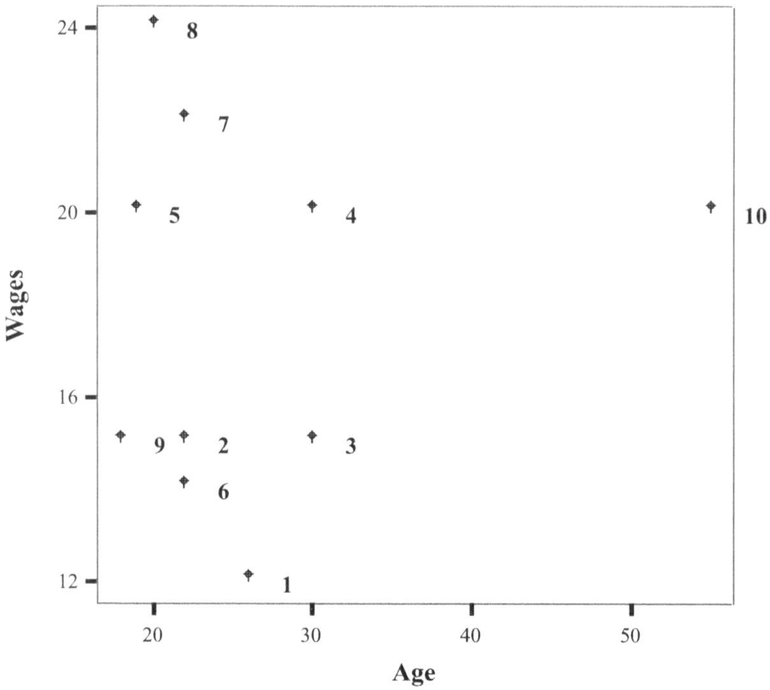

Fig. 1.44 Scatter plot of employee's age and wages

Some Useful Mathcad Built-In Statistics Functions

corr(A, B) – returns the Pearson's r correlation coefficient for the arrays A and B.

cvar(A, B) – returns the covariance of the elements in arrays A and B.

hist(int, A) – returns a vector representing the frequencies with which values in A fall in the intervals represented by int. When int is a vector of intervals in ascending order, the ith element of the returned vector is the number of points in data falling between the ith and $(i + 1)^{th}$ element of int. When int is an integer, it represents the number of subintervals of equal length. Both int and A must be real.

histogram(int, A) – returns a matrix with two columns. When int is an integer, the first column contains the midpoints of int subintervals of the range of equal length and the second column is identical to hist(int, A). When int is a vector of intervals in ascending order, the first column contains midpoints of the intervals represented by the int vector; both int and A must be real. For graphing the result of a frequency analysis in a 2D bar plot showing the distribution of data across the bins, use the histogram function rather than hist, and plot the first column of the result against the second column of the result.

hmean(A, B, C, ...) – returns the harmonic mean of the elements of the arrays or scalars A, B, C, ... All elements must be nonzero.
kurt(A, B, C, ...) – returns the kurtosis of the elements of the arrays or scalars A, B, C, ...
mean(A, B, C, ...) – returns the arithmetic mean of the elements of the arrays or scalars A, B, C, ...
median(A, B, C, ...) – returns the median of the elements of the arrays or scalars A, B, C, ... the value above and below which there are an equal number of values. If there are even numbers of elements, this is the arithmetic mean of the two central values.
mode(A, B, C, ...) – returns the element from the arrays or scalars A, B, C, ... that occurs most often.
skew(A, B, C, ...) – returns the skewness of the elements of the arrays or scalars A, B, C, ...
stdev(A, B, C, ...) – returns the population standard deviation (square root of the variance) of the elements of the arrays or scalars A, B, C, ...
Stdev(A, B, C, ...) – returns the sample standard deviation (square root of the sample variance) of the elements of the arrays or scalars A, B, C, ...
var(A, B, C, ...) – returns the population variance of the elements of the arrays or scalars A, B, C, ...
Var(A, B, C, ...) – returns the sample variance of the elements of the arrays or scalars A, B, C, ...

Regression Functions

I. **Linear regression**

intercept(x, y) – returns the y-intercept of the least-squares regression line for the data points in x and y.
slope(x, y) – returns the slope of the least-squares regression line for the data points in x and y.
stderr(x, y) – returns the standard error associated with linear regression of the elements of y on the elements of x.
line(x, y) – returns the y-intercept and slope of the line that best approximates the data in x and y.
medfit(x, y) – returns the y-intercept and slope of the line that best approximates the data in x and y using median-median regression.

II. **Polynomial regression**

regress(x, y, n) – returns a vector of coefficients for the nth degree least-squares polynomial fit for the data points specified in vectors x and y.
loess(x, y, span) – returns a vector specifying a set of second order polynomials that best fit particular neighborhoods of data points specified in vectors x and y; the

argument span, specifies how large a neighborhood loess considers in performing this local regression.

III. **Generalized regression**.

linfit(x, y, F) – returns a vector containing the coefficients used to create a linear combination of the functions in F that best approximates the data points in vectors x and y; F is a function which returns a vector consisting of the functions to be linearly combined. Elements of x should be in ascending order.

genfit(x, y, g, F) – returns the parameters for the best fit by the (possibly nonlinear) function defined by f to the data points in the vectors x and y. F is a function that returns an element vector containing f and its partial derivatives with respect to each of its n parameters; g is an n-element vector of guess values for the n parameters.

References

Antal MJ Jr, Mok WSL, Richards GN (1990) Mechanism of formation of 5-(hydroxymethyl)-2-furaldehyde from -fructose and sucrose. Carbohydr Res 199(1):91–109

Arabshahi A, Lund D (1985) Considerations in calculating kinetic parameters from experimental data. J Food Process Eng 7(4):239–251

Arena E, Fallico B, Maccarone E (2001) Thermal damage in blood orange juice: kinetics of 5-hydroxymethyl-2-furancarboxaldehyde formation. Int J Food Sci Technol 36(2):145–151

Bakken AP, Hill CG, Amundson CH (1992) Hydrolysis of lactose in skim milk by immobilized beta-galactosidase (*bacillus-circulans*). Biotechnol Bioeng 39(4):408–417

Bard Y (1974) Nonlinear parameter estimation. Academic, New York

Barros RM, Malcata FX (2002) Modeling the kinetics of whey protein hydrolysis brought about by enzymes from *cynara cardunculus*. J Agric Food Chem 50(15):4347–4356

Barros RM, Malcata FX (2004) A kinetic model for hydrolysis of whey proteins by cardosin A extracted from *cynara cardunculus*. Food Chem 88(3):351–359

Bates D, Watts D (1988) Non-linera regression analysis and its aplications. Wiley, New York

BeMiller JN, Whistler RL (1996) Carbohydrates. In: Fennema OR (ed) Food chemistry. Marcel Dekker, Inc, New York

Benker H (2000) Practical use of MathCAD. Solving mathematical problems with a computer algebra system. Springer, New York

Bhattacharyya GK, Johnson RA (1997) Statistical concepts and methods. Wiley, New York

Box GEP, Draper NR (1965) The Bayesian estimation of common parameters from several responses. Biometrika 52(3,4):355–365

Box GEP, Hunter WG, Hunter JS (1978) Statistics for experiments. An introduction to design, data analysis and model building. Wiley, New York

Brandão TRS (2004) Application of non-isothermal methods to the estimation of mass transfer parameters: analysis of the effect of experimental design and data analysis on the precision and accuracy of the estimates. Ph.D. Catholic University of Portugal, Porto

Brands CMJ, van Boekel M (2002) Kinetic modeling of reactions in heated monosaccharide-casein systems. J Agric Food Chem 50(23):6725–6739

Brent M (2006) Engineering with MathCAD: Using MathCAD to create and organize your engineering calculations. Butterworth-Heinemann

Byrkit DR (1987) Statistic today: a comprehensive introduction. The Bebjamin/Cummings Publishing Company

Chapra SC, Canale RP (2002) Numerical methods for engineers with programming and software applications, 3rd edn. McGraw-Hill, New York

Clarke MA, Edye LA, Eggleston G (1997) "Sucrose decomposition in aqueous solution, and losses in sugar manufacture and refining". In. Adv Carbohydr Chem Biochem 52:441–470

Cohen E, Birk Y, Mannheim CH, Saguy IS (1994) Kinetic parameter-estimation for quality change during continuous thermal-processing of grapefruit juice. J Food Sci 59(1):155–158

Davis D, Cosenza RM (1993) Business research for decision making, 3rd edn. Wadsworth Publishing Company, Belmont

Defaye J, Garcia Fernandez JM (1994) Protonic and thermal activation of sucrose and the oligosaccharide composition of caramel. Carbohydr Res 256(2):C1–C4

Defaye J, Garcia Fernandez JM (1995) The oligosaccharide components of caramel. Zuckerindustrie 120(8):700–704

Eggleston G, Trask-Morrel B, Vercellotti JR (1996) Use of differential scaning calorimetry and thermogravimetric analysis to characterize the thermal degradation of crystalline sucrose and dried sucrose-salt residues. J Agric Food Chem 44(10):3319–3325

Fausett L (2002) Numerical methods using MathCAD. Prentice Hall

Glass GV, Stanley JC (1995) Statistical Methods in Education and Psychology, 3rd edn. Allyn & Bacon, Boston

Haralampu SG, Saguy IS, Karel M (1985) Estimation of Arrhenius model parameters using three least squares methods. J Food Process Preserv 9(3):129–143

Hartung J, Elpelt B, Klösener K-H (2005) Statistik Lehr- und Handbuch der angewandten Statistik. Oldenburg

Hoerr SL, Lee SY, Schiffman RF, Horodynski MO, McKelvey L (2006) Beverage consumption of mother–toddler dyads in family with limited incomes. J Paediatr Nurs 21(6):403–411

Johnson ML, Frasier SG (1985) Nonlinear least-squares analysis. Methods Enzymol 117:301–342

Knol JJ, Van Loon WAM, Linssen JPH, Ruck AL, Van Boekel M, Voragen AGJ (2005) Toward a kinetic model for acrylamide formation in a glucose-asparagine reaction system. J Agric Food Chem 53(15):6133–6139

Kroh LW (1994) Caramelisation in food and beverages. Food Chem 51(4):373–379

Kyrianov D (2002) The MathCAD 2001i handbook. Charles River Media, Hingham

Labuza TP, Kamman JF (1983) Reaction kinetics and accelerated tests simulation as function of temperature. In: Saguy I (ed) Computer aided techniques in food technology. Marcel Dekker, Inc, New York

Labuza TP, Tannenbaum SR, Karel M (1970) Water content and stability of low moisture and intermediate-moisture foods. Food Technol 24(5):35–42

Larsen RW (2001) Introduction to MathCAD 2000. Prentice Hall, Upper Saddle River

Lee HS, Nagy S (1988) Relationship of sugar degradation to detrimental changes in citrus juice quality. Food Technol 42(10):91–97

Lowary TL, Richards GN (1988) Effects of impurities on hydrolysis of sucrose in concentrated aqueous solutions. Int Sugar J 90(1077):164–167

Lund DB (1983) Considerations in modeling food processes. Food Technol 37(1):92–94

Malcata FX, Hill CG Jr, Amundson CH (1991) Use of a lipase immobilized in a membrane reactor to hydrolyze the glycerides of butteroil. Biotechnol Bioeng 38(8):853–868

Manley-Harris M, Richards GN (1991) Formation of trisaccharides (kestoses) by pyrolysis of sucrose. Carbohydr Res 219:101–113

Manley-Harris M, Richards GN (1996) Di-D-fructose dianhydrides and related oligomers from thermal treatments of inulin and sucrose. Carbohydr Res 287(2):183–202

Manley-Harris M, Richards GN (1997) Dihexulose dianhydrides. In: Advances in carbohydrate chemistry and biochemistry, pp 207–266

Martins S, Van Boekel M (2003) Kinetic modelling of Amadori N-(1-deoxy-D-fructos-1-yl)-glycine degradation pathways. Part II – Kinetic analysis. Carbohydr Res 338(16):1665–1678

Martins S, Jongen WMF, van Boekel M (2000) A review of Maillard reaction in food and implications to kinetic modelling. Trends Food Sci Technol 11(9–10):364–373

MathSoft Engineering & Education (2001) MathCAD user's guide with reference manual. MathSoft Engineering & Education, Cambridge, MA
Mauch W (1971) Chemical properties of sucrose. Sugar Technol Rev 1:239–290
Mundt S, Wedzicha BL (2005) Role of glucose in the Maillard browning of maltose and glycine: a radiochemical approach. J Agric Food Chem 53(17):6798–6803
Nabais RM, Malcata FX (1997) Modelling of lactic fermentation of carrot slices in salted brines. Food Sci Technol Int 3(4):263–283
Neter J, Knuter MH, Nachtsheim CJ, Wasserman W (1996) Applied linear statistical model. Irwin, Chicago
Peleg M (1992) On the use of the WLF model in polymers and foods. Crit Rev Food Sci Nutr 32(1): 59–66
Priestley MB (1992) Spectral Analysis and Time Series. Academic, London
Ratsimba V, Fernandez JMG, Defaye J, Nigay H, Voilley A (1999) Qualitative and quantitative evaluation of mono- and disaccharides in -fructose, –glucose and sucrose caramels by gas-liquid chromatography-mass spectrometry: Di-fructose dianhydrides as tracers of caramel authenticity. J Chromatogr A 844(1–2):283–293
Rice KE, Watkins J, Hill CG (1999) Hydrolysis of menhaden oil by a Candida cylindracea lipase immobilized in a hollow-fiber reactor. Biotechnol Bioeng 63(1):33–45
Richards GN (1986) Initial steps in thermal degradation of sucrose. Int Sugar J 88(1052):145–148
Rosner BA (1989) Fundamentals of biostatistics. Mass PWS-Kent XV, Boston
Sachs L (2003) Angewandte Statistik. Springer, Heidelberg
Saguy I, Karel M (1980) Modeling of quality deterioration during food processing and storage. Food Technol 34(2):78–85
Seber GAF, Wild CJ (1989) Nonlinear regression. Willey, New York
Simkovic I, Surina I, Vrican M (2003) Primary reactions of sucrose thermal degradation. J Anal Appl Pyrolysis 70(2):493–504
Stewart WE, Caracotsios M, Sorensen JP (1992) Parameter estimation from multiresponse data. AICHE J 38(5):641–650
Sugisawa H (1966) The thermal degradation of sugars: II. The volatile decomposition products of glucose caramel. J Food Sci 31(3):381–385
Suwannaporn P, Linnemann A (2008) Consumer preferences and buying criteria for jasmine rice. J Food Prod Mark 14(4):1–21
Suwannaporn P, Pitiphunpong S, Champangern S (2007) Classification of rice amylose group using discriminant analysis. Starch-Starke 59:171–177
Torres CF, Hill CG (2004) Lipase-catalyzed acidolysis of butter oil with conjugated linoleic acid: a kinetic study involving multiple reuse of the immobilized enzyme. Ind Eng Chem Res 43(14): 3714–3722
Torres CF, Lessard LP, Hill CG (2003a) Lipase-catalyzed esterification of conjugated linoleic acid with sorbitol: a kinetic study. Biotechnol Prog 19(4):1255–1260
Torres CF, Moeljadi M, Hill CG (2003b) Lipase-catalyzed ethanolysis of fish oils: multi-response kinetics. Biotechnol Bioeng 83(3):274–281
van Boekel MAJS (1996) Statistical aspects of kinetic modeling for food science problems. J Food Sci 61(3):447–486
van Boekel MAJS (1998) Effect of heating on Maillard reactions in milk. Food Chem 62(4): 403–414
van Boekel MAJS (1999) Testing of kinetic models: usefulness of the multiresponse approach as applied to chlorophyll degradation in foods. Food Res Int 32(4):261–269
Zs H, Horváth MF, Halász P, Záhonyi R, Kispéter J (2006) Effect of grain size and storage on colour coordinates of paprika grist. J Food Phys 17-18:61–70

Chapter 2
Sampling

Victoria Jideani and Afam I. O. Jideani

2.1 Objectives and Learning Outcomes

After completion of this chapter, the reader should be able to:

1. Define a population and discuss how to acquire representative random samples from a population of interest.
2. Describe the steps involved in a sampling operation.
3. Discuss the purpose of a sampling step in an analysis.
4. Differentiate between a bulk sample and a laboratory sample
5. What are the factors considered in choosing a representative sample?
6. Describe the limitations associated with using operating characteristic (OC) curves and acceptance sampling by attributes.
7. Plan a random sample, given the packaging information for incoming lot.
8. Construct and interpret an OC curve.
9. Read and use the MLT-STD-105D tables.

V. Jideani (✉)
Department of Food Science and Technology, Cape Peninsula University of Technology, Cape Town, South Africa
e-mail: jideaniv@cput.ac.za

A. I. O. Jideani (✉)
Department of Food Science and Technology, University of Venda, Thohoyandou, Limpopo Province, South Africa

G. Schleining et al. (eds.), *Statistics in Food and Biotechnology*, Integrating Food Science and Engineering Knowledge Into the Food Chain 16,
https://doi.org/10.1007/978-3-031-51568-2_2

2.2 Introduction

Sampling is part of daily life. For example, we sample food by taking a bite or two and then deciding if it is too hot, cold or spicy. Without sampling, we would need to eat all of the food before deciding what to do. A supermarket sells prepacked sandwiches. The sandwiches are brought in large batches from a catering firm. The supermarket manager wishes to test the sandwiches to make sure they are fresh and of good quality. She can test them only by unwrapping them and tasting them. After the test it will no longer be possible to sell them. She must, therefore, make a decision as to whether or not the batch is acceptable based on testing a relatively small sample of sandwiches. The term sample is a portion of material selected in some manner to represent a larger body of parent material, lot or population. The nature of sampling will vary if the population is homogenous or heterogeneous. A bowl of soup may be heterogeneous, if all of the solid stuff is at the bottom and the liquid is at the top. Homogeneity is the degree to which a property or a constituent is uniformly distributed throughout a quantity of material. The result obtained from the sample is merely an estimate of the amount or concentration of a constituent or property of the parent material or population. The sample, if it's a good one, must be representative of the whole. To achieve this, the sample should:

1. Represent both the population and its important parts.
2. The sample should be drawn by a probability or random process.
3. The sample should be as small as possible in order to save time and effort.
4. The sample should meet statistical tests for generalization.

Adequacy of a sample depends on the following four attributes: (1) Representativeness of the sample (is it random?); (2) Size of the sample; (3) Variability in the population; and (4) Desired precision of the estimates.

Sampling is that part of statistical practice concerned with the selection of individual observations intended to yield some knowledge about a population of concern for the purposes of statistical inference. The sampling methods discussed in this section assume that the finite population of interest consists of material in bulk form rather than discrete elements, or individuals. Bulk sampling is described as follows:

Bulk materials are continuous and do not consist of populations of discrete, constant, identifiable, unique units or items that may be drawn into the sample. Rather, the ultimate sampling units must be created, during sampling, using some sampling device.

Bulk sampling deals with problems such as sampling a drum of bulk chemicals, sampling a truck load of bags of bean, sampling a tank truckload of milk, etc.

2.3 Designing a Sampling Plan

A sampling plan is a detailed outline of which measurements will be taken at what times, on which material, and in what manner. Sampling plans should be designed so that the resulting data will contain a representative sample of the parameters of interest and allow for all questions, as stated in the goals, to be answered. The steps involved in developing a sampling plan are:

1. Defining the population
2. Identify the parameters to be measured, the range of possible values and the required resolution (Sampling frame)
3. Design a sampling scheme that details how and when samples will be taken (Sampling method).
4. Determining the sample size.
5. Implementing the sampling plan.
6. Sampling and data collecting.
7. Reviewing the sampling process.

2.3.1 Defining the Population

Sampling procedures must be tailored to the nature of the population being evaluated. Establishing the type of population is usually accomplished by observing the process just before the sampling point. Simple random sampling is effective if the population has satisfactory and unsatisfactory products scattered uniformly. If the population is divided into small subgroups, each containing product of uniform quality, then clustered random samples are indicated. If the population consists of layers of good quality and layers containing defects, then stratified random sampling is required. If the type is not easily determined by observation, analysis of groups of samples taken throughout the population will readily define it.

2.3.2 The Sampling Frame

The goal of the investigation is to tell us what characteristic to measure and how. The first step is to carefully scrutinize the goals. This will tell you which response variables need to be sampled and how. Samples fall into two classes: those in which variables are measured and those in which attributes are counted. Quality decisions are based on the degree to which variables depart from some normal or desired figure or the extent to which attributes exceed some specified limit.

Variables are product or process characteristics that can be measured. All variables can be measured using a scale and expressed in increments of inches, milligrams, degrees, percent, and so on.

When it is not feasible to conduct scalar measurements, another type of quality rating must be used. The most practical method of checking a product's process or characteristic is to classify it as acceptable or unacceptable. This is the method of classifying attributes. Attributes, as contrasted with variables are quality characteristics of either a product or process, which may be tabulated as either present or absent, satisfactory or unsatisfactory, within or outside of limits, or right or wrong. Attributes are tabulated whereas variables are measured.

One complications of attribute counting is the need to distinguish between defects and defectives. A defect is a characteristic of either a process or a product which is not in conformance with requirements. A defective is a product containing more defects than allowed.

It may be convenient to reclassify a variable as an attribute. For example, a package of crisps may be considered a defect if it exceeds 3 cm in length. In this case, a gauge exactly 3 cm long would be used to count the package defects in a lot. This type of sampling testing might be used if the purpose of the test were to assure that all of the crisps packages would fit into dispensing machines without jamming. However, if the goal were to control the amount of packaging material used, exact measurements of package length might be required.

Defining the expected ranges of values is useful for screening outliers. Finally, the required resolution for the measurements should be specified. This specification will help guide the choice of measurement equipment and help define the measurement procedures.

2.3.3 Sampling Scheme

A sampling scheme is a detailed description of (a) what sample will be obtained and (b) how this will be done. In other words, the sampling scheme describes how sampling units are selected from the population.

(a) ***What type of sample to collect?***

After determining where to collect samples, the next step in designing a sampling plan is to decide what sample to collect. Three methods are commonly used to obtain samples: grab, composite, and in situ.

The **grab sample**, a portion of the target population removed at a given time and location in space is the most common. Grab sample is easily adapted to any of the sampling schemes. If the target population is fairly uniform in time and space, a set of grab samples collected at random can be used to establish its properties. A systematic grab sample can be used to characterise a target population whose composition varies over time or space.

A **composite sample** consists of a set of grab samples combined to form a single sample. The composite sample is analysed after thorough mixing. In some cases it may be desirable to analyse each grab sample separately and in other cases, there are advantages to working with composite samples.

A significant disadvantage of grab and composite samples is the need to remove a portion of the target population for analysis. Hence, neither sample type can be used to monitor a time-dependent change in the target population continuously. In such cases **in situ sampling**, in which an analytical sensor is placed directly in the target population, allowing continuous monitoring without removing individual grab samples. For example, the pH of grape juice during fermentation can be continually monitored by immersing a pH electrode within the fermentation vat.

(b) ***How samples will be collected***

Once we have selected our response parameters, it would seem to be a rather straightforward exercise to take some measurements, calculate some statistics and draw conclusions. However, many things can go wrong along the way, which can be avoided with careful planning and knowing what to watch for. Two guiding principles will guide the design of our sampling scheme.

1. The first principle is that of **precision**. If the sampling scheme is properly laid out, the difference between our estimate of some parameter of interest and its true value will be due only to random variation. The size of this random variation is measured by a quantity called standard error. The magnitude of the standard error is known as precision. The smaller the standard error, the more precise our estimates are. The precision of any estimate will depend on (a) the inherent variability of the process estimator, (b) the measurement error, (c) the number of independent replications (sample size), and (d) the efficiency of the sampling scheme.
2. The second principle is the **systematic sampling error** (or confounded effects). Systematic sampling error occurs when the levels of one explanatory variable are the same as some unaccounted for explanatory variable. The way to prevent systematic sampling errors (and at the same time increase precision) is through stratification and randomization. Stratification is the process of segmenting our population across levels of some factor to minimize variability within those segments or strata. For instance, if we want to try several different process recipes to see which one is best, we may want to apply each of the recipe to each of the three work shifts. This will ensure that any systematic errors caused by a shift effect is eliminated. This is where the ANOVA designs are particularly useful.

Randomization refers to any process for selecting a sample from the population that involves impartial or unbiased selection. A sample chosen by randomization procedures is called a random sample. The most population types of sampling techniques involving randomization are simple random, stratified, cluster and systematic.

Simple Random Sampling
If a sample is chosen so that all samples of the same size are equally likely to be chosen, then it is called a simple random sample.

The advantages of using random sampling are (1) it helps to reduce or eliminate bias in the way the sampled items are chosen and (2) enables us to make precise statements about the extent to which conclusions drawn from the sample can be applied to the entire population.

Sampling can be done with or without replacement. Sampling **with replacement** means that after each successive item (or integer) is selected for the random sample, the item is "replaced" back into the population and may even be selected again later. In practice, sampling with replacement is rarely used. Instead, the more common sampling is to allow only distinct items from the population in the sample. That is no repeats are allowed and this is called sampling **without replacement**. Unless otherwise stated, we will always assume that random sampling is done without replacement.

Example 2.1
Suppose we want to draw 50 cans of baked beans in order to determine the drained weight from a lot of 1000 cans. A possible method of creating a simple random sample might be to number each of the 1000 cans from 1 to 1000. Next write 1 to 1000 each on different piece of paper. Fold each piece individually and drop all folded paper into a bag, mix and remove 50 of them at random. Obviously, this is an impractical procedure, and only serves to illustrate the principle of simple random sampling. More useful methods would be to: (1) remove a total of 50 cans from the line, 1 at a time, intermittently, and spread out over the period required to produce a lot of size 50; (2) use random numbers.

Example 2.2
Suppose we want to draw 15 cans of baked beans in order to determine the drained weight from a lot of 500 cans. We can use random number table to select the 15 cans.

Solution:

Units of the population from which a sample is required are assigned with an equal number of digits. When the size of the population is less than thousand, three-digit numbers 000, 001, 002,. . ., 999 are assigned. Based of the size of the population and the random number table available to us, we

(continued)

Example 2.2 (continued)

proceed according to our convenience. We can start in any place and may go on in any direction such as column- or row- wise in a random number table.

Random number Table

5688	2376	7403	3746	5109	3879
5249	4595	9238	8753	9788	1578
3541	6858	8674	4856	8560	1267
4919	8155	0943	2311	5535	9257
2204	6599	8531	7888	3466	1843
3918	8806	8407	0831	6442	5743
3095	8567	9131	6420	2322	1902
6229	4600	3518	2700	3586	8464
3226	5426	8875	6825	2089	2672
9529	1942	4860	6302	3638	1085

Since our population is 500 cans and we have given them numbers 001 to 500, we will be locating three digits in the random table. Starting arbitrarily from column 2, row 4, the last three digits of the random number 8155 represents can #155. Since there are 500 cans, the can will be selected. The next random number is 6599 representing can #599. Since this is more than 500, this number will not be selected. The 15 cans to be selected are:

155	131	466
426	311	442
403	420	322
238	302	89
407	109	267

Systematic Random Sampling

This is a variation of the simple random sampling. It is generally known as a 1/k random sampling and is obtained by selecting every k^{th} (sampling interval) unit from the population, beginning with a unit selected at random from the first k units. Systematic sampling is a frequently used method of sampling when a complete list of the population is available.

$$\text{Sampling interval}, k = \frac{N}{n} = \frac{\text{Population size}}{\text{Sample size}} \quad (1)$$

Example 2.3
Referring back to the 1000 cans of baked beans out of which we require 50 cans of baked beans in order to determine the drained weight from a lot of 1000 cans. Our sampling interval from equation 1 will be:

$$k = \frac{1000}{50} = 20$$

Systematic sampling consists of selecting a random number say *i* and every *k*th unit subsequently. Suppose the random number 'i' is 5, then we select 5, 25, 45, 65, 85, 105, . . . i.e. adding the sampling interval (20) to 5 to obtain 25, then adding 20 to 25, etc. The random number 'i' is called random start.

The merits of systematic sampling are (1) simple and convenient; (2) time and work are reduced; (3) accurate results obtained when properly done; (4) it can be used in an infinite population. However, systematic sampling has the following limitations (1) may not represent the whole population and (2) there is a chance of personal bias of the investigators.

Stratified Random Sampling
Stratification means division into groups. In this method, the population is divided into several subgroups or strata. The strata should be so formed that each stratum is homogeneous as far as possible. Then, a simple random sample may be selected from each stratum, and these are combined together to form the required sample from the population. This technique is mainly used to reduce the population heterogeneity and increase the estimates' efficiency.

There are two types of stratified sampling. They are proportional and non-proportional. In proportional sampling equal and proportionate representation is given to subgroups or strata. If the number of items is large, the sample will have a higher size and vice versa. The population size is denoted by N and the sample size is denoted by 'n'. The sample size is allocated to each stratum in such as way that the sample fraction is a constant for each stratum. That is given by n/N = c. Hence, in this method each stratum is represented according to size.

In non-proportionate sample, equal representation is given to all the sub-strata regardless of their existence in the population.

Example 2.4
Referring back to the 1000 cans of baked beans, we require 50 cans of baked beans to determine the drained weight from a lot of 1000 cans. In this case, the cans of baked beans were obtained from two manufacturers: Heinz and

(continued)

Example 2.4 (continued)

Bokomo. We have 600 cans from Heinz and 400 cans from Bokomo. How will you draw the sample using proportional stratified sampling?

Solution:

There are two strata in this case with sizes $N_1 = 600$ and $N_2 = 400$ making a total population $N = N_1 + N_2 = 1000$.

The sample size, n is 50.

If n_1 is the sample size required from Heinz and n_2 the sample size from Bokomo, then the sizes from each brand will be:

$n_1 = \frac{n}{N} \times N_1 = \frac{50}{1000} \times 600 = 30$ cans from Heinz

$n_2 = \frac{n}{N} \times N_2 = \frac{50}{1000} \times 400 = 20$ cans from Bokomo

The sample sizes are 30 cans from Heinz and 20 from Bokomo. The units from each brand will then be selected by simple random sampling.

Cluster Sampling

Cluster sampling is a technique where the entire population is divided into groups, or clusters, and a random sample of these clusters is selected. If the investigator uses all of the population elements in the selected subsets for the sample, the procedure is one-stage cluster sampling. If, on the other hand, a sample of elements is selected probabilistically from the selected subsets, the procedure is known as two-stage cluster sampling.

Example 2.5

In a warehouse containing 200 crates of 24 cans per crate of baked beans, we wish to take a sample size of 50 for drained weight determination.

Solution:

We consider the crates as clusters. Then we could draw a group of random cluster crates from which we draw the samples. There are many possibilities for arriving at a sample size of 50, depending on how many clusters are selected, as illustrated in Table 2.1.

Table 2.1 Examples of choices for cluster sample selection

Cluster size	Sub-sample size	Total sample size
50 crates	1 can from each	50
25 crates	2 cans from each	50
10 crates	5 cans from each	50
5 crates	10 cans from each	50

The variation between clusters is best shown by the use of a 50-crate cluster size. The variation within clusters becomes more apparent as the sub-sample is increased. As with all sampling plans, the goal of the sampling procedure must be established before a rational selection can be made.

2.4 Approaches to Determine the Sample Size

When choosing a sample size, we must consider the following issues (1) What population parameters we want to estimate; (2) Confidence or risk level associated with the information; (3) How much is already known; (4) Level of precision- how precise we want the final estimate to be; (5) Degree of variability of the population; (6) Practicality: how hard is it to collect data.

If our process has been studied before, we can use that prior information to reduce sample sizes. This can be done using prior mean and variance estimates and stratifying the population to reduce variation within groups.

The *level of precision*, sometimes called *sampling error*, is the range in which the true value of the population is estimated to be. This range is often expressed in percentage points (e.g., ±5%). Thus, if a researcher finds that 60% of farmers in the sample have adopted a recommended practice with a precision rate of ±5%, then he or she can conclude that between 55% and 65% of farmers in the population have adopted the practice.

The confidence or risk level is based on ideas encompassed under the Central Limit Theorem. The key idea encompassed in the Central Limit Theorem is that when a population is repeatedly sampled, the average value of the attribute obtained by those samples equals the true population value. Furthermore, the values obtained by these samples are distributed normally about the true value, with some samples having a higher value and some obtaining a lower score than the true population value. In a normal distribution, approximately 95% of the sample values are within two standard deviations of the true population value (e.g., mean). If a 95% confidence level is selected, 95 out of 100 samples will have the true population value within the range of precision specified earlier. There is always a chance that the sample obtained does not represent the true population value. Such samples with extreme values are represented by the shaded areas in Fig. 2.1. This risk is reduced for 99% confidence levels and increased for 90% (or lower) confidence levels.

When choosing sample sizes, we need to select risk values. If the decisions we will make from the sampling activity are very valuable, then we will want low risk values and hence larger sample sizes.

Degree of variability: We take samples to estimates some characteristics of the population of interest. The variance of that estimate is proportional to the inherent variability of the population divided by the sample size:

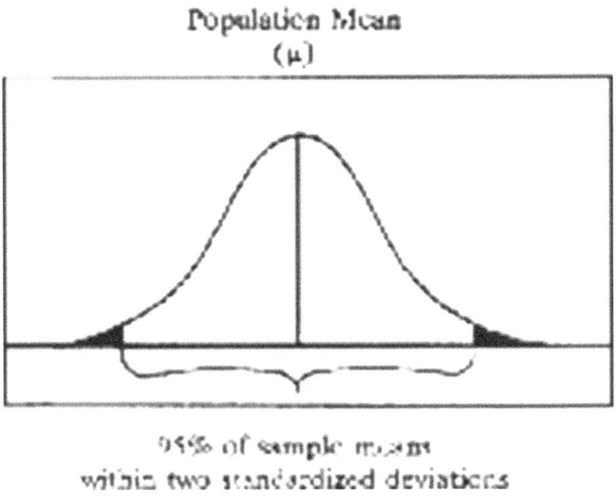

Fig. 2.1 Population mean and the distribution sample means

$$Var(\widehat{p}) \approx {}^{\sigma^2}/_{n}$$

with $\widehat{p}$ denoting the parameter we are trying to estimate. This means that the more heterogeneous a population, the larger the sample size required to obtain a given level of precision. The less variable (more homogeneous) a population, the smaller the sample size. Note that a proportion of 50% indicates a greater variability than either 20% or 80%. This is because 20% and 80% indicate that a large majority do not or do, respectively, have the attribute of interest. Because a proportion of 0.5 indicates the maximum variability in a population, it is often used in determining a more conservative sample size, that is, the sample size may be larger than if the true variability of the population attribute were used.

Practicality of collecting the samples: Of course, the sample size select must make sense. This is where the trade-offs usually occur. We want to take enough observations to obtain reasonably precise estimates of the parameters of interest but we also want to do this within a practical resource budget. The important thing is to quantify the risks associated with the chosen sample size.

Several approaches to determine the sample size exist. These include using a census for small populations, imitating a sample size of similar studies, using published tables, and applying formulas to calculate a sample size. Each strategy is discussed below.

2.4.1 Using a Census for Small Populations

One approach is to use the entire population as the sample. Although cost considerations make this impossible for large populations, a census is attractive for small populations (e.g., 200 or less). A census eliminates sampling error and provides data on all the individuals in the population. Finally, virtually the entire population would have to be sampled in small populations to achieve a desirable level of precision.

2.4.2 *Using a Sample Size of a Similar Study*

Another approach is to use the same sample size as those of studies similar to the one you plan. A review of the literature can provide guidance about "typical" sample sizes which are used.

2.4.3 *Using Published Tables*

A third way to determine sample size is to rely on published tables which provide the sample size that would be necessary for given combinations of precision, confidence levels, and variability.

2.4.4 *Using Formulas to Calculate a Sample Size*

Although tables can provide a useful guide for determining the sample size, you may need to calculate the necessary sample size for a different combination of precision, confidence, and variability levels. The fourth approach to determining sample size is the application of one of several formulas.

Formula for Calculating a Sample Size for Proportions
When sampling proportions, start with a probability statement about the desired precision (δ). We want the probability that the precision of our estimate to be less than we want is α. We would like to set α low, usually 0.1 or less. Using some assumption about the proportion being approximately normally distributed we can obtain an estimate of the required sample size as:

$$n = pq\left(\frac{z_{\alpha/2}}{E^2}\right)^2$$

where z is the ordinate on the normal curve corresponding to the probability value (α = 1 − desired confidence level), E is the desired precision of the estimate or maximum error of the estimate, p is the estimated proportion of an attribute that is present in the population, and q is 1 − p. If no approximation for p is known, one should use p = 0.5.

Example 2.6

Olive food plans to run a new process and sample output for yield (good/bad). The current process has been yielding 55% (p = 0.55, q = 0.45). The decision was to estimate the new process yield to be accurate to within 0.1 at 95% confidence (α = 0.05, z = 1.96). Estimate the sample size to be drawn from the new process.

Solution:

$$n = pq\left(\frac{z_{\alpha/2}}{E}\right)^2 = n = 0.55 \times 0.45\left(\frac{1.96}{0.1}\right)^2 = 95$$

Thus if we draw 95 random parts from the output of the new process and estimate the yield, then we are 95% sure the yield estimate is within 0.10 of the true process yield.

Formula for Calculating a Sample Size for Continuous Variables:

To determine sample size for variables that are continuous the following formula is used

$$n = \left(\frac{z_{\alpha/2}\sigma}{E}\right)^2$$

where z is the ordinate on the normal curve corresponding to the probability value (α = 1 − desired confidence level), E is the desired precision of the estimate or maximum error of estimate, σ is the variance of the population. If σ is unknown, the standard deviation obtained from a sample taken previously can be used.

Example 2.7

A bakery shop owner wishes to find the 95% confidence interval of the true mean cost of a large cheese pizza. How large should the sample be if she wishes to be accurate to within $0.10? A previous study showed that the standard deviation of the prize was $0.16.

Solution:

$$n = \left(\frac{z_{\alpha/2}\sigma}{E}\right)^2 = \left(\frac{1.96 \times 0.16}{0.10}\right)^2 = 9.8$$

Thus she draws 10 random pizza, estimates the prize, then she will be 95% sure the prize estimate is within $0.10 of the true prize.

2.5 Implementing the Sampling Plan

Implementing a sampling plan involves three steps: (1) physically removing the sample from its target population, (2) preserving the sample, and (3) preparing the sample for analysis. Since sampling exposes the target population to contamination, the sampling device must be inert and clean. Samples are often preserved before transporting them to the laboratory for analysis to avoid chemical and physical deterioration.

The initial sample is called the primary or gross sample and may be a single increment drawn from the target population, or a composite of several increments. Often, the gross sample cannot be analysed without further treatment. Processing the sample may be to reduce the sample's particle size, to transfer the sample into a more readily analysable form, or to improve its homogeneity.

2.5.1 Sampling Bulk and Packaged Materials

Bulk materials may consist of a single pile, a shipload of grain, or a portion of the atmosphere or ocean, all characterised by a lack of permanently identifiable units. For convenience in handling, bulk materials may be partially packaged in smaller units called segments (bags, bales, drums) separated in space or in time (moving on a conveyor belt), which can be identified for random sampling. The segments are usually still too large for submission as a laboratory sample. Packaged goods are often packed in larger and larger multiple units (cases, pallets, truckloads) comprised of identifiable units which may be assigned numbers, either actually or conceptually.

Figure 2.2 shows the relationship between the various operations in a sampling scheme up to the sample taken for analysis. Both bulk and packaged materials may be delivered in consignment requiring some analysis. Consignment is a quantity of material transferred on one occasion and covered by a single set of shipping documents. It may consist of one or more lots or portions of lots (Fig. 2.2). The presence of different lots in a consignment is important from the point of view of the sampling plan and the interpretation of the results of the analysis. The lot is the sampling target, by definition all the original material being subject to sampling, i.e., a stockpile, a process stream, a barrel, a lorry load, a train wagon, or a plot in the field, in the forest etc. The term sometimes also denotes the physical form of the material. In bulk materials, the lots are usually too large for submission as a laboratory sample, so the initial physical sample, called an increment, is a unit created by the sampling device (e.g., trier, shovel). Increments of a bulk population correspond to units of a packaged population. Collecting one or more increments or units initially from a population gives the primary sample. The portions may be combined (composited) or kept separate (gross sample). A subsample may be obtained from the primary sample. A subsample may be (a) a portion of the sample obtained by selection or division, (b) an individual unit of the lot taken as part of the

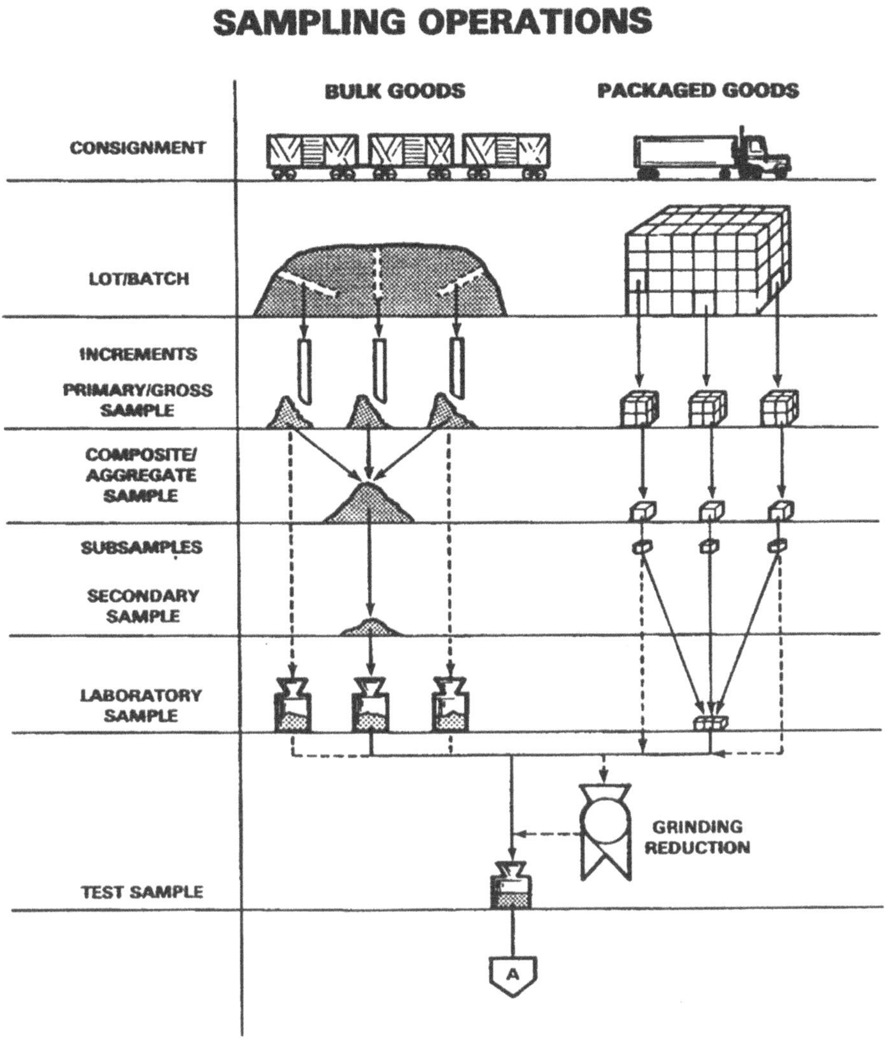

Fig. 2.2 Relationship between the various operations in a sampling scheme up to sample taken for analysis

sample, or (c) the final unit of multistage sampling. The laboratory sample is the sample or subsample(s) sent to or received by the laboratory. When the laboratory sample is further prepared (reduced) by subdividing, mixing, grinding or combinations of these operations, the result is the test sample. When no preparation of the laboratory sample is required, the laboratory sample is the test sample. The laboratory sample is the final sample from the point of view of sample collection but the initial sample from the point of view of the laboratory.

In all cases of sampling bulk materials by using increments or composites, the primary sample as well as the laboratory sample, must be reduced in size (subdivided by riffling, or rolling, coning and quartering, etc.) to a test (or analytical) sample.

Packaged goods are comprised of identifiable units which may be assigned numbers, either actually or conceptually. Numbered units are then selected for the laboratory sample by use of a table of random numbers.

Example 2.8
A lot of 400 canned baked beans is shipped in eight boxes. Each box contains five layers, and each containing 10 rows of 10 cans of baked beans. A sample of 10 cans is to be taken. Design a sample plan.

Solution:

To pick the cans we need to locate at each instance of picking 50 cans, a box, the layer in the box and the particular can in the box (Fig. 2.3). The domain of the box (1–8), the layer in a box (1–5) and cans in a box is (1–100).

887	407	160	903	676
616	419	355	353	227
955	889	145	110	032
117	270	298	915	893
595	713	998	800	417
073	905	121	940	792
853	506	272	913	016
704	694	337	392	057
888	912	700	090	911
413	999	731	385	173
576	379	716	503	820
651	032	636	539	131
445	061	332	756	130
552	882	497	937	314
782	750	147	004	066
172	924	484	723	224
430	341	854	698	569
729	414	394	971	747
003	101	053	584	081
456	379	697	971	545
597	034	687	575	499

Fig. 2.3 A box showing the layer and the cans

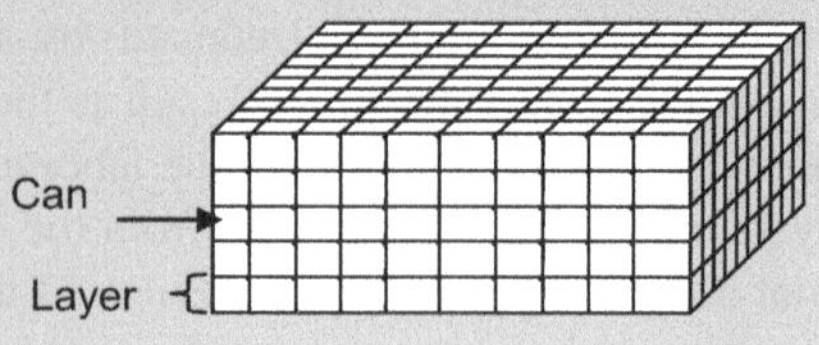

(continued)

Example 2.8 (continued)

We will use a 3-digit random number above to select the 10 cans required. Starting from column 2, row 4 we select the first three random numbers. The first random number will specify the box number with the first digit, The second random number will specify the layer number with the last digit and the third number will specify the can number with the last two digits.

Can number	Random numbers	Box	Layer	Can
1	270, 713, 905,	2	3	5
2	506, 694, 912	5	4	12
3	~~999~~, 379, 032, 061	3	2	61
4	882, ~~750~~, 924, 341	8	4	41
5	414, 101, 379	4	1	79
6	~~034~~, 160, 355, 145	1	5	45
7	298, 998, 121	2	8	21
8	272, 337, ~~700~~, 731	2	7	31
9	716, 636, 332	7	6	32
10	497, 147, 484	4	7	84

The 10 cans that will be selected are shown above. The first can (number 1) will be selected from box number 2, layer 3 and the 5th can. The crossed out numbers are not in the domain for either the box, layer or can.

2.5.2 *Acceptance Sampling Plans for Attribute Data*

Acceptance sampling is performed on goods that have already been produced to determine what percentage of products conform to specifications. These products may be items received from another company and evaluated by the receiving departments, or they may be components that have passed through a processing step and are evaluated by company personnel either in production or later in the warehousing function. If all items in the lot are checked, the process is called 100% inspection. Acceptance sampling is the 'middle of the road' approach between no inspection and 100% inspection. Acceptance sampling is employed when one or several of the following hold: (1) testing is destructive, (2) the cost of 100% inspection is very high and (3) 100% inspection takes too long. An Acceptance Sampling Plan defines how many samples must be taken to verify the lot. However, Acceptance Sampling isn't to be considered a replacement or alternative to Statistical Process Control. Statistical Process Control focuses on improving the process to improve future products, while Acceptance Sampling focuses on evaluation of products that have already been produced to ensure their quality.

There are two classifications of acceptance sampling plans: (1) by variables, which makes accept/reject decisions based on measurement values; (2) by attributes which make accept/reject decisions based on the number of defectives within a lot (2-class plan). In some instances, a marginally defective grouping is permitted to make some allowance for variation in the distribution of organisms in the food and the imprecision associated with the analytical method. Such a scheme is referred to as 3-class sampling plan and is commonly used for microbiological colony count data. The main purpose of acceptance sampling is to decide whether or not the lot is likely to be acceptable, not to estimate the quality of the lot. There are many methods available for acceptance testing, but Acceptance Sampling plans which take into account the Acceptable Quality Level (AQL) and Lot Tolerance Percent Defective (LTPD) are considered to be the most statistically valid.

The Acceptable Quality Level (AQL)

The AQL is a percent defective that is the base line requirement for the quality of the producer's product. The producer would like to design a sampling plan such that there is a *high probability of accepting* a lot that has a defect level less than or equal to the AQL. The acceptable quality level is generally defined as the percent defectives that the plan will accept 95% of the time. Lots that are at or better than the AQL will be accepted 95% (P_a) of the time. If the lot fails, we can say with 95% confidence that the lot quality level is worse than the AQL. Likewise, we can say that a lot at AQL that is acceptable has a 5% ($1 - P_a$) chance of being rejected. This probability of rejecting the acceptable lot is called the Producer's Risk (α) or type I error (Fig. 2.4). This is the probability, for a given (n, c) sampling plan, of rejecting a lot that has a defect level equal to the AQL. The producer suffers when this occurs, because a lot with acceptable quality was rejected. Typical values for α range from 0.2 to 0.01.

When a reference quality level has been established and agreed upon by the responsible parties (such as a consumer and supplier/producer) to be specified for the parameter of sampling procedures, and it is associated with an expected high probability of acceptance, then the probability of rejection of such quality can be considered a risk to the producer. Better control of such a producer's risk can be exercised in the process of designing the sampling inspection plans.

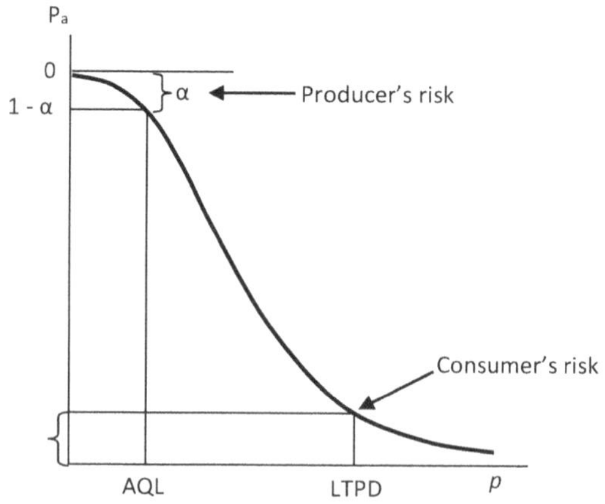

Fig. 2.4 OC curve indicating the producer's risk α and consumer's risk β

Table 2.2 Risks associated with attribute sampling

	Lot Quality (Center)	
Decision	Good (AQL)	Bad (LTPD)
Accept	Producer's Confidence $(1 - \alpha)$	Consumer's point **TYPE II ERROR** Consumer's risk (β)
Reject	Producer's point **TYPE I ERROR** Producer's Risk $= \alpha$	Consumer's confidence $(1 - \beta)$

Lot Tolerance Percent Defective (LTPD)

The LTPD is a designated high defect level that would be unacceptable to the consumer. The consumer would like the sampling plan to have a *low probability of accepting* a lot with a defect level as high as the LTPD. The lot tolerance percent defective is generally defined as percent defective that the plan will reject 90% of the time. We can say that a lot, at or worse than the LTPD, will be rejected 90% of the time. If the lot passes, we can say with 90% confidence that the lot quality is better than the LTPD. We could also say that a lot at the LTPD that is defective has a 10% chance of being accepted. This probability of accepting a defective lot is called the Consumer's Risk (β) or Type II error (Fig. 2.4, Table 2.2). This is the probability, for a given (n,c) sampling plan, of accepting a lot with a defect level equal to the LTPD. The consumer suffers when this occurs, because a lot with unacceptable quality was accepted. The symbol β is commonly used for the Type II error and typical values range from 0.2 to 0.01.

When a limiting or tolerance quality level has been established and agreed upon by the responsible parties (such as a consumer and producer) to be specified for the parameter of a sampling inspection procedure and it is associated with an expected low probability of acceptance, then the probability of acceptance of such quality can be considered a risk to the consumer. Hence, P_a associated with an LTPD for any given sampling plan is a form of a consumer's risk.

The OC curves are a means of quantifying alpha and beta risks for a given attribute sampling plan. The P_a value obtained assumes that the distribution of defectives among a lot is random; either the underlying process is in control, or the product was well mixed before being divided into lots. The samples must be selected randomly from the entire lot. The alpha risk is $1 - P_a$. The shapes of the OC curves is affected by the sample size (n) and accept number (c) parameters. An OC curve should be steep in the region of most interest (between the AQL and the LTPD), which is accomplished by varying n and c. If c remains constant, increasing the sample size n causes the OC curve to be more vertical. While holding n constant, decreasing c (the maximum number of defective units) also makes the slope more vertical, moving closer to the origin. Increasing both the accept number and sample size will bring the curve closer to the ideal shape, with better discrimination. The size of the lot that the sample is taken from has relatively little effect on the quality of protection.

Two-Class Plan

In the two-class scheme, samples are classified as acceptable or defective. A two-class or single sampling scheme is defined by two numbers: (1) n = the number of sample units to be tested from a lot; and (2) c = the maximum allowable number of sample units which may exceed m before the lot is rejected. Tables based on the binomial and Poisson distributions showing the size of sample and the acceptable number of defectives for different values of AQL and LTPD have been published, e.g. the Military standard MIL-STD 105D. Examples of these tables are shown in Tables 2.3 and 2.4. These are attribute sampling plans that provide *n* and *c* value for a specified AQL. Rejected lots are not 100% impacted if this standard is used. In these tables the AQL is fixed while the LTPD vary according to lot size so that larger lots receive more sampling and therefore have a lower risk of being accepted with a high proportion of poor quality products. For instance, it can be seen from Table 2.3 that for a lot containing from 281 to 500 units, 50 sample units should be tested. Assuming an AQL of 1.5%, the lot would be accepted if not more than two units out of the 50 were found to be defective- this critical acceptance number is termed *c*. At such a level there would still be a chance that defective items would occur in the acceptable lot to the extent that 1.7% defective items would be accepted with a probability of 0.95 (equivalent to 5% producer's risk) and that 10% defectives would be accepted with a probability of 0.10 (equivalent to 10% purchaser's or consumer's risk).

Some variations within the sampling procedure included the classification of critical, major, and minor defects. Also, inspection is sub-classified as normal, tightened and reduced, and rules dictate when to switch from one classification to another. If the quality history has been good, fewer items are inspected under

Table 2.3 Sample size code letters for MIL-STD-105D

	Special inspection levels				General inspection levels		
Lot size	S-1	S-2	S-3	S-4	I	II	III
2–8	A	A	A	A	A	A	B
9–15	A	A	A	A	A	B	C
16–25	A	A	B	B	B	C	D
26–50	A	B	B	C	C	D	E
51–90	B	B	C	C	C	E	F
91–150	B	B	C	D	D	F	G
151–280	B	C	D	E	E	G	H
281–500	B	C	D	E	F	H	J
501–1200	C	C	E	F	G	J	K
1201–3200	C	D	E	G	H	K	L
3201–10000	C	D	F	G	J	L	M
10001–35000	C	D	F	H	K	M	N
35001–150000	D	E	G	J	L	N	P
150001–500000	D	E	G	J	M	P	Q
500001– and over	D	E	H	K	N	Q	R

Table 2.4 Master table for normal inspection (single sampling) for MIL-STD-105D

Sample size code letter	Sample size	0.010	0.015	0.025	0.040	0.065	0.10	0.15	0.25	0.40	0.65	1.0	1.5	2.5	4.0	6.5	10	15	25	40
		Acceptable quality levels (AQL)																		
		Ac Re	Ac Re	Ac Re	Ac Re	Ac Re	Ac Re	Ac Re	Ac Re	Ac Re	Ac Re	Ac Re	Ac Re	Ac Re	Ac Re	Ac Re	Ac Re	Ac Re	Ac Re	Ac Re
A	2														↓	0 1		↓	1 2	2 3
B	3													↓	0 1	↑	↓	1 2	2 3	3 4
C	5												↓	0 1	↑	↓	1 2	2 3	3 4	5 6
D	8											↓	0 1	↑	↓	1 2	2 3	3 4	5 6	7 8
E	13										↓	0 1	↑	↓	1 2	2 3	3 4	5 6	7 8	10 11
F	20									↓	0 1	↑	↓	1 2	2 3	3 4	5 6	7 8	10 11	14 15
G	32								↓	0 1	↑	↓	1 2	2 3	3 4	5 6	7 8	10 11	14 15	21 22
H	50							↓	0 1	↑	↓	1 2	2 3	3 4	5 6	7 8	10 11	14 15	21 22	↑
J	80						↓	0 1	↑	↓	1 2	2 3	3 4	5 6	7 8	10 11	14 15	21 22	↑	
K	125					↓	0 1	↑	↓	1 2	2 3	3 4	5 6	7 8	10 11	14 15	21 22	↑		
L	200				↓	0 1	↑	↓	1 2	2 3	3 4	5 6	7 8	10 11	14 15	21 22	↑			
M	315			↓	0 1	↑	↓	1 2	2 3	3 4	5 6	7 8	10 11	14 15	21 22	↑				
N	500		↓	0 1	↑	↓	1 2	2 3	3 4	5 6	7 8	10 11	14 15	21 22	↑					
P	800	↓	0 1	↑	↓	1 2	2 3	3 4	5 6	7 8	10 11	14 15	21 22	↑						
Q	1250	0 1	↑	↓	1 2	2 3	3 4	5 6	7 8	10 11	14 15	21 22	↑							
R	2000			1 2	2 3	3 4	5 6	7 8	10 11	14 15	21 22									

↓ = Use first sampling plan below the arrow. If sample size equals, or exceeds, lot or batch size, do 100% inspection

↑= Use first sampling plan above arrow

Ac Acceptance number, *Re* Rejection number

reduced inspection, while poor-quality history indicates the need for tightened acceptance criteria, so *c* is decreased. The goal is to reduce inspection whenever possible. The various sample sizes are designated by code letters, listed in Table 2.3. The relationship between the lot size and the sample size determines the inspection level. Three inspection levels are usually shown: I, II and III (Table 2.3). Level II is used unless otherwise specified. Level I gives less discriminating and level III gives more discriminating sampling plans. Four additional levels, S-1, S-2, S-3 and S-4 are used when small samples are necessary or when large sampling risks may be taken. The code letter and sample size are then used with a master table such as Table 2.4 to determine the acceptance and rejection numbers. MIL-STD-105D is the most commonly used attribute sampling plan standard and we shall illustrate its use with an example.

In most standards, double and multiple sampling plans are given. For a double sampling plan two samples n1 and n2, are allowed, and two acceptance numbers listed. If the number of defective items d in n1 is $d \leq c1$, accept the lot; and if $d > c2$, reject the lot. But if $c1 < d \leq c2$, take a second sample of size n2. If the total defectives in $n1 + n2$ is $d \leq c2$, accept the lot; but if $d > c2$, reject it. Double sampling usually requires that fewer items be inspected on the average than single sampling requires.

Example 2.9

Use MLT-STD-105D to determine the sample size and the acceptance number with the following criteria:

- Single sampling
- Normal inspection
- Inspection level II
- Lot size 3000
- AQL 6.5%

Solution:

The code letter is obtained from Table 2.3. Find the row with lot size 1201 to 3200. Follow it across to the inspection level II column and read the code letter, K. Single sampling for normal inspection is found in Table 2.4. The row corresponding to the code letter K shows that a sample of 125 is to be taken. Follow the row across to the AQL = 6.5 column. The numbers at the intersection of the row and column are 14 and 15.

A sample of 125 is randomly selected; and if the number of defects is 14 or less the lot is accepted. If 15 or more defectives are found, the lot is rejected.

Tables 2.3 and 2.4 are only samples from MIL-STD-105D. The actual standard includes plans for single, double, and multiple sampling for normal, tightened and reduced inspection.

In microbiological sampling a sample is described as defective, if it is shown to contain more than a specified number of organisms or, in cases where a presence or absence test is applied, the target organism is detected. For 2-class plans to be applicable in microbiological data three numbers are required: (a) n = the number of sample units to be tested from a lot; m = the count above which the sample is regarded as defective (i.e. microbial limit). This term would not appear in schemes employing a presence/absence test since a positive result is sufficient for the sample to be defective; c = the maximum allowable number of sample units which may exceed m before the lot is rejected.

Such schemes simply classify sample units according to the test result. For example, if m is 10^4 cfu/g, those samples giving counts of 10^2, 9×10^3 and 1.2×10^4 would be considered acceptable, acceptable and defective respectively. Despite the fact that the first sample had a count almost 2 log cycles lower than the second and the difference in count between the second and third samples is relatively small.

If ten sample units were analysed and the aerobic plate count (APC) of eight are equal to or less than m and two are greater than m (i.e. c = 2), what decision can one make regarding the acceptability of the lot? If two defective sample units (i.e. c = 2) are allowed of the ten that are examined, then the entire lot of a food product that has eight acceptable and two defective units would be considered as acceptable. However, if there are three defective units (c = 3), the lot would be rejected.

Three-Class Plans

In the same way that 2-class plans are constructed for various AQL and LTPD levels based on binomial or poisson distributions, 3-class plans can be derived from the trinomial distribution. Three-class attributes sampling plans introduce a further category and divide samples into three classes: acceptable, marginally acceptable and unacceptable. The use of this extra classification of marginally acceptable means that three-class attribute plans are not used with presence or absence tests but only with microbiological count data. A three-class plan is defined by four numbers: (1) n = the number of sample units to be tested from a lot; (2) M = a count which if exceeded by any of the test samples would lead to rejection of the lot; (3) m = the count which separates good quality from marginal quality and which most test samples should not exceed; (4) c = the maximum number of test samples which may fall into the marginally acceptable category before the lot is rejected. The lower level (m) is the target level below which colony counts should fall most of the time; the upper level (M) is that level which should never be exceeded. Hence marginal defectives will have counts intermediate between m and M.

Microbiological criteria using attributes sampling plans for different foods have been produced by a number of organisations and some examples of these are give in Table 2.5. As is clear, the stringency of the sampling plan varies with the type of food and the organism being sought.

Table 2.5 Attributes sampling plans

Product	Organisms	Plan class	n	m	M	c	Source
Ice cream	APC	3	5	10^5	10^6	2	Canada
	Coliforms	3	5	10	10^3	1	
Dried milk	APC	3	5	5×10^4	2×10^3	2	International Dairy Federation
	Coliforms	3	5	10	100	1	
	Salmonella	2	15	0	–	0	
					10^7	3	
Frozen raw crustaceans	APC	3	5	10^6	500	3	ICMSF
	E. coli	3	5	11	10^4	2	
	Staph. aureus[a]	3	5	10^3	–	0	
	Salmonella[a]	2	5	0	10^3	1	
	V. parahaemolyticus	3	5	10^2	10^7	2	
Frozen cooked crustaceans	APC	3	5	5×10^3	10^7	2	ICMSF
	E. coli	3	5	11	500	2	
	Staph. aureus[a]	2	5	10^3	–	0	
	Salmonella[a]	2	10	0	–	0	
	V. parahaemolyticus	3	5	10^2	10^3	1	
Frozen fruits & vegetables pH > 4.5	*E. coli*	3	5	10^2	10^3	2	ICMSF

Source: Adams and Moss (1999)

APC Aerobic plate count; *ICMSF*, International Commission on Microbiological Specifications for Foods

[a]Additional tests; M and m values are expressed as cfu/g

Example 2.10

The following aerobic plate count was obtained from analysing 5 samples, each of which was a ready-to-eat snack from processors.

	Aerobic count (cfu/g)		
Sample	A	B	C
1	1.3×10^6	1.25×10^6	1.7×10^5
2	4.97×10^5	1.95×10^5	6.9×10^8
3	1.2×10^6	3.05×10^3	4.2×10^6
4	1.5×10^8	4.77×10^1	2.99×10^6
5	1.33×10^8	5.6×10^5	3.6×10^6

Using a three-class plan where $n = 5$, $m = 10^5$, $M = 10^6$ and $c = 2$, determine whether the batch of food is acceptable or not.

(continued)

Example 2.10 (continued)
Solution:

Sample	Processor A	Processor B	Processor C
1	Marginally accept	Marginally accept	Accept
2	Accept	Accept	Accept
3	Marginally accept	Accept	Marginally accept
4	Reject	Accept	Marginally accept
5	Reject	Accept	Accept

For processor A, two of the samples exceeded M, a count which, if exceeded by any of the test samples, would lead to rejection of the lot; therefore, the lot for processor A is rejected.

For processor B, the lot will be accepted since c, the maximum number of test samples which may fall into the marginally acceptable category before the lot is rejected is less than 2.

For processor C, the lot will be rejected since c = 2.

Operating Characteristic Curve

It is natural to always know that every bad lot would be rejected and that every good lot would be accepted. However, since we are taking a sample of the lot, there is always a risk of making the wrong decision. Sampling plans are used to control that risk. A statistically valid sampling plan tells us the probability of accepting bad lots and the probability of rejecting good lots.

The ability of a sampling plan to discriminate between good and bad lots is best explained by an Operating Characteristic (OC) curve. The OC curve is a graph that shows the probability of acceptance of a lot (P_a) on the Y-axis versus all possible fraction or percent defective values, p on the X-axis. Fig. 2.5 details the shape of an operating characteristic (OC) curve and an ideal OC curve.

In an ideal OC curve, we might want to accept all lots $\leq p_0$ defective and reject all lots that are $> p_0$ defective. With this ideal (no risks) curve, all batches with $< p_0$ defective incoming quality level would have a probability of acceptance (P_a) of 1.0. All lots with $> p_0$ defective would have P_a of 0.

Since there will always be some risks, a more typical looking OC curve looks like the one shown in Fig. 2.5. It is based on the Poisson distribution (with the defective rate $< 10\%$ and n is relatively large compared to N). The steeper the OC curve, the better its discriminating power. An ideal curve, vertical at some desirable fraction defective level p_0 (Fig. 2.5), is obtainable only for error-free inspection.

The acceptance number *c* and sample size *n* are the most important factors in defining the OC curve. Decreasing the acceptance number is preferred over increasing sample size. The larger the sample size the steeper is the OC curve, i.e. it becomes more discriminating between good and bad lots.

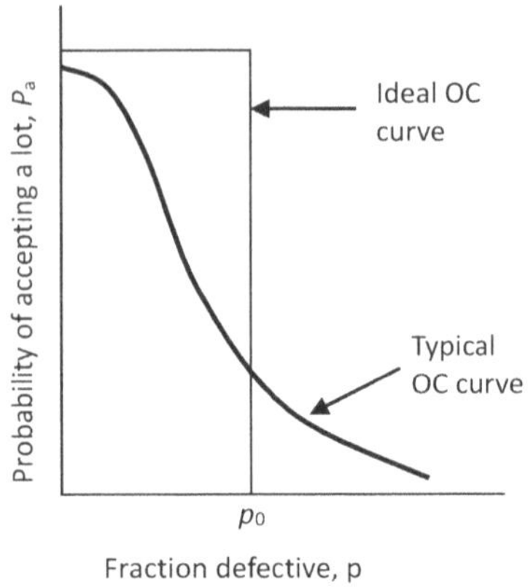

Fig. 2.5 General shape of an operating characteristic (OC) curve and ideal OC curve

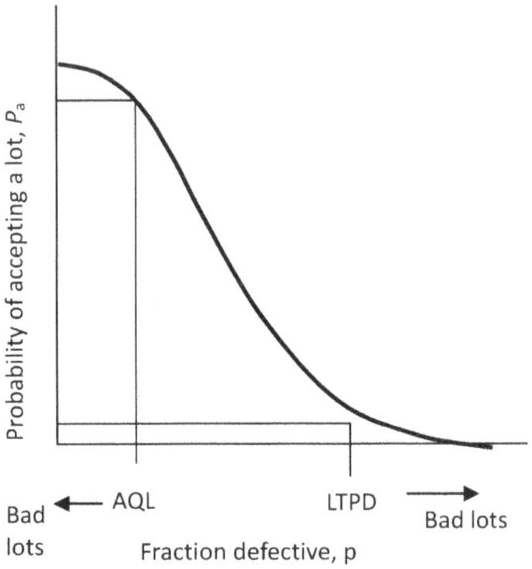

Fig. 2.6 Typical OC curve showing the location of the Acceptable Quality Level (AQL) and the Lot Tolerance Percent Defective (LTPD)

The OC curve is usually summarised by two points on the curve: the Acceptable Quality Level (AQL) and the Lot Tolerance Percent Defective (LTPD) (Fig. 2.6).

OC curves for 3-class plans can be considered to form a geometric surface on a 3-dimensional plot of P against the percentage of lot defectives and the percentage of lot marginal defectives.

Constructing an OC Curve

The probability values are correctly computed using the hypergeometric distribution. However, if n is small compared to N and if p is small, the Poisson distribution P(d; np), where np is the main number of defectives in n items, gives a good approximation. The following steps are used to compute and graph the OC curve values, given n and c.

1. List the several p values that cover the expected range of defective fractions.
2. Compute the Poisson mean $\lambda = np$ for each p value.
3. Set $c = x'$ and use the Poisson Table (Appendix 1) to determine $P_a = P(d \leq x')$.

4. Plot p versus P_a.
5. Construct a smooth, continuous curve through the points. More p values might be needed to obtain a well-shaped OC curve.

Example 2.11 illustrates these steps.

Example 2.11
The line inspection department of Cola company planned to use a sampling plan based on attributes of sample size n = 100 and an acceptance number of c = 2 for fill height of their cola product.

(a) Produce the OC curve for this plan.
(b) If in a lot 10% of the bottles are defective, what is the probability that the lot will be rejected?

Solution:

The steps outlined above are used, and the results are given in Table 2.6.

1. Steps 1 & 2: Values of fraction defective to cover the expected range of defectives are chosen. Let us choose 0.005 to 0.070 (0.5 to 7%) defectives for our case. These are listed with the Poisson means $\lambda = np = 100p$ since n = 100 in this case.

   Step 3: From the Cummulative Poisson distribution table (Appendix) the values of P_a for c =x' = 2 are recorded for each *np* value.
   Steps 4 & 5: The values are plotted in Figure and a smooth curve approximated through them (Table 2.6).

Table 2.6 Computations for the OC curve for the sampling plan n = 100, c = 2, Example

p	$n \times p$	P_a
0	0	1.0
0.01	1	0.920
0.02	2	0.677
0.03	3	0.423
0.04	4	0.238
0.05	5	0.125
0.06	6	0.062
0.07	7	0.030
0.08	8	0.014
0.09	9	0.006
0.10	10	0.003
0.11	11	0.001
0.12	12	0.001

(continued)

Example 2.11 (continued)

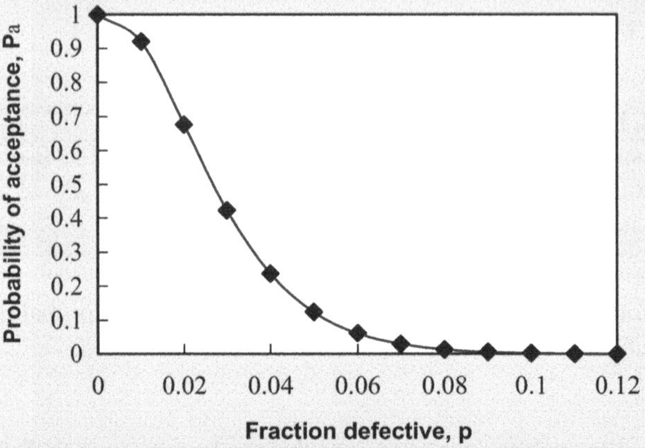

Figure OC curve for the sampling plans n = 100, c = 2

2. From Figure and Table the probability of accepting a lot with 10% (0.1) defective is 0.003. Therefore the probability of rejecting the lot is 1 – 0.003 = 0.997.

If we had not OC curve we could also answer question (2) by calculating the $\lambda = np$ of the 10% defect which is 100(0.1) = 10. Then we read off the probability on the Poisson table at $\lambda = 10$ and $c = 2$, which is 0.003, subtracting from 1 gives the probability of rejection.

Rework (b) above for lots that are 3%, and 5% defective.

Example 2.12

A large wholesale outlet uses the plan in Example 2.11 of n = 100, c = 2 to inspect incoming pre-packed frozen peas. The vendor-buyer contract requires that lots of 0.8% or less defective be accepted and 5% or more defective be rejected. Determine the producer's risk (α) and the consumer's risk (β) for this plan.

Solution:

Given: AQL = 0.008; LTPD = 0.05

For AQL, the $\lambda = np = 100(0.008) = 0.8$. From the Poisson table 0.8 against $c = 2$ gives a probability of $P_a = 0.953$. Recall from Table that the probability of AQL is $1 - \alpha$. Therefore, $1 - \alpha = 0.953$, solving this equation gives $\alpha = 0.047$. Therefore the chance that a good quality lot will be rejected is 5%.

For LTPD, the $\lambda = np = 100(0.05) = 5$. From the Poisson table 5 against $c = 2$ gives a probability of $P_a = 0.125$. Recall from Table that the probability of LTPD is β. Therefore, $\beta = 0.125$. Therefore the chance that a poor quality lot will be accepted is 12.5%.

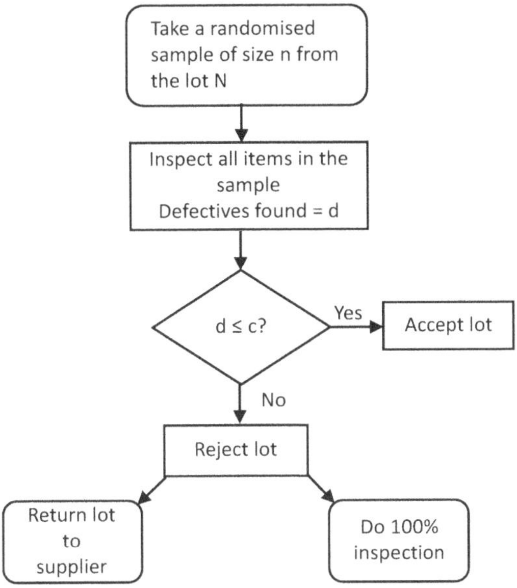

Fig. 2.7 Procedure for attributes single sampling

Designing a Single Acceptance Sampling Plans

The single sampling plan is basic to all acceptance sampling. Applicable to all sampling situations, the single sampling plan has become the benchmark against which other sampling plans are judged. It is employed in inspection by counting the number of defects found in the sample (Poisson distribution) or evaluating the proportion defective from processes or large lots (binomial distribution) or from individual lots (hypergeometric distribution). Single sampling is undoubtedly the most used of any sampling procedure.

The implementation of a single sampling plan is very simple. It involves taking a random sample of size n from a lot of size N. The sample may be intended to represent the lot itself (Type A sampling) or the process used to produce the lot (Type B sampling). The number of defectives (or defects) d found is compared to an acceptance number c. If the number found is less than or equal to c, the lot is accepted. If the number found is greater than c, the lot is rejected. The operation of the plan is illustrated in Fig. 2.7.

The design of the sampling procedure is essential to determine a set of n and c values that come close to satisfying the specification of two points on the desired OC curve for the derived sampling plan. Specifically, the two points are (1) AQL fraction and its associated P_a value $(1 - \alpha)$, i.e. (AQL, $1 - \alpha$) and (2) LTPD fraction and its associated P_a value (β), i.e. (LTPD, β). The objective is a single plan that has an OC curve shown in Fig. 2.4.

Example 2.13

A single plan is desired that will have a high probability of acceptance in the order of 0.95 [a producer's risk α of (1-0.95) of 0.05] for fraction nonconforming of 0.005 (AQL) and simultaneously provide consumer protection by a low probability of acceptance of approximately 0.05 (a consumer's risk β of 0.05) for fraction nonconforming of approximately 0.030 (LTPD).

What needed to be done is to determine the acceptance number 'c' and sample size,'n' of the sampling plan.

1. The first step is to determine the desired or design operating ratio R = LTPD/AQL, which for the case being considered is:

$$R = 0.030/0.005 = 6$$

2. Enter the table below (excerpt from operating ratio table), find the value for "c" by selecting the value in the "n(AQL)" column that is equal to or just greater than the ratio above. So, c = 2.

c	LTPD/AQL	n AQL
0	44.890	0.052
1	10.946	0.355
2	6.509	0.818
3	4.890	1.366
4	4.057	1.970
5	3.549	2.613
6	3.206	3.286
7	2.957	3.981
8	2.768	4.695
9	2.618	5.426

3. Since, c = 2, the n(AQL) for this value is 0.818.

$$n(AQL) = 0.818 \text{ and since } AQL = 0.005$$
$$n = \frac{0.818}{0.005} = 163.6$$

4. Since n must be an integer, rounding up to the next higher integer obtains the acceptance sampling plan, n = 164, c = 2.

Take a random sample of 164 units from the lot. Reject the lot if more than 2 units are defective.

Practice Exercise

1. Compute the probability of acceptance P_a the following sampling plans:

   (a) n = 50, c = 2, p = 0.3
   (b) n = 150, c = 10, p = 0.08
   (c) n = 15, c = 0, p = 0.5

2. (a) Plot the OC curves for n = 75, c = 2 and n = 50, c =2 on the same graph.
   (b) Compare the consumer's risk values for 8% defective lots.
3. A Tomato puree factory inspects incoming fresh tomatoes using the sampling plan n = 80, c = 3 with an AQL = 2.5%. Compute

   (a) Probability of accepting a lot which has a percent defective value equal to the AQL.
   (b) Producer's risk of this plan for p = 0.015.
   (c) The percent defective of a lot which has a 50% chance of being rejected.
   (d) The percent defective for a consumer's risk of 0.10.of AQL = 2,5 % to inspect the sodium content of

4. Use MIL-STD-105D to find single sampling plans for the inspection of bottles of fruit drink which are received a the weekly rate of 150 cases of 24 bottles each. Use level II, normal inspection and an AQL of (a) 2.5% and (b) 10%.
5. The following facts are stated: MIL-STD-105D; normal inspection; Level II; single sampling plans; and an AQL of 1.5%. At Grand foods, 1500 parts are received each week; and at Picky foods, 2000 are received. (a) What sampling plans should be used? (b) Which plan has the smaller producer's risk?

2.6 Variable Sampling

A major disadvantage of attributes sampling schemes is that they make no assumptions about the distribution of measurable parameters within a "lot". In schemes based purely on inspection for faults, this has no consequence. The application of attribute schemes based on the acceptability or otherwise of estimates of variable parameters will result in excessive risk of rejection of otherwise acceptable materials.

Acceptance sampling can be done by measuring a variable rather than classifying an item as conforming or non-conforming. Variables such as thickness, strength or weight might be measured. The primary advantage of a variable sampling scheme is that the proper use of quantitative data should permit "better decision processes" and lead to "economics benefits. The primary disadvantages are the need for analysis of laboratory data, rather than a "go- no go" decision; and the requirement that the data obtained for the variable being measured conforms to a normal frequency distribution. With the advent of computers, the first disadvantage can be overcome readily;

and provided that the analytical data can be "normalised", the second disadvantage can also be overcome.

A typical plan would be to take a sample of size n and reject the batch if the mean measurement, $\bar{x}$, is less than k or greater than k depending on the scenario.

Application of Variable Sampling in Food Microbiology

In variable sampling, it is assumed that microorganisms are distributed log-normally. That is, the logarithms of the counts from different samples conform to a normal distribution. The percentage, d, of units in a "lot" which exceeds some specific critical level can be used to define whether the 'lot' is unsatisfactory. If the critical level is defined in terms of the critical log colony count (V), it is possible to determine whether for a set of n samples, the percentage of values determined exceeds the permissible level (d).

The shape of a normal distribution curve is determined by the parameters μ, the population mean which determines the maximum height of the curve, and σ, the standard deviation of the distribution which determines its spread. For any log count V in a log-normal distribution, a certain proportion of counts will lie above V determined by:

$$\frac{(V-\mu)}{\sigma}=K$$

where K is known as the standardised normal deviate. Rearranging we get:

$$\mu+K\sigma=V \tag{5.2}$$

If V is chosen to represent a log count related to a safety or quality limit and K determines the acceptable proportion of samples in excess of V then a lot would be acceptable if:

$$\mu+K\sigma\leq V$$

and unacceptable if:

$$\mu+K\sigma>V$$

In practice V is likely to be very close to the logarithm of M, used in three-class attribute plans.

Since we do not know μ or σ, we must use estimates derived from our testing, $\bar{x}$, the mean log count, and s, the sample standard deviation. K is replaced with a value k, derived from Standard Tables, which makes allowance for our imprecision in estimating K and chosen to give a desired lowest probability for rejection of a lot having an unacceptable proportion of counts greater than V. This gives us the condition for rejection:

$$\bar{x} + k_1 s > V$$

Some k_1 values are presented in Table 2.7. If we decrease the desired minimum probability of rejection for a given proportion exceeding V, i.e. decrease the stringency of the plan, then k_1 decreases. Application of a variables plan for control purposes will give a lower producer's risk than the equivalent attributes scheme.

It is possible to apply the variables plan as a guideline to Good Manufacturing Practice (GMP). In this case the criterion is:

$$\bar{x} + k_2 s > v \quad (5.6)$$

where k_2 is derived from Tables and gives a certain minimum probability of acceptance provided less than a certain proportion exceeds v a lower limit value characteristic of product produced under conditions of Good Manufacturing Practice (Table 2.7). The value v will be very similar to the logarithm of m used in the three-class attributes plan. If a lot were to fail a GMP criterion it would not lead to rejection of the lot but would alert the manufacturer to an apparent failure of GMP.

Example 2.14

Three batches of product give the following log counts:

	A	B	C
	5.48	5.55	5.95
	5.23	5.30	6.12
	5.81	6.10	6.20
	5.97	5.75	5.65
	5.46	6.01	6.21
$\bar{x}$	5.59	5.74	6.03
s	0.30	0.33	0.23

The critical safety/quality limit is 10^6 cfu/g ($V = 6$), but under conditions of GMP a level below 10^5 is usually attainable ($v = 5$). We are testing 8 samples from the lot and wish to be 95% certain to reject lots where more than 10% of samples exceed a count of 10^6.

Solution:

For safety/quality limit V = 6 and d = 0.1with n = 8 with 95% confidence.

From Table 2.7, $k_1 = 2.6$ and our specification becomes:

If $\bar{x} + 2.6s > 6$ then reject.

For GMP quality limit v = 5 and d = 0.1 with n = 8 with 95% confidence. From Table $k_2 = 0.66$, our GMP criterion becomes:

(continued)

Example 2.14 (continued)

Table 2.7 k values for setting specifications for variables sampling

Safety/quality $\bar{x} + k_1 s > V$									
Probability of rejection, (P)	Proportion exceeding *V (d)*	Number. of replicate samples (n)							
		3	4	5	6	7	8	9	10
0.99	0.1			5.4	4.4	3.9	3.5	3.2	3.0
	0.2			4.0	3.3	2.9	2.6	2.4	2.2
	0.3			3.0	2.5	2.2	2.0	1.8	1.7
	0.4			2.3	1.9	1.6	1.5	1.4	1.3
	0.5		2.3	1.7	1.4	1.2	1.1	1.0	0.9
0.95	0.05	7.7	5.1	4.2	3.7	3.4	3.2	3.0	2.9
	0.10	6.2	4.2	3.4	3.0	2.8	2.6	2.4	2.4
	0.30	3.3	2.3	1.9	1.6	1.5	1.4	1.3	1.3
	0.50	1.7	1.2	0.95	0.82	0.73	0.67	0.62	0.58
0.90	0.10	4.3	3.2	2.7	2.5	2.3	2.2	2.1	2.1
	0.25	2.6	2.0	1.7	1.5	1.4	1.4	1.3	1.3
GMP criterion $\bar{x} + k_2 s < v$									
	Proportion exceeding *v*								
0.95	0.10	0.33	0.44	0.52	0.57	0.62	0.66	0.69	0.71
	0.20	−0.13	0.02	0.11	0.17	0.22	0.26	0.29	0.31
	0.30	0.58	−0.36	−0.24	−0.16	−0.10	−0.06	−0.02	0.00
0.90	0.10	0.53	0.62	0.68	0.72	0.75	0.78	0.81	0.83
	0.20	0.11	0.21	0.27	0.32	0.35	0.38	0.41	0.43
	0.30	0.26	−0.13	−0.05	0.01	0.04	0.07	0.10	0.12

Source: Adams and Moss (1999)
GMP Good Manufacturing Practice

(continued)

Example 2.14 (continued)
If $\bar{x} + 0.66s < 5$ then accept.

Then for lot A:

$\bar{x} + 2.6s > 6$ i.e. 5.59 + (0.26×0.30) = 5.67 < 6 (Safety criterion) and
$\bar{x} + 0.66s < 5$ i.e. 5.59 + (0.66 × 0.30) = 5.79 > 5 (GMP criterion)

Hence, lot A is accepted based on the safety criterion but rejected on GMP criterion.

For lot B:

$\bar{x} + 2.6s > 6$ i.e. 5.74 + (0.26×0.33) = 5.83 < 6 (Safety criterion) and
$\bar{x} + 0.66s < 5$ i.e. 5.74 + (0.66 × 0.33) = 5.96 > 5 (GMP criterion)

Hence, lot B is accepted based on the safety criterion but rejected on GMP criterion.

For lot C:

$\bar{x} + 2.6s > 6$ i.e. 6.03 + (0.26×0.23) = 6.09 > 6 (Safety criterion) and
$\bar{x} + 0.66s < 5$ i.e. 6.03 + (0.66 × 0.23) = 6.18 > 5 (GMP criterion)

Hence, lot C is rejected based on both the safety and GMP criteria.

Appendix

Tables of the Poisson Cumulative Distribution

The table below gives the probability of that a Poisson random variable X with mean = λ is less than or equal to x. That is, the table gives

$$P(X \le x) = \sum_{r=0}^{x} \lambda^r \frac{e^{-\lambda}}{r!}$$

λ=	0.1	0.2	0.3	0.4	0.5	0.6	0.7	0.8	0.9	1.0	1.2	1.4	1.6	1.8
x= 0	0.9048	0.8187	0.7408	0.6703	0.6065	0.5488	0.4966	0.4493	0.4066	0.3679	0.3012	0.2466	0.2019	0.1653
1	0.9953	0.9825	0.9631	0.9384	0.9098	0.8781	0.8442	0.8088	0.7725	0.7358	0.6626	0.5918	0.5249	0.4628
2	0.9998	0.9989	0.9964	0.9921	0.9856	0.9769	0.9659	0.9526	0.9371	0.9197	0.8795	0.8335	0.7834	0.7306
3	1.0000	0.9999	0.9997	0.9992	0.9982	0.9966	0.9942	0.9909	0.9865	0.9810	0.9662	0.9463	0.9212	0.8913
4	1.0000	1.0000	1.0000	0.9999	0.9998	0.9996	0.9992	0.9986	0.9977	0.9963	0.9923	0.9857	0.9763	0.9636
5	1.0000	1.0000	1.0000	1.0000	1.0000	1.0000	0.9999	0.9998	0.9997	0.9994	0.9985	0.9968	0.9940	0.9896
6	1.0000	1.0000	1.0000	1.0000	1.0000	1.0000	1.0000	1.0000	1.0000	0.9999	0.9997	0.9994	0.9987	0.9974
7	1.0000	1.0000	1.0000	1.0000	1.0000	1.0000	1.0000	1.0000	1.0000	1.0000	1.0000	0.9999	0.9997	0.9994
8	1.0000	1.0000	1.0000	1.0000	1.0000	1.0000	1.0000	1.0000	1.0000	1.0000	1.0000	1.0000	1.0000	0.9999
9	1.0000	1.0000	1.0000	1.0000	1.0000	1.0000	1.0000	1.0000	1.0000	1.0000	1.0000	1.0000	1.0000	1.0000
λ=	2.0	2.2	2.4	2.6	2.8	3.0	3.2	3.4	3.6	3.8	4.0	4.5	5.0	5.5
x= 0	0.1353	0.1108	0.0907	0.0743	0.0608	0.0498	0.0408	0.0334	0.0273	0.0224	0.0183	0.0111	0.0067	0.0041
1	0.4060	0.3546	0.3084	0.2674	0.2311	0.1991	0.1712	0.1468	0.1257	0.1074	0.0916	0.0611	0.0404	0.0266
2	0.6767	0.6227	0.5697	0.5184	0.4695	0.4232	0.3799	0.3397	0.3027	0.2689	0.2381	0.1736	0.1247	0.0884
3	0.8571	0.8194	0.7787	0.7360	0.6919	0.6472	0.6025	0.5584	0.5152	0.4735	0.4335	0.3423	0.2650	0.2017
4	0.9473	0.9275	0.9041	0.8774	0.8477	0.8153	0.7806	0.7442	0.7064	0.6678	0.6288	0.5321	0.4405	0.3575
5	0.9834	0.9751	0.9643	0.9510	0.9349	0.9161	0.8946	0.8705	0.8441	0.8156	0.7851	0.7029	0.6160	0.5289
6	0.9955	0.9925	0.9884	0.9828	0.9756	0.9665	0.9554	0.9421	0.9267	0.9091	0.8893	0.8311	0.7622	0.6860
7	0.9989	0.9980	0.9967	0.9947	0.9919	0.9881	0.9832	0.9769	0.9692	0.9599	0.9489	0.9134	0.8666	0.8095
8	0.9998	0.9995	0.9991	0.9985	0.9976	0.9962	0.9943	0.9917	0.9883	0.9840	0.9786	0.9597	0.9319	0.8944
9	1.0000	0.9999	0.9998	0.9996	0.9993	0.9989	0.9982	0.9973	0.9960	0.9942	0.9919	0.9829	0.9682	0.9462
10	1.0000	1.0000	1.0000	0.9999	0.9998	0.9997	0.9995	0.9992	0.9987	0.9981	0.9972	0.9933	0.9863	0.9747
11	1.0000	1.0000	1.0000	1.0000	1.0000	0.9999	0.9999	0.9998	0.9996	0.9994	0.9991	0.9976	0.9945	0.9890
12	1.0000	1.0000	1.0000	1.0000	1.0000	1.0000	1.0000	0.9999	0.9999	0.9998	0.9997	0.9992	0.9980	0.9955
13	1.0000	1.0000	1.0000	1.0000	1.0000	1.0000	1.0000	1.0000	1.0000	1.0000	0.9999	0.9997	0.9993	0.9983
14	1.0000	1.0000	1.0000	1.0000	1.0000	1.0000	1.0000	1.0000	1.0000	1.0000	1.0000	0.9999	0.9998	0.9994
15	1.0000	1.0000	1.0000	1.0000	1.0000	1.0000	1.0000	1.0000	1.0000	1.0000	1.0000	1.0000	0.9999	0.9998
16	1.0000	1.0000	1.0000	1.0000	1.0000	1.0000	1.0000	1.0000	1.0000	1.0000	1.0000	1.0000	1.0000	0.9999
17	1.0000	1.0000	1.0000	1.0000	1.0000	1.0000	1.0000	1.0000	1.0000	1.0000	1.0000	1.0000	1.0000	1.0000

References

Adams MR, Moss MO (1999) Food microbiology. The Royal Society of Chemistry Publication, Cambridge

Devore JL, Farnum NR (1999) Applied statistics for engineers and scientists. Duxbury Press, Pacific Grove

Harvey D (2000) Modern analytical chemistry. McGraw-Hill Publishing Company, New York

Hubbard MR (1996) Statistical quality control for the food industry, 2nd edn. Chapman and Hall, New York

Jarvis B (1989) Statistical aspects of the microbiological analysis of foods, vol 21. Elsevier, New York

Krajewski LJ, Ritzman LP (2005) Operations management, processes and value chains, 7th edn. Prentice Hall, Hoboken

Ledland Blank PE (1980) Statistical procedures for engineering, management and science. McGraw-Hill Book Company, New York

Levinson WA (2006) How to design sample plans on a computer, 6 April. http://www.qualitydigest.com/july99/html/body_samplan.html

Montgomery DC (2005) Introduction to statistical quality control, 5th edn. Wiley

Sampling In IS 540:Research Methods. http://web.utk.edu/~wrobinso/540.html

Schilling EG (1982) Acceptance sampling in quality control, vol 42. CRC Press

Smith GM (2001) Statistical process control and quality improvement, 4th edn. Prentice Hall, Hoboken

Smith G (1991) Statistical process control and quality improvement. Prentice Hill, Columbus

Taylor WA (1992) Guide to acceptance sampling. Taylor Enterprises, Libertyville

Wadsworth HM Jr (1990) Handbook of statistical methods for engineers and scientists. McGraw-Hill Publishing Company, New York

Wadsworth HM, Stephens KS, Godfrey AB (2002) Modern methods for quality control and improvement, 2nd edn. Wiley

Chapter 3
Statistic for Food Product and Process Development

Maria Margarida Cortez Vieira, Sócrates Quispe-Condori, Prisana Suwannaporn, Davide Ballabio, Fabio Gosetti, Saverio Mannino, and Viviana Consonni

3.1 Development of a Vegetable, Fish, and Meat Burger Using Mixture Design

Maria Margarida Cortez Vieira

M. M. C. Vieira (✉)
Instituto Superior de Engenharia, Universidade do Algarve, Campus of Penha and MED—Mediterranean Institute for Agriculture, Environment and Development, CHANGE—Global and Sustainability Institute, Faculty of Science and Technology, Universidade do Algarve, Campus de Gambelas, Faro, Portugal
e-mail: mvieira@ualg.pt

S. Quispe-Condori
School of Food Engineering, Faculty of Engineering and Architecture, Universidad Peruana Unión, Lima, Perú
e-mail: socrates@upeu.edu.pe

P. Suwannaporn
Faculty of Medical and Health Food Technology, Chulabhorn Royal Academy, Bangkok, Thailand

BioScience, Faculty of Science, Kasetsart University, Bangkok, Thailand
e-mail: prisana.s@ku.th

D. Ballabio · F. Gosetti · V. Consonni
Milano Chemometrics and QSAR Research Group, Department of Earth and Environmental Sciences, University of Milano-Bicocca, Milan, Italy

S. Mannino
Department of Food Science and Technologies, University of Milan, Milan, Italy

G. Schleining et al. (eds.), *Statistics in Food and Biotechnology*, Integrating Food Science and Engineering Knowledge Into the Food Chain 16,
https://doi.org/10.1007/978-3-031-51568-2_3

3.1.1 Objectives and Learning Outcomes

The objective of this work was to study and develop a new kind of hamburger including meat, fish, and vegetables. After completion of this chapter the reader should be able to know how to develop a new product with statistical analysis, experimental design, and sensory analysis.

3.1.2 Introduction

DOE (Design of Experiments) comprises a systematic series of tests, in which purposeful changes are made to the input factors, so that you may identify causes for significant changes in the output responses. Multiple independent variables can be studied and their responses at the same time and always explained by a mathematical model. Consequently, there is a reduction in costs, as a smaller number of experiences are needed in a shorter time and with fewer resources.

Factorial Design
When running an experiment, often there is a need to study the effect of several factors or independent variables at the same time on a given parameter (response or dependent variable). Factorial designs allow the inclusion in each complete trial or replicate, of all possible level combinations of these factors. (Montgomery 2017). The effect of a factor is the response produced by each factor level which is called the main effect. If the difference in response between the levels of one factor is not the same at all levels of the other factors that is because we are in the presence of an interaction between the factors. Factorial designs allow finding interactions by crossing two (or more) independent variables with one another, in a way that there are observations for each combination of levels of the two independent variables.

When in a study there are k factors to be studied at two levels, the number of experiments is given by 2^k. To reduce costs and time, a fractional factorial design can be used which is a fraction of a full factorial design. By providing a very reduced number of experiments, this kind of design is very useful when starting a new research project, with a high degree of uncertainty regarding which factors are likely to affect the results (factor screening) reason why they are widely used when developing a new product (Ferreira et al. 2019). If only two factors are considered, A and B, each run at two levels (e.g. high and low), the design is called a "2^2 factorial design". In a two-level factorial design, we can define the average effect of a factor as the change in the response produced by a change in the level of that factor averaged over the levels of the other factor. Also, the symbols (3.1), a, b, and ab represent the total of the response observation at all n replicates taken at the treatment combination, as exemplified in Fig. 3.1

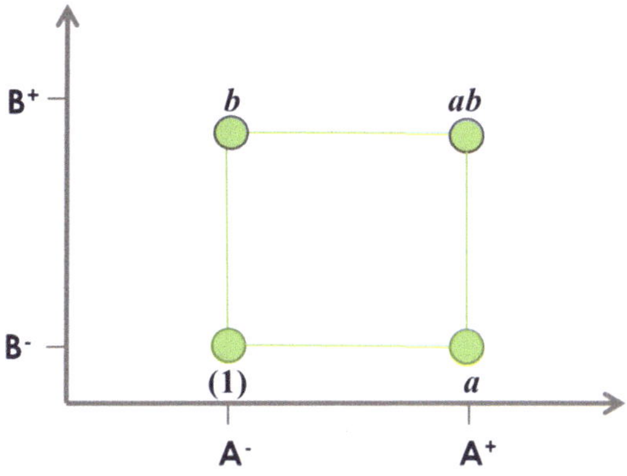

Fig. 3.1 The 2^2 factorial design

Now the effect of A at the low level of B is [a – (1)]/n, and the effect of A at the high level of B is [ab – b]/n. Averaging these two quantities yields the main effect of A, as shown in Eq. 3.1.

$$A = \frac{1}{2n}\{[a - (1) + (ab - b)]\} \tag{3.1}$$

As for the main effect of B, the same reasoning is followed, the average of the effect of B at the low level of A is [b – (1)]/n, and at the high level of A, [ab – a]/n is represented by Eq. 3.2.

$$B = \frac{1}{2n}\{[b - (1) + (ab - a)]\} \tag{3.2}$$

A factorial design allows the detection of interactions between factors when they exist. When the two factors A and B do not interact with each other to affect the dependent variable (the effect), the interaction effect AB is by definition the average of the difference between the effect of A at the high level of B and the effect of A at the low level of B (Montgomery 2017). This interaction AB can also be defined as the difference between the effect of B at the high level of A and the effect of B at the low level of A, leading also to Eq. 3.3

$$AB = \frac{1}{2n}\{[ab - b] - (a - (1))]\} \tag{3.3}$$

If the response is plotted against factor A for both levels of factor B and there is no interaction between B and A, the B^+ and B^- lines are parallel (Fig. 3.2a) but when there is an interaction, the B lines will cross each other (Fig. 3.2b)

Response Surface Methodology

As defined by Montgomery (2017), response surface methodology, or RSM, includes several mathematical and statistical techniques that help in the modeling and analysis of problems that involve a response of interest (Y) that is dependent on

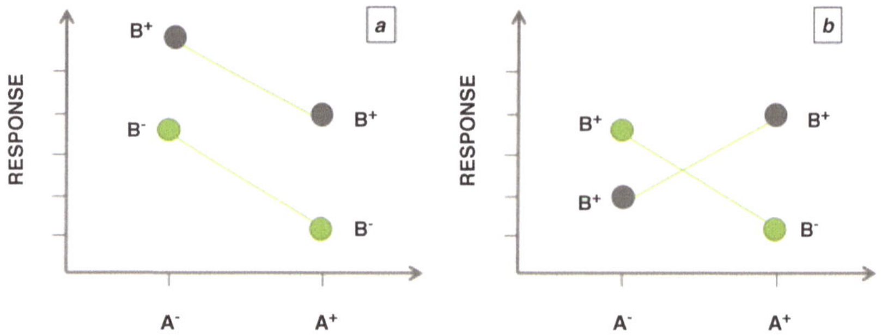

Fig. 3.2 Two-factor factorial experiment without interaction (**a**) and with interaction (**b**)

several independent variables ($x_1, x_2, \ldots, x_n$) and where the final objective is to find its optimum value. Response surfaces are usually represented graphically, either a surface or as a contour of the response surface, which are lines of constant response drawn in the x1, x2 plane, each corresponding to a certain height of the response surface. The function that usually defines a response surface is usually unknown and to find a relationship, usually a polynomial, that is the most closer to the true one between y and the set of independent variables, a low-order is used. And if a curvature is present on the surface then a higher order of the polynomial function is required.

Mixture Design

Mixture design is a good tool when dealing with food formulation studies (Squeo et al. 2021). Any quality parameter evaluated (response) will depend not only on the total properties of the food in evaluation but also on its proportions (Myers and Montgomery 1995). This technique, combined with sensory analysis, presents a relevant tool to design new food products which together with response surface methodology, have been used in the optimization of food formulations, such as fish patties (Cornell 1990) or fruit juice blends (Huor et al. 1980; Vieira and Silva 2004; Branger et al. 1999). The use of design of experiments allows the exploration of various operation conditions to optimize raw materials, optimize processes, or production cost reduction (Hubbard 1990). In mixture designs, the factors are the components of a mixture and therefore they cannot be independent. Considering n components on a mixture and respective fractions, x_n, then (Eq. 3.4)

$$0 < x_i < 1, \text{with } i = 0 \text{ to } n \text{ and } x_1 + x_2 + \ldots + x_n = 1 \tag{3.4}$$

In Fig. 3.3a, it can be seen the graphic representation of a mixture with only two components. In this case, a line segment defined by $x_1 + x_2 = 1$ will include all combinations of these two components. In Fig. 3.3b a mixture of three components is represented as a triangle surface, where the vertices represent mixtures that are composed of 100% of a single component. When this does not happen, but the constrained region forms a simplex triangle due to the components being constrained

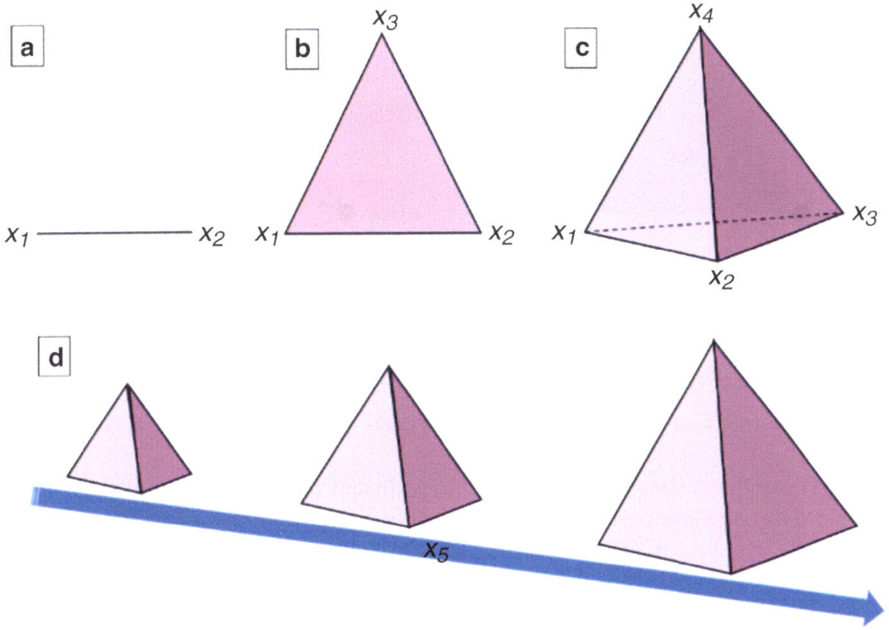

Fig. 3.3 Mixture designs with: (**a**) 2 components, (**b**) 3 components, (**c**) 4 components, and (**d**) 5 components.

to equal ranges, this geometry also occurs (Anderson and Whitcomb 1999). Figure 3.3c represents a 4-component mixture and in Fig. 3.3d, a 5-component design is also presented. These designs, combined with sensory analysis, presents a relevant tool to design new food products which together with RSM, have been used in the optimization of food formulations, such as beef patties (Los et al. 2020) or fruit juice blends (Lawless et al. 2013; Branger et al. 1999; Vieira and Silva 2004) or even the best mixture of antimicrobial components on an active packaging film (Rubilar et al. 2013).

1. Mixture design selection and evaluation

To select an ideal number of experiments it is nowadays feasible to apply matrix-based algorithms due to the availability of user-friendly software such as DESIGN Expert from STATEASE, JMP from SAS, or XLSTAT from AddinSoft. One of the most popular is the D-optimal design. To apply this design, a general procedure has to be followed (Anderson and Whitcomb 1999). The order of the polynomial, needed to obtain a prediction model for the response to be studied, is first chosen. It is often assumed that the responses will fit a second or third-order polynomial of Scheffé (Eqs. 3.5 and 3.6) (a special cubic polynomial), to explain the behavior of the responses.

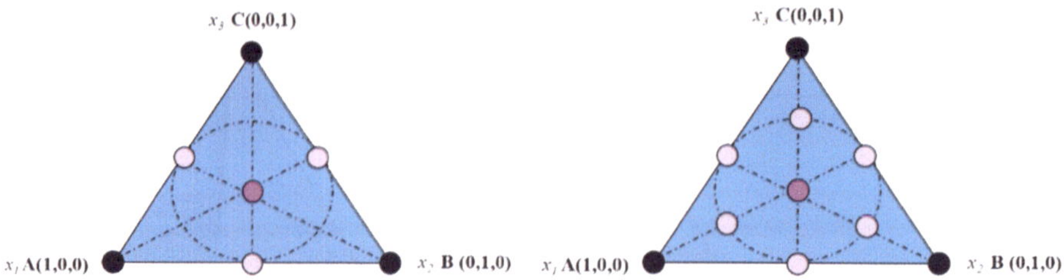

Fig. 3.4 The simplex centroid and the axial simplex design

$$Y_i = \beta_1 x_1 + \beta_2 x_2 + \beta_3 x_3 + \beta_{12} x_1 x_2 + \beta_{23} x_2 x_3 + \beta_{13} x_1 x_3 + \varepsilon \quad (3.5)$$

$$Y_i = \beta_1 x_1 + \beta_2 x_2 + \beta_3 x_3 + \beta_{12} x_1 x_2 + \beta_{23} x_2 x_3 + \beta_{13} x_1 x_3 + \beta_{123} x_1 x_2 x_3 + \varepsilon \quad (3.6)$$

Where Y_i is the response variable of the evaluated parameter i (quality attribute evaluated through sensory analysis), $x_{1,}$ x_2 and x_3 are three components' proportions and β_1, β_2, β_3, β_{12}, β_{23}, β_{13} and β_{123} are the estimated coefficients (Myers and Montgomery 1995) and ε is the associated pure error.

A minimum number of points, needed to fit the model, should be selected. Several designs can be chosen, the simplex centroid design the axial simplex are the most common (Fig. 3.4).

Blocking may be needed to avoid any interference due to variations from a set of experiments to another such as to reduce the possible effects of within-day or day-to-day due to differences in weather or any events extraneous to the experiments being carried out (Draper et al. 1993). Next, cycles of design evaluation-design modification should be performed to increase the number of degrees of freedom to evaluate pure error, maximize the matrix of the coefficients' determinant and minimize the trace of the inverse of the same matrix, which can be visualized by a contour plot of the standard error that is minimized in the region of interest, the central region, as well as the distribution of the experimental points on the design as it can be observed in Fig. 3.14 on the case study presented in this chapter.

2. Optimization of mixtures

In an optimization problem, several factors are involved:

1. an objective function, which in product development is usually identified as the maximization of the average mark given in the sensory analysis of a particular response.
2. decision variables, adjustable and with an independent nature. These are variables that affect the value of the objective function. In product development, they are specific to the product to be developed, e.g. temperature of the oven or griding time, etc).
3. a set of constraints allowing the decision variables to take certain values, but not others, eg. the cost of raw materials

4. a mathematical model that allows the prediction of the objective function by constraining the decision variables, and
5. the optimization technique to be chosen

The desirability function d (Eq. 3.7) (Derringer and Suich 1980; Anderson and Witcomb 1996) is the only objective function developed in the optimization of multiple responses, by combining the desirability ranges, d_i, of the responses studied.

$$d_i = \left[\frac{Y_i - Low_i}{High_i - Low_i}\right] \tag{3.7}$$

3.1.3 Problem to Solve and Case Study

On the scope of the Food Product and Processes Development course, in the Bachelor's degree of Food Engineering of the University of Algarve, the students were asked to propose an innovative product to be developed by them throughout the semester. One group of students was challenged with the following requirements: the product should be innovative and bring some benefit to the very actual problem: child obesity. They proposed to develop a new kind of hamburger that besides meat should also include fish and vegetables. The choice of this product was based on the fact that fish and vegetables are less appreciated by children and it could be a way to introduce these two foods on their diet in a form of a hamburger that would still contain meat (though in a lower percentage) to keep its taste but much less intense. A market survey was carried out indicating that the future consumer, preferred spinach or leek as a vegetable. The fish chosen was hake, based on the fact that it is a low cost and easy to manipulate fish that gives higher yields when deboned and skinned. As for the kind of meat, pork and chicken were chosen by the market survey. Which ingredients or process conditions would be powerful enough to influence and, in which way, the panellists' preference and which quantities of ingredients or process conditions were the best in the new product to be developed.

3.1.4 Procedure

The development of this new product involved four main steps in three different laboratories (Fig. 3.5):

1. Design of the experiments in the computer laboratory. This step is very important once well-designed experiments allow the exploration of various operation

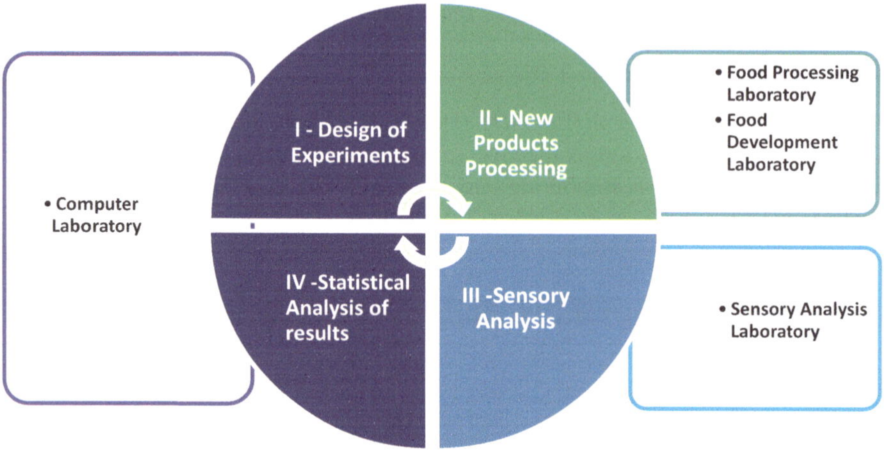

Fig. 3.5 Scheme of steps and laboratories involvement in product development

conditions to optimize raw materials, optimize processes, or production cost reduction (Hubbard 1990).

2. Processing the several products in the food processing laboratory according to the different formulations imposed by the experimental design.
3. Sensory analysis of the several products carried out in the sensory laboratory.
4. Statistical evaluation of the results, (average of the marks given by the panelists for each response and formulation obtained in the sensory analysis).

Development of a Prototype

To start the development of the new product, a prototype was first created, based on a thorough literature review carried on a meat hamburger recipe. The recipe that could better adapt to the students' ideas, was chosen. Then, a part of the meat in its composition was replaced by chicken and hake. Next, a descriptive sensory analysis was carried out within the team (composed of four people) to obtain the best sensory descriptors for this kind of product (Meilgaard et al. 1999). The best descriptors were: color, aggregation, juiciness, and taste.

Study on the Effect of Some Ingredients/Processes Steps on the Panelist's Opinion

To assess which ingredients and its effect on the product's formulation such as kind of meat and vegetable, percentage of hake and mincing time that should be used in the recipe of the triburger, an incomplete (1/2) factorial design 2^4 factorial design was created, being the factors: A, grinding degree, B, quantity of Hake (%), C, kind of meat and D, kind of vegetable, as presented in Table 6.1. As responses, Y_1—Color; Y_2—Aggregation; Y_3—Juiciness, and Y_4—Taste, were evaluated by a sensory panel to which a questionnaire was given (Fig. 3.6) to fill in after tasting each sample.

Questionnaire

Name: ______________________ **Date:** ________ **Hour:** _______

Type of sample: fish, meat and vegetables

Studied Characteristics: Color, Aggregation, Juiciness and Taste.

Instructions

Please write the samples' codes on the questionnaire when evaluating them

Between sample tasting, please wait 30 s and wash your mouth with water;

Evaluate visually color and aggregation in the hamburger

Evaluate sample juiciness relatively to the humidity mouth feel;

Evaluate all parameters by giving marks according to the following table:

1	2	3	4	5
Bad	Not good	Aceptable	Good	Very good

PARAMETERS	Sample #			
Color				
Aggregation				
Juiciness				
Taste				

Thank you!

Fig. 3.6 Questionnaire for factorial design 2^4 sensory analysis

To evaluate the different products, a sensory panel was needed. Initially, 20 people from both sexes with ages ranging from 20 to 46, not trained, were invited to take part on the sensory panel, but due to the non-availability of some of them or uneasiness to perform this task, a selection of the panelist was carried out to avoid erratic results. The data obtained on the sensory analysis, (panelists' marks) were always averaged. These averages were then used as the correspondent response value for each recipe, Y1, Y2, Y3, and Y4, and inserted in the factorial design (software DX13) to perform the statistical analysis (Table 3.1).

To visualize better the most significant effects of each factor on the panellists' opinion, semi-normal probability plots were produced where the absolute value of the effect was shown. In this graph, the effects were selected and interactions were determined by lining up the straight line with the factors that are closer to zero. The ones that are more distant from the origin situated on the right side of the straight line are the factors that most contribute to influencing the panelist's opinion. Some factors may not effect on their own, (e.g. A or B) but they have to be considered if taking part in an interaction (ex: AB). From Fig. 3.7 and Table 3.2, it was concluded that the factors that had an effect on the panelists' opinion about the color of the

Table 3.1 Incomplete ½ Factorial design to evaluate the effect of 4 factors on the panelist's opinion.

	A	B	C	D	Y_1	Y_2	Y_3	Y_4
Run	Texture (grinding degree)	Hake (%)	Meat	Vegetable	Color	Aggregation	Juiciness	Taste
1	5	15	Pork	Spinach	3.6	3.6	3.6	3.9
2	5	15	Chicken	Leek	2.4	3.6	3.3	2.8
3	40	25	Chicken	Leek	2.7	3.6	3.2	2.8
4	40	25	Pork	Spinach	3.4	3.8	3.7	3.7
5	5	25	Pork	Leek	3.6	3.7	3.6	3.8
6	5	25	Chicken	Spinach	3.5	3.6	3.9	3.5
7	40	15	Pork	Leek	3.3	3.5	3.3	3.1
8	40	15	Chicken	Spinach	3.9	3.7	3.5	3.5

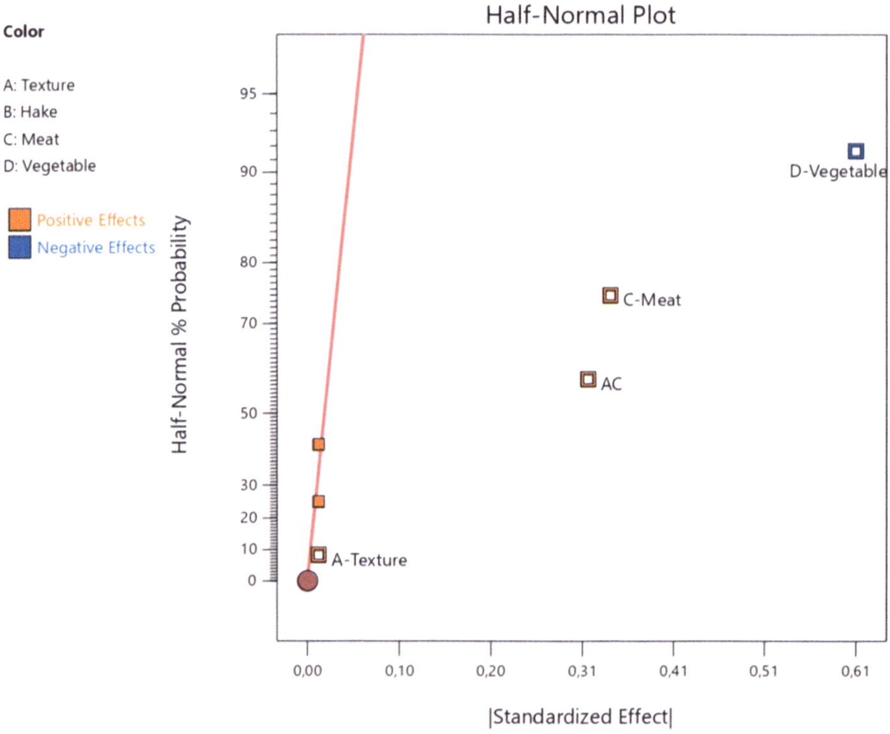

Fig. 3.7 Semi-normal plot of the effects of the factors on the panelists' opinion regarding taste

Table 3.2 Effect list showing the contribution of the factors on the panelists' opinion regarding taste

	Term	Effect	SumSqr	% Contribution
Require	Intercept			
Model	A-Texture	0.01	0.00	0.0
Model	B-Hake	0.01	0.00	0.0
Model	C-Meat	0.34	0.23	13.0
Model	D-Vegetable	−0.61	0.75	42.8
Model	AB	0.54	0.58	33.0
Model	AC	0.31	0.20	11.1
Error	AD	0.01	0.00	0.0

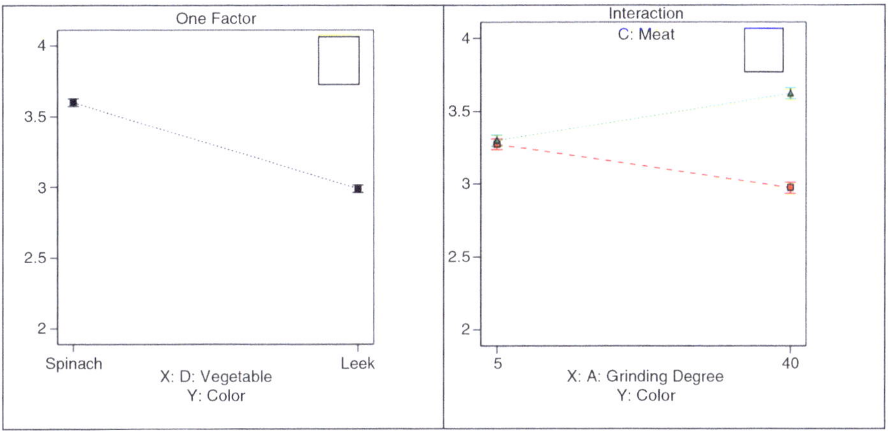

Fig. 3.8 (**a**) Effect of the kind of vegetable, on the opinion of the panelists about the hamburger's color (spinach is preferred). (**b**) Effect of the interaction of the factors Texture and Meat on the opinion of the panelists about the hamburger's color. (•) Chicken and (▲) Pork

triburger are factors C-Meat, D-Vegetable, and the interactions AB Texture/Hake and AC Texture/Meat.

In Fig. 3.8a, it is shown that spinach gives a better color to the burger than leek and from Fig. 3.8b, the interaction AC is presented and it can be concluded that when chicken is the meat used a less ground product is preferred but the most appreciated product is a more ground product with pork.

Regarding response Juiciness, by observing Fig. 3.9 and Table 3.3, it was concluded that the factors that affected the panelists' opinion factors were A-Grinding Degree, B-Hake, C-Meat, D-Vegetable, and the interactions AB Texture/Hake. In Fig. 3.10, it can be seen that the panelists chose spinach as the vegetable that would impart more juiciness to the triburger and that more Hake is preferred.

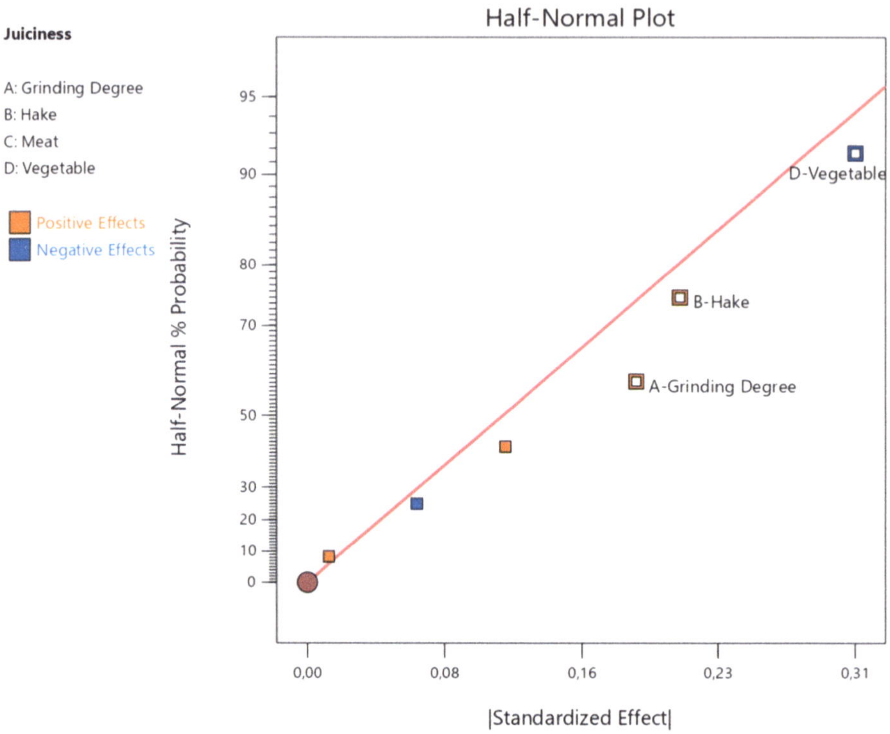

Fig. 3.9 Semi-normal plot of the effects list showing the contribution of the factors on the panelists' opinion regarding Juiciness

Table 3.3 Effect list showing the contribution of the factors on the panelists' opinion regarding Juiciness

	Term	Effect	SumSqr	% Contribution
Require	Intercept			
Model	A-Texture	0.19	0.07	16.5
Model	B-Hake	0.21	0.09	21.1
Model	C-Meat	0.11	0.03	5.91
Model	D-Vegetable	−0.31	0.20	45.7
Model	AB	0.14	0.04	8.9
Error	AC	−0.06	0.01	1.8
Error	AD	0.01	0.00	0.1

When analyzing response Taste, by observing Fig. 3.11 and Table 3.4 it was concluded that the factors that affected the panelists' opinion factors were C-Meat, D-Vegetable.

According to the sensory analysis performed the kind of meat chosen was pork probably due to the higher fat content which gives a better taste to the final product

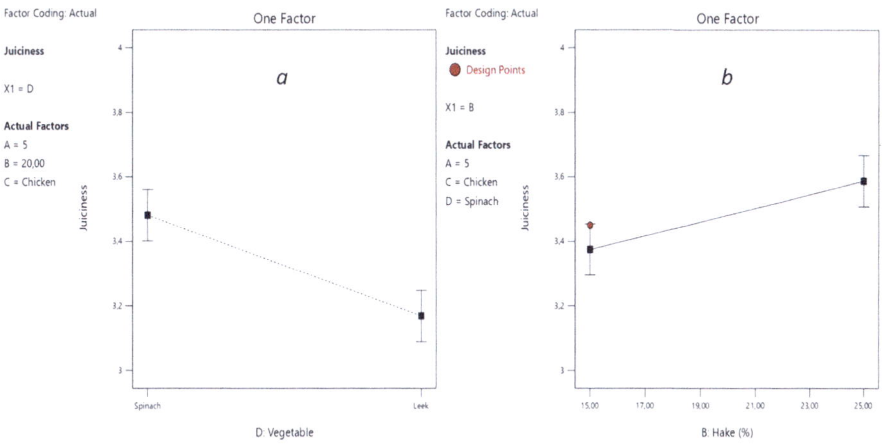

Fig. 3.10 (**a**) Effect of the vegetable, on the opinion of the panelists about the hamburger's juiciness (spinach is preferred). (**b**) Effect of the % of the Hake on the opinion of the panelists about the hamburger's juiciness (more hake is preferred)

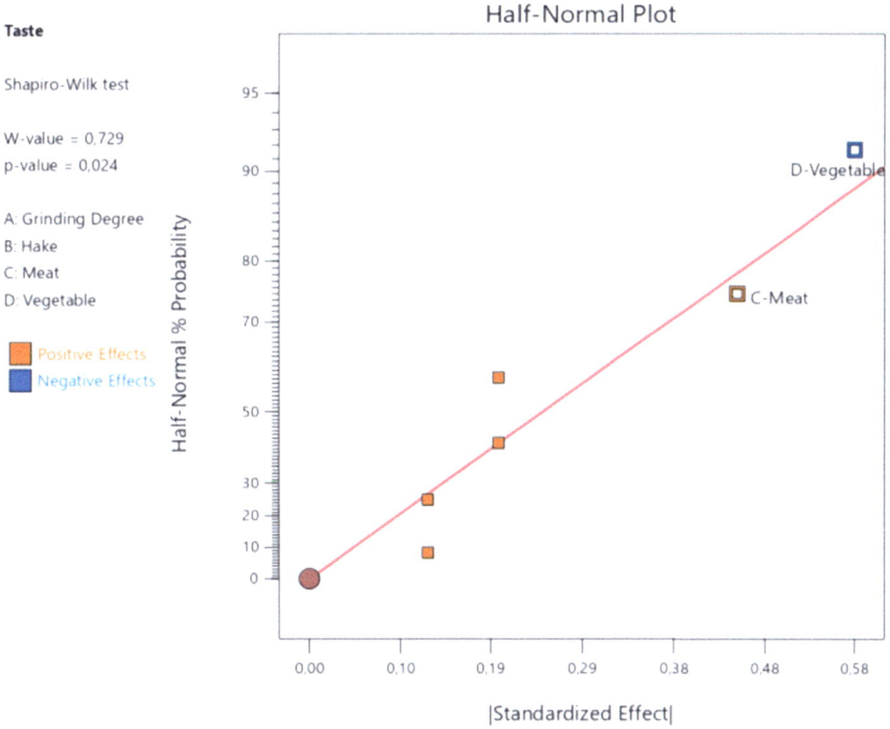

Fig. 3.11 Semi-normal plot of the effects list showing the contribution of the factors on the panelists' opinion regarding taste

Table 3.4 Effects list showing the contribution of the factors on the panelists' opinion regarding taste

	Term	Effect	SumSqr	% Contribution
Require	Intercept			
Error	A-Texture	0.20	0.08	6.2
Error	B-Hake	0.13	0.03	2.4
Model	C-Meat	0.45	0.41	31.4
Model	D-Vegetable	−0.58	0.66	51.3
Aliased	AB	Aliased		
Error	AC	0.20	0.08	6.2
Error	AD	0.13	0.03	2.4

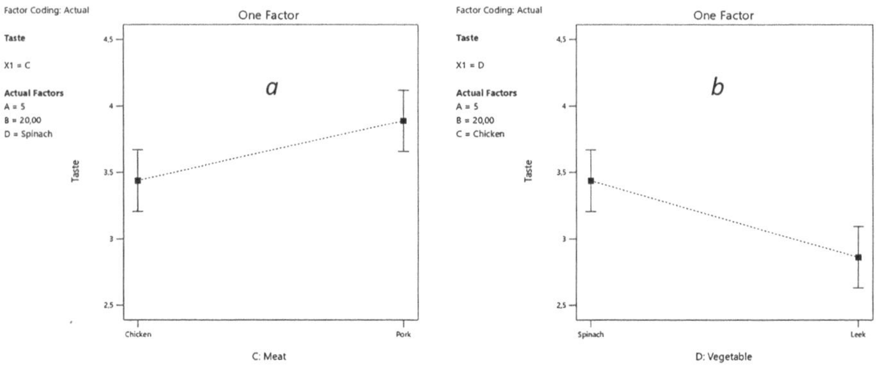

Fig. 3.12 (**a**) Effect of the kind of meat, on the opinion of the panelists about the hamburger's taste (pork is preferred). (**b**) Effect of the kind of vegetable on the opinion of the panelists about the hamburger's taste (spinach is preferred)

(Fig. 3.12a). The panelists preferred spinach to leek (Fig. 3.12b), which may be due to the intense taste of the latter, imparting an unpleasant taste to the final product.

Optimization of the Formulation

1. Design Selection

The software DX13 (DX13, 2021) was used. A third-order polynomial of Scheffé was chosen (Eq. 3.5). In Table 3.5, the proportion ranges for the three ingredients are shown, being by x_1, x_2, and x_3 are the three hamburger components' proportions and β_1, β_2, β_3, β_{12}, β_{23}, β_{13} and β_{123} are the estimated coefficients (Myers and Montgomery 1995). The quality attributes or response variables, Y_i, were evaluated through sensory analysis.

A simplex centroid design, augmented with axial blends, was initially chosen with a candidate set of 10 points: 3 vertices of the pre-defined region, 3 axial points, 3 centers of edges, and 1 overall centroid. As the response variables would be

Table 3.5 Quantity limits (%) of the ingredients used in the Mixture design

Ingredients	Limits (%)	
	Minimum	Maximum
Pork meat	40	60
Hake	20	40
Spinach	20	40

Table 3.6 Mixture design

Run #	Block	Hake (%)	Spinach (%)	Pork (%)	Juiciness	Taste
1	1	20	30	50	4.06	3.59
2	1	40	20	40	4	
3	1	30	20	50	3.88	3.82
4	1	30	30	40	3.88	3.12
5	1	20	40	40	3.88	3.29
6	2	26.7	26.7	46. 7		3.29
7	2	23.3	33.3	43.3	4.19	3.71
8	2	26.7	26.7	46. 7	3.94	3.53
9	2	26.7	26. 7	46. 7		3.53
10	2	33.3	23.3	43.3	4.31	3.82
11	3	20	20	60	3.44	3.29
12	3	23.3	23.3	53.3	3.63	3.53
13	3	33.3	23.3	43.3	3.81	3.41
14	3	23.3	23.3	53.3	3.69	3.41
15	3	23.3	33.3	43.3	3.94	3.41

evaluated by panelists, the design runs would be done 1/3 each day, and 3 blocks were selected, to avoid day-to-day variation. The midpoints between the centroid and the corners of the triangular region (axial check blends) were then added to the design, to obtain some more information, in case some unusual behavior occurred. The vertices, the centroid, and the center of edges were replicated, to estimate the pure error. The final design was obtained (Table 3.6 and Fig. 3.13).

A sensory panel with 20 elements was initially used on the mixture design but due to the absence of some elements this panel was reduced to 17 elements, 65% women and 35% men. From the previous experiment, the question of exactly which would be the best quantities of hake, spinach, and pork was raised. These three ingredients were selected to take part in a mixture design, and therefore in the hamburger formulation only this part was studied, the % of the other ingredients was kept constant in all formulations. The total quantity of their lower and higher percentages in the hamburger formulation were defined for the three of them, 15 formulations were tested, (DX13 2013), by varying the proportions of pork, hake, and spinach (Table 3.6).

2. **Model Evaluation**

To obtain the models, besides eliminating some panelists as referred before, the values that presented some discrepancy (outliers) were eliminated. The search for the

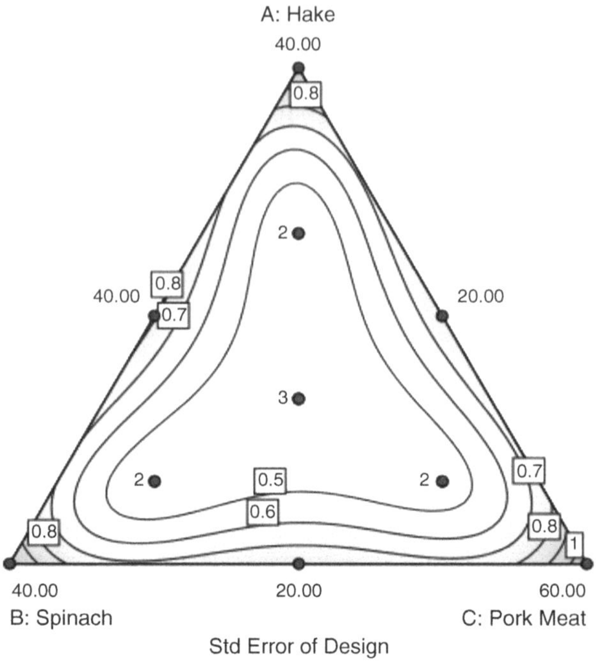

Fig. 3.13 Contour plot of the standard error of the mixture design

responses' prediction models that could best fit the experimental data, was then performed in the following way: for each response, an analysis of the sequential F-tests, lack of fit test results., $R^2_{adjusted}$ (only increases if necessary terms are added to the model) and the $R^2_{predicted}$ (which indicates how well is the model able to predict data), indicated whether the quadratic terms would contribute significantly ($p < 0.001$) to the model, as well as the special cubic term ($\beta_{123}x_1x_2x_3$). Models where the residual error significantly exceeded the pure error from replication (significant lack of fit) were rejected. Only Juiciness and Taste presented significant models as shown in Table 3.7.

As it can be observed in Fig. 3.14, where the surfaces and contour plots are shown for both juiciness (a) and taste (b) the red region represents the higher preference by the panelists and the light blue the lower preference, meaning that the hamburger was juicier at a lower percentage (%) of Hake. As for taste, the panelists revealed to prefer samples with a lower percentage (%) of spinach or a lower percentage (%) of hake. These antagonistic results present some difficulty to optimize a formulation or in other words to find a region in this triangular map that satisfies both requirements of having a good taste and being juicy.

3. **Recipe Optimization**

In search for the optimum formulation, the maximum and minimum (desirability range, $d_{i,}$ Eq. 3.7) for both responses, were imposed in terms of the sensory mark of

Table 3.7 Significant mathematical models

Response	Equation	Type of model	Model	Lack of fit	R^2	R^2_{adj}	R^2_{pred}
Juiciness	$-1.3773 \cdot A - 1.4783 \cdot B - 0.5970 \cdot C + 0.0871 \cdot A \cdot B + 0.0464 \cdot A \cdot C + 0.0487 \cdot B \cdot C - 2.19 \times 10^{-3} \cdot A \cdot B \cdot C$	Cubic vs. Quadratic	F = 8.41 Prob > F = 0.0293 Significant	F = 2.96 Prob > F = 0.3980 Non-significant	0.93	0.82	0.62
Taste	$-2.1341 \cdot A - 2.0308 \cdot B - 0.8789 \cdot C + 0.1236 \cdot A \cdot B + 0.0699 \cdot A \cdot C + 0.0675 \cdot B \cdot C - 2.13 \times 10^{-3} \cdot A \cdot B \cdot C$	Cubic vs. Quadratic	F = 7.41 Prob > F = 0.0220 Significant	F = 0.11 Prob > F = 0.9021 Non-Significant	0.90	0.78	0.64

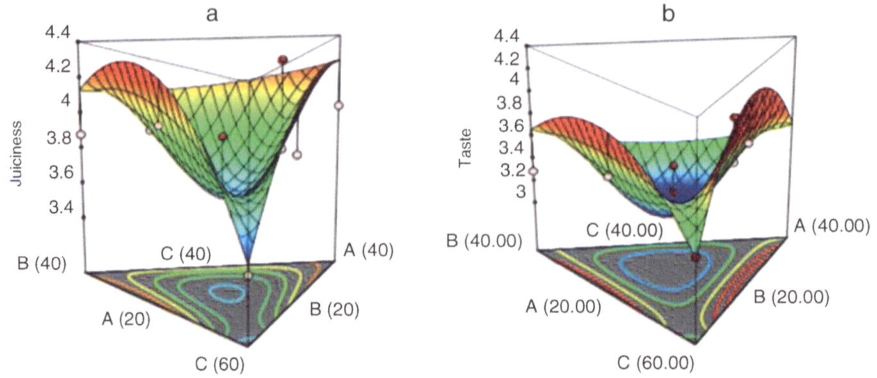

Fig. 3.14 Contour plot and response surface of hamburger preference in what concerns (**a**) juiciness and (**b**) taste

Table 3.8 Imposed limits for each studied response to obtain the optimum response optimization.

Response	Limits		Importance
	Mínimum	Maximum	
Juiciness	4	5	+++
Taste	3	5	+++++

the hedonic scale used in the sensory questionnaire (Fig. 3.6), through a transformation of the estimates Y_i, after goals were established for each response (Derringer and Suich 1980; Anderson and Whitcomb 1996). The desirability function D (Eq. 3.7) was obtained, by combining the desirability ranges, d_i, of the 2 responses (Table 3.8). The search algorithm was then applied to reach the optimum. This process also used the software DX13. The optimum composition was found using numerical optimization. After data interpretation, four optimum formulations were obtained, and the one with a higher desirability D = 0.435, was selected, in which the ingredients' proportions were the following (Figs. 3.15 and 3.16): 47.3% of Pork, 20.0% of Hake, 32.7% of Spinach.

4. **Optimum formulation validation**

To validate the optimum formulation a triangular test was then carried out. Besides Formulation A, which was the optimum found in this piece of work, another formulation (B) was chosen from the contour plot in an area that was not very appreciated by the panelists. To the panelists, 3 samples were then given, two equal and one different from these two formulations A and B, which are presented in Table 3.9.

Regarding the preference of the panelists for these formulations, the results are presented in Fig. 3.17. It can be stated that among the 15 panelists, 10 preferred formulation A, the optimum formulation obtained, and 3 preferred the other formulation chosen to validate the results.

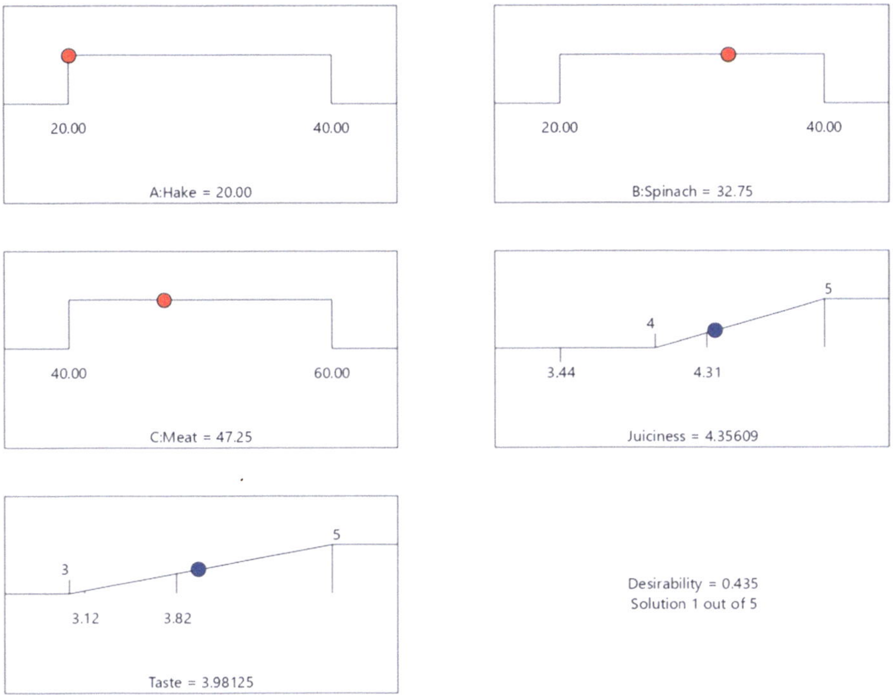

Fig. 3.15 Ramps showing the optimum found and the predicted values for the sensory parameters analyzed after imposing the restrictions

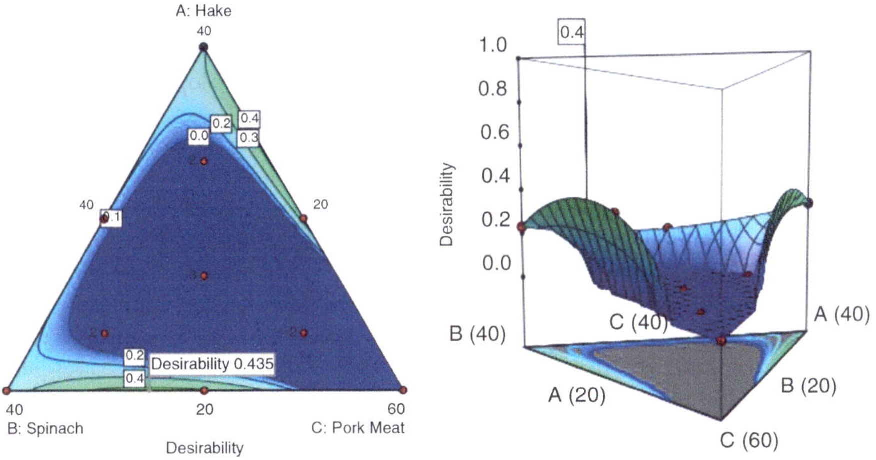

Fig. 3.16 Contour plot and Response Surface obtained for the optimization

Table 3.9 Proportion in %, of the ingredients in the (A) optimum formula and the (B) non-optimum formula

Ingredients	Formulations	
	A	B
Pork	47.3	45.2
Hake	20.0	27.2
Spinach	32.7	27.6

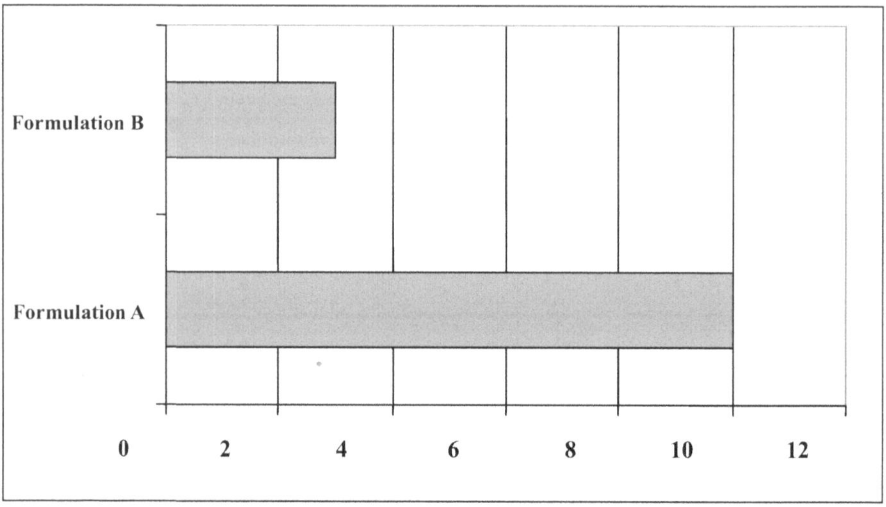

Fig. 3.17 Preference of formulations A (the optimum found statistically) and formulation B

3.1.5 Conclusions

A formulation of a hamburger, denominated triburger, was developed and optimized using statistical designs of experiments. First, a factorial design was used to screen the variables that most influenced the panelists and it was concluded that regarding color, spinach, pork, and to a lesser extent the grinding time contributed to a better evaluation of color. As for Aggregation, the panelists couldn't distinguish the samples indicating that the factors studied didn't influence this response. In what concerns Juiciness, hake at 25%, spinach, pork and to a lesser extent grinding time, imparted more juice to the product and for Taste, pork, meat, and spinach gave a better taste. With these conclusions, a mixture design was next used having as main objective to optimize the hamburger formulation. Only Juiciness and Taste presented significant models. After data interpretation, from four optimum formulations obtained, the one with a higher desirability D = 0.435, was selected, where the composition of meat was reduced to only 47.3% and included 20.0% of hake and 32.7% of spinach. This formulation was validated as when comparing it with a different formulation, the majority of the panelists were able to detect the differences and 77% preferred the optimum formulation.

3.2 Optimization of the Clarifying Process of Apple Juice Using Response Surface Design

Sócrates Quispe-Condori

3.2.1 Objectives and Learning Outcomes

The main goal is to present a sequential methodology for the analysis of second order models in Response Surface Design for process optimization studies. The learning outcomes are:

- To understand and use response surface design to solve problems in scientific research and industrial application.
- To present the usual procedures and tools of statistical analysis in order to obtain valid conclusions.

3.2.2 Introduction

In food processing research, response surface design is used to establish a functional relationship between responses or dependent variables (y) and process factors or independent variables (x_1, x_2, x_3 … a_n), through first, second, or higher order polynomials (Myers and Montgomery 2002). Since the actual physical or chemical relationship is not explained by these models, it is considered an empiric model or black box math model.

- First order linear model without factor interactions

$$y = \beta_0 + \sum_{i=1}^{k} \beta_i x_i + \varepsilon$$

- First order linear model with factor interactions

$$y = \beta_0 + \sum_{i=1}^{k} \beta_i x_i + \sum_{i=1}^{k-1} \sum_{j=i+1}^{k} \beta_{ij} x_i x_j + \varepsilon$$

- Second order or quadratic model

$$y = \beta_0 + \sum_{i=1}^{k} \beta_i x_i + \sum_{i=1}^{k} \beta_{ii} x_{ii}^2 + \sum_{i=1}^{k-1} \sum_{j=i+1}^{k} \beta_{ij} x_i x_j + \varepsilon$$

Where: ε is the error term,
β_i are the linear effects,
β_{ii} are the quadratic effects,
β_{ij} is the linear-by-linear interactions.

The coefficients of these mathematical models are calculated using the collected experimental data. The minimum number of experimental data to fit the model must be equal to the number of unknown parameters.

The linear model produces response surface which is a hyperplane, and the addition of interaction terms allows hyperplane deformations. On the other hand, the quadratic terms of the second order model capture the potential curvature in the relationship between factors and response variables. The response surface can show different patterns, such as maximum, minimum, saddle, or rising ridge, and consequently determines an optimum response location.

There are different experimental designs that can be used to calculate the number of experiments; however, the challenge is to choose an experimental design that represent properly a particular response surface.

Experimental Designs for Fitting Response Surfaces

An appropriated experimental design gives a reasonable distribution of data points in the interest regions and allows the investigation of the suitable estimated regression equation. Moreover, it must allow a sequential formulation of higher-order designs in order to guarantee simplicity in the estimation of model parameters.

A response surface design is a subgroup of design of experiments (DOE) that allows to better understand the relationships between factors and responses, as well to optimize the responses. In general, a response surface design must be selected if the response surface of product, process, or system have second order polynomial behavior. Response surface design must have at least three levels for each independent variable (inferior, center, and superior level), and it is augmented with a group of axial points (also called star points) that let estimates curvature. However, compared to full factorial design, response surface design requires fewer number of experiments, with a consequent lower cost for researching. The most known response surface designs are: Central Composite Design, Box-Behnken Design, and Uniform Shell Design (or Doehlert Design).

1. **Central Composite Design (CCD)** (Box and Wilson 1951)

A Box-Wilson Central Composite Design, commonly called Central Composite Design is constituted of:

- A set of points of the full factorial design 2^k or the fractional factorial design 2^{k-p}
- A group of axial or star points along the axis. Each factor is selected in its major or minor levels (to a distance $\pm\alpha$ from the center of the design) and the other factors in the central level or central point.
- A number of central points.

The total number of design points needed (N) is determined by the equation

$$N = 2^k + 2k + C_0$$

where k is the number of factors and C_0 is the number of center points

There are three variations of Central Composite Design (CCD). The type of CCD used (position of the factorial points and stars) is determined by the number of factors and by the desired properties of the design (orthogonality and rotatability). The CCD's geometrically design for two and three factors, indicating the values of the factorial design levels and the corresponding star points, are shown in Fig. 3.18.

- **CCD-Circumscribed**

This design requires 5 levels of each factor. The design consists of factorial points, which is the product of the intervals (−1, +1), star points along the axis, and out of the factorial space and central points in the origin. The precise value of α depends on the number of factors, according to the equation $\alpha = \pm 2^{k/4}$.

- **CCD-Inscribed**

For process in which the specified levels for the factors have restrictions, the star points can be used for the factor settings ±1 and creates a factorial or fractional factorial design within those limits. This new arrangement is called CCD-Inscribed. This design also requires 5 levels of each factor.

- **CCD-Face Centered**

This design has 3 levels for each factor, unlike the previous ones that have 5 levels for each factor. Thus, this design has the star points in the space faces of the design region, so $\alpha = \pm 1$.

2. **Box- Behnken Design** (Box and Behnken 1960)

The Box-Behnken Design is an independent quadratic design that does not contain a factorial design or fractional factorial. This design is rotatable (or almost rotatable) but have limited capacity for orthogonal blocking compared to Central Composite Design. Box-Behnken design uses only 3 levels for each factor. The combinations of the levels of the factors are in the central points of the axis of the design space and in the center. For example, the Box-Behnken Design for three factors involves three blocks, 2 factors in each one of them are varied through the 4 possible high and low combinations. It is essential to include central points, where all the factors are in their central values. The geometric representation of this design is shown in the Fig. 3.19. Thus, this design does not contain points in the vertex or extreme area of the

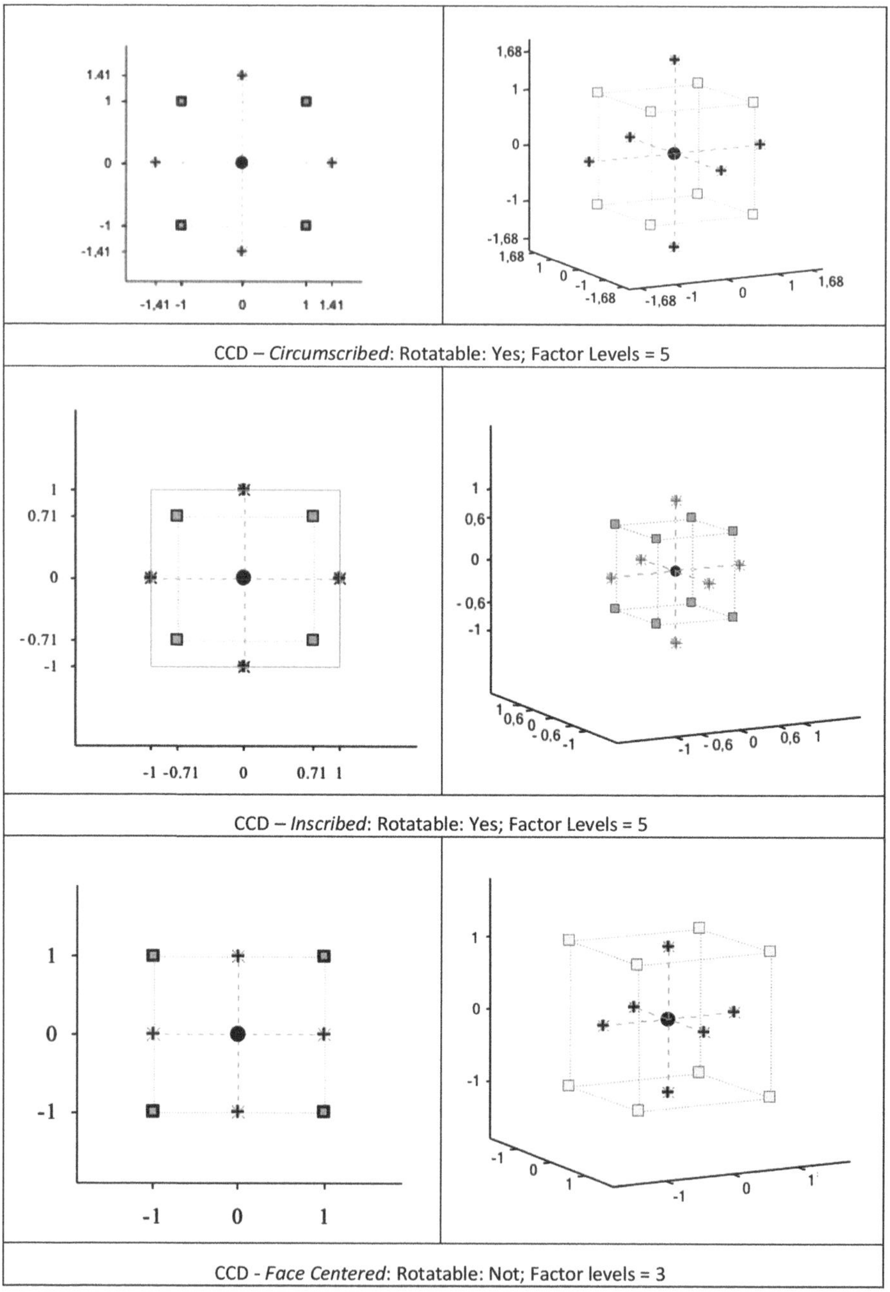

Fig. 3.18 Geometric representation of the Central Composite Design (CCD)

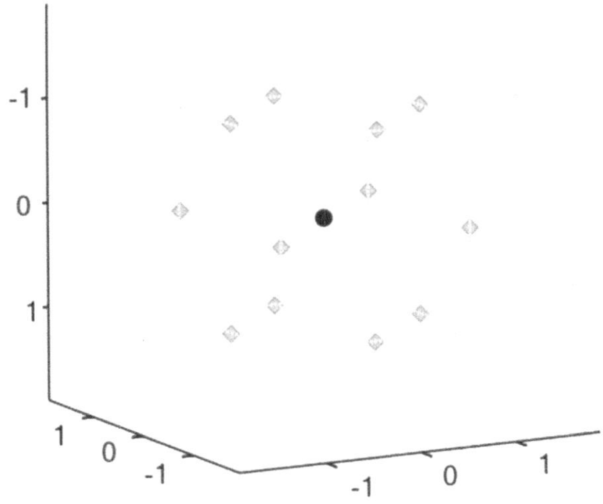

Fig. 3.19 A Box-Behnken Design for three factors

experimental region. The advantage is when the cube square points represent combinations of the levels of factors which are expensive or impossible to test because of physical process limitations. The number of experiments required (N) is defined by this equation:

$$N = 2k(k - 1) + C_0,$$

where k is the number of factors and C_0 is the number of center points.

3. **Uniform Shell Design (Doehlert Design)** (Doehlert 1970)

The Doehlert design describes a spherical experimental domain, and it attempts uniformity in space filling. Although this matrix is neither orthogonal nor rotatable, it does not significantly diverge from the required quality for effectiveness (Ferreira et al. 2004). The Doehlert design for two factors consists of one central point and six points forming a regular hexagon (Fig. 3.20a). For three factors, it can be viewed in different ways depending on the geometric structure selected (Fig. 3.20b). In Doehlert design the level numbers are not the same for all factors. In a two-factor Doehlert design, one factor is studied at five levels while the other is studied at only three levels. The number of experiments required (N) is given by the equation:

$$N = k^2 + k + C_0,$$

where k is the number of factors and C_0 is the number of center points

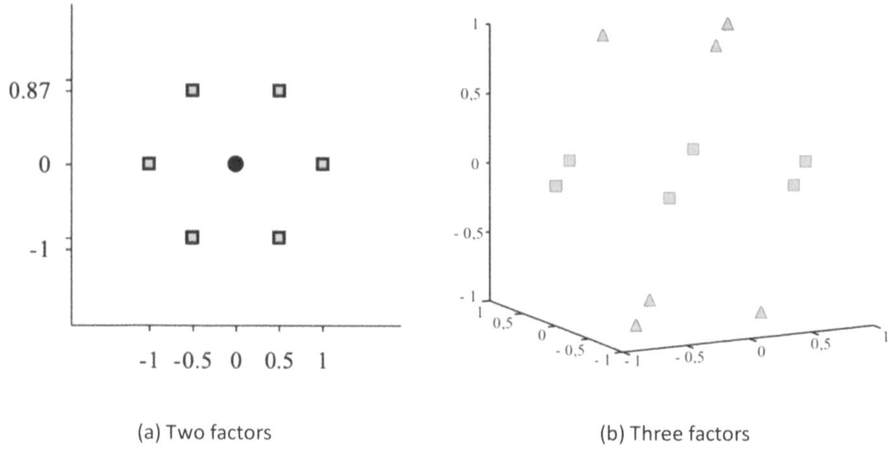

Fig. 3.20 Uniform shell designs for two and three factors.

3.2.3 *Problem to Solve*

An adequate process optimization demands the accomplishment of a systematic sequence of procedures, that includes formulation of experimental design (screening and planning), carrying out experiments, and statistical analysis of the results. The main goal will always be to calculate the best values of process parameters (levels of independent variables), in order to obtain the optimal results or responses (value of independent variable). It should be noted that in process optimization, a better result means having a maximum or minimum, depending on the type of process. For this chapter, the problem to be solved is related to the search of the best parameters values to decrease turbidity of clarified apple juice process (Pinto-Castro 1995).

3.2.4 *Selection and Coding of Variables*

The experimentation strategy for the appropriate application of the response surface design is shown in Fig. 3.21. After throwing many trivial factors, the experimental sequence should enter the breakthrough phase where interactions become evident. This requires two-level full factorial designs. If curvature is detected, the processes must be optimized. The step where response surface design come into play was indicated in Fig. 3.21.

For the selection and coding of variables, the researcher's expertise, experience, and knowledge about the studied process is very important. The step includes:

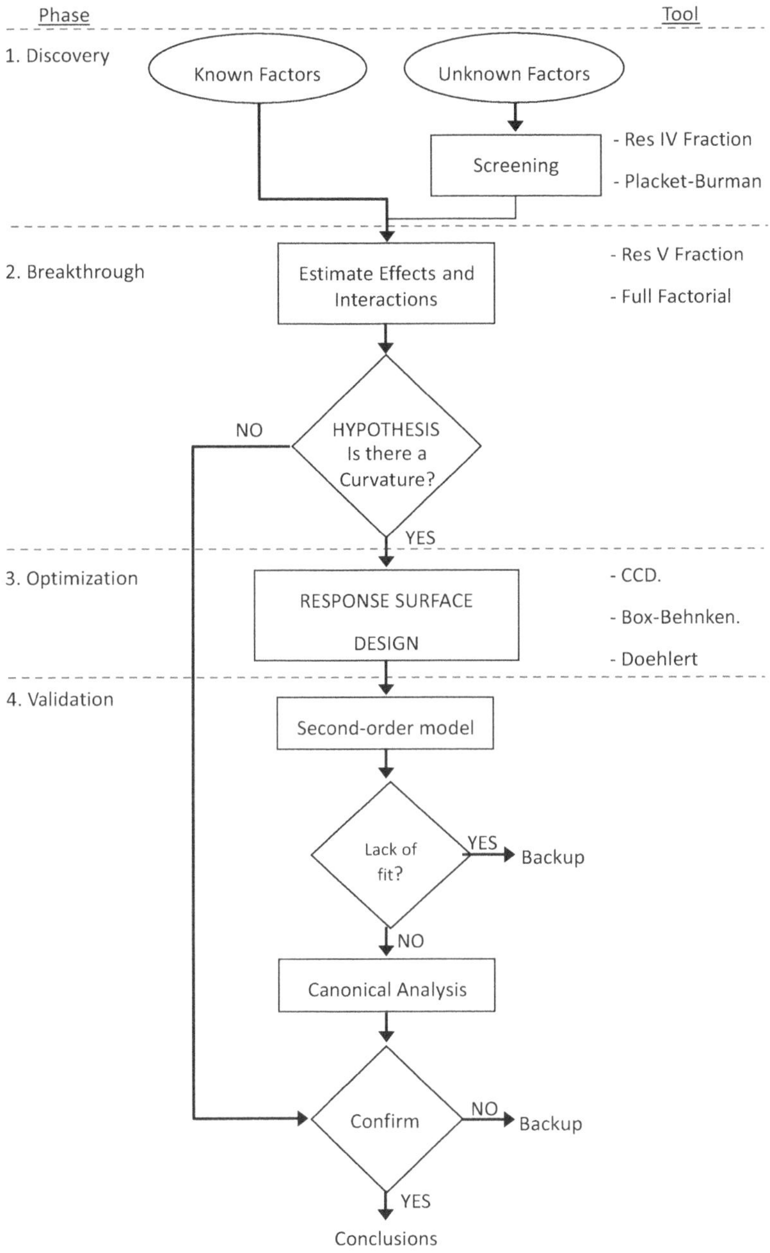

Fig. 3.21 Strategy for process optimization using response surface design. (Adapted from Anderson and Whitcomb 2005)

1. Determine the levels of the natural variables.

The definition of the factors, as well as their levels, depends on the nature of the experiment, and the region of interest to be modeled (Box and Draper 1987).

If the effect of the independent variables on a given response is unknown, the discovery and breakthrough stages will be followed (as presented in Fig. 3.21), to find variables that significantly affect the response. Fractional factorial, Placket-Burman and factorial designs can be used at these stages.

On the other hand, if the researcher knows the effects of independent variables on the response, it is not necessary to perform these first two stages. In this case, the researcher can define the independent variables that will be studied in the process optimization, as well as the upper and lower values of these variables (ξ)). The intervals of values of independent variables will define what is called region of interest.

2. Coding of natural variables.

Values of independent variables (ξ) must be transformed in a coded scale from -1 a $+1$ (the low and high levels, respectively). The coded variables allow a better comparison between them, and it is recommended that the codification between the levels should be equidistant. The codification of the variables is given by the following equation:

$$x_i = \frac{\xi_i - \overline{\xi}}{S/2}$$

Where ξ_i is the value of the independent variable $\overline{\xi}$ is the average of the values of independent variable and S is the difference between superior and inferior levels.

In the optimization step the following procedure could be followed (Fig. 3.21):

3.2.5 Selecting an Experimental Design

The most efficient design should be determined before planning the optimization process. The appropriate design selection will ensure that experimental data can represent efficiently the response surface. Some of the features that should be determined when selecting a response surface design are listed below (Montgomery 2013):

- Provides a reasonable distribution of data points throughout the region of interest,
- Allows model adequacy including lack of fit,
- Allows design of higher order to be built up sequentially,
- Provides an internal estimate of error,
- It does not require a large number of trials,
- It does not require too many independent variables.

Table 3.10 Comparison of efficiency of Central Composite Design (CCD), Box- Behnken Design (BBD) and Doehlert Design (DD) (Ferreira et al. 2004)

		Number of experiments (N)			Efficiency (p/N)		
Factors (k)	Number of coefficients (p)	CCD	BBD	DD	CCD	BBD	DD
2	6	9	–	7	0.67	–	0.86
3	10	15	13	13	0.67	0.77	0.77
4	15	25	25	21	0.60	0.60	0.71
5	21	43	41	31	0.49	0.61	0.68
6	28	77	61	43	0.39	0.46	0.65
7	36	143	85	57	0.25	0.42	0.63
8	45	273	113	73	0.16	0.40	0.62

Moreover, a response surface design could be selected by using its efficiency value (amount of coefficients/number of experiments). A comparison of the efficiency of central composite, Box–Behnken, and Doehlert design are presented in Table 3.10.

The surface response matrices can be formulated based on the number of factors and levels of each factor. Software programs (Design Expert, Statistica, Minitab, R, etc.) can be used to create experimental matrices. To apply a response surface design, the coded factor levels must be scaled to the actual values of the process variables (experimental factors). However, the most important consideration is that the design shows consistent performance in the following criteria:

- Perform the experiment in orthogonal blocks. The condition of orthogonality implies that the estimations of the parameters in the model are independent among them and, also, independent of the block effect.
- Rotate the design. A design is rotatable if the precision in the estimation of the surface is the same for all equidistant points from the central point, independently of the direction where it is found; it means, that the surface can be rotated in their central point and the precision of the estimation is the same.

3.2.6 *Carrying out the Experiment*

Although this step seems more straightforward, it is also essential crucial to follow an appropriate experimental plan. This must consider randomization, replication, and blocking. It may be beneficial to perform some preliminary experiments, such that the vital data is not affected during the execution of the experiments.

3.2.7 Fitting the Experimental Data to a Response Surface Model

Any data set in a multidimensional space that exhibits a trend, can be fitted by a mathematical model that represents the response surface. Often the mathematical model is selected through a fitting process, known as least-squares method. This fitting can be done also using statistical Software.

If the researcher does not have enough information about the model for the response surface, the fitting of the model to the experimental data is usually done using a first-order equation. When the first-order model shows a lack of fit for the experimental data, higher-order models, such as the quadratic model, must be analyzed. The advantage of higher-order models is that it incorporates terms related to curvature, which allows that "optimal" values can be identified for the studied region.

3.2.8 Analysis of Variance (ANOVA) And Test of Lack of Fit

Analysis of variance (ANOVA) is performed to assess the significance of the statistical model. Table 3.11 presents the Analysis of Variance with all the statistics used to analyze response surface design.

$$SS_{TO} = \sum_{i=1}^{N} y_i^2 - \frac{\left(\sum_{i=1}^{N} y_i\right)^2}{N}$$

$$SS_E = \sum_{i=1}^{m} \sum_{j=1}^{n_i} \left(y_{ij} - \widehat{y}_i\right)^2 = SS_{LOF} + SS_{PE}$$

$$SS_{LOF} = \sum_{i=1}^{m} n_i \left(\overline{y}_i - \widehat{y}_i\right)^2$$

Table 3.11 Analysis of variance table for response surface design

Source	d.f.	SS	MS = SS/d.f.	*F*-statistic
Regression	p	SS_R	MS_R	MS_R/MS_E
Error or residual	$N - p - 1$	$SS_E = SS_{LOF} + SS_{PE}$	MS_E	
Lack of fit	$m - p - 1$	SS_{LOF}	MS_{LOF}	MS_{LOF}/MS_{PE}
Pure error	$N - m$	SS_{PE}	MS_{PE}	
Total	$N - 1$	$SS_{TO} = SS_R + SS_{LOF} + SS_{PE}$		

Where: d.f. = Degrees of freedom, PE = Pure error, SS = Sum of squares, MS = Mean square, N = Number of observations, m = Number of unique experiments, p = Number of coefficients in the model, TO = Total, E = Error, R = Regression, LOF = Lack of fit

$$SS_{PE} = \sum_{i=1}^{m} \sum_{j=1}^{n_i} \left(y_{ij} - \bar{y}_i\right)^2$$

$$SS_R = SS_{TO} - SS_E$$

To test lack of fit it is necessary to compare $F_{\text{statistics}}$ (F_o) versus $F_{\text{critical value}}$ ($F_{\alpha,\text{m-p,N-m}}$):

- If $F_o < F_{\alpha,\text{m-p,N-m}}$ Evidence of no lack of fit. The model fits the experimental data.
- If $F_o > F_{\alpha,\text{m-p,N-m}}$ Evidence of lack of fit. It's necessary to choose another model.

3.2.9 Location Model

After confirming the validity of the second-order model in representing the experimental data, the Canonical polynomial is determined for the optimization of the surface through the exploration of the math optimum, called Stationary Point (Montgomery 2013). When writing the second order model in matrix notation, it gets:

$$\widehat{y} = \widehat{\beta}_0 + \mathrm{x^t b} + \mathrm{x^t B x}$$

Where

$$\mathbf{x} = \begin{bmatrix} x_1 \\ x_2 \\ \vdots \\ x_k \end{bmatrix} \quad \mathbf{b} = \begin{bmatrix} \widehat{\beta}_1 \\ \widehat{\beta}_2 \\ \vdots \\ \widehat{\beta}_k \end{bmatrix} \quad \text{and} \quad \mathbf{B} = \begin{bmatrix} \widehat{\beta}_{11} & \widehat{\beta}_{12}/2 & \dots & \widehat{\beta}_{1k}/2 \\ \widehat{\beta}_{12}/2 & \widehat{\beta}_{22} & & \widehat{\beta}_{2k}/2 \\ \vdots & \vdots & \ddots & \\ \widehat{\beta}_{1k}/2 & & & \widehat{\beta}_{kk} \end{bmatrix}$$

The derived $\widehat{y}$ with respect to the elements of the vector x equaled with 0 is:

$$\mathrm{b} + 2\,\mathrm{B\,x} = 0$$

The Stationary point will be:

$$\mathrm{x_s} = -\frac{1}{2}\,\mathrm{B}^{-1}\mathrm{b}$$

The predicted response in the stationary point can be found as:

$$\widehat{y}_s = \widehat{\beta}_0 + \frac{1}{2}\mathrm{x_s}^{\mathrm{t}}\mathrm{b}$$

3.2.10 Characterization of the Response Surface

After finding the stationary point, it is necessary to characterize the response surface in the proximity of this point, in order to determine if the stationary point is the point of a maximum, minimum or saddle response. There are two ways to characterize a response surface:

1. Indirect Analysis: Surface and contour plots

The simplest way to characterize a response surface is examining the tridimensional plot of the response with its corresponding bidimensional contour plot. Tridimensional plots represent the effects of multiple factors on the response, as well as the interactions of the factors, and their effect on the response.

2. Direct Analysis: Canonical Model

After obtaining the quadratic model which fit the response surface, it is convenient to do a canonical analysis of the resultant model in order to interpret in a simple way the geometry of the estimated surface in the experimentation region and to get information that can be useful in order to suggest possible theoretical explanations of the problem. The procedure consists of transforming the quadratic model in a new system of coordinates with the origin in the stationary point $\mathbf{x_s}$. This equation is known as the canonical form of the model:

$$\widehat{y} = \widehat{y}_s + \lambda_1 w_1^2 + \lambda_2 w_2^2 + \lambda_3 w_3^2 + \ldots + \lambda_k w_k^2$$

Where: w_i: Transformed independent variables, λ_i: Model constants.

The nature of the response surface is determined from the signs and magnitudes of λ_i. Thus:

- If $\lambda_i > 0 \rightarrow \mathbf{x_s}$ = Point of minimum response.
- If $\lambda_i < 0 \rightarrow \mathbf{x_s}$ = Point of maximum response.
- If λ_i has different signs $\rightarrow \mathbf{x_s}$ = Saddle point.

3.2.11 Case Study

In order to find the optimum level of turbidity of the clarified apple juice, Pinto-Castro (1995) studied the effect of variables that control the clarification process using a response surface design. From preliminary results using the Placket Burman Design, the researcher has been identified that the main parameters that affect the turbidity of the apple juice were Gelatin concentration (ξ_1) and Flocculation time (ξ_2).

1. **Selection and coding of variables**

For the parameters that affect the turbidity of apple juice, the lower and upper levels of the actual and coded values are shown in the Table 3.12.

2. **Selecting the experimental design**

Because of its rotatability and orthogonality, it was decided to use CCD – Circumscribed for the optimization of the clarification process of apple juice. So, the value of α for two variables ($k = 2$) is $\alpha = \pm 1.4142$. Using this coded value it's possible to calculate the new actual values of the factors. The units, the actual values and the coded levels of the independent variables are presented in Table 3.13.

The response variable analyzed was the turbidity of apple juice, measured in Nephelometric Turbidity Units (NTU). As mentioned above, CCD – Circumscribed in the experimental design is composed by 2^2 factorial points, four axial points and three replicates of the central point, as shown in the Table 3.14.

Table 3.12 Factor levels used for the clarification of apple juice

	Coded level (Natural values)	
Factors	−1 (lower value)	1 (upper value)
ξ_1: Gelatin concentration (g/Hl)	15	25
ξ_2: Flocculation time (min)	10	20

Table 3.13 Values of the coded and actual factor levels for the clarification of apple juice using a CCD - Circumscribed

	Coded level and actual values				
Factors	−1,4142	−1	0	+1	+1,4142
ξ_1: Gelatin concentration (g/Hl)	12.9	15	20	25	27
ξ_2: Flocculation time (min)	7.9	10	15	20	22

Table 3.14 Rotatable central composite design arrangement

	Coded level of variables (Actual value)			
	$x_1(\xi_1)$		$x_2(\xi_2)$	
1	−1	(15)	−1	(10)
2	−1	(15)	+1	(20)
3	+1	(25)	−1	(10)
4	+1	(25)	+1	(20)
5	−1.4142	(12.9)	0	(15)
6	+1.4142	(27)	0	(15)
7	0	(20)	−1.4142	(7.9)
8	0	(20)	+1.4142	(22)
9 (c)	0	(20)	0	(15)
10 (c)	0	(20)	0	(15)
11 (c)	0	(20)	0	(15)

For this design the following quadratic model will be used:

$$y = \beta_0 + \beta_1 x_1 + \beta_2 x_2 + \beta_{11} x_1^2 + \beta_{22} x_2^2 + \beta_{12} x_1 x_2 + \varepsilon$$

3. **Carrying out the experiment**

Table 3.15 shows the sequence on which the experiments were done, as well as the results of turbidity for the clarified apple juice.

4. **Fitting the experimental data to a response surface model:**

The independent and dependent variables were analyzed to obtain a regression equation for Turbidity (y). In Table 3.16 the values of the regression coefficients are presented, as well as the levels of significance (p-value). It was observed that the lineal coefficients for the gelatin concentration x_1 (p-value = 0,000520) and

Table 3.15 Arrangement of the experiments and Response

Run	Coded level of variables (Actual value)				
	x_1 (ξ_1)		x_2 (ξ_2)		Turbidity (NTU)
9 (c)	0	(20)	0	(15)	190
1	−1	(15)	−1	(10)	250
2	−1	(15)	+1	(20)	178
4	+1	(25)	+1	(20)	274
3	+1	(25)	−1	(10)	275
10 (c)	0	(20)	0	(15)	194
8	0	(20)	+1.4142	(22)	221
5	−1.4142	(12.9)	0	(15)	215
6	+1.4142	(27)	0	(15)	277
7	0	(20)	−1.4142	(7.9)	282

Table 3.16 Values of regression coefficients calculated for the Turbidity of apple juice

Independent variable	Regression Coefficient (β)	Standard Error	t-value	Significance level (p-value)
Constant	192.0001	5.114712	37.53879	0.000003
Linear				
x_1	26.0853	2.557368	10.20006	0.000520***
x_2	−19.9085	2.557368	− 7.78475	0.001468**
Quadratic				
x_1^2	25.8752	3.383106	7.64835	0.001570**
x_2^2	28.6252	3.383106	8.46123	0.001069**
Interaction				
$x_1 x_2$	17,7500	3,616648	4,90786	0,007999**

*** Significant at 0.001; ** Significant at 0.01

flocculation time x_2 (p-value = 0,001468) have a highly significant influence on the turbidity of the clarified apple juice. From the regression coefficients presented in Table 3.16, the equation that represents the response surface in function of the coded variables is:

$$y = 192,0001 + 26,0853x_1 - 19,9085x_2 - 25,8752{x_1}^2 - 28,6252{x_2}^2 + 17,75x_1x_2$$

Transforming of the coded variables into actual variables will yield the following function. This empirical formula allows predicting the turbidity response in regions that were not studied in the design.

$$y = 840,0142 - 46,83323\xi_1 - 52,53197\xi_2 + 1,03501\xi_1^2 + 1,14501\xi_2^2 + 0,71\xi_1\xi_2$$

5. **Analysis of Variance (ANOVA)**

In Table 3.17 the analysis of variance for the turbidity of the clarified apple juice was presented. For the statistical analysis of the regression model, it proceeded to decompose the term of total error in the pure error and the error of the lack of fit of the model (Table 3.18), which represents a test of goodness of fit for response surface models (Montgomery 2013). As observed in Table 3.18, the fitted mathematical model doesn't show a significant lack of fit ($F_o = 8.3868 < F_{0.05,4,1} = 224.560$.

Table 3.17 Analysis of variance (ANOVA) of independent variables for clarification of apple juice

Source	Sum of squares (SS)	Degree freedom (df)	Mean squares	*F*-statistic	*p*-value
x_1	5443.49	1	5443.491	104.0411	0.000520
x_2	3170.75	1	3170.746	60.6023	0.001468
${x_1}^2$	3060.61	1	3060.610	58.4973	0.001570
${x_2}^2$	3745.75	1	3745.755	71.5924	0.001069
$x_1\ x_2$	1260.25	1	1260.250	24.0871	0.007999
Error	209.28	4	52.321		
Total SS	14866.40	9			

Table 3.18 Analysis of variance (ANOVA) for the fitted quadratic polynomial model

Source	Sum of squares (SS)	Degree freedom (df)	Mean squares	*F*-statistic	*p*-value
x_1	5443.49	1	5443.491	680.4364	0.024393
x_2	3170.75	1	3170.746	396.3432	0.031951
${x_1}^2$	3060.61	1	3060.610	382.5763	0.032519
${x_2}^2$	3745.75	1	3745.755	468.2193	0.029400
$x_1\ x_2$	1260.25	1	1260.250	157.5313	0.050615
Lack of fit	201.28	3	67.094	8.3868	0.247342
Pure Error	8.00	1	8.000		
Total SS	14866.40	9			

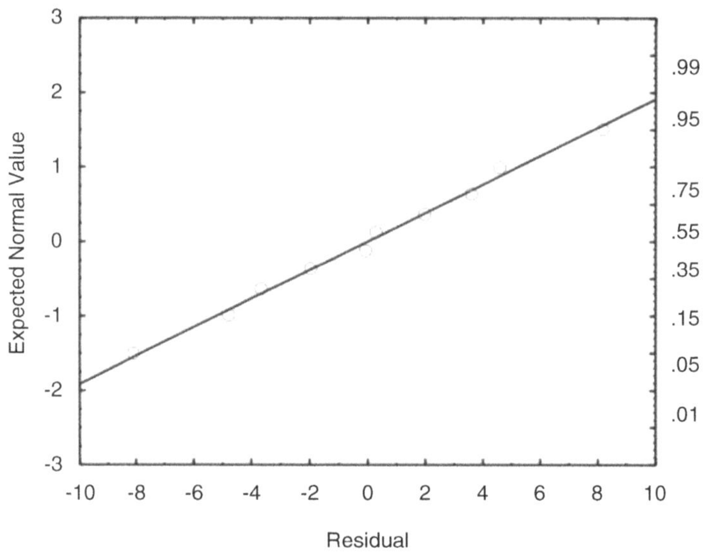

Fig. 3.22 Plotting of the expected normal values of residuals versus residuals

Also, the *p*-value for F-statistic is very small, so it's possible to conclude that the mathematical model fit adequately the experimental data in the surface response.

A plotting of the expected normal values versus residuals (Fig. 3.22) showed that data were very close to the straight line and situated at both sides of it. Then, the surface fit was fairly good.

6. **Location model**

Considering the values of the regression coefficients presented in Table 3.16, it was found the following matrices to determine the optimum response surface.

$$b = \begin{bmatrix} 26.0853 \\ -19.9085 \end{bmatrix}$$

$$B = \begin{bmatrix} 25.8752 & 8.875 \\ 8.875 & 28.6252 \end{bmatrix}$$

For the previous equation, the stationary point is:

$$x_s = -\frac{1}{2}\begin{bmatrix} 25.8752 & 8.875 \\ 8.875 & 28.6252 \end{bmatrix}^{-1}\begin{bmatrix} 26.0853 \\ -19.9085 \end{bmatrix}$$

$$\mathrm{X_S} = \begin{bmatrix} -0.697508 \\ 0.563999 \end{bmatrix}$$

Where $x_{1,\ s} =$ - 0.697508; $x_{2,\ s} = 0,\ 563999$

These values are the optimal parameters in order to reduce the turbidity of the clarified apple juice. When using the equation for the predicted response in the stationary point, it was found:

$$\widehat{y}_s = 192,0001 + \frac{1}{2}\begin{bmatrix} -0.697508 \\ 0.563999 \end{bmatrix}^t \cdot \begin{bmatrix} 26.0853 \\ -19.9085 \end{bmatrix}$$

$$\widehat{y}_s = 177,2886$$

In terms of the actual variables, the stationary point is:

$$\xi_{1,s} = 16.51 \times 10^{-2} \mathrm{g/L};\ \xi_{2,s} = 17.82\ \mathrm{min}$$

7. **Characterization of the response surface: indirect and direct analysis**

Figure 3.23 shows the tridimensional response surface of the significant interactions between the gelatin concentration and flocculation time. The response surface contour plots showing the gelatin concentration and flocculation time on turbidity are presented in Fig. 3.24. The response surface contour plot clearly showed that the turbidity of the clarified apple juice could be minimized at 16.51×10^{-2} g/L of gelatin concentration and 17.82 min of flocculation time.

For the direct analysis, it is necessary to express the fitted model in the canonic way. The values λ_1 y λ_2 are the roots in the equation of determinants:

$$|\mathrm{B}-\lambda \mathrm{I}| = 0$$

$$\begin{bmatrix} 25.8752 - \lambda & 8.875 \\ 8.875 & 28.6252 - \lambda \end{bmatrix} = 0$$

This is reduced to:

$$\lambda^2 - 54.49\lambda + 661.72 = 0$$

The roots of this quadratic equation are:

$$\lambda_1 = 36.22$$

$$\lambda_2 = 18.22$$

So, the canonical form of the adjusted model that characterized the response surface was:

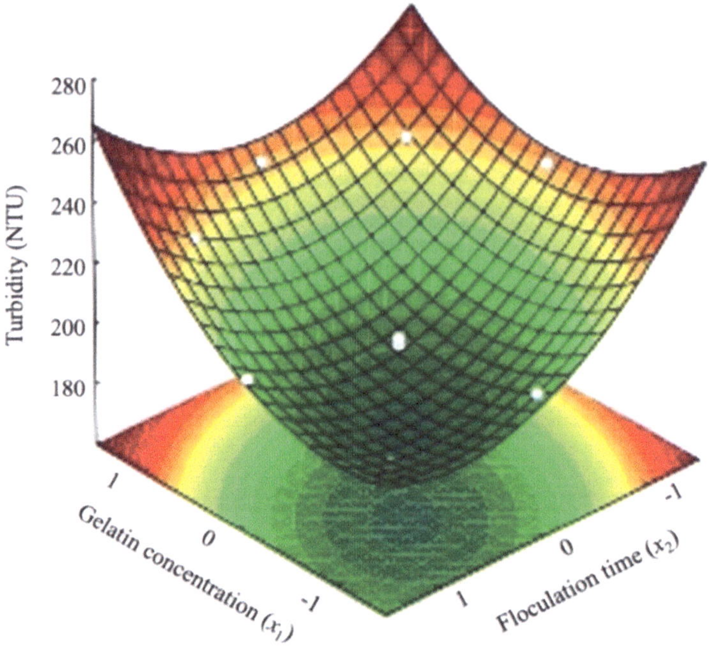

Fig. 3.23 Estimated response surfaces in the central composite design experiment obtained by plotting the Gelatin Concentration and Flocculation Time for Turbidity

Fig. 3.24 Contour plot for Turbidity (NTU) as a function of Gelatin concentration and Flocculation time

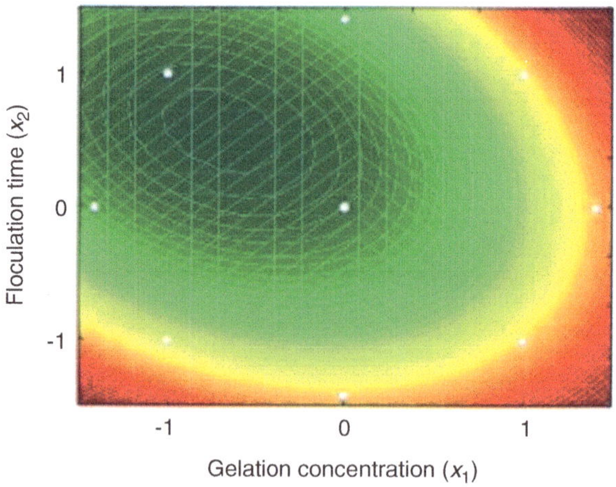

$$\widehat{y} = 177.2886 + 36.22w_1^2 + 18.26w_2^2$$

It is evident that the values λ_1 and λ_2, when both are positive, they indicate that the stationary point was a minimum point, as shown in Fig. 3.23.

3.3 Classification of Rice Amylose Group Using Discriminant Analysis

Prisana Suwannaporn

3.3.1 Objectives and Learning Outcomes

Discriminant Analysis was applied to determine set of independent variables which can significantly differentiate among two or more groups of study units. Moreover it can tell which variables are contributing more to the discrimination function. Specific set of variables that provides the most separation among the groups were defined and finally if the discrimination function is fitted, it can correctly reassign each study unit into its proper group by using Z score (Hair et al. 1995). Discriminant function is presented as followings:

$$\underset{\text{(non metric)}}{Y} = \underset{\text{(metric)}}{X_1 + X_2 + X_3 + \ldots\ldots + X_n} \tag{3.8}$$

$$Z = W_1X_1 + W_2X_2 + \ldots\ldots + W_nX_n \tag{3.9}$$

Where:Z= Discriminant Score, W_i= Discriminant weight,X_i= Independent variable

Group Centroid

The centroid indicates the most typical location of any individual from a particular group. The comparison of the group centroid shows how far apart the groups by testing the significance of the distance between the centroids. By computing the distribution of the discriminant scores of each group. If the overlap in the distribution is small, the discriminant function can separate group well. If the overlap in the distribution is large, the function is poorly discriminated between groups. Number of discriminant function equals to number of group – 1. In Fig. 3.25, we use only 1 discriminant function to separate 2 groups (A and B). Shaded area represent probabilities of misclassify unit from A and B. If A and B are highly overlapped (high similarity), then probabilities of misclassify unit A from B and vice versa is also high. The best discriminant function should separate group well with low misclassification (Fig. 3.25)

Cutting Score Determination

The cutting score is the criterion (score) against which each individual's discriminant score is judged to determine into which group the individual should be classified (Hair et al. 1995). Use t-test to determine the significance of classification accuracy. Statistic software will compute Discriminant Score (Z score) which followed

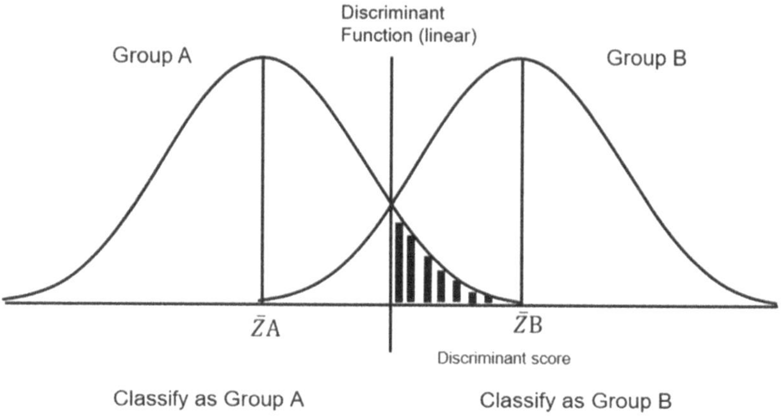

Fig. 3.25 Discriminant function, discriminant score (cutting score) of two groups (A and B) of equal sample size. (Source: Modified from Hair et al. 1995)

the statistical rule of maximizing the variance between the group and minimizing the variance within group. If the variance between the groups is large relative to the variance within group, then the discriminant function separates the group well. If equal size, optimal cutting score (Z) will be the halfway between two group centroids. Discriminant Score (Z score) can be calculated as followings;

$$Z_{cu} = \frac{N_B Z_A + N_A Z_B}{N_A + N_B} \tag{3.10}$$

Where: Z_{cu} = critical cutting score, Z_A = centroid for group A, Z_B = centroid for group B

3.3.2 *Introduction*

Primary objective of Discriminant Analysis is to identify the group to which the object (e.g. person, firm, and product) belongs. Discriminant analysis is a powerful tool uses for eliminate attributes that are poorly discriminated and retain only the best discriminators hence promote grouping prediction by use only few important attributes. Discriminant analysis involves in deriving linear combination of two (or more) independent variables which can discriminate best between groups. By setting the weight for each variable to maximize the between group variance relative to within group variance

Discriminant Analysis is sensitive to assumptions in normality of the independent variables, linear relationships, unequal covariance metric, multicollinearity among independent variables and small or unequal sample size. If each group size is unequal, randomly sampling should be done until same size within each group is obtained. Since larger group size has a disproportionately higher chance of being classified. Stepwise procedure can be applied in order to reduce multicollinearity problem.

3.3.3 Problem to Solve

Researcher seeks for a rapid and accurate method for detecting rice amylose group (low, medium and high amylose) for exporting purpose. Pasting properties variables (peak viscosity, trough, final viscosity, breakdown and setback) and sensory texture attributes (cohesion of bolus, adhesion to lips, hardness, cohesion of mass, roughness of mass, tooth pack and tooth pull) are measured in order to investigate whether they can be used as parameters to differentiated rice amylose group (Suwannaporn et al. 2007).

3.3.4 Procedure

Step 1. Define Dependent and Independent Variables

Dependent variables should be dichotomous or multichotomous. In this study, amylose group is dependent variable (non-metric variable). There are three groups; low, medium and high. Result in Table 3.19 shows that discrimination function can separate three groups well enough. However, if poor discrimination result is obtained then "polar extreme approach" that are low and high amylose group by excluding the middle group, will be more appropriated for better group separation. Independent variables are those parameters of pasting properties and sensory texture attributes mentioned above (metric variables).

If sample size is large, it can divide into two subsamples, one for estimate the discriminant function another for validation purpose (such as 60–40 or 75–25 split). But if sample size is small (as always encountering in scientific study), same set of samples can be used both for analysis and validation. By applying the "leave one out" principle to estimate k − 1 samples and eliminating one observation at a time.

Step 2. Model and Assess Overall Fit

There are two discriminant analysis methods (1) Simultaneous estimation or Direct method, by compute discrimination function with all independent variables are considered altogether (2) Stepwise estimation, by entering independent variables one at a time on the basis of their discriminating power. Use the latter method when there is relatively large number of independent variables. In this study, Stepwise estimation is applied since there are many independent variables with small sample size. So we try to extracted only variables that most effect on amylose content of rice.

Table 3.19 Discriminant coefficient (or discriminant function) of three amylose groups rice using paste viscosity and sensory texture attributes

Pasting property	Discriminant coefficients		
	Low amylose	Medium amylose	High amylose
Peak viscosity	0.332	−0.125	−1.236
Trough	−0.118	0.756	3.037
Constant	−24.963	−25.626	−85.713
Correctly predict (%)	66.7	100.0	100.0

Significance of the discriminant function (from Wilk's Lambda) is 0.000
Source: Suwannaporn et al. (2007)

There are three groups so two discriminant functions are obtained (number of groups − 1). The first function which accounts for 99.7% of variance is shown in Table 3.19. The first function consists of only two variables, peak viscosity and trough. Only two variables are enough to correctly predict low, medium and high amylose content rice at 66.7%, 100% and 100% respectively. As a consequence, the second discriminant function which accounts for only 0.3% of variance is negligible.

Table 3.19 shows "Discriminant weights or Discriminant coefficient" which are sign and weight represent the relative contribution of its associated variable to the function. The larger weight the higher discriminating power. In low amylose rice, "peak viscosity" is the good discriminator. Whereas, "trough" has higher discriminating power in medium and high amylose rice. Software program calculated altogether the "discriminant loadings" which is a measure of the linear correlation between each independent variable and the discrimination function or variance that the independent variable shares with the discrimination function. It is interpreted like factor loading. Discriminant loadings are more valid than weight to consider discrimination power. Statistical significance for testing the significant of discrimination function between group overall was Wilks' lambda, Hotelling's trace, Pillai's (Mahalanobis D^2 and Rao's V for Stepwise method). Its significance should lower than 0.05 to be considered a reliable function. However, if there are more than two groups, each of the estimated discrimination functions should also statistically significance. In this study, Wilks' lambda is used to test the significant of discrimination function. It is significant at 0.000.

Step 3. Validation

Statistically significance does not tell how well the function predicts. In discrimiant analysis we use "hit ratio" which is similar to R^2 in Regression analysis to assess overall fit of the function. Hit Ratio is the percentage that discriminant function can correctly classify. It tells how well the function can classify the units. The "hit ratio" in Table 3.19 using "leave one out" showed that discriminant function 1 consists of only peak viscosity and trough can correctly predict low, medium and high amylose content rice at 66.7%, 100% and 100% respectively. Add in more variables will increase "hit ratio" of the discriminant function however it depends on researcher decision to trade off with economic consideration and practically implication. For example it may not practical or economic if we have to test many parameters in order to add up a little prediction power.

Stage 4. Discriminant Loadings Plot

A graphically classification of interested samples can be plotted. Using discriminant score in function 1 (X axis) and discriminant score in function 2 (Y axis) to plot each data (sample) as shown in Fig. 3.26. Result shows that function 1 (vertical dimension) can separated three groups quite well. But function 2 (horizontal dimension) is poorly discriminated these three groups. Pay attention to function 1, low and medium amylose group do not separate well since they are quite close together. But low and medium amylose groups are clearly separated from high amylose group since they are far apart (back to explanation in Fig. 3.26). It can be concluded that function 1 which are mainly peak viscosity and trough can be used for predicting rice amylose group. It can correctly predict low, medium, and high amylose content rice at 66.7%, 100% and 100% respectively. As a result, rice exporters can employ the more easy and rapid pasting measurement to predict rice amylose group instead of complicated and error prone chemical analysis of amylose. However, this method may not apply for low and medium amylose rice due to its low prediction accuracy.

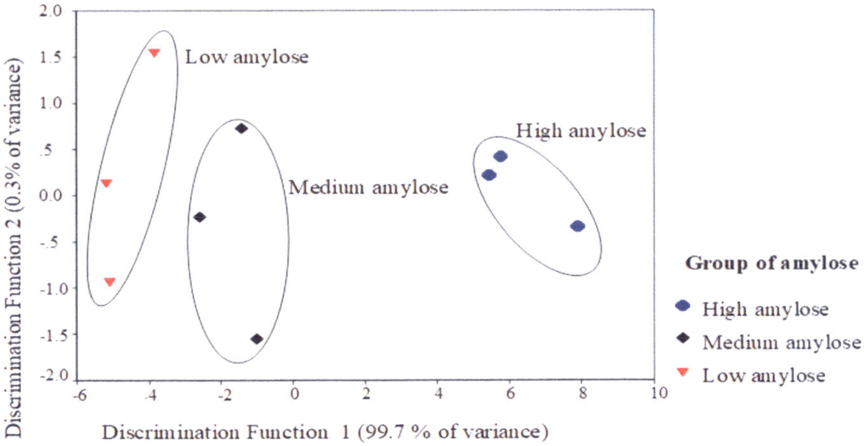

Fig. 3.26 Score plot on the plane of the two discriminant functions with three amylose group rice. (Source : Suwannaporn et al. 2007)

3.4 Multivariate Classification of Olive Oils Using Matlab

Davide Ballabio, Fabio Gosetti, and Saverio Mannino

3.4.1 Objectives and Learning Outcomes

The aim of this chapter is to show how to build multivariate classification models by using the functions included in the MATLAB Statistics Toolbox. Beside this option, many toolboxes in the MATLAB environment exist for applying classification model. However, this case study can help the user to understand step by step the application of these approaches to real case studies. In this framework, Discriminant Analysis and Principal Component Analysis are used in order to classify the geographical origin of olive oils samples on the basis of chemical measurements.

3.4.2 Introduction

Classification methods are fundamental multivariate statistical techniques aimed to find mathematical models able to recognize the membership of each sample to its proper class on the basis of a set of measurements. Once a classification model has been built, the membership of unknown samples can be predicted. Thus, classification techniques can be defined as quantitative methods for the modeling of qualitative responses, i.e. classification methods find mathematical relationships between a set of descriptive variables (e.g. chemical measurements) and a qualitative variable, i.e. the membership to a defined class.

Classification methods (also called supervised pattern recognition methods) are increasingly used in several fields, such as chemistry, process monitoring, medical sciences, pharmaceutical chemistry, and food science. For example, food products acquire a higher value when their geographical origin is controlled and assured. As a consequence, it is clear how the development of reliable methods for detecting authenticity is becoming very important and several efforts have been made to authenticate the origin of food products, with different chemical and physical parameters and on several food matrices, by means of multivariate classification.

Among traditional classification methods, Discriminant Analysis is probably the most known method (Brereton 2003; Fisher 1936; McLachlan 1992). Canonical Variates Analysis (CVA) separates samples into classes by minimizing the within-class variance and maximizing the between-class variance. So, the aim of CVA is to find directions (i.e. linear combinations of the original variables) in the data space that maximize the ratio of the between-class to within-class variance. These directions are called discriminant functions or canonical variates and are in number equal to the number of classes minus one. Then, a sample (**x**) is assigned to the class for which is minimum the discriminant score $d_g(\mathrm{x})$, defined as:

$$d_g(\mathrm{x}) = \left(\mathrm{x} - \overline{\mathrm{x}}_{\mathrm{g}}\right)^{\mathrm{T}} \mathrm{S}_{\mathrm{g}}^{-1} \left(\mathrm{x} - \overline{\mathrm{x}}_{\mathrm{g}}\right) + \ln \mid \mathrm{S}_{\mathrm{g}} \mid - 2\ln\left(P_g\right) \tag{3.11}$$

where **x** is the p-dimensional vector containing the p variables describing the sample, $\overline{\mathrm{x}}_{\mathrm{g}}$ and $\mathbf{S_g}$ are the centroid and the covariance matrix of the g-th class, respectively. P_g is the prior class probability and is usually defined as $P_g = 1/G$ or $P_g = n_g/n$, where G is the number of considered classes, n is the number of samples and n_g is the number of samples belonging to the g-th class.

Quadratic Discriminant Analysis (QDA) is a probabilistic parametric classification technique and is based on the classification rule described above. It is able to separate the class regions by quadratic boundaries. A special case, referred to as Linear Discriminant Analysis (LDA), occurs if all the class covariance matrices are assumed to be identical and separates the class regions by linear boundaries.

Both QDA and LDA are expected to work well if the class conditional densities are approximately normal, i.e. data are multinormally distributed. When the number of samples is small compared to the dimension of the measurement space (the number of p variables), the inversion of covariance matrices $\mathbf{S_g}$ became difficult. Consequently, when applying LDA, the number of samples must be greater than the number of variables, while QDA requires the number of samples of each class to be greater than the number of variables. In order to overcome these problems, a basic approach is to reduce the number of considered variables, by means of variable selection techniques (Jennrich 1977). A second approach, that will be shown in the next paragraphs of this chapter, can be based on a method for dimensionality reduction, such as Principal Component Analysis (PCA) (Jackson 1991; Jolliffe 1986; Wold et al. 1987); in fact, the classification model can be built by considering only the significant scores calculated by means of PCA.

Nowadays, statistical software packages contain procedures referred to by various names such as Linear Discriminant Analysis and Canonical Variates Analysis. One of the most popular software for numerical computing is MATLAB® (The MathWorks Inc., Natick, Ma, USA) that offers also the possibility to directly handle the programming language of the mathematical algorithms. Basically, MATLAB allows easy matrix manipulation and plotting of data, but also implementation of algorithms and creation of user interfaces. In fact, lists of commands can be written in a text file, as a script, extending the available commands. MATLAB is built around the MATLAB language, also called M-code. However, several additional MATLAB toolboxes can be installed. One of these toolboxes is called “Statistics Toolbox”: it provides tools for running specific mathematical and statistical models, such as Principal Components Analysis and Discriminant Analysis.

3.4.3 Problem to Solve

As stated before, classification methods appear as optimal tools for facing issues where a qualitative response is studied and modeled. For example, consider a consortium that wants to characterize a high quality food product on the basis of different chemical parameters, in order to assure geographical origin and uniqueness to the product. Multivariate classification can be used in order to distinguish the considered food product from products belonging to other geographical areas. In this model, each sample can be associated to a class on the basis of its provenience; when the model will be applied on unknown samples, each new sample will be assigned to one of the considered geographical groups.

This kind of issue will be applied on a dataset of olive oils (Ballabio et al. 2006a, b; Cosio et al. 2006) that includes 53 samples of extra virgin olive oils (EVOOs); 36 samples belong to a small production area, located in the lake of Garda (north of Italy) and distinguished with a European Protected Denomination of Origin trademark since 1998; on the other hand, 17 samples belong to other different regions of Italy and Spain. The sampling included also 19 commercial EVOOs: 6 samples labeled as Garda and 13 samples collected on the market, produced with unknown cultivars. For our purposes, all the samples have been divided into 2 classes (Garda and not-Garda). Each olive oil sample is described by 22 values of chemical measurements (variables) that correspond to the parameterized signals collected with an electronic nose. Consequently, the data can be arranged as a numerical table (matrix) with 72 (53 + 19) rows and 22 columns. Each entry x_{ij} of this data matrix represents the j-th chemical measurement on the i-th olive oil sample see Table 3.22 in the Appendix.

The final aim of this classification approach is the discrimination of Garda and not-Garda olive oil samples on the basis of electronic nose sensor measurements. In order to evaluate the predictive power of the classification model, i.e. the ability of the classification model to correctly assign unknown samples, all the 19 commercial olive oil samples will be used as external test set. Consequently, the classification model will be built on the basis of the 53 olive oil samples (training set) and then applied on the 19 commercial oils (test set). The model validation and the control of the classification predictive performances are extremely important, since the quality of a classification model must be assessed on the basis of its capability to be applied in the future on unknown samples.

3.4.4 Procedure

The data set will be initially classified by means of Linear Discriminant Analysis (LDA). Then, Quadratic Discriminant Analysis (QDA) will be applied. Since the not-Garda class includes just 17 samples described by 22 variables and QDA requires the number of samples of each class to be greater than the number of variables, it will be necessary to apply Principal Component Analysis in order to reduce the data size. All the calculations will be performed with the MATLAB Statistics Toolbox.

Linear Discriminant Analysis

First of all, the data (which is provided as supplementary material of this chapter) has to be imported in the MATLAB workspace with the following matrices and vectors:

X: is a numerical matrix made of 53 rows and 22 columns, representing the training set;

class: is a numerical vector made of 53 values, each representing the membership of the corresponding sample of X to one of the two defined classes (1 for not-Garda samples, 2 for Garda samples);

X_test: is a numerical matrix made of 19 rows and 22 columns, representing the test set;

class_test: is a numerical vector made of 19 values, each representing the membership of the corresponding sample of X_test to one of the two defined classes (1 for not-Garda samples, 2 for Garda samples).

As second step, the prior class probability (P_g) must be defined. As stated before, it can be calculated as $P_g = 1/G$ or $P_g = n_g/n$, i.e. in the MATLAB command window for this case study:

```
Pa = [0.5 0.5];
ng1 = length(find(class==1))/size(X,1);
ng2 = length(find(class==2))/size(X,1);
Pb = [ng1 ng2];
```

where ng1 is the proportion of samples belonging to class 1 (not-Garda), ng2 is the proportion of samples belonging to class 2 (Garda); consequently Pa corresponds to $P_g = 1/G$, since G is equal to 2; Pb corresponds to $P_g = n_g/n$. In the following steps we'll always consider Pb as the prior class probability.

Since LDA is scale independent, data are not pre-processed and the classification model based on Linear Discriminant Analysis can be directly applied:

```
C = classify (X,X,class,'linear',Pb);
```

where classify is the function of the MATLAB Statistic Toolbox, X is the training set, class is the vector defining the membership of the training samples, 'linear' specifies the type of discriminant function, and Pb is the prior class probability previously calculated. The output of this function is stored in a vector named C. This numerical vector contains the assignations to one of the two defined classes calculated by the classification model; consequently, C is made of 53 values, each containing the assignation of the corresponding sample of X to class 1 (not-Garda) or class 2 (Garda).

A basic approach to evaluate the errors in the model assignations is to compare the real class vector (class) and the calculated one (C):

e = abs(C − class);
accuracy = 100 − 100*length(find(e > 0))/size(X,1);

where accuracy is the percentage of correctly classified training samples. Beside accuracy, other classification measure can be used, such as sensitivity, specificity and balanced accuracy (Ballabio et al. 2018), especially when dealing with unbalanced classes. In our case study, we have obtained accuracy = 100%, i.e. LDA is able to perfectly separate the olive oil training samples. But, as stated above, in order to get the classification predictive power, we have to apply the model on the external test set:

C_test = classify(X_test,X,class,'linear',Pb);
As before, C_test is made of 19 values, each containing the assignation of the corresponding sample of the test set (X_test) to class 1 or class 2. Then, we can evaluate the percentage of correct classifications obtained on the test set (accuracy_test):
e_test = abs(C_test − class_test);
accuracy_test = 100 − 100*length(find(e_test>0))/size(X_test,1);

In the case study under analysis, only one test sample is misclassified, leading to accuracy on the test set equal to 1 − 1/19 (94.73%). The procedure for calculating the number of errors in the test set assignments is resumed in Table 3.20, by comparing the vectors representing the true class (class_test), the assigned class (C_test) and the absolute values of their difference (e_test).

Quadratic Discriminant Analysis Coupled with Principal Component Analysis Data matrices (X and X_test), class vectors (class and class_test) and class prior probabilities (Pa and Pb) are the same used in the previous application. If Quadratic Discriminant Analysis is directly applied on the training set:

C = classify(X,X,class,'quadratic',Pb);

MATLAB returns the following error: "the covariance matrix of each group in TRAINING must be positive definite". This is due to the fact that QDA requires the number of samples of each class to be greater than the number of variables, while the not-Garda class includes just 17 samples described by 22 variables. Thus, by using Principal Component Analysis, we can reduce the data size and then apply QDA on the calculated PCA scores. PCA must be calculated on the training set samples, while test samples must be successively projected into the component space in order to assure that test samples do not affect the model building.

Since PCA is scale dependent, we have to autoscale the data in order to make the chemical measurements (the variables) comparable (Krzanowski 1988). Basically the autoscaling procedure consists in subtracting mean values and dividing by standard deviations. Means and standard deviations relative to the training set are used to autoscale also the external test set:

m = mean(X);
s = std(X);

Table 3.20 Vector representing the true class (class-test), the assigned class (C-test) and the absolute values of their difference (e-test) for the test samples

test set samples	class_test	C_test	e_test
1	2	2	0
2	2	2	0
3	2	2	0
4	2	2	0
5	2	2	0
6	2	2	0
7	1	1	0
8	1	1	0
9	1	1	0
10	1	1	0
11	1	1	0
12	1	1	0
13	1	1	0
14	1	1	0
15	1	1	0
16	1	1	0
17	1	1	0
18	1	1	0
19	1	2	1

```
for i = 1:size(X,1); for j = 1:size(X,2); X_scal(i,j) = (X(i,j) – m(j))/s(j); end; end;
for i = 1:size(X_test,1); for j = 1:size(X_test,2); X_test_scal(i,j) = (X_test(i,j) – m
  (j))/s(j); end; end;
```

where X_scal is the numerical matrix containing the autoscaled values for the training set (53 rows and 22 columns) and X_test_scal is the autoscaled test set (19 rows and 22 columns).

Once the data have been autoscaled, we can apply PCA by using the corresponding MATLAB function:

```
[~,E,L] = svd(X_scal,0);
E = diag(E).^2/(size(X_scal,1) – 1);
T = X_scal*L;
```

where L, T and E are the loading matrix, the score matrix and the eigenvalues, respectively. By looking at the eigenvalues (Table 3.21), we can decide how many component to retain. When data are autoscaled, one criterion is to preserve all those components that have eigenvalue greater than 1 (Meloun et al. 2000; Todeschini 1997; Wasim and Brereton 2004). Following this approach, we can retain 3 components and reduce the data space from 22 dimensions to just 3 dimensions:

```
T = T(:,1:3);
T_test = X_test_scal*L(:,1:3);
```

Table 3.21 Eigenvalues

component	eigenvalue	component	eigenvalue
1	15.29	12	0.03
2	3.06	13	0.02
3	1.93	14	0.01
4	0.59	15	0.01
5	0.36	16	0.00
6	0.28	17	0.00
7	0.15	18	0.00
8	0.11	19	0.00
9	0.06	20	0.00
10	0.05	21	0.00
11	0.03	22	0.00

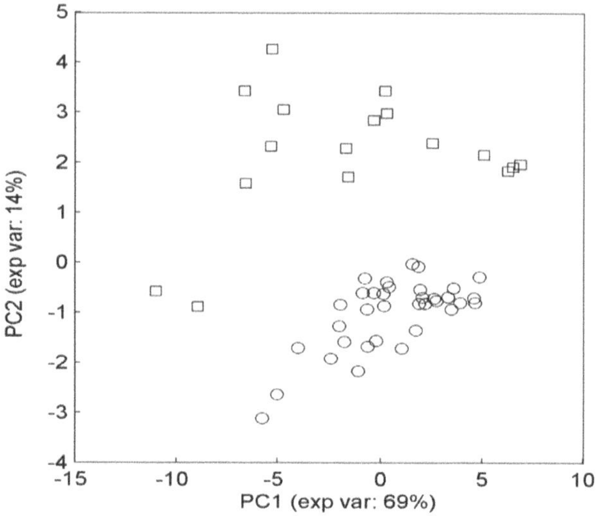

Fig. 3.27 PCA score plot. Garda samples are drawn as circles, not-Garda samples as squares

where T is the score matrix of the training set (53 rows and 3 columns) and T_test is the score matrix of the test set (19 rows and 3 columns), obtained by projecting the test set samples in the training set component space.

The scatter plot of the first two principal scores (Fig. 3.27) of the trainings samples can be easily obtained by typing on the MATLAB command window the following code:

```
figure
hold on
plot(T(find(class==1),1),T(find(class==1),2),'ks')
plot(T(find(class==2),1),T(find(class==2),2),'ko')
```

The two classes are drawn with different symbols ('s' squares for class 1, not-Garda samples, 'o' circles for class 2, Garda samples). In this way we can get an idea on the data structure and on how the classes are placed in the component space. Then, Quadratic Discriminant Analysis can be directly applied on the score matrix relative to the training set:

C = classify(T,T,class,'quadratic',Pb);

where C is the numerical vector containing the class assignations calculated by QDA. As before, the errors in the model assignations and accuracy can be easily calculated:

e = abs(C – class);
accuracy = 100 – 100*length(find(e>0))/size(S,1);

In our case study, we get accuracy equal to 100%, i.e. QDA is able to perfectly discriminate the olive oil training samples on the basis of their component scores. Finally, we can test the classification model by applying it on the test set scores and calculating the corresponding error rate:

C_test = classify(T_test,T,class,'quadratic',Pb);
e_test = abs(C_test – class_test);
accuracy_test = 100 – 100*length(find(e_test>0))/size(X_test,1);

Differently from the previous approach (LDA on the original data), in this case we get no classification error on the test set (accuracy_test = 100%).

3.5 Validation of Multivariate Models

Davide Ballabio, Viviana Consonni, and Saverio Mannino

3.5.1 Objective and Learning Outcomes

In the present chapter, the most common validation techniques for multivariate models are described with the aid of practical examples that focus on the estimation of the predictive capability of regression models built on simulated data.

3.5.2 Introduction

When evaluating the quality of multivariate models (both regression and classification models), the fitting performance of the models should be always taken into account to understand if models are able to correctly describe the relationship between dependent and independent variables. Moreover, it is also fundamental to assess the predictive capability of the model. In fact, since the main aim of multivariate models is their application for the calculation of unknown responses, great attention must be paid to their predictive capability.

A multivariate model can contemporaneously have a good fitting behaviour (i.e., it can correctly describe the relationship between data and response) and poor predictive capabilities (i.e., it is not able to correctly predict the responses of unknown samples). If this happens, the model can be considered useless, since it cannot be later applied for predicting new samples, making its application limited to data description. Therefore, the evaluation of predictive capabilities of multivariate models is extremely important and can be carried out by means of different validation techniques that constitute an essential tool for the evaluation of the model validity by multivariate methods. Validation techniques are used to objectively give a measure of their capability to provide reliable predictions of the modelled response for new samples where the response is unknown (Myers 1986; Rawlings 1988).

3.5.3 Fitting and Predictive Capabilities of Multivariate Models

When regression is taken into account, one of the mandatory conditions for the validity of a multivariate model is that the coefficient of determination (R^2) is as close as possible to one or that the Root-Mean-Square Error (RMSE) is as close as possible to zero (see next paragraphs for formulas). Nevertheless, this condition (fitting quality) is not enough to ensure the model validity, because multivariate models can improve their fit (larger R^2 or small *RMSE*) increasing the number of variables in the models; moreover, the fitting parameter does not account for the true capability of the model to provide with reliable predictions for unknown samples.

It has been demonstrated that the fitting performance of a multivariate regression model always increases if new variables are added, even if these variables are not correlated to the modelled response. On the other hand, the predictive performance of the model increases only if valuable variables are added to the model, otherwise it should decrease. As an example, a regression model was built by means of Ordinary Least Square regression (OLS) with the simulated data shown in Table 3.22. The data set consists of 20 samples, each described by 3 independent variables (X1, X2, X3) and an experimental response (Y). The fitting performance of the OLS regression model calculated between the independent variables (X1, X2, X3) and the

Table 3.22 Simulated data for regression

sample	X1	X2	X3	Y
1	0.4966	0.7948	0.5828	1.9347
2	0.8998	0.9568	0.4235	2.5055
3	0.8216	0.5226	0.5155	2.2177
4	0.6449	0.8801	0.3340	2.3054
5	0.8180	0.1730	0.4329	1.5604
6	0.6602	0.9798	0.2260	1.9933
7	0.3420	0.2715	0.5798	1.6260
8	0.2897	0.2523	0.7604	1.4186
9	0.3412	0.8757	0.5298	2.1492
10	0.5341	0.7373	0.6405	2.3661
11	0.7271	0.1365	0.2091	1.1886
12	0.3093	0.0118	0.3798	0.8205
13	0.8385	0.8939	0.7833	2.5406
14	0.5681	0.1991	0.6809	1.4872
15	0.3704	0.2987	0.4611	1.4506
16	0.7027	0.6614	0.5678	2.0275
17	0.5466	0.2844	0.7942	2.0471
18	0.4449	0.4692	0.0592	1.0602
19	0.6946	0.0648	0.6029	1.4476
20	0.6213	0.9883	0.0503	2.1571

response (Y) is evaluated in terms of determination coefficient (R^2) equal to 0.90 and *RMSE* equal to 0.149. The predictive performance of the same model is the determination coefficient calculated in validation by means of the leave-one-out procedure (see next paragraph for further details) equal to 0.85, while *RMSE* in cross-validation (namely, *RMSECV*) is equal to 0.186. This means that the regression model can have an acceptable fitting performance, while the validation capability is lower, but comparable in this case.

Afterwards, 7 random variables were created and added to the regression model as independent variables, one at a time. Therefore, the OLS model was recalculated first with 4 variables (the three original ones and the first random variable), then with 5 variables (the three original ones and the first two random variables), and so on. Fitting and predictive performance calculated for each model are shown in Fig. 3.28 (R^2 and R^2_{cv}) and Fig. 3.29 (*RMSE* and *RMSECV*) as a function of the number of independent variables included in the model. As stated before, R^2 calculated in fitting never decreases if variables (of any type) are added to the model, reaching the value of 0.93 when all the random variables have been included. Obviously, *RMSE* has the opposite behaviour and always decreases when variables are added.

This clearly shows that fitting performance never decreases, even if the added information is random and consequently not related to the modelled response Y. On the other side, the predictive performance correctly decreases when random variables are added to the original three variables.

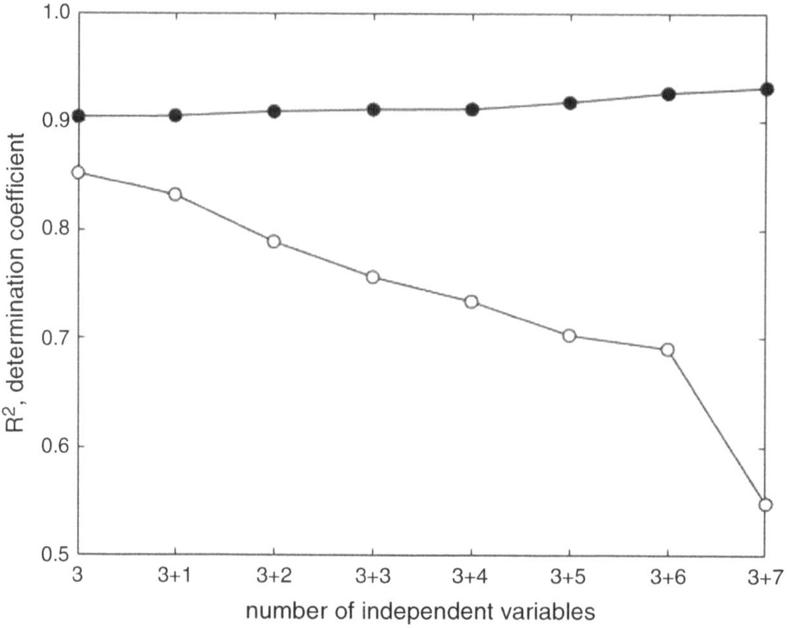

Fig. 3.28 Plot of the determination coefficient in fitting (R^2, black) and cross-validation (R^2_{cv}, white) as a function of the number of independent variables included in the model. Random variables were added to the three original ones

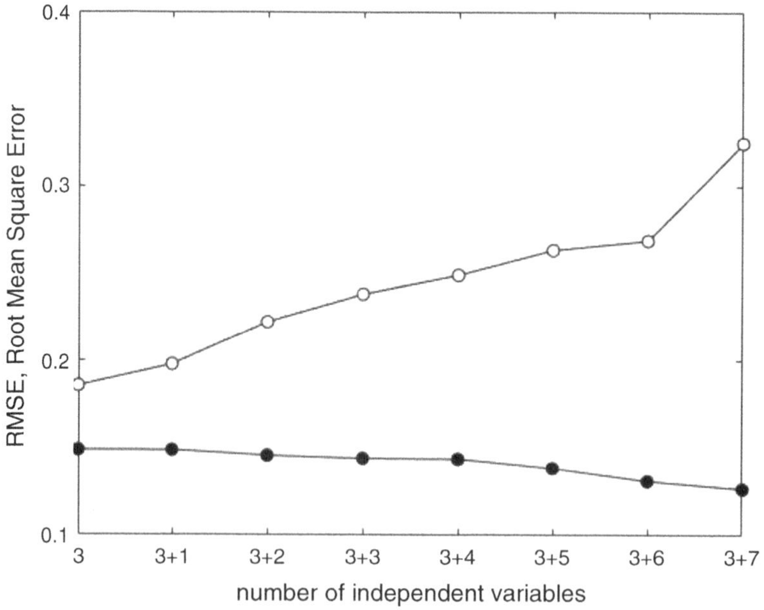

Fig. 3.29 Plot of Root-Mean-Square Error in fitting (*RMSE*, black) and validation (*RMSECV*, white) as a function of the number of independent variables included in the model. Random variables were added to the three original ones

As shown in Fig. 3.28, the coefficient of determination calculated by the leave-one-out validation procedure significantly decreases from 0.85 to 0.55 (*RMSECV* increases from 0.186 to 0.325), showing how important is to consider both the fitting and predictive performance for the final evaluation of a multivariate regression model.

The same conditions can be easily extended to classification models. Obviously, the prediction ability of classification models is evaluated by different parameters from those of regression methods, since the modelled response is qualitative and not quantitative. The most common parameters are the percentage of overall correctly classified objects or the percentage of correctly classified objects of a category of interest (Frank and Todeschini 1994). Anyway, as in regression applications, it is important to evaluate and test both fitting and predictive capabilities of classification models, for the same reasons stated above.

3.5.4 Methods of Validation

Different statistical techniques have been proposed to estimate the predictive ability of a model. The most common validation methods are listed below.

Cross-Validation

To carry out cross-validation, a number of reduced data sets is created by removing one sample or a small group of samples from the original data set, so that each sample is taken away once and only once (Efron 1983; Osten 1988). For each reduced data set, the model is calculated, and the responses for the removed samples are predicted from the obtained model. Then, the final predictive performance is calculated by comparing the predicted responses and the experimental ones.

The simplest cross-validation technique is the leave-one-out procedure, where each sample is taken away from the data set, one at a time. Then the model is built (not considering the left out sample) and the response of the removed sample is predicted by using the obtained model. All the samples are sequentially removed and predicted and therefore, given n samples, n reduced models are calculated. Leave-one-out is one of the most common cross-validation procedures, since its deletion scheme is not based on random selection and the predictive ability of different models can be perfectly compared. However, it can provide over optimistic results (Golbraikh and Tropsha 2002), especially when the number of samples is large. This is due to the small perturbation applied on the model calculation when only one sample is left out.

Consequently, when the number of objects is not too small, other cross-validation procedures can be adopted. Alternatively, instead of deleting one sample at a time, at each step a group of samples can be left out. The samples are divided into different groups (cancellation groups) and one group at a time is removed from the training set, the model is built without the left out samples and the responses of the removed samples are predicted.

Table 3.23 Example of sample splitting in cancellation groups with two different approaches: venetian blinds and contiguous blocks

Samples	Cancellation group Venetian blinds	Cancellation group Contiguous blocks
1	1	1
2	2	1
3	3	1
4	4	1
5	1	1
6	2	2
7	3	2
8	4	2
9	1	2
10	2	2
11	3	3
12	4	3
13	1	3
14	2	3
15	3	3
16	4	4
17	1	4
18	2	4
19	3	4
20	4	4

This procedure is then repeated for each cancellation group, and finally the predicted responses are compared to the experimental ones. The number of cancellation groups to be used is defined by the user: it can range from 2 to n (in this case, it coincides with the leave-one-out procedure), where n is the total number of samples that constitute the data set. Moreover, the user chooses the way for selecting the cancellation group at each step. The most common methods are random selection, venetian blinds and contiguous blocks. A visual example of these last two procedures is shown in Table 3.23 for a data set comprised of 20 samples.

External Validation

One problem with cross-validation is that the multivariate model is tested internally, that is, using the same data as to fit the model. In addition to internal cross-validation, one can test the model predictive ability on an independent data set, often referred to as test set or external test set. Consequently, the external validation procedure requires the samples to be divided into two different sets: the training and external test set. The training set is used to build the multivariate model, while the external test set does not participate in the model calibration (i.e., it can be assumed to be a set of unknown samples for the calculated model). Then, the fitted model is applied on the test set to calculate the predicted capability of the model by comparing the predicted with the experimental responses for the test samples.

When selecting the samples to be retained in the test set, some issues must be taken into account, such as the number of samples to be used. Of course, in order to apply this kind of validation, significant numbers of samples are needed, since a portion of these samples is not used for the calculation of the model. Then, the percentage of samples to be included in the test set is a key factor that must be defined, and it can change on the basis of the total number of available samples. Generally, the greater the number of samples the greater the percentage of samples that can be included in the test set; however, as a general rule, a percentage comprised between 10% and 50% is usually selected.

Another issue is how to properly select the test set samples. Since these samples must be representative of the whole set of data, the test set can be used reliably if the splitting is performed by choosing the samples by a well-stated criterion, such as a criterion based on cluster analysis or experimental design (Marengo and Todeschini 1992). Another common procedure is the Kennard-Stone algorithm (Kennard and Stone 1969), that ensures that representative samples, internal to the data domain, are selected in the test set. However, when the total number of samples in the data set is significantly large, the test set samples should be randomly selected.

Moreover, when dealing with classification, the distribution of samples retained in the external test set has to be considered on the basis of the analysed classes. Consider a data set where the samples are divided into three classes (A, B and C). If the entire class A is selected in the test set (all the samples belonging to A are used to test the model), the validation result will be unsuccessful. In fact, the model will be built without samples of the removed class (the model will not consider class A) and consequently will not recognize objects belonging to that class.

As general rules, a correct test set selection in classification should (a) retain samples of all the considered classes and (b) represent the classes in the training and test sets with the same proportion. In the previous example, if classes A, B and C are represented by the 40%, 40% and 20% of the training set, respectively, the test set should represent the same proportions of the three classes.

In general, even if external validation should be unbiased since (unlike cross-validation) no information related to the training set is used, it can provide unreliable results as a consequence of bad designs in the test set selection. Moreover, the optimal size and representativeness of the test set can be an issue. A proposed solution is to use a large number of samples, but this is not always feasible, since it can be expensive and time consuming. Basically, unique and standardized approaches of validating multivariate models do not exist, but it is fundamental to have some insight into the aims of an experiment before deciding how important the structure of the training set is to the success of the model in future predictions (Brereton 2007).

Training/Evaluation Set Splitting

This validation technique is based on the splitting of the data set into training and evaluation sets. The splitting is performed by random selection of the samples, then the multivariate model is obtained from the training set and the predictive performance is calculated on the evaluation set. As the results can be strongly dependent on the splitting of the data, the splitting is repeated hundreds or thousands of times and the prediction results are finally averaged on the repeated evaluation sets (Boggia et al. 1997).

Bootstrap
Let the data set be comprised of n samples; this validation procedure divides the original data set into training and evaluation sets with a random selection, but the original size of the data set is preserved in the training set, by selecting n samples with repetition. In this way, the training set consists of repeated samples, while the evaluation set of samples randomly selected (Efron 1982). As in the previous approach, the model is calculated on the training set and then the responses are predicted on the evaluation set. Since the selection is random, the bootstrap procedure is repeated thousands of time and the average predictive power is calculated.

Y-Scrambling
This validation method is used to evaluate regression models affected by chance correlation, that is, models with independent variables that are randomly correlated to the response variable. The test is performed by calculating the fitting or predictive quality of the model randomly modifying the sequence of the response vector, that is, by assigning each sample a response randomly selected from the true responses (Lindgren et al. 1996). If the original model is not affected by chance correlation, the quality of the original model is significantly greater than the quality of the model calculated on the randomized response vector. Since the method is based on random permutation, it is repeated hundreds of times and the final results are averaged.

Data Scaling in Validation Procedures
Some regression and classification methods (for example Partial Least Squares and Principal Component Regression) are scale dependent, that is, the results are affected by the scaling metric of the data set. In these cases, a particular data pre-treatment is required prior to the model validation. Since the test set samples are not involved into the model calculation, the data scaling parameters are calculated considering only the training data and then applied for the scaling of the samples belonging to the test set. The same procedure must be used on each cancellation group when cross-validation is performed.

As an example, consider to select the last five samples of the data set shown in Table 3.22 as the test samples and the first fifteen as the training samples. In order to use a scale dependent multivariate method, we can decide to apply auto-scaling on the data, in order to make the variables comparable. The auto-scaling procedure is calculated in the following way:

$$x'_{ij} = \frac{x_{ij} - \overline{x}_j}{s_j}$$

where x_{ij} is the value of the jth variable for the ith sample, x'_{ij} is the auto-scaled value, $\overline{x}_j$ and s_j are the mean and standard deviation of the jth variable in the training set,

Table 3.24 Example of autoscaling on training samples

	original variables			Auto-scaled variables		
Training sample	X1	X2	X3	X1	X2	X3
1	0.4966	0.7948	0.5828	−0.3781	0.7469	0.4583
2	0.8998	0.9568	0.4235	1.5068	1.2078	−0.4523
3	0.8216	0.5226	0.5155	1.1416	−0.0275	0.0737
4	0.6449	0.8801	0.3340	0.3155	0.9897	−0.9642
5	0.8180	0.1730	0.4329	1.1245	−1.0222	−0.3985
6	0.6602	0.9798	0.2260	0.3871	1.2730	−1.5816
7	0.3420	0.2715	0.5798	−1.1007	−0.7420	0.4413
8	0.2897	0.2523	0.7604	−1.3449	−0.7964	1.4734
9	0.3412	0.8757	0.5298	−1.1043	0.9771	0.1555
10	0.5341	0.7373	0.6405	−0.2026	0.5833	0.7884
11	0.7271	0.1365	0.2091	0.6997	−1.1258	−1.6781
12	0.3093	0.0118	0.3798	−1.2535	−1.4807	−0.7020
13	0.8385	0.8939	0.7833	1.2204	1.0288	1.6047
14	0.5681	0.1991	0.6809	−0.0437	−0.9477	1.0189
15	0.3704	0.2987	0.4611	−0.9677	−0.6644	−0.2374
mean	0.5774	0.5323	0.5026	0.0000	0.0000	0.0000
SD	0.2139	0.3515	0.1749	1.0000	1.0000	1.0000

Table 3.25 Example of autoscaling on test samples

	original variables			Auto-scaled variables		
Test sample	X1	X2	X3	X1	X2	X3
16	0.7027	0.6614	0.5678	0.5858	0.3675	0.3728
17	0.5466	0.2844	0.7942	−0.1443	−0.7051	1.6669
18	0.4449	0.4692	0.0592	−0.6196	−0.1793	−2.5350
19	0.6946	0.0648	0.6029	0.5476	−1.3299	0.5731
20	0.6213	0.9883	0.0503	0.2051	1.2974	−2.5859

respectively. In Tables 3.24 and 3.25, the results of the auto-scaling procedure are shown. Consider now the value of the first variable (X1) for the last sample (ID 20) of the test set: the auto-scaled value is calculated by using the mean and standard deviation obtained on the training set:

$$x_{20,1} = \frac{0.6213 - 0.5774}{0.2139} = 0.2051$$

On the other hand, if training and test samples were scaled all together, without considering their partition, a different result would be erroneously obtained. In this case, the test set samples would affect the data scaling and indirectly also the model calculation.

3.5.4.1 Problem to Solve

In the next paragraph, different validation techniques are applied on the simulated data set shown in Table 3.22 in order to estimate the predictive capabilities of multivariate models built by means of Ordinary Least Square regression (OLS).

3.5.4.2 Procedure

Initially, the regression predictive performance can be evaluated by means of the leave-one-out procedure. As described before, this procedure takes away each sample from the data set, one at a time; then the model is built with the remaining $n - 1$ samples (where n is the total number of available samples) and the response of the left out sample is predicted. Since the OLS method is scale independent, we can avoid the data pre-treatment for scaling training and test samples.

Therefore, considering the simulated data, the leave-one-out procedure starts by excluding the first sample and calculates the following OLS regression model with the remaining 19 samples:

$$Y = 0.3002 + 0.6994^{*}X1 + 1.2180^{*}X2 + 1.0063^{*}X3$$

Afterwards, the response of the first sample can be predicted by using the regression equation:

$$\hat{y}_1 = 0.3002 + 0.6994^{*}0.4966 + 1.2180^{*}0.7948 + 1.0063^{*}0.5828 = 2.2020$$

As second step, the first sample is included in the training set, while the second sample is excluded and the regression model is recalculated:

$$Y = 0.2982 + 0.7547^{*}X1 + 1.1717^{*}X2 + 0.9659^{*}X3$$

Obviously, the second model is slightly different with respect to the first one, since it is calculated on a different set of samples. Again, it is possible to calculate the predicted response for the left out sample:

$$\hat{y}_2 = 0.2982 + 0.7547^{*}0.8998 + 1.1717^{*}0.9568 + 0.9659^{*}0.4235 = 2.5075$$

The same procedure is sequentially repeated for all the 20 samples and all the predicted responses, that is, the responses calculated with the leave-one-out procedure, are then compared with the experimental responses. By this comparison, the coefficient of determination in validation (R^2_{cv}, also known as Q^2) can be calculated as:

$$R^2_{cv} = 1 - \frac{\sum_{i=1}^{n} (y_i - \hat{y}_i)^2}{\sum_{i=1}^{n} (y_i - \bar{y})^2}$$

where $\bar{y}$ is the mean of the experimental responses, n is the number of samples, y_i and $\hat{y}_i$ are the experimental and the predicted responses for the ith sample, respectively.

The coefficient of determination is the most common correlation-based metric for model predictive ability evaluation: it is independent of the measurement scale of the response and has an upper bound of 1 allowing to directly perceive how good a model is (the closer to 1, the better). The predictive ability of multivariate models can be also quantified through the usage of measures of dispersion, like the Root-Mean-Square Error in Prediction (RMSEP), which is a function of the prediction residual sum of squares, or the Standard Error of Predictions (SEP). These measures, in fact, provide simple and unambiguous estimates of the model prediction error and are easy to interpret (the lower, the better) (Alexander et al. 2015). The Root-Mean-Square Error in Prediction (RMSEP) calculated by the cross-validation procedure is also known as Root-Mean-Square Error in Cross-Validation (*RMSECV*) and defined as:

$$RMSECV = \sqrt{\frac{\sum_{i=1}^{n} (y_i - \hat{y}_i)^2}{n}}$$

Considering that $\bar{y}$ is equal to 1.8152, the steps for the calculation of numerator and denominator of R^2_{cv} and *RMSECV* are shown in Table 3.26. Then, the values of R^2_{cv} and *RMSECV* can be easily calculated as:

$$R^2_{cv} = 1 - \frac{0.6900}{4.6675} = 0.85$$

$$RMSECV = \sqrt{\frac{0.6900}{20}} = 0.1857$$

Now we can proceed with the evaluation of the predictive performance of the same OLS model by using cancellation groups of larger size. Initially, we must decide the number of cancellation groups to be used: in this example, since we have 20 samples, we can easily divide the samples into 4 groups. Then, the method for partitioning the samples must be decided too; in this first step, the venetian blind procedure can be used. Therefore, the validation procedure removes the samples of the first cancellation group, calculates the regression model on the remaining samples and, as usual, applies the model on the left-out samples in order to predict their responses.

In our example, the first cancellation group is composed by five samples (1, 5, 9, 13, and 17) as previously described in Table 3.23. The OLS regression model calculated with the remaining 15 samples is the following:

Table 3.26 Experimental (y) and predicted ($\hat{y}$) responses calculated with the leave-one-out procedure on the data set in Table 3.22. $\bar{y}$ is the average response equal to 1.8152

sample	y	$\hat{y}$	$(y-\hat{y})^2$	$(y-\bar{y})^2$
1	1.9347	2.2020	0.0714	0.0143
2	2.5055	2.5075	0.0000	0.4765
3	2.2177	1.9970	0.0487	0.1620
4	2.3054	2.1161	0.0358	0.2403
5	1.5604	1.5283	0.0010	0.0649
6	1.9933	2.2014	0.0433	0.0317
7	1.6260	1.4017	0.0503	0.0358
8	1.4186	1.5892	0.0291	0.1573
9	2.1492	2.0743	0.0056	0.1116
10	2.3661	2.1587	0.0430	0.3035
11	1.1886	1.2174	0.0008	0.3926
12	0.8205	0.9446	0.0154	0.9894
13	2.5406	2.8237	0.0801	0.5262
14	1.4872	1.6371	0.0225	0.1076
15	1.4506	1.3623	0.0078	0.1329
16	2.0275	2.1636	0.0185	0.0451
17	2.0471	1.7640	0.0801	0.0538
18	1.0602	1.3146	0.0647	0.5700
19	1.4476	1.4890	0.0017	0.1351
20	2.1571	1.8926	0.0700	0.1169
		sum	0.6900	4.6675

Table 3.27 Independent variables (X1, X2, X3) and predicted responses ($\hat{y}$) for the first cancellation group partitioned according to the venetian blinds approach

sample	X1	X2	X3	$\hat{y}$
1	0.4966	0.7948	0.5828	2.2283
5	0.8180	0.1730	0.4329	1.5003
9	0.3412	0.8757	0.5298	2.1611
13	0.8385	0.8939	0.7833	2.8264
17	0.5466	0.2844	0.7942	1.8254

$$Y = 0.2026 + 0.7494^{*}X1 + 1.302^{*}X2 + 1.061^{*}X3$$

Afterwards, the responses of the left out samples can be predicted by using this regression equation (Table 3.27). The same procedure is then repeated for each of the remaining cancellation groups, the predicted responses are compared with the experimental ones (Table 3.28), so that the coefficient of determination in cross-validation and *RMSECV* are calculated. In this case, R^2_{cv} with 4 cancellation groups partitioned by means of the venetian blind approach is equal to 0.82 and *RMSECV* is equal to 0.204.

The procedure for the validation by means of contiguous blocks is similar to the previous one, but the splitting of the samples in the cancellation groups is different, as shown in Table 3.23. Therefore, the samples selected in the first cancellation group are 1, 2, 3, 4, and 5, the samples in the second group are 6, 7, 8, 9, and 10, and

Table 3.28 Experimental (y) and predicted ($\hat{y}$) responses calculated with 4 cancellation groups partitioned by means of venetian blinds on the data set shown in Table 3.22

sample	y	$\hat{y}$	$(y-\hat{y})^2$	$(y-\bar{y})^2$
1	1.9347	2.2283	0.0862	0.0143
2	2.5055	2.5312	0.0007	0.4765
3	2.2177	2.0010	0.0470	0.1620
4	2.3054	2.0489	0.0658	0.2403
5	1.5604	1.5003	0.0036	0.0649
6	1.9933	2.2361	0.0590	0.0317
7	1.6260	1.3885	0.0564	0.0358
8	1.4186	1.6724	0.0644	0.1573
9	2.1492	2.1611	0.0001	0.1116
10	2.3661	2.1714	0.0379	0.3035
11	1.1886	1.1683	0.0004	0.3926
12	0.8205	0.9798	0.0254	0.9894
13	2.5406	2.8264	0.0817	0.5262
14	1.4872	1.6025	0.0133	0.1076
15	1.4506	1.3305	0.0144	0.1329
16	2.0275	2.1371	0.0120	0.0451
17	2.0471	1.8254	0.0492	0.0538
18	1.0602	1.3633	0.0919	0.5700
19	1.4476	1.4313	0.0003	0.1351
20	2.1571	1.8080	0.1219	0.1169
		sum	0.8314	4.6675

Table 3.29 Resume of fitting and validation results obtained on the simulated data of Table 3.22

fitting	R^2	*RMSE*
all samples in the model	0.90	0.149
validation	R^2_{cv}	*RMSECV*
leave-one-out	0.85	0.186
4 venetian blinds	0.82	0.204
4 contiguous blocks	0.86	0.180
2 venetian blinds	0.71	0.258
2 contiguous blocks	0.87	0.169
bootstrap with 1000 repetitions	0.84	
bootstrap with 5000 repetitions	0.84	

so on. In the analysed data set, R^2_{cv} with 4 cancellation groups partitioned by means of contiguous blocks is equal to 0.86 (*RMSECV* equal to 0.180).

One can also split the samples in a different number of cancellation groups: smaller the number of groups, more severe the validation procedure due to a larger perturbation in the model calculation.

Table 3.29 shows the validation results obtained with different numbers of cancellation groups and other validation strategies. R^2 calculated in fitting (i.e., considering all the samples in the training set) is the highest value and describes the fitting performance of the OLS model. All the validation results are lower. In particular, the leave-one-out procedure gives the second highest value, confirming that in some cases it can provide overoptimistic evaluations of the predictive performance, especially when the number of samples is large. Since for this simulated data set the number of samples is small, this can somehow justify the fact that R^2_{cv} calculated with four contiguous blocks is higher than R^2 calculated in leave-one-out. Then, R^2_{cv} obtained with a reduced number of cancellation groups is lower if venetian blinds are used, while with the contiguous blocks the results are comparable. Finally, the bootstrap method gives the same results with different number of repetitions, in accordance with the other validation procedures. The same conclusions can be drawn by looking at the *RMSE* and *RMSECV* values of Table 3.29.

Obviously, the final quality and acceptability of the model should be judged on the basis of its final aim. So, in this example, produced on simulated data, we can just say that the predictive performance of the OLS model is acceptable.

As stated at the beginning of this chapter, validation methods are used in order to calculate the predictive capability of a multivariate model. By comparing fitting and predictive capabilities, it is possible to evaluate if the model is reliable, stable and not affected by the presence of overfitting, that is, the model capability in describing the data (fitting) is significantly higher than its predictive capability. The larger the difference between fitting and predictive performance, the greater the probability that the model is affected by overfitting (i.e., the model correctly describes the relationship between the response and the independent variables, but it is not able to correctly predict unknown samples).

Basically, there is no a general optimal procedure for evaluating the predictive ability of multivariate models. The most appropriate procedure has to be selected case-by-case on the basis of the data characteristics (e.g., the number of included samples) and the final objectives of the model application. However, as shown in this chapter, the model validation is a fundamental step for the evaluation of the effective model quality and should be always considered in order to interpret (and show) the feasibility of the model.

Appendix

Table 3.22 Data obtained on samples of olive oils

sample	set	V01	V02	V03	V04	V05	V06	V07	V08	V09	V10	V11	V12	V13	V14	V15	V16	V17	V18	V19	V20	V21	V22	class
S01	training	43.80	16.98	85.81	1.56	30.89	47.77	9.80	65.93	0.80	14.70	233.91	197.05	245.08	147.71	63.37	70.87	6.82	204.35	220.27	292.63	217.74	215.37	1
S02	training	51.11	15.71	97.19	1.58	33.11	62.62	9.10	71.57	0.97	16.24	245.26	200.25	250.21	151.10	57.35	65.67	7.94	218.53	317.97	301.56	219.68	219.71	1
S03	training	29.49	10.15	60.05	0.52	20.99	31.50	5.99	45.88	0.40	9.41	189.35	170.48	192.15	111.18	47.40	50.91	4.53	174.74	221.97	259.16	155.54	172.40	2
S04	training	27.83	9.67	57.64	0.49	21.01	28.91	5.82	43.23	0.31	9.86	124.47	131.02	113.63	80.35	46.08	48.58	3.29	102.34	191.87	183.07	96.48	119.93	2
S05	training	30.41	10.67	62.69	0.45	22.30	31.30	6.38	47.01	0.13	9.87	154.77	150.88	149.49	95.14	49.26	52.03	3.68	140.87	214.98	220.28	123.06	144.13	2
S06	training	34.65	11.96	69.53	0.62	24.98	36.76	6.98	51.98	0.49	10.83	170.45	158.29	169.14	104.42	51.81	55.01	4.18	152.75	214.82	233.82	139.15	157.30	2
S07	training	33.07	11.41	66.00	0.68	23.18	36.13	6.71	50.98	0.54	10.99	205.04	179.96	200.14	122.95	51.13	55.87	5.53	179.48	211.13	261.96	173.49	187.36	2
S08	training	32.65	10.96	66.17	0.59	23.88	34.22	6.45	49.06	0.46	10.24	177.22	160.26	155.33	103.49	48.92	53.55	3.83	155.92	236.62	247.84	145.23	157.71	2
S09	training	33.49	11.33	68.72	0.69	26.57	36.01	6.64	50.82	0.49	11.95	138.94	139.47	131.01	87.53	48.79	51.71	3.37	111.51	217.52	194.79	107.18	130.67	2
S10	training	31.82	11.51	64.84	0.90	23.71	34.54	6.78	49.53	0.52	11.40	186.33	168.51	188.90	111.20	50.49	55.06	4.50	161.20	212.43	250.83	154.76	170.82	2
S11	training	33.21	10.77	65.65	0.50	22.87	36.37	6.49	49.71	0.51	10.37	192.36	170.90	193.56	113.71	49.09	52.65	4.93	178.37	223.74	258.37	157.41	174.07	2
S12	training	31.73	10.42	63.28	0.36	21.86	34.20	6.28	47.90	0.50	9.86	218.23	186.35	196.56	131.08	49.43	58.02	6.17	183.35	217.26	284.84	190.05	200.70	2
S13	training	27.52	9.96	55.92	0.41	19.79	28.02	5.64	42.68	0.26	9.08	165.73	156.79	155.79	100.44	47.82	50.82	3.47	147.55	193.07	234.19	132.58	154.14	2
S14	training	27.52	9.96	55.92	0.41	19.79	28.02	5.64	42.68	0.26	9.08	165.73	156.79	155.79	100.44	47.82	50.82	3.47	147.55	193.07	234.19	132.58	154.14	2
S15	training	27.86	9.30	55.53	0.08	18.65	27.97	5.50	41.42	0.39	8.42	139.80	139.28	120.52	88.67	46.38	49.29	3.18	118.58	190.12	200.06	109.14	134.12	2
S16	training	27.55	9.37	55.28	0.23	18.82	27.27	5.50	41.33	0.28	8.25	140.11	140.29	118.19	88.94	46.92	49.52	3.14	122.56	189.37	202.85	110.12	133.77	2
S17	training	29.89	9.81	58.97	0.13	19.85	30.00	5.89	43.50	0.32	8.72	147.91	144.69	126.53	93.25	48.12	51.12	3.31	131.34	198.00	211.72	116.69	140.15	2
S18	training	32.64	10.83	63.38	0.81	23.19	34.22	6.28	48.18	0.36	10.73	197.44	174.54	175.04	120.03	49.75	54.84	4.54	161.90	224.69	252.18	161.69	182.88	2
S19	training	28.34	9.60	56.49	0.24	19.47	28.34	5.61	42.29	0.28	8.58	147.70	144.66	132.32	92.64	47.16	49.97	3.24	128.45	196.30	211.63	116.29	140.32	2
S20	training	27.95	9.85	56.52	0.35	19.51	28.17	5.81	42.67	0.34	8.84	154.09	149.33	141.10	95.71	48.15	50.84	3.35	137.53	192.00	220.40	123.29	145.15	2
S21	training	32.76	10.79	65.68	0.38	22.26	32.89	6.56	48.49	0.55	10.04	152.02	146.63	132.48	95.87	49.28	52.93	3.02	129.66	291.99	217.61	117.62	143.80	2
S22	training	27.61	9.69	55.82	0.35	19.23	27.66	5.68	42.07	0.38	8.78	150.39	147.97	130.30	93.78	47.66	50.11	3.29	139.63	203.19	217.66	118.36	141.28	2
S23	training	28.52	10.19	57.25	0.55	19.96	28.92	5.84	43.84	0.45	9.24	171.40	160.61	162.21	103.83	48.60	51.62	3.65	154.71	193.20	238.23	138.30	158.72	2
S24	training	27.54	9.69	55.65	0.38	19.24	27.65	5.68	42.15	0.43	8.72	172.64	156.71	140.92	101.19	47.70	53.23	3.34	142.85	194.13	238.83	141.46	156.04	2

(continued)

Table 3.22 (continued)

sample	set	V01	V02	V03	V04	V05	V06	V07	V08	V09	V10	V11	V12	V13	V14	V15	V16	V17	V18	V19	V20	V21	V22	class
S25	training	28.89	10.57	57.84	0.52	19.85	29.14	6.27	44.48	0.40	9.29	165.70	155.56	156.47	101.52	49.99	52.83	4.23	147.98	183.51	229.40	134.32	153.45	2
S26	training	28.13	10.11	56.49	0.48	18.95	28.01	6.06	43.25	0.24	8.74	186.24	161.31	153.87	108.11	49.59	57.54	4.24	144.11	181.74	247.79	160.46	166.73	2
S27	training	30.79	10.90	61.63	0.42	19.36	31.63	6.59	48.27	0.44	9.29	182.79	163.77	178.33	111.29	50.85	54.56	4.77	163.14	255.61	247.86	149.93	168.61	2
S28	training	29.09	10.21	58.06	0.41	19.13	29.31	6.27	44.46	0.41	8.84	171.93	157.08	154.70	103.62	49.64	53.57	4.32	149.40	195.52	235.43	139.98	157.44	2
S29	training	31.44	11.36	62.21	0.60	21.15	31.76	6.71	47.80	0.45	9.92	179.84	164.47	170.71	110.43	52.57	55.72	4.57	167.62	192.36	246.84	149.23	165.27	2
S30	training	43.69	11.01	68.19	0.55	25.33	45.09	6.51	54.94	0.64	10.45	232.65	193.57	237.11	146.02	56.11	62.68	7.24	202.46	201.70	268.38	220.68	213.65	2
S31	training	37.24	10.16	62.35	0.33	24.76	35.90	5.98	48.60	0.57	10.03	202.23	168.83	171.18	121.58	53.21	64.52	4.38	142.26	196.35	241.58	189.73	183.65	2
S32	training	34.09	10.26	61.47	0.32	24.11	32.69	6.05	47.92	0.36	10.02	170.60	158.65	156.84	106.57	52.96	56.42	4.00	144.73	173.78	218.21	143.46	158.64	2
S33	training	37.31	9.71	61.07	0.60	26.26	35.84	5.66	48.45	0.50	10.39	189.90	169.14	180.05	115.73	51.41	55.49	4.11	145.29	204.43	225.42	159.48	174.63	2
S34	training	46.80	11.84	72.85	0.77	30.45	44.12	6.95	59.47	0.52	12.55	206.19	171.49	177.24	124.97	55.98	66.16	3.75	140.52	317.77	246.60	181.52	185.39	2
S35	training	41.40	10.55	66.27	0.76	29.77	38.84	6.14	52.42	0.50	11.43	182.60	159.16	159.84	110.70	52.88	59.80	3.53	119.68	206.41	217.96	153.69	166.15	2
S36	training	36.71	10.69	64.03	0.58	24.94	37.58	6.16	51.34	0.50	10.67	212.20	183.12	216.56	130.64	54.02	58.99	5.61	179.19	192.04	251.45	187.37	195.78	2
S37	training	39.44	10.36	64.75	0.50	24.77	41.39	6.10	54.41	0.47	10.73	253.64	208.92	240.26	162.55	56.30	64.04	8.60	219.46	201.53	287.52	244.54	232.81	2
S38	training	40.82	11.01	66.35	0.82	30.00	39.69	6.24	52.66	0.61	12.61	193.72	167.61	167.82	116.76	54.07	61.67	3.93	134.64	205.68	228.10	165.99	175.09	2
S39	training	24.88	9.70	51.13	2.16	18.35	24.65	7.04	40.06	0.59	8.79	137.56	131.62	120.89	73.90	39.50	44.33	2.86	120.82	195.20	211.50	111.70	120.22	1
S40	training	24.05	9.07	49.30	1.58	18.54	24.43	6.63	38.88	0.52	8.33	124.34	123.13	125.52	70.01	38.16	42.02	2.75	97.63	190.15	192.25	99.07	114.09	1
S41	training	22.71	9.23	47.24	1.49	20.88	23.01	6.36	36.99	0.26	8.59	125.78	123.13	115.18	67.64	37.00	40.03	2.53	123.49	200.00	206.70	96.93	108.16	1
S42	training	21.84	8.60	45.62	1.23	20.11	22.39	6.23	35.98	0.30	8.16	126.00	123.30	121.83	67.11	35.58	38.45	2.40	121.44	201.34	208.63	97.17	109.42	1
S43	training	32.28	13.73	67.86	4.15	28.16	36.27	9.64	55.97	1.22	12.70	240.68	202.93	260.37	130.31	47.76	55.86	7.97	223.57	260.93	319.03	210.51	208.95	1
S44	training	32.65	14.91	70.39	3.25	30.93	34.34	10.74	55.84	1.04	12.83	209.42	187.38	212.54	112.51	51.00	61.32	4.36	180.00	255.06	287.01	175.50	180.85	1
S45	training	31.04	15.78	68.31	5.09	32.21	33.09	10.24	54.66	1.04	13.66	216.03	189.01	230.67	116.75	48.73	55.85	5.95	198.32	235.06	295.75	182.35	187.69	1
S46	training	39.14	12.93	77.46	2.31	25.65	43.53	10.53	61.28	0.90	11.98	235.66	199.10	257.08	127.09	48.69	58.02	6.95	209.26	273.54	314.72	203.74	204.04	1
S47	training	33.40	13.24	69.33	2.98	26.52	36.49	9.94	56.20	0.93	11.95	230.49	197.72	248.85	123.11	48.54	56.58	6.46	211.35	242.33	310.55	197.44	198.29	1
S48	training	25.91	11.49	55.32	3.29	23.55	27.36	7.12	43.96	0.83	10.25	188.83	169.58	199.28	99.71	41.87	47.38	4.49	164.31	228.54	268.63	144.95	164.47	1
S49	training	26.20	12.12	56.48	3.45	24.32	26.66	7.50	45.07	0.79	10.52	193.77	175.17	199.82	104.05	43.99	49.69	4.51	167.25	223.69	269.71	151.17	170.58	1
S50	training	27.85	12.63	61.01	3.57	24.84	28.28	8.67	48.03	0.93	10.97	168.86	159.72	181.08	91.26	44.37	50.12	4.10	145.69	219.10	248.29	130.93	148.10	1
S51	training	24.71	10.86	53.37	2.72	22.57	25.14	6.90	41.99	0.57	9.37	162.63	155.41	171.37	88.90	42.15	46.45	3.60	140.13	225.81	240.74	120.73	144.99	1

S52	training	28.97	12.69	61.49	2.29	26.84	30.16	8.29	48.15	0.67	10.67	200.41	178.92	202.89	108.97	47.78	56.04	4.36	167.78	236.83	273.66	160.57	176.31	1
S53	training	32.81	13.74	69.77	2.58	28.77	34.03	10.39	54.70	0.67	11.79	174.57	165.57	189.78	96.48	49.55	56.10	3.73	146.26	228.87	256.04	139.87	153.31	1
T01	test	35.40	8.19	56.75	0.71	11.74	31.77	4.87	43.85	0.42	7.92	176.24	155.21	157.00	96.92	48.65	53.87	3.82	104.47	198.79	236.83	144.81	153.10	2
T02	test	29.73	9.56	52.96	1.14	21.54	29.73	5.09	44.56	0.44	10.16	198.45	174.87	195.01	112.04	49.13	52.43	4.91	147.08	222.41	260.42	161.00	176.98	2
T03	test	30.22	9.69	53.81	0.95	21.66	29.76	5.24	43.84	0.35	9.78	183.28	165.22	185.59	105.07	50.32	53.15	4.24	123.89	205.18	244.30	148.10	165.49	2
T04	test	31.61	9.51	55.06	1.03	20.33	30.91	5.17	44.49	0.47	9.53	173.89	158.68	178.14	101.05	49.71	52.14	4.16	115.72	235.62	235.19	138.34	158.83	2
T05	test	31.03	10.00	55.97	0.91	22.04	30.36	5.41	45.38	0.46	9.88	151.86	144.52	157.36	91.67	50.59	52.51	3.80	99.28	261.69	213.57	119.36	141.33	2
T06	test	30.56	9.74	54.06	1.08	22.05	29.85	5.35	43.70	0.55	9.93	167.03	155.01	174.37	97.57	49.96	51.76	4.11	120.05	199.43	232.97	132.54	152.46	2
T07	test	25.27	10.87	54.47	1.90	23.90	25.90	7.24	42.29	0.68	9.38	172.30	160.80	175.68	91.98	44.11	48.94	3.50	156.08	210.33	256.02	130.71	148.88	1
T08	test	29.92	13.03	63.00	2.08	27.10	31.27	8.73	49.95	0.78	10.84	208.16	184.29	216.76	114.38	50.04	55.57	4.62	191.77	235.75	286.89	168.11	182.32	1
T09	test	25.41	11.23	54.83	2.08	24.18	26.22	7.24	43.75	0.52	9.78	186.41	170.32	191.59	99.85	43.89	49.18	4.06	160.65	227.82	262.48	143.53	163.49	1
T10	test	26.65	9.85	53.45	2.19	17.74	26.92	7.20	42.53	0.58	8.76	177.88	156.80	170.53	93.36	39.71	47.59	3.67	150.56	210.71	254.87	154.18	158.58	1
T11	test	27.38	10.45	55.09	2.80	18.07	27.57	7.49	43.83	0.81	9.63	173.84	156.48	150.51	89.79	41.34	49.33	3.24	146.06	202.28	252.24	148.68	150.21	1
T12	test	29.34	12.19	60.27	2.30	26.28	30.97	8.64	47.74	0.51	11.12	192.14	165.89	194.42	103.19	44.41	48.82	4.15	183.76	225.30	275.45	167.68	169.56	1
T13	test	25.22	10.87	52.43	2.27	23.61	26.45	7.45	41.83	0.39	10.14	163.25	147.58	162.02	84.78	40.79	45.10	3.14	153.49	227.44	251.14	135.75	140.36	1
T14	test	26.91	10.78	54.39	1.84	23.82	29.45	7.50	43.60	0.39	10.52	182.02	158.05	191.92	95.00	40.24	44.47	4.05	167.82	209.14	268.24	157.04	159.99	1
T15	test	25.54	8.98	54.07	1.47	18.08	26.93	6.91	44.09	0.56	8.15	215.95	185.31	219.47	107.97	39.24	47.87	5.06	186.89	236.99	295.47	175.23	181.01	1
T16	test	28.55	12.40	60.97	2.98	25.34	29.73	8.40	48.30	0.77	10.79	188.51	171.73	199.01	101.78	45.46	51.77	4.46	163.25	229.31	266.80	149.38	165.15	1
T17	test	26.31	10.84	55.55	2.31	20.50	26.94	7.69	43.85	0.51	9.63	161.82	146.48	155.34	88.34	40.05	45.07	3.60	127.80	254.78	243.50	134.94	147.84	1
T18	test	29.22	10.90	57.88	2.40	20.70	30.20	8.00	46.62	0.65	9.85	192.91	167.30	205.22	101.96	43.36	48.18	4.23	170.83	220.26	275.95	167.56	169.92	1
T19	test	22.67	9.18	46.95	1.81	21.31	23.60	6.31	37.63	0.53	8.88	191.81	165.71	188.82	97.04	37.53	43.92	4.32	166.37	213.44	290.34	164.51	167.47	1

References

Alexander D, Tropsha A, Winkler DA (2015) Beware of R(2): simple, unambiguous assessment of the prediction accuracy of QSAR and QSPR models. J Chem Inf Model 55:1316–1322

Anderson MJ, Whitcomb PJ (1996) Optimization of paint formulations made easy with computer-aided design of experiments for mixtures. J Coatings Technol 68(858):71–75

Anderson MJ, Whitcomb PJ (1999) Computer-aided tools for optimal mixture design. PCI Paint Coatings Ind Magaz 15(11):68

Anderson MJ, Whitcomb PW (2005) RSM simplified, optimizing processes using response surface methods for design of experiments. Productivity Press, New York. 292p

Ballabio D, Cosio MS, Mannino S, Todeschini R (2006a) A chemometric approach based on a novel similarity/diversity measure for the characterization and selection of electronic nose sensors. Anal Chim Acta 578:170–177

Ballabio D, Mauri A, Todeschini R, Buratti S (2006b) Geographical classification of wine and olive oil by means of CAIMAN (Classification And Influence Matrix Analysis). Anal Chim Acta 570: 249–258

Ballabio D, Grisoni F, Todeschini R (2018) Multivariate comparison of classification performance measures. Chemom Intell Lab Syst 174:33–44

Boggia R, Forina M, Fossa P, Mosti L (1997) Chemometric study and validation strategies in the structure-activity relationship of new cardiotonic agents. Quant Struct Activ Relationship 16: 201–213

Box G, Behnken D (1960) Some new three level designs for the study of quantitative variables. Technometrics 2:455–475

Box GEP, Draper NR (1987) Empirical model building and response surfaces. Wiley, New York

Box GEP, Wilson KB (1951) On the experimental attainment of optimum conditions. J R Stat Soc Ser B 13:1–45

Branger EB, Sims CA, Schmidt RH, O'Keefe SF, Cornell JA (1999) Sensory characteristics of cottage cheese whey and grapefruit juice blends and changes during processing. J Food Sci 64(1):180–184

Brereton RG (2003) Chemometrics data analysis for the laboratory and chemical plant. Wiley, Chichester

Brereton RG (2007) Applied chemometrics for scientists. Wiley, Chichester

Cornell JA (1990) Experiments with mixtures: designs, models and the analysis of mixture data, 2nd edn. Wiley, New York

Cosio MS, Ballabio D, Benedetti S, Gigliotti C (2006) Geographical origin and authentication of extra virgin olive oils by an electronic nose in combination with artificial neural networks. Anal Chim Acta 567:202–210

Derringer G, Suich R (1980) Simultaneous optimization of several response variables. J Qual Technol 12(4):214–219

Doehlert DH (1970) Uniform shell designs. Appl Stat 19(3):231–239

Draper NR, Prescott P, Lewis SM, Dean AM, John PWM, Tuck MG (1993). Mixture designs for four components in orthogonal blocks. Technometrics 35:268–276

Efron B (1982) The jackknife, the bootstrap and other resampling methods. Society for Industrial and Applied mathematics, Philadelphia

Efron B (1983) Estimating the error rate of a prediction rule: improvement on cross-validation. J Am Stat Assoc 78:316–331

Ferreira SLC, dos Santos WNL, Quintella CM, Neto BB, Bosque-Sendra JM (2004) Doehlert matrix: a chemometric tool for analytical chemistry-review. Talanta 63(4):1061–1067

Ferreira SLC, Silva Junior MM, Felix CSA, da Silva DLF, Santos AS, Santos Neto JH et al (2019) Multivariate optimization techniques in food analysis - a review. Food Chem 273:3–8

Fisher RA (1936) The use of multiple measurements in taxonomic problems. Ann Eugenics 7:179–188

Frank IE, Todeschini R (1994) The data analysis handbook. Elsevier, Amsterdam

Golbraikh A, Tropsha A (2002) Beware of $q2$! J Mol Graph Model 20:269–276
Hair JF, Anderson RE, Tatham RL, Black WC (1995) Multivariate data analysis with readings, 4th edn. Prentice-Hall, Upper Saddle River
Hubbard MR (1990) Statistical quality control for the food industry. Van Nostrand Reinhold, New York
Huor SS, Ahmed EM, Rao PV, Cornell JA (1980) Formulation and sensory evaluation of a fruit punch containing watermelon juice. J Food Sci 45:809–813
Jackson JE (1991) A user's guide to principal components. Wiley, New York
Jennrich RJ (1977) Stepwise discriminant analysis. In: Enslein K, Ralston A, Wilf HF (eds) Statistical methods for digital computers. Wiley, New York
Jolliffe IT (1986) Principal component analysis. Springer, New York
Kennard RW, Stone LA (1969) Computer aided design of experiments. Technometrics 11:137–148
Krzanowski WJ (1988) Principles of multivariate analysis. A user's perspective. Oxford University Press, Oxford
Lawless LJR, Threlfall RT, Meullenet JF, Howard LR (2013) Applying a mixture design for consumer optimization of black cherry, concord grape and pomegranate juice blends. J Sens Stud 28:102–112
Lindgren F, Hansen B, Karcher W, Sjostrom M, Eriksson L (1996) Model validation by permutation tests: applications to variable selection. J Chemom 10:521–532
Los PR, Marson GV, Dutcosky SD, Nogueira A, Marinho MT, Simoes DRS (2020) Optimization of beef patties produced with vegetable oils: a mixture design approach and sensory evaluation. Food Sci Technol 40:12–20
Marengo E, Todeschini R (1992) A new algorithm for optimal, distance-based experimental design. Chemom Intell Lab Syst 16:37–44
McLachlan G (1992) Discriminant analysis and statistical pattern recognition. Wiley, New York
Meilgaard M, Civille GV, Carr BT (1999) Descriptive analysis techniques. In: Sensory evaluation techniques. CRC Press, Boca Raton, pp 161–142
Meloun M, Capek J, Miksik P, Brereton RG (2000) Critical comparison of methods predicting the number of components in spectroscopic data. Anal Chim Acta 423:51–68
Montgomery DC (2013) Design and analysis of experiments, 8th edn. Wiley, New York
Montgomery DC (2017) Design and analysis of experiments. Wiley, Hoboken
Myers RH (1986) Classical and modern regression with applications. Duxbury Press, Boston
Myers RH, Montgomery DC (1995) Experiments with mixtures. In: Response surface methodology, process and product optimization using designed experiments. Willey Interdisciplinary Science, pp 535–623
Myers RH, Montgomery DC (2002) Response surface methodology: product and process optimization using designed experiments, 2nd edn. Wiley, New York
Osten DW (1988) Selection of optimal regression models via cross-validation. J Chemom 2:39–48
Pinto-Castro J (1995) Aplicación de la metodología de superficie de respuesta en la optimización del proceso de clarificación de jugo de manzana. Tesis Ingeniero en Industrias Alimentarias. Universidad Nacional Agraria La Molina. Lima, Perú. 140p
Rawlings JO (1988) Applied regression analysis. Wadsworth & Brooks/Cole, Pacific Grove
Rubilar JF, Cruz RMS, Khmelinskii I, Vieira MC (2013) Effect of antioxidant and optimal antimicrobial mixtures of carvacrol, grape seed extract and chitosan on different spoilage microorganisms and their application as coatings on different food matrices. Int J Food Stud 2:22–38
Squeo G, De Angelis D, Leardi R, Summo C, Caponio F (2021) Background, applications and issues of the experimental designs for mixture in the food sector. Foods 10:1128
Suwannaporn P, Pitiphunpong S, Champangern S (2007) Classification of rice amylose group using discriminant analysis. Starch-Starke 59:171–177
Todeschini R (1997) Data correlation, number of significant principal components and shape of molecules. The K correlation index. Anal Chim Acta 348:419–430

Vieira MC, Silva CLM (2004) Optimization of a Cupuacu (Theobroma grandiflorum) nectar formulation. J Food Process Eng 27(3):181–196
Wasim M, Brereton RG (2004) Determination of the number of significant components in liquid chromatography nuclear magnetic resonance spectroscopy. Chemom Intell Lab Syst 72:133–151
Wold S, Esbensen KH, Geladi P (1987) Principal component analysis. Chemom Intell Lab Syst 2: 37–52

Chapter 4
Statistical Process Control (SPC)

Daniela Borda, Lester Wilson, Iuliana Banu, Iulia Bleoanca, Davide Ballabio, Saverio Mannino, Ivanka Zheleva, Todor Todorov, Hüseyin Bozkurt, and Fahrettin Göğüş

4.1 Introduction and Implementation of Statistical Process Control

Daniela Borda and Lester Wilson

4.1.1 Objectives and Learning Outcomes

The objectives of this chapter is to explain how a procedure can be implemented to keep a process under control and to produce products within the specification limits.

After completion of this chapter the reader should be able to understand the principles of SPC and to implement SPC.

D. Borda (✉) · I. Banu · I. Bleoanca
Faculty of Food Science and Engineering, Dunarea de Jos University of Galati, Galati, Romania
e-mail: Daniela.Borda@ugal.ro

L. Wilson
Department of Food science and Human Nutrition, Iowa State University, Ames, IA, USA

D. Ballabio
University of Milano Bicocca, Milan, Italy

S. Mannino
Department of Food Science and Technologies, University of Milan, Milan, Italy

I. Zheleva · T. Todorov
University of Rousse, Ruse, Bulgaria

H. Bozkurt · F. Göğüş
Department of Food Engineering, University of Gaziantep, Gaziantep, Turkey

G. Schleining et al. (eds.), *Statistics in Food and Biotechnology*, Integrating Food Science and Engineering Knowledge Into the Food Chain 16,
https://doi.org/10.1007/978-3-031-51568-2_4

4.1.2 Introduction

Without statistical process control (SPC), the basis for decisions regarding quality improvement are either empiric, or after-the-fact product inspections. SPC provides a scientific basis for decisions regarding ways for process improvement. A process control system is essentially a feedback system that links process outcomes with process inputs.

Food quality and safety has always been important to consumers and the food industry. We assume a food is safe and then make a decision about the quality (degree of excellence, uniformity, personal expectations, acceptability, etc. about food. Historically, specifications were developed, guidelines and test methods set, etc., and conformance to these standards were based upon end product inspection using attribute (yes/no) sampling. This approach has many problems:

- The product is in its final package (can, carton, bag), and if it fails inspection you have to find ways to recover the product (rework) or destroy it if it is a safety issue.
- It is risky to ship the product right off the line, as in Just In Time (JIT) operations. You do not know if it is good enough to ship without sampling and testing (time wasted).
- This process is 'reactive' (after the fact), instead of 'proactive' (fix it before the problem occurs, often referred to as "Do it right the first time."
- Final product inspection is costly, product is often destroyed in testing, and its labor intensive, and it does not improve the quality of the product.

A newer approach to manage food quality is to control the process using statistical methods. Work by Deming, Juran, Crosby, Ishikawa, Shewhart, Freigenbaum, Hubbard, Taguchi, Peters, Taylor, Conway, and others helped to develop a holistic, managerial, statistically proactive method of maintaining and improving the quality of foods, Statistical Process Control. SPC is part of a Total Quality Management (TQM) program (Vasconcellos 2003). By including the employees, education/training programs, knowing the actual process, controlling the process with real-time measurements, not only the quality and safety of the food can be improved, but also it is possible to improve productivity and lower costs. Thus by controlling the process and reducing the assignable causes of variation, it is possible to improve the quality and safety of the food before it is in the final finished form.

There are seven basic tools that are used to set up a SPC program.

1. Flow charts to define the actual process from start to finish.
2. Cause and Effect diagrams, often referred to as 'Fishbone' or Ishikawa diagram' or 'tree branch' diagrams, to determine potential causes of problems and thus variability.
3. A Pareto Analysis is used to rank the importance of different defects based upon their importance (frequency or cost). This helps identify which problem needs be solved first.

4. Histograms can be used to determine the shape of the distribution (frequency over time), extent of variation, and process mean.
5. Run charts can be used to see how a process is performing over time.
6. Scatter plots, correlations analysis, and regression can identify possible relationships between two variables.
7. Control charts, a statistical type of run chart than can be used to monitor and control the process through real-time measurements and decision-making.

By using these techniques, assignable causes can be identified and reduced. This will help stabilize the mean and the variation around the mean, thus allowing the control charts to monitor the process (Sect. 4.2). When the process is under control, the capability of the process can be determined (Sect. 4.3). If the process is not capable of meeting the product specification, the process needs to be changed, or the specification needs to be changed, or the company should cease to produce the product.

The remainder of this chapter will address the reduction of variability and the basis behind control charts. Section 4.2 will look at applications of process control charts, and Sect. 4.3 will analyze process capability.

When analyzing process stability some elements should be considered:

1. The samples taken must be representative of the products being run at that time interval, as your decision in SPC is based upon these samples.
2. Process must deliver batches of qualitatively homogenous products;
3. Process holds inner inherent, normal variations;
4. Process is influenced by a complex and diverse range of external factors, which interact between themselves;
5. Any component of the system varies in time.

The variation of a process associates statistical principles to the process management and divides total variation into inherent variation (common causes) and variation generated by special causes (assignable causes).

The variations of the elements contained by the process can be considered relatively small when (Petrescu and Voda 2002a, b):

(i) the total variability of the process is small and randomly distributed, but within acceptable limits;
(ii) variability of the products' characteristics is small and the products can be considered homogenous statistical population;
(iii) the conformance of the products is satisfactory;
(iv) process displays a statistical control state, or is **in statistical control**.

The assignable variations in a system are present when one or many elements are suddenly changing, or a new element is present in the system, and then:

(i) the variability of the process increases;
(ii) the variability in the products quality is very high;
(iii) products cannot be considered as part of the same statistical population;
(iv) conformance is very poor;

(v) process has the traits of an out of control process and is unpredictable and unstable.

To have an in-control process means to control two elements: centering and precision. There is no such a thing as fixed values for long time, thus Shewhart's great idea was to let the two elements "move" within acceptable limits of variation, based on the three sigma approach.

A Shewhart chart is a graphical representation of a temporal evolution of a process for which the statistical parameter θ is estimate with $\theta\ (X_{i1}, X_{i2}, X_{i3}, \ldots, X_{in})$, using m samples $(i = \overline{1, m})$ that contains n elements each $\{X_{i1}, X_{i2}, \ldots, X_{in}\}$, drawn from the process.

4.1.3 Steps of Implementation

The main steps of implementation of a statistical process control are (Fig. 4.1):

1. Preliminary test batch analysis—evaluates the process readiness for statistical process control methods using a small number of measurements. A preliminary chart is drawn in this step.
2. The second step is a complex one involving analyses of data using control charts, checking the stability, capability, and optimizing the results. It is part of the so-called off-line control.
3. On-line control is for a capable process when is necessary to monitor the evolution of the system studied.

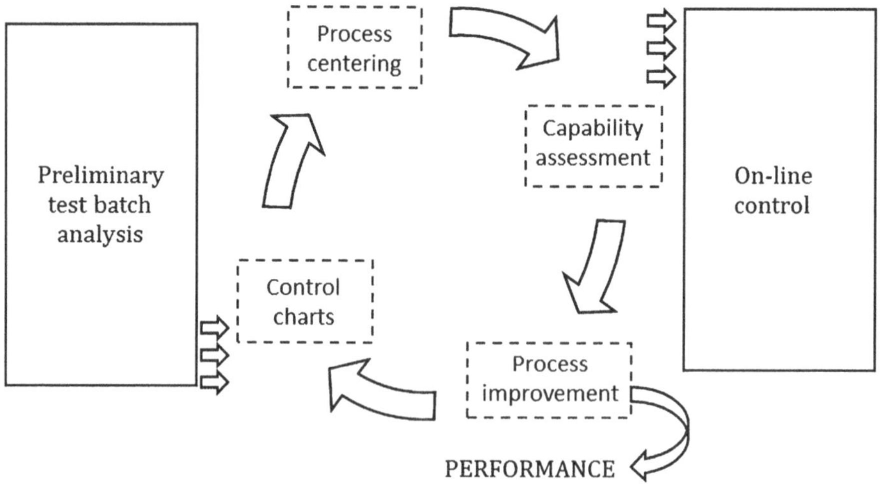

Fig. 4.1 Steps in statistical process control

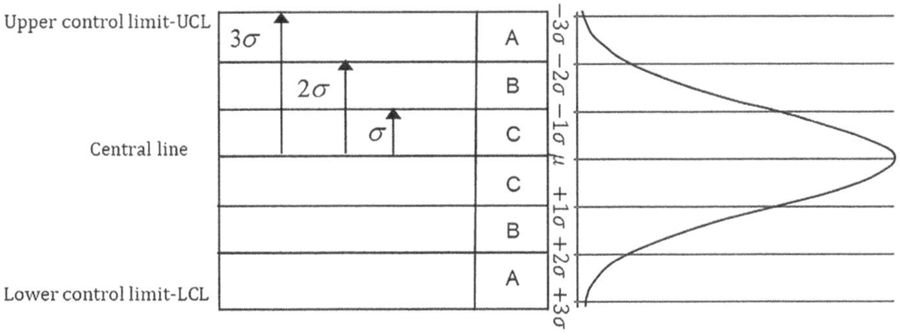

Fig. 4.2 Control chart area ($\pm 3\sigma = 99.7\%$, $\pm 2\sigma = 95.44\%$, $\pm\sigma = 68.26\%$)

To get an accurate representation in Shewhart control charts it is necessary to have a minimum 20 samples with 4 or 5 elements in each subgroup The chart area is divided into three main zones: ± A, B and C (Fig. 4.2). Shewhart recommended 4 or 5 elements in the subgroups to easy manually compute the average, but later on it has proved the effectiveness of the calculations dramatically decreases when (1) the subgroup is made of less than 4 values due to the incapacity of the subgroups to express the process evolution, or (2) when there are more than 5 values and the effect might be a covering effect induced by the average, which do not allow the individual values to be accurately displayed (Voda 2001).

This $\pm 3\sigma$ limits shows that 99.7% of subgroup values are within the control limits, if the process is in statistical control. Often is useful to draw the $\pm 2\sigma$ interval (the upper and lower B lines: 95,44%) and use those lines as warning limits for a tendency of losing control.

A uniform process should display:

(i) all the points of the chart between the control limits;
(ii) points uniformly distributed in all the charts' area;
(iii) the highest points density should be in C and B areas and a reduced density in zone A;
(iv) absence of bias in a certain area;
(v) absence of bias in case of several samples.

If an abnormal distribution of points is displayed than corrective action should be taken. It is important to underline that statistical control only can help noticing and requiring the application of corrective actions, but it cannot tell how these measures should be applied.

Another fact that should be taken into account is that Shewhart charts are made only for one characteristic and if the process needs to be monitored for more than one characteristic than it is necessary to draw as many charts as characteristics are needed.

The practitioners should clearly distinguish between the control limits and specifications. While control limits are the "natural" process tolerances, calculated with the process data, specifications are given tolerances settled before controlling the process. Two types of errors are met when applying Shewhart charts:

Type I –when the process is under control, but one point is randomly out of control limits. The result is the wrong conclusion that process is out of control and as a consequence an useless effort to find the cause of the problem.

Type II –when the process is out of control, but a random point is within the control limits. This conclusion generates a rising of the costs with nonconformities. Three factors are significantly contributing to this type of errors: the wideness of control limits, the rate of the deviation compared to the control limits and sample size. Shewhart charts only take into account the type I errors and the error implied is 0.3% for the 3σ control limits.

Sampling Frequency
It is difficult to set specific rules for the frequency of sampling to generate good control charts. As a rule of thumb the sampling interval (SI) could be estimated as mean time between failures (MTB) devided by 6 (SI = MTBF/6). For example if about all 90 min a failure happens, then SI = 90/6 = 15 min. Normally it is desirable to get 20–25 subgroups made out of 4 or 5 elements each. To be able to implement an effective process control it is necessary first to generate a preliminary chart. Recognizing the abnormal variations in the chart and identifying the unstable processes is the key for a successful application of statistical process control.

Unstable Path
The abnormal assignable variation is registered when one or more of the situations listed below (SR ISO 825/1999) appears (Fig. 4.3):

(i) one point is outside upper control limit (UCL), or lower control limit (LCL), beyond zone A;
(ii) two out of three points in a row are in zone A, or beyond of this area;
(iii) four out of five points in a row are in zone B, or on the same side of the central line;
(iv) eight points in a row are on both sides of central line with none in zone C;
(v) nine points in a row in zone C, on the same side of the central line;
(vi) fifteen consecutive points up and down in zone C;
(vii) fourteen points in a row alternatively up and down;
(viii) six points in a row steadily ascending or descending;

Other abnormal variations that can generate unstable processes can be classified in simple or complex variations such as: sudden drifts, up and down trends, recurring cycles, and unstable mixtures.

A Shewhart chart gives an overview of the process centering (localization) using the average and process precision using the range or standard deviation.

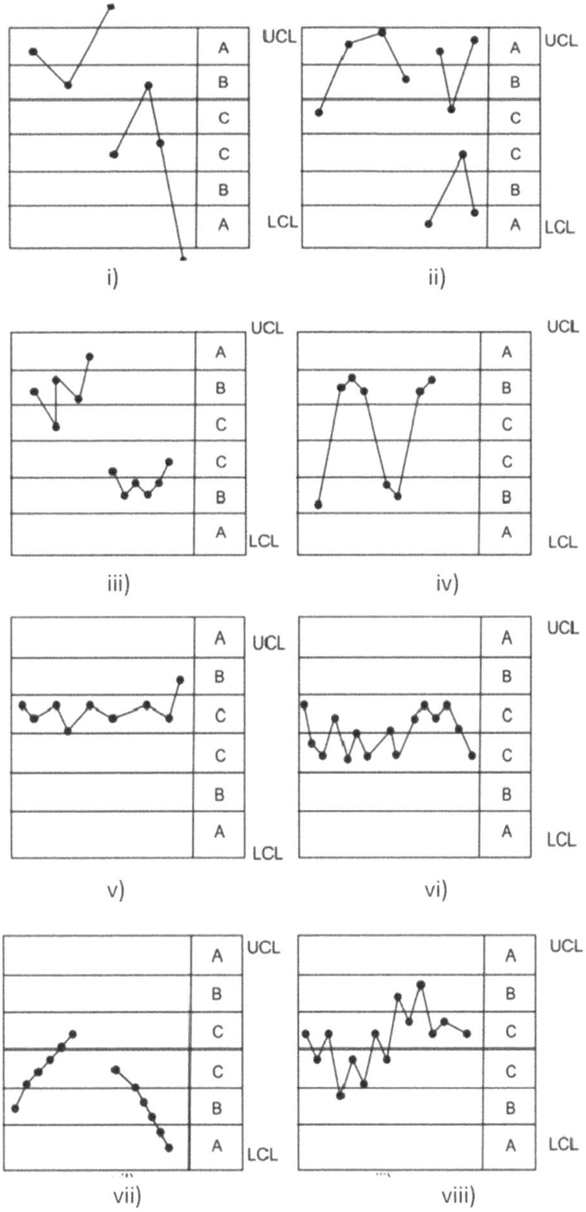

Fig. 4.3 Examples of unstable situations

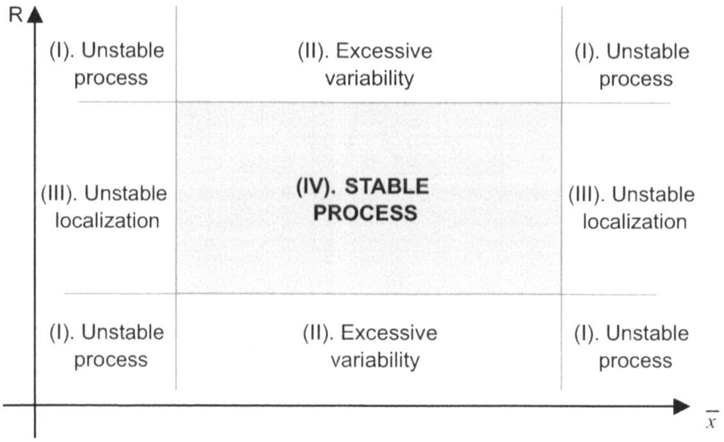

Fig. 4.4 Decision areas of the control charts

For example an average and range chart can highlight the decisional areas (Voda 1998) (Fig. 4.4). Area I -indicates the global instability of the process, area II- too large variations in the process, area III- problems with process centering, IV-the stability area.

4.2 Application of Process Control Charts

Daniela Borda, Iuliana Banu, and Iulia Bleoanca

4.2.1 Objectives and Learning Outcomes

The objectives of this chapter is to explain how a process could be monitored and assessed to be under control.

After completion of this chapter the reader should be able to apply the right type of control chart.

4.2.2 Introduction

The general strategy of prevention is nowadays the key to success in production activities. Certainly, food quality and safety management require specific tools to keep processes in control and monitor the critical limits and the technological parameters. Control charts are statistical tools mainly designed to control processes.

Shewhart charts were developed by Walter Shewhart during 1920s while he was working with Bell Telephone Laboratories. The power of Shewhart charts lies in its ability to separate assignable causes of quality variation from the inevitable ones (Grant and Leavenworth 1988).

The assignable variation is produced by special causes, while inherent variation is produced by common causes. A process is considered *in statistical control* when the variability is generated only by common causes (SR ISO 8258+C1).

The Shewhart charts can be applied in cases with:

(a) known standard deviation, or
(b) unknown standard deviation.

Types of Shewhart charts for variable and attributive data:

(a) Variable charts:

- Average and range or average and standard deviation control charts;
- Charts for individual measurements (X and moving range R);
- Charts for median (Me) and range (R).

(b) attributive charts:

- Control charts for percentage defective (p) or for number of defectives (np);
- Charts for nonconformities per unit control (c) or average occurrence per unit (u).

Using a graphical representation to display and compare information obtained from consecutive samples, charts compare in fact the real status of a process with the limits settled after estimation of inherent process variability.

The variable charts offer information on: basic variability of quality characteristics, consistency of performance and average level of quality characteristics.

The charts for fraction rejected/defectives, even less sensitive than ($\overline{X}$, R) charts have the advantage of using data already available for nonconformities to evaluate suppliers, products or processes.

Although U and C charts are less applied than the others, the special cases in which the nonconformities can be assessed by unit make them valuable tools.

Control charts allow estimation of the *in control status* of a process or to determine if the process tends to reach a control state for a certain level and afterwards to maintain control and a high uniformity level.

Control charts are considered the basis of statistical quality control and they are ideal tools for discovering ways to improve product quality and process control (Hubbard 2003).

Variable charts are trouble shooters whenever is difficult to meet quality specification in terms of variables (Grant and Leavenworth 1988).

Variable Charts

The mean, range and standard deviation are the commonly used statistics to analyze the data.

Arguments for applying variable charts:

- most processes and their results have measurable characteristics (time, temperature, pressure, weight, % fat, etc.);
- a measured value (accurate and precise) contains a higher amount of information compared to an attribute (yes/no) one;
- although variable sampling is more expensive than attribute sample, subgroups are smaller for the variable charts (SR ISO 8258+C1).

Control charts may describe the process variability inter-samples and locate the process average. The average (X-bar) and range (R) control charts are the best known and widely applied Shewhart charts.

Table 4.1 shows the formulas to calculate Shewhart's statistics in different cases.

Table 4.1 Control limits for Shewhart charts (Constants are indicated in Appendix Table 4.20)

Statistics	With unknown standard deviation Central line UCL and LCL		With known standard deviation Central line UCL and LCL	
$\overline{X}$ R	$\overline{\overline{X}}$ $\overline{R}$	$\overline{\overline{X}} \pm A_2 \cdot \overline{R}$ $D_4 \cdot \overline{R}, D_3 \cdot \overline{R}$	$X_0\, or\, \mu$ $R_0\, or\, d_2 \cdot \sigma_0$	$X_0 \pm A \cdot \sigma_0$ $D_2 \cdot \sigma_0, D_1 \cdot \sigma_0$
$\overline{X}$ s	$\overline{\overline{X}}$ $\overline{s}$	$\overline{\overline{X}} \pm A_3 \cdot \overline{s}$ $B_4 \cdot \overline{s}, B_3 \cdot \overline{s}$	$X_0\, or\, \mu$ $s_0 or\, c_4 \cdot \sigma_0$	$X_0 \pm A \cdot \sigma_0$ $B_6 \cdot \sigma_0, B_5 \cdot \sigma_0$

Average and Range Control Charts

The X, R charts can be used as a tool for improving monitoring of the product quality variation and ensure process is under control. The X and R charts are individual, directly related graphs, that plot the average of samples ($\overline{X}$) and the variation in each sample, *range* (R), over time.

Control Charts for Individual Measurements (X Charts)

The X charts or moving range charts are used when it is not feasible to draw rational samples or subgroups. The charts are designed to evaluate the central tendency of a process over time.

There are many possible reasons why moving range (MR-bar) charts are preferred: observations may be too expensive to get (destructive testing) or output may be too homogeneous over short time intervals (Pyzdek 2003).

There are no subgroups in the individual charts therefore X charts rely on variation between two successive range values.

Moving range (MR-bar) is the absolute value obtained when subtracting two successive values.

When working with X charts there is need to take into account:

- the fact that individual chart are not as sensitive as ($\bar{x}$, R) ones;
- the analysis of the charts should be very carefully performed when the process distribution is not normal (SR ISO 8258+C1).

Table 4.2 shows the formulas to calculate X charts.

Attributive Control Charts

The main advantages of atribute charts are: easy to get, cheap and often does not need qualified personnel to draw the samples. Table 4.3 shows the formulas to calculate Shewhart attribute charts.

Table 4.2 Control limits for X charts

Statistics	With unknown standard deviation Central line UCL and LCL		With known standard deviation Central line UCL and LCL	
Individual values, X	$\overline{X}$	$\overline{X} \pm E_2 \cdot \overline{R}$	$X_0 \, or \, \mu$	$X_0 \pm 3 \cdot \sigma_0$
Moving average, R	$\overline{R}$	$D_4 \cdot \overline{R}, D_3 \cdot \overline{R}$	$R_0 \, or \, d_2 \cdot \sigma_0$	$D_2 \cdot \sigma_0, D_1 \cdot \sigma_0$

Note: $E_2 = \frac{3}{d_2}$; Constants are indicated in Appendix Table 4.20

Table 4.3 Control limits for Shewhart attribute charts

Statistics	Without standard values Central line UCL and LCL		With standard values Central line UCL and LCL	
p	$\overline{p}$	$\overline{p} \pm 3\sqrt{\frac{\overline{p}(1-\overline{p})}{n}}$	p_0	$p_0 \pm 3\sqrt{\frac{p_0(1-p_0)}{n}}$
np	$n\overline{p}$	$n\overline{p} \pm 3\sqrt{n\overline{p}(1-\overline{p})}$	np_0	$np_0 \pm 3\sqrt{np_0(1-p_0)}$
c	$\overline{c}$	$\overline{c} \pm 3\sqrt{\overline{c}}$	c_o	$c_o \pm 3\sqrt{c_0}$
u	$\overline{u}$	$\overline{u} \pm 3\sqrt{\frac{\overline{u}}{n}}$	u_0	$u_0 \pm 3\sqrt{\frac{u_0}{n}}$

Charts for Nonconformities

The *p*-chart is generally applied for attributes data, when the classification of a product is done based on the comparison with a standard (Montgomery and Runger 2003).

The name of this type of control chart comes from *proportions*, due to the fraction or the percentage defective used for constructing the chart.

4.2.3 *Using Average/Range Charts with Unknown Standard Deviation to Control the Weight of Bread*

In a bread-making plant the weight of the bread is an important quality parameter. To determine the variation in process parameters the dough-dividing apparatus was verified in terms of automatic dough-weight. Due to the weight-loss during the technological process (dough-dividing), for a 300 g bread the dough should be divided in 335 g ± 3% pieces. Data on dough-weight (in grams) are given in the Table 4.4 for 20 samples of five parts each.

Calculate the control limits for X and R charts, construct the chart and plot the data in order to determine whether the process is in statistical control.

Table 4.4 Dough weigh for 25 subgroups

Subgroup number	X_1	X_2	X_3	X_4	X_5	$\bar{x}$	R
1	330	322	345	321	345	332.60	24
2	345	325	344	326	352	338.40	27
3	346	329	344	319	325	332.60	27
4	343	341	343	302	338	333.40	41
5	339	333	359	339	345	343.00	26
6	325	326	362	337	337	337.40	37
7	324	337	356	358	324	339.80	34
8	345	329	357	320	325	335.20	37
9	356	328	343	334	329	338.00	28
10	329	336	356	347	320	337.60	36
11	326	349	345	326	354	340.00	28
12	325	359	325	329	351	337.80	34
13	334	362	364	343	347	350.00	30
14	336	356	350	322	345	341.80	34
15	328	345	328	328	326	331.00	19
16	361	368	346	322	334	346.20	46
17	341	343	341	345	345	343.00	4
18	335	328	335	345	342	337.00	17
19	329	326	327	321	346	329.80	25
20	337	369	347	360	331	348.80	38
						$\overline{\overline{X}} = 338.67$	$\overline{R} = 29.60$

Start with calculating the average for each subgroup.

$$\overline{X} = \frac{\sum X}{n}$$

For the first subgroup $\overline{X} = \frac{330+322+345+321+345}{5} = 332.60$.

The average of the subgroup average ($\overline{\overline{X}}$) is computed by summing the values for $\overline{X}$ and dividing it by the number of subgroups (k). In this case:

$$\overline{\overline{X}} = \frac{332.60 + 338.40 + \ldots + 348.80}{20} = \frac{6773.40}{20} = 338.67$$

For each subgroup the range (R) is calculated, by subtracting the smallest value from the largest value in the subgroup. Here, $R_1 = 345 - 321 = 24$.

An average range ($\overline{R}$) is then calculated, by summing the values for R and dividing them by the number of the subgroups (k):

$$\overline{R} = \frac{24 + 27 + \ldots + 38}{20} = \frac{592}{20} = 29.60$$

Control limits (*UCL*- upper control limit and *LCL*- lower control limit) are computed separately, for $\overline{X}$ and $\overline{R}$ charts, using equations indicated Table 4.1.

For $\overline{X}$- chart: $LCL_{\overline{x}} = \overline{\overline{X}} - A_2 \cdot \overline{R}$ $\quad UCL_{\overline{x}} = \overline{\overline{X}} + A_2 \cdot \overline{R}$
For $\overline{R}$- chart: $LCL_{\overline{R}} = D_3 \cdot \overline{R}$ $UCL_{\overline{R}} = D_4 \cdot \overline{R}$

A_2, D_3, D_4 are coefficients that can be read from Appendices Table 3.2.3.

For $n = 5$, $A_2 = 0.577$, $D_3 = 0$, $D_4 = 2.114$, the results are:

$$UCL_{\overline{R}} = 2.115 \times 29.60 = 62.60 \qquad UCL_{\overline{x}} = 338.67 + 0.577 \times 29.60 = 355.75$$
$$LCL_{\overline{R}} = 0 \times 29.60 = 0 \qquad LCL_{\overline{x}} = 338.67 - 0.577 \times 29.60 = 321.59$$

At this point the $\overline{X}$ and R charts can be drawn. The charts will have a center line located at $\overline{\overline{X}}$ and $\overline{R}$ respectively, and two boundaries, depicted by the upper (UCL) and lower (LCL) control limits.

The charts show there is no special causes affecting the process, so it can be said that the dough-dividing process is under statistical control (Fig. 4.5).

The lower control limit is not plotted when it is zero, as it is the case for dough (Fig. 4.6). The main causes of variation in process parameters could be dough consistency, the type of bread or the equipment imprecision. The range variation close to LCL can be viewed as an early warning sign that fluctuations in the process might occur and need to be prevented.

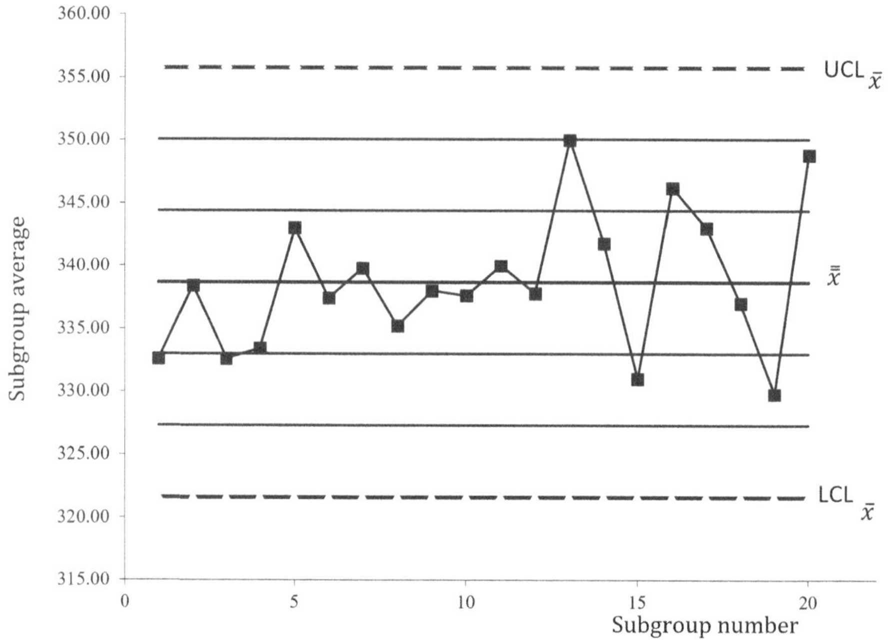

Fig. 4.5 X chart for dough-weight

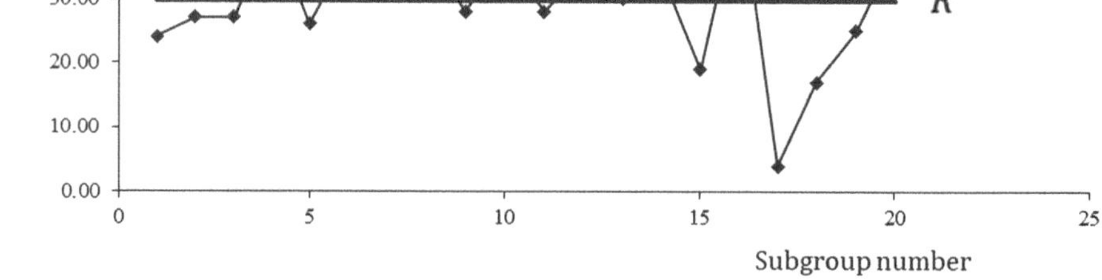

Fig. 4.6 Range chart for dough-weight

Table 4.5 Cocoa powder weigh

Subgroup	X_1	X_2	X_3	X_4	Average	Range
1	199.9	199.8	203.0	201.3	201.00	3.2
2	200.5	201.0	200.0	202.0	200.88	2.0
3	198.9	200.7	200.8	205.0	201.35	6.1
4	200.0	199.4	200.6	201.0	200.25	1.6
5	201.0	202.5	199.7	200.0	200.80	2.8
6	205.0	201.0	200.0	199.5	201.38	5.5
7	200.0	201.0	201.1	204.0	201.53	4.0
8	198.6	200.0	200.6	199.9	199.78	2.0
9	200.0	202.0	201.1	198.0	200.28	4.0
10	199.2	200.0	198.0	202.5	199.93	4.5
11	203.3	199.0	202.4	199.1	200.47	4.3
12	201.8	202.8	200.0	203.0	201.90	3.0
13	201.1	202.0	201.1	202.0	201.55	0.9
14	198.7	203.7	202.4	201.6	201.60	5.0
15	199.5	202.7	201.4	202.0	201.40	3.2
16	203.1	198.9	199.5	205.0	201.63	6.1
17	204.4	201.1	203.0	199.3	201.95	5.1
18	203.4	201.0	202.1	198.8	201.33	4.6
19	200.0	204.0	202.2	203.1	202.33	4.0
20	199.7	199.8	203.0	202.5	201.25	3.3
21	203.5	203.0	199.5	201.0	201.75	4.0
22	198.0	202.7	198.9	203.5	200.78	5.5

4.2.4 Using Average/Range Charts with Known Standard Deviation to Control the Medium Filling Weight of Cocoa Powder Bags

An importer wants to control packaging of cocoa powder in order to check the supplier compliance with a medium weight of 200 g. The process standard deviation is 1.6 determined from similar previous packaging processes.

$\bar{x}$ **and R** chart should be plotted. Central line $X_o = 200$. A number of 22 samples are drawn with 4 observations each.

The average and range are computed (Table 4.5) for each subgroup and then plotted on the control chart.

Subgroup equations for average and range charts are:

$$\overline{X} = \frac{\text{sum of subgroup measurements}}{\text{subgroup size}}$$

$$R = \text{largest value in subgroup} - \text{smallest value in subgroup}$$

Control limits for $\overline{X}$ chart:

$$UCL = X_0 + A \cdot \sigma_0 = 200 + 1.6 \times 1.500 = 202.4$$

$$LCL = X_0 - A \cdot \sigma_0 = 200 - 1.6 \times 1.500 = 197.6$$

Control limits for R chart:

$$R_0 = d_2 \cdot \sigma_0 = 2.059 \times 1.6 = 3.29$$

$$UCL = D_2 \cdot \sigma_0 = 4.598 \times 1.6 = 7.36$$

$$LCL = D_1 \cdot \sigma_0 = 0 \times 1.6 = 0 \text{ (for n} \leq 7 \text{ the limit is not plotted)}$$

The chart demonstrates that process is out of control because there are more than 4 points on the same side of the central line (stratified; low probability of occurring). The cause of this out-of-control deviation must be identified and corrected.

Obviously, the supplier does not meet the specifications and needs to readjust the packaging process due to the variations in mean that tend to bring the process out of control (Fig. 4.7).

Moreover, the process shows a trend up that might indicate the lack of precision in packaging and a gradual deterioration of the equipment (Fig. 4.8).

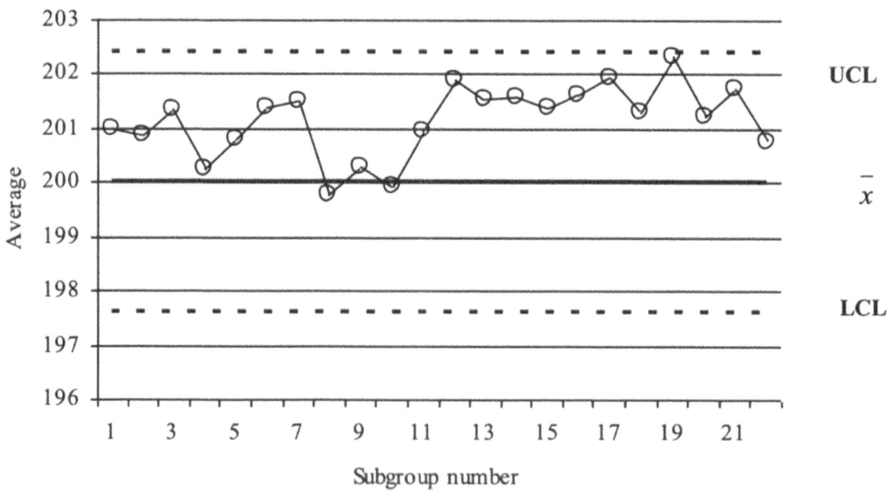

Fig. 4.7 X chart for cocoa powder packages

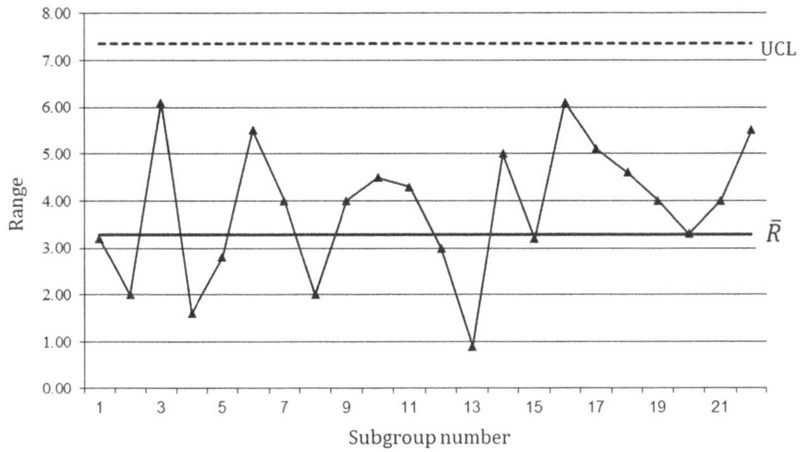

Fig. 4.8 R chart for cocoa powder packages

4.2.5 Using Average/Standard Deviation Charts with Unknown Standard Deviation to Control the Temperature of Tomato Juice Batch Sterilization

The effectiveness of tomato juice batch sterilization in a tubular heat exchanger for 2 min at 119–121 °C is given by monitoring temperature and flow of tomato juice.

Time is covered by the length of the maintaining tube at sterilization temperature, which allows inactivation of *Bacillus coagulans.*

To audit the temperature a control chart for average and standard deviation is provided.

The increase in temperature above the upper control limit will lead to a decrease in quality, while values below the lower control limit would indicate a critical control point out of the critical limits.

The formulas needed for plotting an – X and S chart are:

$$\overline{X}_i = \frac{1}{n_i}\sum_{i=1}^{i} X_i \qquad \overline{\overline{X}} = \frac{1}{k}\sum_{i=1}^{k}\overline{X}_i = 119,94$$

where k-number of subgroups and n-number of observations in subgroup (Table 4.6).

The standard deviation for each group:

$$\bar{s} = \frac{1}{k}\sum_{i=1}^{k} s_i = 0,56 \qquad s_i = \sqrt{\frac{\sum_{i=1}^{n}(x_i - -x)}{n-1}}$$

Table 4.6 Sterilizing temperature of tomato juice

Sample number	X_1	X_2	X_3	X_4	X_5	X_6	X_7	X_8	X_9
1	119.0	119.0	118.5	119.6	120.5	119.0	120.1	119.0	120.2
2	120.0	120.4	120.5	119.7	119.7	120.1	119.5	120.0	119.7
3	119.5	120.7	119.7	119.3	120.0	120.4	119.5	119.8	119.7
4	120.5	120.4	120.7	119.8	119.4	120.5	119.7	120.3	120.4
5	120.5	119.0	119.7	120.5	121.0	120.4	119.0	120.3	120.0
6	119.6	119.7	120.0	120.3	119.0	119.5	120.0	119.8	120.0
7	119.0	119.0	119.5	120.4	120.5	119.0	120.0	120.7	120.8
8	120.3	120.4	120.6	120.4	119.5	120.4	119.7	120.5	120.7
9	120.4	119.5	119.7	119.9	120.0	119.6	120.5	119.4	119.3
10	119.5	120.0	121.1	121.0	119.5	119.5	121.0	120.0	120.0
11	119.8	119.8	120.4	121.0	120.6	120.0	119.9	119.8	120.0
12	118.9	119.6	120.4	120.5	120.7	119	118.9	120.0	120.5
13	118.9	119.6	119.7	120.4	121.0	119.5	120.5	119.8	120.0
14	121.0	119.9	118.9	120.2	119.0	120.5	119.0	119.9	120.8
15	119.2	120.8	120.5	119.6	119.0	119.9	120.0	119.6	120.0
$\overline{\overline{x}}$	119.94								
$\overline{s}$	0.56								
UCLx	120.52								
LCLx	119.36								
UCLs	0.991								
LCLs	0.134								

To calculate the control limits the formulas below are applied:

$$LCL_x = \overline{\overline{X}} - A_3 \cdot \overline{s} = 119.36 \qquad UCL_x = \overline{\overline{X}} + A_3 \cdot \overline{s} = 120.52$$

$$UCL_S = B_4 \cdot \overline{s} = 0.99 \qquad LCL_s = B_3 \cdot \overline{s} = 0.134$$

$$A_3 = 1.032, B_3 = 0.239, B_4 = 1.761$$

The average and standard deviation charts indicate that the process is under control and the CCP ensure the effectiveness of sterilization process (Figs. 4.9 and 4.10).

4.2.6 Using Moving Range Charts with Unknown Standard Deviation to Control the Fat Percentage of Milk Powder

Table 4.7 indicates the results of the fat percentage in powder milk for 10 batches. For each batch one single sample was drawn and the percentage of the fat in powder milk resulted from whole milk should be 26%. Inside the batch the variation is very small, and the manufacturing of powder milk is normally distributed.

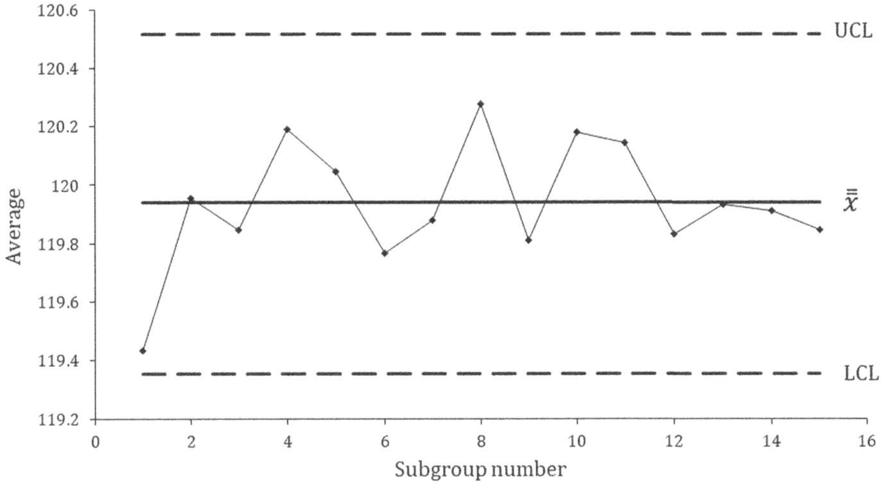

Fig. 4.9 X bar for sterilization of tomato juice

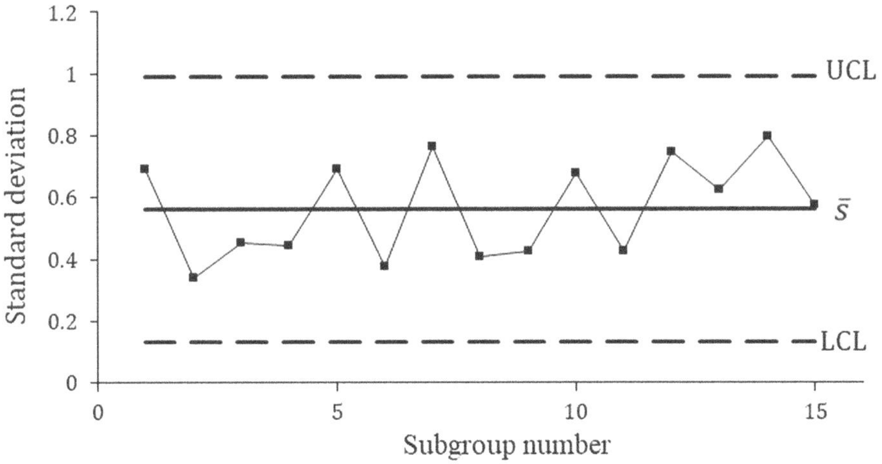

Fig. 4.10 Standard deviation chart for sterilization of tomato juice

Table 4.7 Fat percentage for 10 samples of milk powder

Batch number	1	2	3	4	5	6	7	8	9	10
X: Fat (%)	26.2	26.8	26.0	25.9	26.4	25.8	25.5	26.7	26.4	26.0
R: Moving Range		0.6	0.8	0.1	0.5	0.6	0.3	1.2	0.3	0.4

$$\overline{X} = \frac{\sum_{i=1}^{10} X_i}{10} = 26.17$$

$$\overline{R} = \frac{\sum_{i=1}^{10} R_i}{10} = 0.53$$

Control limits for moving range when n = 2 are:

$$UCL = D_4 \cdot \overline{R} = 3.267 \times 0.53 = 1.742$$

$$LCL = D_3 \cdot \overline{R} = 0 \times 0.53 = 1.742 \quad \text{(for } n \leq 7 \text{ the lower control limit is not plotted)}$$

Control limits for individuals X when n=2 are:

$$UCL = \overline{X} + E_2 \cdot \overline{R} = 26.17 + (2.66 \times 0.53) = 27.59$$
$$LCL = \overline{X} - E_2 \cdot \overline{R} = 26.17 - (2.66 \times 0.53) = 24.75$$

The charts represented in Figs. 4.11 and 4.12 demonstrate that process is under control.

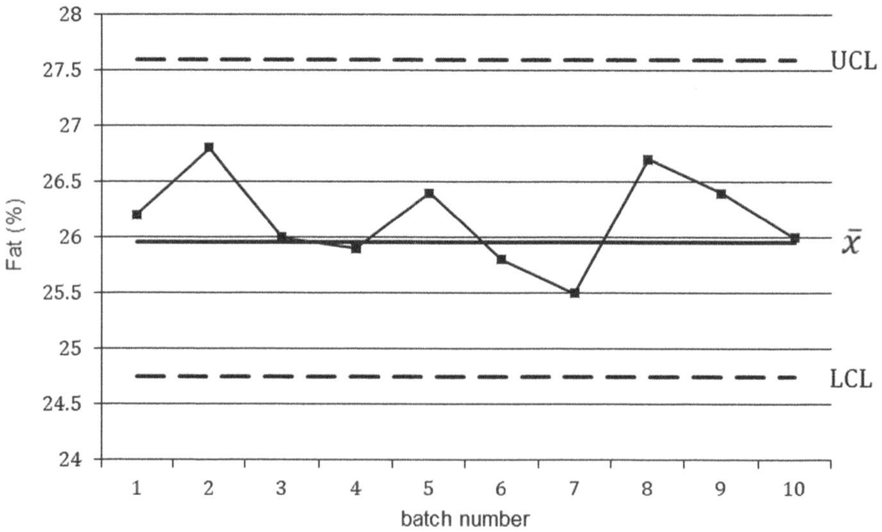

Fig. 4.11 Individual X bar for fat content of milk powder

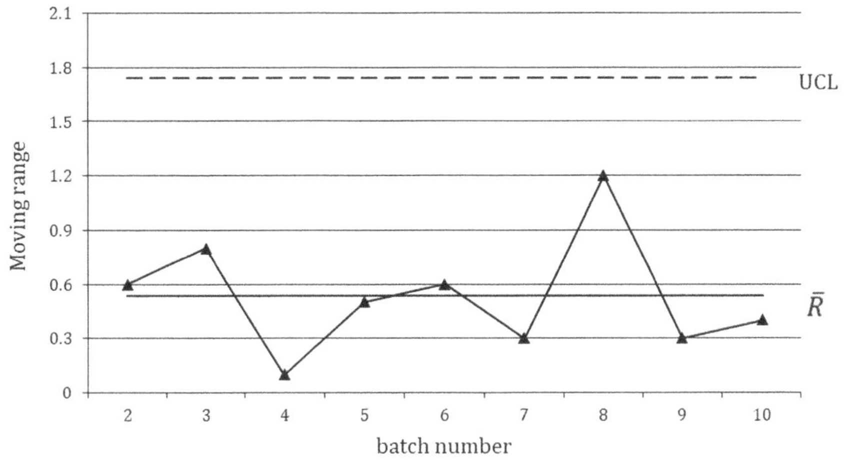

Fig. 4.12 Moving range chart for milk powder

4.2.7 Using p-Charts to Control the Accuracy of Pipetting

Using accurate techniques with pipettes is one of the key factors in obtaining reliable assay results in a laboratory. The major types of defective pipette measurements proved to be caused by several frequent errors, so special attention should be given to the following:

- well fitting of the correct disposable tip, before using the pipette;
- holding the pipette vertically or to an angle less then 20° from the vertical axis;
- immersing the tip into the sample liquid to a depth of at least 2–4 mm;
- possible air bubbles within the tip during intake;
- speed and smoothness during repression and releasing the push- button plunger;
- heat- transfer from the human hand to the pipette during prolonged, unnecessary holding of the pipette.

The easiest way for measuring the accuracy of pipetting is by weighing the volume of liquid delivered by a pipette at a selected setting. In this case, even though weight follows a normal (variable) distribution, the values can be then interpreted considering the tolerance, declared by the producer in the technical booklet. Let's take the example of a 1000 μL pipette with 0.14% accuracy.

During an annual check-up for correct pipetting technique in a microbiology accredited lab, a survey was conducted for 20 days, on 100 measurements/day. The number of defective measurements per subgroup is presented in Table 4.8.

First, we need to calculate the fraction defective in each subgroup (p), by dividing the rejects in subgroup (np) to the size of subgroups (n):

Table 4.8 Data indicating the defective measurements in pipetting

Soubgroups	Subgroup size n	Number of defective measurements in subgroup np	Fraction defective p	①	②	③	④
①	②	③	④	11	100	6	0.06
1	100	3	0.03	12	100	3	0.03
2	100	5	0.05	13	100	4	0.04
3	100	3	0.03	14	100	6	0.06
4	100	3	0.03	15	100	5	0.05
5	100	6	0.06	16	100	5	0.05
6	100	4	0.04	17	100	4	0.04
7	100	5	0.05	18	100	3	0.03
8	100	5	0.05	19	100	5	0.05
9	100	6	0.06	20	100	6	0.06
10	100	3	0.03	**Total**	**2000**	**90**	

$$p = \frac{np}{n}$$

For the 1st subgroup, for example, $np = 3$, which means two measurements were found defective from the first 100 weighing.

$$p = \frac{3}{100} = 0.03$$

The next parameter to calculate is the process average ($\overline{p}$), which is the total number of rejects divided to the total number of measurements in the subgroup:

$$\overline{p} = \frac{\sum np}{\sum n} = \frac{90}{2000} = 0.045$$

The control limits will be calculated with the following equations:

Since this is the first time when measurements are checked the control limits are calculated for this process and they should be considered *trial limits*. If there are data points outside the limits, special causes should be identified and eliminated. Afterwards, the limits would be recalculated without considering the special-cause data. In this case the control limits are

$$UCL_p = \overline{p} + 3\sqrt{\frac{\overline{p}(1-\overline{p})}{n}} \qquad LCL_p = \overline{p} - 3\sqrt{\frac{\overline{p}(1-\overline{p})}{n}}$$

$$\mathrm{UCL_p} = 0.045 + 3\sqrt{\frac{0.045(1-0.045)}{100}} = 0.107$$

$$\mathrm{LCL_p} = 0.045 - 3\sqrt{\frac{0.045(1-0.045)}{100}} = -0.017$$

Since in reality the defective fraction (p) can not have values below zero, it will be set at zero. The p-chart can be plotted by using the data points representing the defective fraction (p). The chart shows there are no special causes affecting the process, so it can be said that pipetting process is in statistical control (Fig. 4.13).

Keep in mind that the sample size "n" is very important in p and np charts. The larger the sample, the narrower the distance between the lower and upper control limits (tighter control). Conversely, the smaller the "n", the larger that distance will be, meaning that a much larger percentage or number of defectives will be accepted.

4.2.8 Using n,p Charts to Control the Number of Disintegrated Fruits in Apricot Cans

An apricot canning plant has a long-established quality control system using percentage defective control charts to evaluate the quality level of production. However, wishing to start a quality improvement program they find out that grouping all of the nonconformities into one single category, specific quality problems are not highlighted. In the past, many jars containing fruits with peeled skin, disintegrates fruits, too big or too small fruits, attacked by insects and/or birds, unclear syrup were simple classified as defectives.

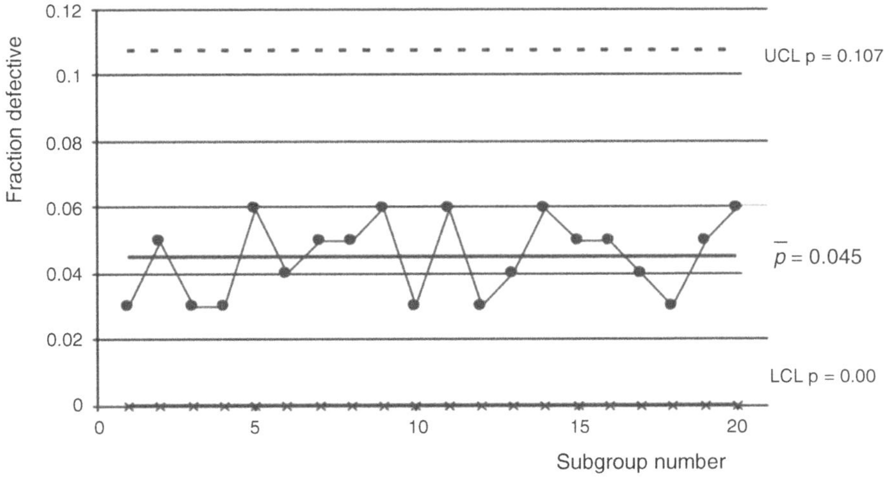

Fig. 4.13 Fraction defective control chart (p—chart), constant lot size

Table 4.9 Data for defective canned apricot

Day	Sample size (n)	Number of defectives np
1	400	10
2	400	12
3	400	10
4	400	14
5	400	15
6	400	9
7	400	10
8	400	15
9	400	23
10	400	4
N = 10	Σ n = 4000	Σ np = 122

For the purpose to set a new quality program, the company has designated the presence of disintegrates fruits as a major nonconformity. They are keen to reduce or eliminate this nonconformity.

A constant sample size of 400 jars drawn constantly within each production run shall be examined for disintegrates fruits, and np chart for number of defective jars will be plotted.

The formulas used for the np chart are:

$$n\overline{p} = \frac{122}{10} = 12.2$$

The first 10 days data are shown in Table 4.9.

$$-p = \frac{12.2}{400} = 0.0305$$

and the control limits are:

$$UCL = n\overline{p} + 3\sqrt{n\overline{p}(1-\overline{p})} = 12.2 + 3\sqrt{12.2(1-0.035)} \qquad UCL = 22.50$$

$$lCL = n\overline{p} - 3\sqrt{n\overline{p}(1-\overline{p})} = 12.2 - 3\sqrt{12.2(1-0.035)} \qquad LCL = 1.88$$

As depicted in Fig. 4.14 disintegrated fruits center line is 12.2. On the ninth day a rise in nonconformities number above the upper control limit is registered. The cause of the variation should be identified and eliminated. Further, after a few more days of production a new chart should be prepared with new closer limits. Any time a point falls outside the limits, the process should be analyzed and improved. The future evolution of the process can indicate if there is a trend of the process to get out of control or if remains in control with improved performances.

In the above example, a constant sample size was chosen, as this makes the control chart much easier to use and interpret. However, there are times when sample size can vary (boxes of fish or shellfish from fishing boats each day, or fresh produce

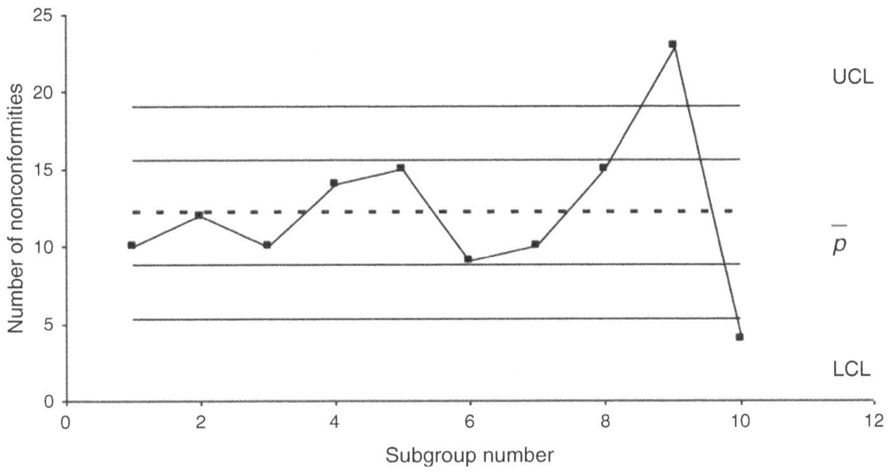

Fig. 4.14 Number of defectives (np) chart

harvested each day, etc.). In this case the sample (boxes) of product will vary every day, as will the samples taken. A np chart can be made, but the upper and lower control limit will vary with each sample size, thus presenting uneven/boxy (up and down) LCL and UCL. This changes the risk that you take each day (for example 30% defectives in control one day and 12% in control the next). As noted in the p-chart section, keep in mind that the sample size 'n' is very important in p and np charts.

4.2.9 Using c Charts to Control the Number of Nonconformities of Apples

The c charts examine the products for nonconformities in subgroups of constant size. C charts are statistical tools to evaluate the nonconformities in a process. Usually the subgroup is considered of size one (a batch of ice-cream mix, a beef carcass, a bin of blended tea, a mixed fruit gift basket etc.).

The subgroup size is constant in the sense that the different subgroups have substantially equal opportunities for the occurrence of nonconformities (Grant and Leavenworth 1988). This type of charts applies Poisson distribution. When the opportunity for occurrence of nonconformities changes from subgroup to subgroup, the u charts, for nonconformities per unit may be used. These charts are less applied in food industry.

Let us apply the c chart for a lorry of apples brought from a certain supplier in a canning plant. The nonconformities usually found during quality control evaluation are: bruises, molded apples, foreign matter, differences in color and ripening stage, fruits too small, etc.

Table 4.10 The daily number of nonconformities

Subgroups	Date	Nonconformities	①	②	③
①	②	③	13	1.05.08	3
1	15.04.08	3	14	2.05.08	3
2	16.04.08	4	15	5.05.08	1
3	17.04.08	2	16	6.05.08	0
4	18.04.08	3	17	7.05.08	1
5	21.04.08	1	18	8.05.08	3
6	22.07.08	0	19	9.05.08	4
7	23.04.08	2	20	12.05.08	2
8	24.04.08	2	21	13.05.08	3
9	25.04.08	2	22	14.05.08	0
10	26.04.08	4	23	15.05.08	0
11	27.04.08	1	24	16.05.08	5
12	30.04.08	3	Total		52

Assuming one lorry with apples is received every day, a survey was conducted for 24 days on the production line to asses the quality of the supplier and the data presented in Table 4.10 were obtained:

To calculate the central line, which represents the medium number of nonconformities, it is necessary to divide the total number of nonconformities to the number of subgroups.

$$\overline{c} = \frac{\Sigma c}{N} = \frac{52}{24} = 2.17$$

$$UCL = \overline{c} + 3\sqrt{\overline{c}} \quad UCL = 2.17 + 3\sqrt{2.17} = 6.97$$

$$LCL = \overline{c} - 3\sqrt{\overline{c}} \quad LCL = 2.17 - 3\sqrt{2.17} = -2.25$$

Whenever LCL gives a negative value, no lower control limit is used and the chart will exhibit only the upper control limit. Firstly, a trial chart is drawn.

As the first 24 points are inside the settled limits the central line or trial $\overline{c}$ can be used for further auditing supplier's activity (Fig. 4.15).

4.2.10 Using u Charts to Control the Number of Nonconformities of Black Pepper Packages

A small spice packaging plant is working on demand for caterers, restaurants and local market by packing pepper, mustard seeds, spice mixtures for pickles and soups, and other similar products.

Depending on demands and company working plan, the black pepper is packed in 20 g packages, with one or two machines of same type CEH1 and CEH2.

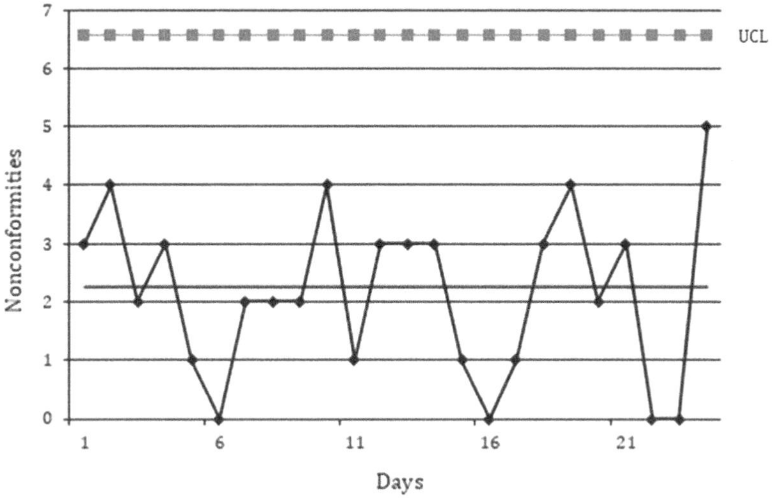

Fig. 4.15 C chart for nonconformities

The nonconformities observed on the packages at the end of the line, before the collection in boxes are: unreadable label, dots of ink on the packages, unclear durability, partially damaged packages and all sorts of different small blemishes. In the u-charts central line represents the average number of nonconformities per unit ($\overline{u}$ *or* u_0) (Table 4.3).

The subgroup size varies (n_i) and, as a consequence, the control limits are variable. If a medium number of nonconformities ($\overline{u}$) was determined in a trial test then:

$$UCL = \overline{u} + \frac{3\sqrt{\overline{u}}}{\sqrt{n_i}} \quad LCL = \overline{u} - \frac{3\sqrt{\overline{u}}}{\sqrt{n_i}}$$

where $\overline{u} = \frac{\sum c_i}{\sum n_i} = \frac{total\ nonconformities}{total\ units\ inspected}$.

Considering the nonconformities registered during an 8 days period (Table 4.11) we want to find out which of the two machines works properly. The medium number of nonconformities was determined in a previous trial, $u_0 = 0.32$.

From Fig. 4.16. we can notice that most of the points are falling between the control limits. However, CEH 1 has one point, the last one, higher than the upper control limit, so it requires adjustments.

Moreover, to demonstrate that process is in statistical control the computed value of nonconformities per unit $\overline{u} = \frac{3032}{9600} = 0.315$ is very close to the value given in the first trials (0.32).

Table 4.11 Control limits for black pepper packaging machines

Date	Machine	Units	Nonconformities	Nonconformities per unit	$3 * \sqrt{\frac{u_0}{n_i}}$	UCL	LCL
15.01	CEH 1	200	53	0.27	0.23	0.55	0.09
	CEH2	500	175	0.35	0.13	0.45	0.19
16.01	CEH 1	1000	300	0.30	0.10	0.42	0.22
	CEH2	200	60	0.30	0.22	0.54	0.10
17.01	CEH 1	500	160	0.32	0.13	0.45	0.19
	CEH2	1000	320	0.32	0.09	0.41	0.23
18.01	CEH 1	500	145	0.29	0.14	0.46	0.18
	CEH2	500	155	0.31	0.14	0.46	0.18
19.01	CEH 1	1500	450	0.30	0.08	0.40	0.24
	CEH2	200	50	0.25	0.24	0.56	0.08
22.01	CEH 1	100	30	0.30	0.31	0.63	0.01
	CEH2	1500	490	0.33	0.08	0.40	0.24
23.01	CEH 1	500	175	0.35	0.13	0.45	0.19
	CEH2	500	115	0.23	0.16	0.48	0.16
24.01	CEH 1	300	154	**0.51**	0.14	0.46	0.18
	CEH2	600	200	0.33	0.12	0.44	0.20
	TOTAL	9600	3032				

4.2.11 *Using CuSum Charts to Control the Weight of Butter After Packaging*

The CuSum chart for sample average is applicable for stable processes that are known to be normally distributed. The method was developed in 1954 by E.S. Page and the technique incorporates information taken from the latest sample into that taken from a sequence of prior samples (Grant and Leavenworth 1988).

A V-mask is placed over the graph of cumulative sums and the latest cumulative sum points are examined to see if they fall below or above the V-mask. As long as the previous point lies between the sides of the V, the process is in control. CuSum charts are more sensitive than Shewhart charts in detecting shifts. It is necessary to specify from the beginning the shift that must be detected in terms of multiple standard deviation of the sample average (δ times $\sigma/\sqrt{n}$). Most users of CuSum procedure prefer tabular chart over the V-mask, which proved to be too difficult to apply into practice. To generate the method we need to use the parameters h -slope and k- the shift to be detected.

The data presented in Table 4.12 represent the weight in grams of butter after packaging, at the end of production line. 15 samples of 4 packages were drawn every quarter of hour, scaled and placed back on the line.

The process mean target is aimed at T = 200 g. The known process standard deviation is 1 and therefore $\sigma/\sqrt{n}$ is

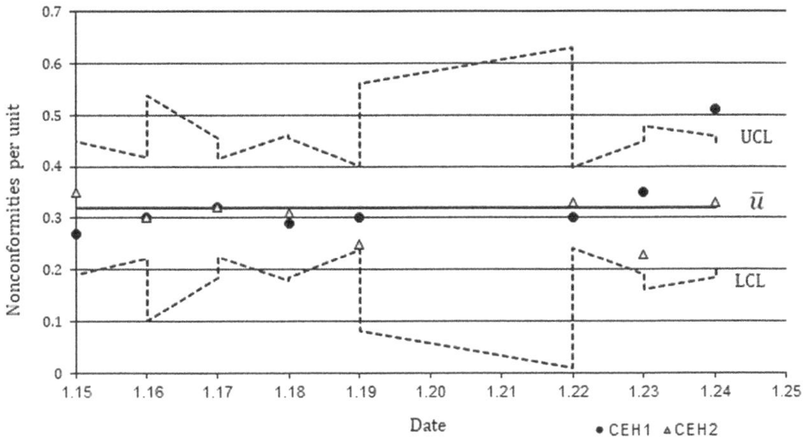

Fig. 4.16 U chart for nonconformities

Table 4.12 Butter packages weight

Sample	Weight [g]			
1	201.0	200.5	200.0	199.5
2	200.5	200.0	201.0	199.8
3	200.0	202.0	199.2	201.5
4	199.7	201.3	197.0	200.8
5	198.7	199.0	200.3	201.0
6	199.5	200.3	201.0	200.9
7	200.6	199.1	199.0	200.8
8	200.4	199.4	199.0	199.5
9	200.6	204.0	200.0	201.0
10	200.1	202.0	200.3	200.8
11	205.0	203.0	203.0	202.0
12	199.0	198.0	200.5	200.0
13	200.0	201.0	201.5	199.7
14	199.8	199.9	200.0	200.6
15	190.0	201.3	194.0	200.5

$1/\sqrt{4} = 0.5$, and considering k = 1.25 and h = 0.995 where k is the size of the shift planned to be detected and h is the number of standard deviations between the center line and the control limits. The h-value set the value at which an out-of-control case occurs.

$U_i = max(0, \ \ U_{i-1} - T + \widehat{\mu}_0 - k)$ where $\widehat{\mu}_0$ is the estimated mean of the subgroup, T is the target value and k is the size of the shift

$$L_i = \max(0, \quad L_{i-1} + T - \widehat{\mu}_0 - k)$$

When either U_i or L_i exceeds h, the process is out of control. V mask slope $= k \cdot \frac{\sigma}{\sqrt{m}} = 0.625$ where m = 4 is subgroup size.

The values are calculated in Table 4.13.

When U_i equals two (Fig. 4.17), for the 11th sample, the process registers a value out of control. After eliminating the cause of the variation, calculation with CuSum method can restart.

Table 4.13 Ui and Li values calculated for CuSum method

Average	$\widehat{\mu}_0 - T$	$\widehat{\mu}_0 - T - k$	Ui	$T - \widehat{\mu}_0$	$T - \widehat{\mu}_0 - k$	Li	CuSum $\sum(\widehat{\mu}_i - T)$
200.250	0.250	−1	0.000	−0.250	−1.500	0.000	0.250
200.325	0.325	−0.925	0.000	−0.325	−1.575	0.000	0.575
200.675	0.675	−0.575	0.000	−0.675	−1.925	0.000	1.250
199.700	−0.300	−1.550	0.000	0.300	−0.950	0.000	0.950
199.750	−0.250	−1.500	0.000	0.250	−1.000	0.000	0.700
200.425	0.425	−0.825	0.000	−0.425	−1.675	0.000	1.125
199.875	−0.125	−1.375	0.000	0.125	−1.125	0.000	1.000
199.575	−0.425	−1.675	0.000	0.425	−0.825	0.000	0.575
201.400	1.400	0.150	0.150	−1.400	−2.650	0.000	1.975
200.800	0.800	−0.450	0.000	−0.800	−2.050	0.000	2.775
203.250	3.250	2.000	**2.000**	−3.250	−4.500	0.000	6.025
199.375	−0.625	−1.875	0.125	0.625	−0.625	0.000	5.400
200.550	0.550	−0.700	0.000	−0.550	−1.800	0.000	5.950
200.075	0.075	−1.175	0.000	−0.075	−1.325	0.000	6.025
196.450	−3.550	−4.800	0.000	3.550	2.300	2.300	3.975

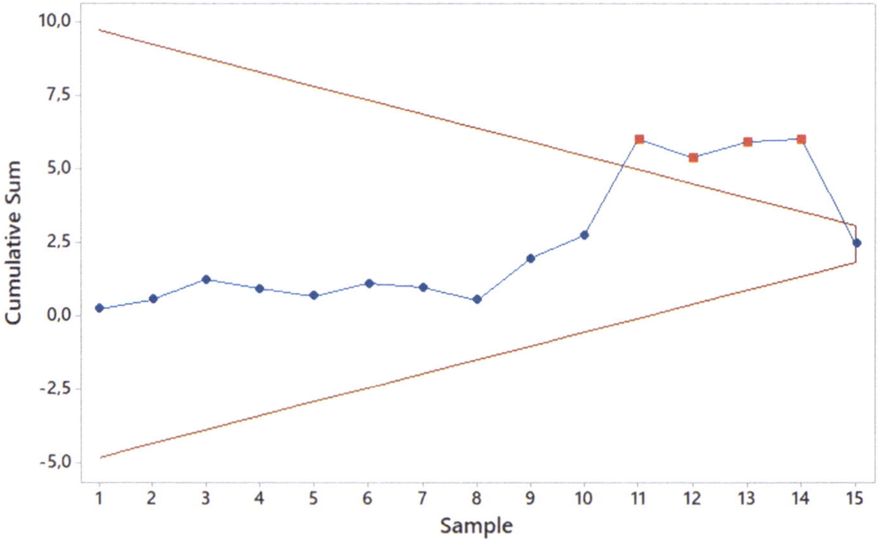

Fig. 4.17 CuSum chart

4.3 Analysing Process Capability

Daniela Borda, Iulia Bleoanca, and Iuliana Banu

4.3.1 Objectives and Learning Outcomes

The objectives of this chapter is to explain how a process can be evaluated to deliver products within the specification limits.

After completion of this chapter the reader should be able to calculate the capability indexes to characterize processes' performance and to evaluate the process capability, its potential and finding ways to improve its performance.

4.3.2 Introduction

Fiengenbaum's definition of capability refers to "process capability that is quality performance capability of the process with given factors, under normal, in-control conditions". From all classical definitions only this one mentions performance as an approach to understand capability and probably this is the reason why it takes a large number of numerical indexes to describe it.

Capability studies show that even processes with a perfect drawn ($\overline{X}$, R) chart, without any out of control variations may demonstrate an unwanted behavior if the UCL/LCL and X-double bar (precision and central tendency) does not fit within specifications (process not capable of meeting the specifications).

However, we might have situations when an excellent centered process within tolerance limits [USL, LSL], right in the middle, with an aimed center value equal with $T = (LSL + USL)/2$ can have a large number of defective fractions if variability is too large.

Understanding processes variability and potential can give information on the ways in which process can be improved. The statistical stability is not characterized only thorough the control charts but also by calculating indexes to check the process centring and the report among its natural limits and the ones imposed by specifications.

Performance is what separates the common processes from the very best ones. A capable process is a desirable one but it is not easy to define what makes it the best and how to achieve this major objective. Maybe this is why capability had until now so many definitions and many indexes to express its complexity.

"Process capability is the measured, inherent reproducibility of production turned out by a process" said Juran.

"Process capability is quality-performance capability of a process with given factors under normal, in-control conditions" was the definition given by Fiengenbaum, just to have two of the most known definitions of the process capability. Finding the best ways to express what performance the process holds and its potential is very important in finding ways to improve.

Statistical performance is determined by three factors:

- process stability -validated by control charts-most often ($\overline{X}$, R) chart;
- process potential-validated by the index C_p;
- process capability- validated by the capability index C_{pk}.

Cp is the process capability index that measures the ratio between the accepted process variability and the real process variability, or specification limits and process control limits.

If the natural variability given by the interval between the upper control limit and lower control limit [UCL, LCL] is included in specifications interval given by the upper specification limit and lower specification limit [USL, LSL], then process has the capability to display a variability smaller than the one required. We can write it as [UCL, LCL] ⊂[USL, LSL] and graphically represent this in Fig. 4.18.

$$C_p = \frac{USL - LSL}{UCL - LCL} = \frac{USL - LSL}{6\,\sigma}$$

$\widehat{\sigma}$-is the estimation of process standard deviation

The number found by calculating the process capability index determines whether the process is capable of meeting the specifications (Table 4.14).

If we carefully look at the index, we might notice that the formula includes a measure of spread but not a measure of location of the quality characteristic. Therefore, we can come across situations like the ones depicted below case a) or b), where $C_p \geq 1$ but we have nonconformities (Fig. 4.19).

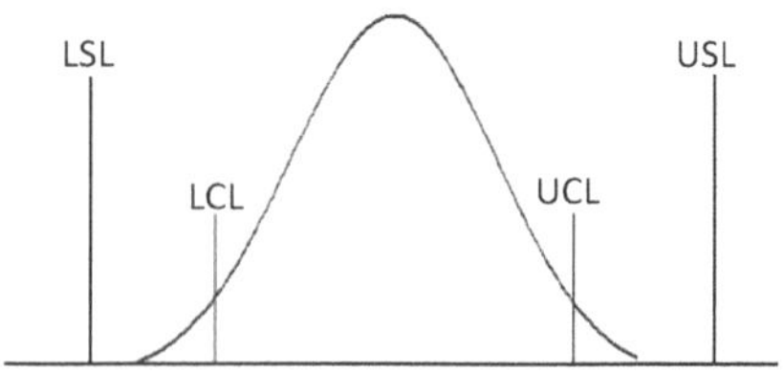

Fig. 4.18 A capable process

Table 4.14 Interpretation of process capability index

Capability index	Conclusion
≥1.33	Process is satisfactory
1÷1.33	Process is capable
≤1	Process is unsatisfactory

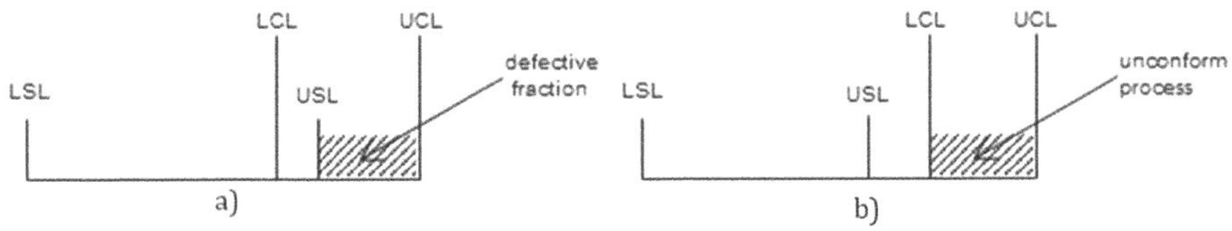

Fig. 4.19 Cases with variations outside control limits and a good C_p values

To describe the statistical performance one coefficient (C_p) is not enough. In order to have a better image of the process behaviour two indicators of the zone capability were introduced: $C_{pk\text{inf}}$ and $C_{pk\text{sup}}$.

$$C_{pk\,\text{inf}} = \frac{|\bar{\bar{x}} - LSL|}{3\,\sigma},\quad C_{pk\text{sup}} = \frac{|\bar{\bar{x}} - USL|}{3\,\sigma}$$

$C_{pk\text{inf}}$ and $C_{pk\text{sup}}$ show the process variability related to the specification limits USL and LSL. The value considered as reference is the lowest value from the indexes:

$$C_{pk} = \min\{C_{pk\,\text{inf}}, C_{pk\text{sup}}\}$$

To have at least a low performance process it is required that C_{pk} should be equal with 1, if not higher than 1.33, as it is expected for better performances.

Montgomery et al. (1994) suggests, depending on the type of process, recommended values for C_p and C_{pk}, such as for a process in progress Cp should be equal to 1.33 and C_{pk} to 1.25; while for a new process the recommended C_p and C_{pk} values are 1.5 respectively 1.45.

The new six sigma philosophy implemented by Motorla and Ford requires an Cpk of minimum 2, which means a drastic reduction of variability to 3.4 ppm nonconformities.

To quantify the performance of the process the value for C_{pk} and C_p will be compared with a specified level of performance (SPL), with the minimum requirement SPL $\geq$ 1.

If we are to represent the graph using for the reference system (Cp, Cpk) coordinates, than we obtain for a minimal SPL

Zone A –the process is poor for both its potential capability and process capability are smaller than 1.

Zone B- the performance of the process is only partially covered, because only the potential capability is adequate.

Zone C-the performance of the process is adequate (SPL >1) (Fig. 4.20).

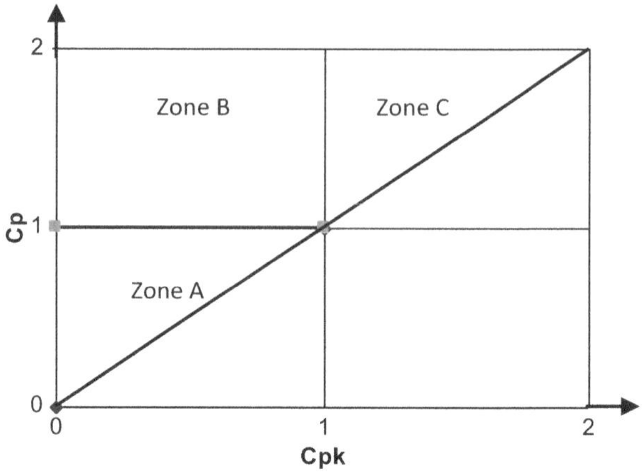

Fig. 4.20 The process level of performance

For zone C we can consider the following cases:

(i) if $1 \leq SPL \leq 1.33$—the performance of the process is low
(ii) if $1 \leq SPL \leq 1.67$ the process performance is good
(iii) if $SPL > 1.67$ the performance achieved by the process is very good

To explain why although there are no values outside the specified tolerance, in the sample the process performance is low and knowing that the sample is drawn from a normally distributed population, we can calculate the defective fraction using Z (or Z_U and Z_L).

$$z_U = \frac{USL - \overline{\overline{x}}}{\sigma}, \quad z_L = \frac{\overline{\overline{x}} - LSL}{\sigma}$$

$$p_i = Prob\ (X < LSL) = Prob\left(U < \frac{\overline{\overline{x}} - LSL}{\widehat{\sigma}}\right)$$

$$p_s = Prob\ (X > USL) = Prob\left(U > \frac{\overline{\overline{x}} - LSL}{\widehat{\sigma}}\right)$$

In general, the bigger z is, the better. A value of at least +3 is required to assure that 0.1% or less defective is produced.

Another capability index is C_{pm} introduced by Taguchi which express the relationship with an optimum target value denominated T. The widely used equation for C_{pm} was introduced by Chan et al. in 1988 and written as:

$$C_{pm} = \frac{USL - LSL}{6\sqrt{\sigma^2 + (\mu - T)^2}} = \frac{\frac{USL - LSL}{6\sigma}}{\sqrt{1 + \frac{(\mu - T^2)}{\sigma^2}}} = \frac{C_p}{\sqrt{1 + V^2}},$$

where $V = \frac{T - \mu}{\sigma}$ and $T = \frac{USL + LSL}{2}$

When we use the estimation of C_{pm} the formula is:

$$\underset{pm}{C} = \frac{C_p}{\sqrt{1+V^2}}$$

and $V = \frac{\bar{\bar{x}}-T}{\sigma}$ and $\sigma_1 = \sqrt{\frac{1}{n-1}\sum (x_i - T)^2}$
where $C^*_{pm\ inf} = \frac{T-LSL}{3\sigma_1}$ and $C^*_{pm\ sup} = \frac{USL-T}{3\sigma_1}$

In practice there is also used an estimation of Pearn-Kotz-Johnson (1992) index:

$$\underset{pmk}{C} = \min\left\{\frac{USL-\bar{\bar{x}}}{3\sqrt{s^2+\left(\bar{\bar{x}}-T\right)^2}}, \frac{\bar{\bar{x}}-LSL}{3\sqrt{s^2+\left(\bar{\bar{x}}-T\right)^2}}\right\}$$

where T is the target value

If $T = \mu$ than $C_{pmk} = C_{pk}$

An estimation of the Luceňo (1996) index can be computed:

$\underset{pc}{C} = \frac{USL-LSL}{6\sqrt{\frac{\pi}{2}} x \frac{1}{n} x \sum |x_i - m|}$ where $m = \frac{USL+LSL}{2}$ and $6\sqrt{\frac{\pi}{2}} \approx 7.52$ is correction for the cases when the variable X is not normally distributed.

4.3.3 Calculating Process Capability Indices Cp and Cpk for the Length of Pickled Cornichons

A canning plant manufactures pickle cornichons (small cucumbers) and the length of a cornichon should vary between 5 ÷ 9 cm. The capability of the process needs to be evaluated.

Samples were collected from 25 subgroups of consecutively produced unit, measured and placed back on the line.

The results are given in Table 4.15. together with average and range computed for each subgroup.

For $\overline{X}$- chart: $UCL_{\bar{x}} = \overline{\overline{X}} + A_2 \cdot \overline{R}$ $\quad LCL_{\bar{x}} = \overline{\overline{X}} - A_2 \cdot \overline{R}$

For $\overline{R}$- chart: $UCL_{\overline{R}} = D_4 \cdot \overline{R}$ $\quad LCL_{\overline{R}} = D_3 \cdot \overline{R}$

A_2, D_3, D_4 are coefficients that can be read from Appendix Table 4.20.

For n = 4, $A_2 = 0.729$, $D_3 = 0$, $D_4 = 2.282$.

Applying the coefficients, the results are:

$$LCL_{\bar{x}} = 5.89 - 0.729 \times 0.97 = 5.18 \quad UCL_{\bar{x}} = 5.89 + 0.729\ x\ 0.97 = 6.60$$

Table 4.15 Cornichons' length

S	X1	X2	X3	X4	Average	Range	S	X1	X2	X3	X4	Average	Range
1	4.9	6.0	6.5	6.4	5.95	1.6	14	5.5	5.8	5.9	6.0	5.80	0.5
2	4.7	6.0	6.3	5.8	5.70	1.6	15	6.0	5.4	5.6	6.0	5.75	0.6
3	5.8	6.0	6.6	6.3	6.18	0.8	16	4.9	6.0	6.4	6.0	5.83	1.5
4	6.0	5.5	5.9	6.1	5.88	0.6	17	5.2	5.9	6.1	6.2	5.85	1.0
5	5.0	6.4	6.5	6.0	5.98	1.5	18	5.4	5.3	5.4	5.9	5.50	0.6
6	5.8	5.5	5.0	6.0	5.58	1.0	19	5.7	6.0	6.4	6.3	6.10	0.7
7	5.8	6.0	6.3	6.1	6.05	0.5	20	5.4	5.0	6.4	6.0	5.70	1.4
8	6.0	5.1	5.7	6.1	5.73	1.0	21	5.7	5.9	6.3	6.0	5.98	0.6
9	5.5	6.5	7.0	6.2	6.30	1.5	22	5.6	5.5	6.0	6.1	5.80	0.6
10	6.0	6.3	7.0	5.4	6.18	1.6	23	6.0	6.3	5.6	6.0	5.98	0.7
11	5.7	5.8	6.0	6.3	5.95	0.6	24	5.5	5.6	5.7	6.0	5.70	0.5
12	5.2	5.9	6.1	6.2	5.85	1.0	25	4.9	6.4	6.5	6.2	6.00	1.6
13	5.7	6.0	6.4	6.0	6.03	0.7						**5.89**	**0.97**

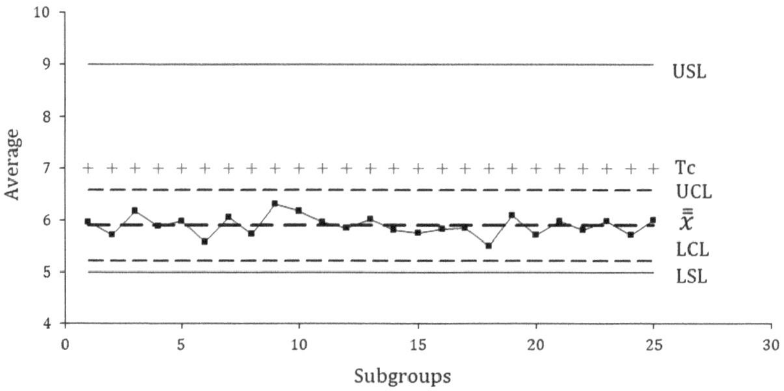

Fig. 4.21 X chart for cornichons

$$UCL_{\overline{R}} = 2.282\ x\ 0.97 = 2.21 \quad LCL_{\overline{R}} = 0 \times 0.97 = 0$$

The average and range charts are presented in Figs. 4.21 and 4.22. Both average and range charts show a in control variation of the length.

σ can be estimated: $\sigma = \frac{\overline{R}}{d_2}$ $\sigma = \frac{0.97}{2.059} = 0.471$, where $d_2 = 2.059$ (Appendix Table 4.20)

The capability index $C_p = \frac{USL - LSL}{6\sigma} = \frac{9-5}{6 \times 0.471} = 1.41$

The process capability and chart for average and range indicates a good process, and if we calculate the natural spread of the process we obtain:

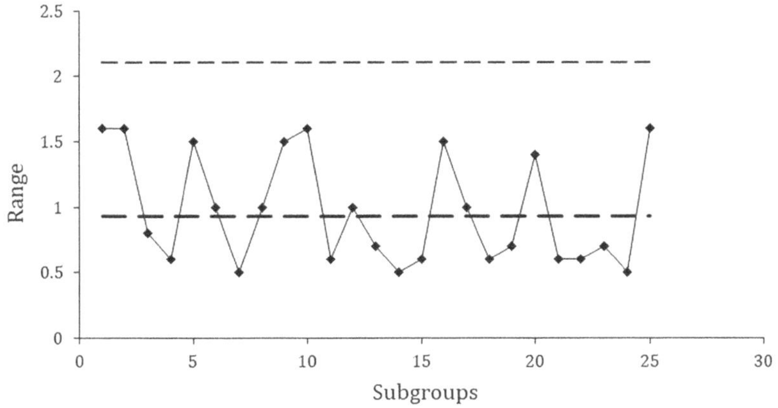

Fig. 4.22 R chart for cornichon

$$UCL = \bar{\bar{x}} + 3 \cdot \hat{\sigma} = 5.89 + 3\,x\,0.472 = 7.31$$

$$LCL = \bar{\bar{x}} - 3 \cdot \hat{\sigma} = 5.89 - 3\,x\,0.472 = 4.74$$

However, it can be noticed tolerance are too large compared with the measured values and the central tolerance Tc is equal 7 which differs from the calculated mean.

4.3.4 Calculating Process Capability Indices Cp, Cpk and Cpm, Cpmk for the Weight of Bags of Mixed Frozen Vegetables

A frozen vegetable mix is packed in 400 g weight bags. According to its internal specification the allowed weight variation of a bag is ± 10 g. To check the process capability 50 samples in subgroups of five are drawn from a batch and scaled. The values are presented below (Table 4.16):

$\bar{\bar{x}} = 401$, $\bar{s} = 3.17$ and $\sigma = \frac{\bar{s}}{c_4}$ where $c_4 = 0.94$ (Appendix Table 4.20)

$$\sigma = \frac{3.17}{0.94} = 3.37, \; C_p = \frac{USL - LSL}{6\,\sigma} = \frac{410 - 390}{6 \times 3.37} = 0.99,$$

$$C_{pk\,\text{inf}} = \frac{\left|\bar{\bar{x}} - LSL\right|}{3\,\sigma} = \frac{401 - 390}{3x3.37} = 1.09$$

Table 4.16 Weight of vegetable bags

	S_1	S_2	S_3	S_4	S_5	S_6	S_7	S_8	S_9	S_{10}
	404	406	405	403	399	400	395	401	400	405
	399	398	402	405	402	399	400	403	400	396
	404	405	401	401	400	405	400	400	398	406
	398	401	399	402	405	403	399	405	395	405
	399	400	400	391	404	398	399	400	405	400
$\bar{x}$	400.8	402.0	401.4	400.4	402.0	401.0	398.6	401.8	399.6	402.4
s	2.95	3.39	2.30	5.46	2.55	2.92	2.07	2.17	3.65	4.28

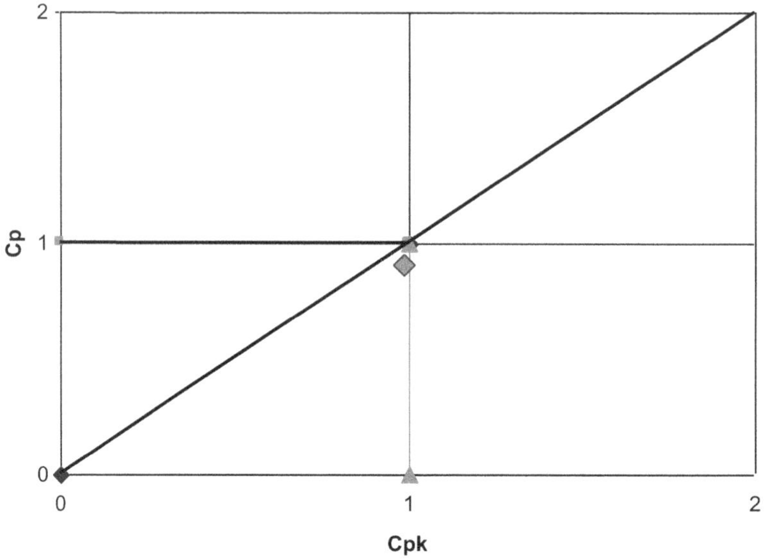

Fig. 4.23 The performance of the vegetable mix packaging process

$$C_{pk\sup} = \frac{|\bar{\bar{x}} - USL|}{3\ \sigma} = \frac{410 - 401}{3x3.37} = 0.89$$

$$C_{pk} = \min\{C_{pk\inf}, C_{pk\sup}\} = \min\{1.09; 0.89\} = 0.89$$

If we place the point {0.89;0.99} in the C_{pk}, C_p graph we notice that the point is under the acceptance diagonal, although very close to the minimal allowed performance (Fig. 4.23).

To explain why although there are no values outside the specified tolerance, in the sample the process performance is low and knowing that the sample is drawn from a normally distributed population, we can calculate the defective fraction using Z (or Z_U and Z_L).

$$p_i = Prob(U < z_L) \ z_L = \frac{401 - 390}{3.37} = 3.26$$

In this case the defective fraction (p_i) should be less than 0.1%.

Using the Gauss-Laplace distribution (Appendix Table 4.21) we can calculate p_i:

$$p_i = 0.5 - \mathrm{A} = 0.5 - 0.4994 = 0.0006 = 0.06\% \ \text{(A-read from appendices 2).}$$

The situation is different for the USL:

$$p_s = Prob\left(U > \frac{410 - 401}{3.37}\right) \ z_U = \frac{410 - 104}{3.37} = 2.67$$

$$p_s = 0.5 - A = 0.5 - 0.4961 = 0.0039 = 0.39\%$$

Here z_U is less than 3, so the percentage of defective (p_s) is higher (0.39%) than p_i (0.06%).

Considering this information, the process variation should be reduced in the upper limit to bring the capability indexes in the performance zone.

4.4 Multicriteria Decision Methods for Process Analytical Technology (PAT)

Davide Ballabio and Saverio Mannino

4.4.1 *Objectives and Learning Outcomes*

An example based on a real case study will illustrate the practical use of a multi-criteria decision methodology. The Multiple Criteria Decisions considered are the indices of desirability and utility.

After completion of this chapter the reader should be able to understand the principles of PAT and to implement PAT.

4.4.2 *Introduction*

There are various types of analytical methods applied to food that can provide information concerning physical and chemical characteristics of food. However, these methods are often complicated and time consuming while the food industry has general need for methods suitable for processing and quality control that are simple, rapid and able to provide real time data and information on processed food.

Nowadays there is a change of strategies in food control: from the final product to the process. The concepts of quality by design have assuming greater importance by using of computerized data gathering and evaluation techniques, and process and product monitoring methods through on/in line sensors. This approach, now defined Process Analytical Technology (PAT), can provide products of better quality at lower costs as any shifts from optimal or defined processing conditions could be promptly pointed out and the process modified according to.

According to FDA website, PAT is defined as: "a system for designing, analysing, and controlling manufacturing through timely measurements (i.e., during processing) of critical quality and performance attributes of raw and in-process materials and processes with the goal of ensuring final product quality". The goal of PAT is "to understand and control the manufacturing process, which is consistent with our current drug quality system: quality cannot be tested into products, it should be built-in or should by design". The main PAT objective is to increase process knowledge and therefore ensure a consistent pre-defined product using data from on-line and real time analysis.

It is well known that the application of the concepts of quality, as required by ISO 9001:2000, require management tools that are useful to support not only the design but also the control systems implemented. The process approach provides -by the use of control systems and decision-making strategies- to intervene appropriately when the system is out of limits.

There are numerous technical and operational management described in the literature and the seven tools of quality management are the most known. Recently, the use of decision-making strategies based on Multiple Criteria Decision (1) (Multicriteria Decison Making), have proved useful management tool for the control of production processes. These methods make it possible to reach the optimal choice among the many possible, based on the construction of a ranking of the different operational possibilities. In order to achieve this ranking is, however, necessary to identify for each individual criterion conditions that are considered optimal, acceptable or unacceptable and the ways in which you can switch from one situation to another.

4.4.3 Controlling the Production Processs Instant Sports Drinks

A company produces an energy supplement salt of the type "instant sports drink" according to the flowchart shown in Fig. 4.24. The process as seen mainly consists in a grinding step, which follows the mixing, the granulation and the final packaging.

Manufacturer in accordance with the requirements of its Quality Management System has defined the tests and inspections to be carried out along the production process in order to monitor some important features of the product. The controls are designed to assess the correspondence of the product specifications laid down and operate corrections from the input so that this correspondence is kept continuously up to the finished product.

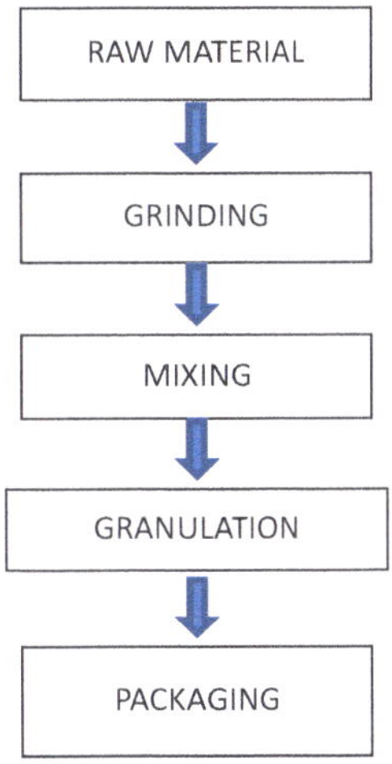

Fig. 4.24 Production steps

Table 4.17 Parameters of the study

Parameter	Units	Minimum	Optimum	Maximum
Sodium	mg/100	675	750	825
Potassium	mg/100	135	150	165
Magnesium	mg/100	38	45	52
Vitamin C	mg/100	48	60	84
Acidity	%	3	4	5
Weight loss	%	0	0	2.1

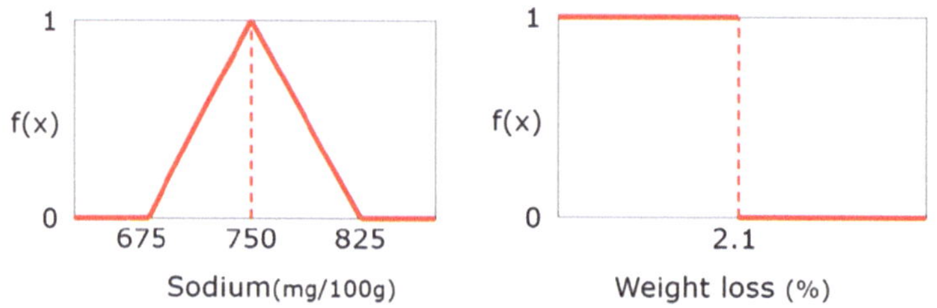

Fig. 4.25 Functions for sodium and weight loss

In practice, these checks consist on the determination of six different parameters immediately after the mixing step of the granules. The parameters controlled, for which are provided for a range of compliance and a value of optimality, are described in Table 4.17.

Only when the product complies with all specification limits can move on to the next stage of packaging and placing on the market. Otherwise is optimized by further mixing processes.

It is clear that at this level it is necessary to have fast, safe and easy tools to interpret, which provide decision-making to optimize the manufacturing process, while identifying the main causes of non- conformity of the mixed products. This would lead to an increase in the number of semi-finished products that meet the quality specifications with subsequent decrease in the costs of materials and production time.

The methodology based on the index of desirability and utility proved to be a useful tool.

Indices of Desirability and Utility

Below is presented in five successive points a simplified construction of these indices relatively to the analysis under study:

1. For each of the six parameters considered (x), depending on the values of optimality and interval of conformity set Table 4.17), a function f(x) is built. For instance, for sodium and for weight loss are defined the functions represented in Fig. 4.25. In practice, as regards the sodium, the optimal amount established by the company is 750 mg/100 g with lower specification limit of 675 mg/100 g and

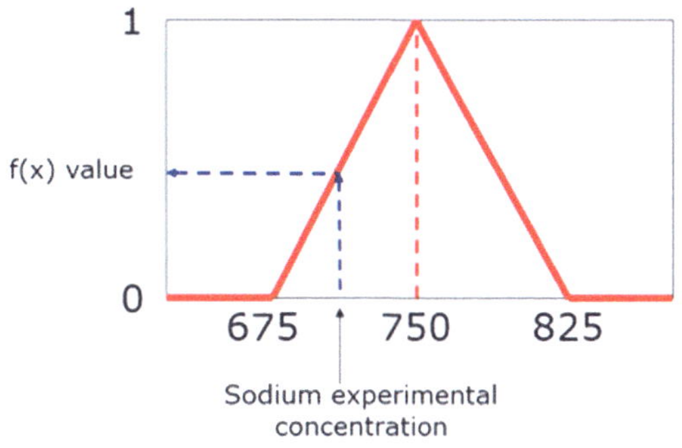

Fig. 4.26 Triangular function for sodium

specification limit higher than 825 mg/100 g, while for Weight Loss there is only the upper limit of 2.1%.

2. For each product, depending on the value of the parameter considered, the corresponding value of f(x) is calculated. For example, considering the sodium and the triangular function constructed above, it continues in the manner shown in Fig. 4.26.

   A sample with a concentration of Sodium equal to 750 mg/100 g (value of optimality) assumes a corresponding value of f(x) equal to 1. A sample with a concentration of Sodium outside the range of compliance (greater than 825 or less than 675 mg/100 g) assumes a corresponding value of f(x) equal to 0. In all other cases, the more the Sodium concentration approaches the optimal value, the more the function f(x) takes values close to 1.
3. This procedure is carried out on all six parameters of Table 4.17. For each are then associated six values of f (x) between 0 and 1.
4. The index of desirability (D) is than calculated according to the following formula:

$$D_i = \sqrt[p]{f_{i1} \cdot f_{i2} \cdots f_{ip}}$$

where f_{ip} is the value of the function f(x) of the product i-th for the p parameter, p is the number of parameters considered (in this case $p = 6$). The index of desirability has the following properties:

$D_i = 0$ if the i -th product has at least a f_{ip} equal to 0 (i.e., it moves outside of the conformity for at least one parameter)

$D_i = 1$ if the i-th product f_{ip} is always equal to 1 (i.e., for all parameters the product meets the value of optimality)

5. The utility index (U) is calculated for each product using the following equation:

$$U_i = (f_{i1} + f_{i2} + \cdots + f_{ip})/p$$

where f_{ip} is the value of the function f (x) of the product for the pth - parameter p and, p is the number of parameters considered. The index of utility satisfies the following properties:

$U_i = 0$ if the i-th product f_{ip} is always equal to 0 (i.e., moves outside of the compliance for all parameters)
$U_i = 1$ if the i-th product f_{ip} is always equal to 1 (i.e., for all parameters the product meets the value of optimality)

In the case under study, 365 products obtained in the mixing step in a time range of 2 years were determined by the values of the six parameters (Sodium, Potassium, Magnesium, Vitamin C, Acidity and Weight Loss) identified by the product specifications. On the basis of these analyses and considering specifications listed above in Table 4.17, for each product were calculated the indices of desirability and utilities. To optimize the screening of all 365 products is therefore possible to realize a chart reporting on the x axis values of desirability and on the Y-axis values of utility of each production (Fig. 4.27).

What indications may suggest the graph of Fig. 4.27 to the person in charge of the production line that has to assess the quality system implemented?

1. First of all, by analysing the graph it is clear that only a small part of the productions meet compliance (82 out of 365 products, corresponding to 22% of total). Productions that meet the limit for compliance for all the six parameters considered (labelled in the diagram as "eligible") have all values of desirability greater than 0 and are represented by points that lie on the right side of the graph. On the contrary productions that do not meet the limit of Compliance for at least

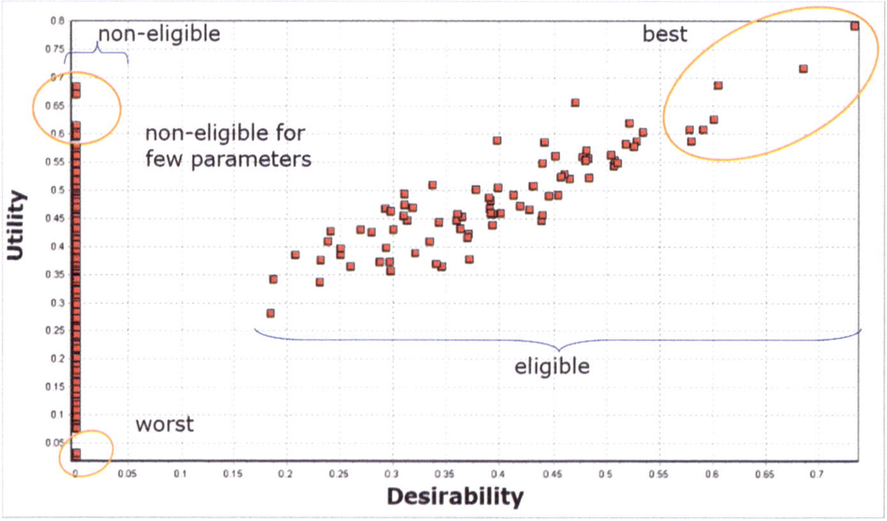

Fig. 4.27 Chart reporting for the study

one parameter ("non-eligible") all have a value equal desirability 0. They are represented by the points, which are located in the left part of the graph.
2. The production manager, after having established that most of the production has provided unsatisfactory results from the qualitative point of view, with a consequent and evident economic loss, can first identify easily which are the products which are closest to the values of optimality (labelled as "best" in the graph of Fig. 4.27). They are characterized by high values of desirability and utility (close to 1), and then represented by points that are located at the top - right of the graph. The standard of production of these samples can be repeated to increase the performance of the process. In contrast, samples that the more you move away from the values of optimality ("worst") are characterized by values close to 0 desirability and utility: they are in fact represented by the points that are located in the bottom—left of the graph.
3. Another opportunity to increase the performance of the production line would be the introduction of some changes in the manufacturing of some products. But which ones? The products, that "are not eligible for a few parameters", have a value of desirability of 0, but a high value for the utility. This corresponds to products that come from the range of compliance for one or a few parameters, while the remaining assume values near optimality and therefore are positioned in the upper—left portion of the graph. These tools allow you to easily locate, retrieve them and improve them, rather than acting on eligible products but with a lower utility value (corresponding to the points that are located in the centre of the graph), or products that always fall within the ranges of conformity, but always far away from the values of optimality set in the specifications. In the latter case it should make any changes at all (or nearly all) the six parameters considered, while in the first case, reworking products that are not generally successful in the range of conformity of a single parameter, it will be sufficient to act on the latter (with significant savings).
4. Finally, it is easy to calculate the number of products that do not comply with the intervals of conformity for each parameter p considered, identifying all samples characterized by an $f_p = 0$. In this way, the parameters that are monitored total cause in the production line the majority of non-conforming products. By using the histogram in Fig. 4.28, which refers to the case study in question, it can be deduced, as are necessary interventions and monitoring more closely on the production line for potassium and magnesium. These in fact are the parameters that cause the greatest amount of products do not comply with the technical specifications (including 146 samples fall outside the range of Potassium and 136 in accordance with one of the Magnesium).

In conclusion, what are the advantages of using multi-criteria decision-making for the control of production processes? Surely their speed of application, ease of interpretation, the ability to provide a tool for the interpretation of complex systems and the possibility to obtain a global screening of the products, by means of which implement the next programming and optimization of corrective actions on processes productive (Fig. 4.28).

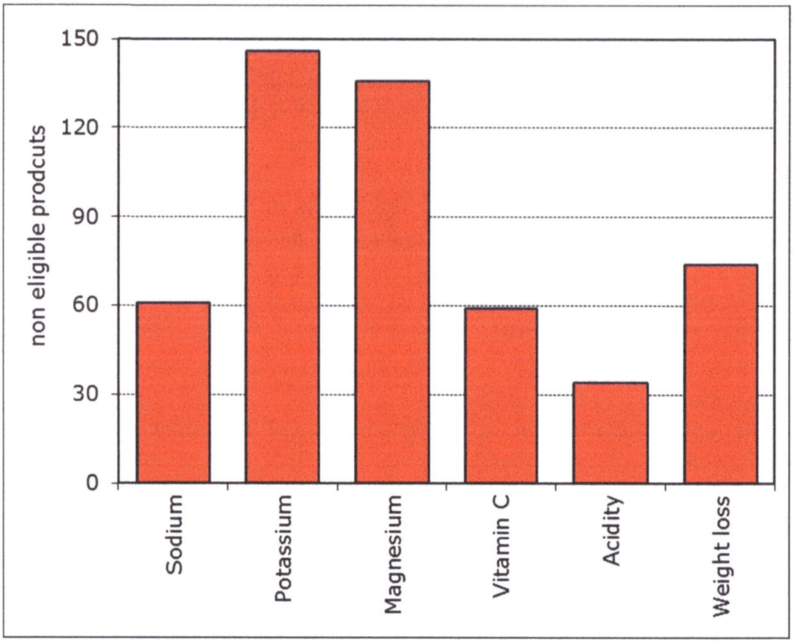

Fig. 4.28 Number of non eligible products for each considered parameter

In this case only 22% of the analysed products meet all ranges of conformity set and only 7 of these assume values close to optimality. If the production system was analysed according to the six-sigma metric (see Tip), which takes into account the number of defects and the number of opportunities to defect, the process would be classified on an annual basis as 2.64 sigma.

If the process was rather at the level of 6sigma, and with fewer defects, and the cost per defect could be evaluated in 25 euro, there would be a saving of around €70,000 in a year.

Tip
Six Sigma

The fundamental objective of the Six Sigma methodology is the implementation of a measurement-based strategy that focuses on process improvement and variation reduction through the application of two Six-Sigma methodologies. DMAIC and DMADV. The Six Sigma DMAIC process (Define, Measure, Improve, and Control) is an important system for existing process falling below specification and looking for incremental improvement. The Six Sigma

(continued)

DMADV process (Define, Measure, Analyse, Design, and Verify) is an improvement system used to develop new processes or products at Six Sigma quality levels.

The term sigma is a Greek alphabet letter (σ) used to describe variability. In Six Sigma, the common measurement index is DPMO (Defects Per Million Operations) and can include anything from a component, piece of material, or line of code, to an administrative form, time frame or distance. A sigma quality level offers an indicator of how often defects are likely to occur, where a higher sigma quality level indicates a process that is less likely to create defects. Consequently, as sigma level of quality increases, product reliability improves the need for testing and inspection diminishes, work in progress declines, cycle time goes down, costs go down, and customer satisfaction goes up.

In other words, as the process sigma value increases from zero to six, the variation of the process around the mean value decreases. With a high enough value of process sigma, the process approaches zero variation and is known as 'zero defects.'

To have a more comprehensive understanding about sigma quality level, visit the following website: http://www.isixsigma.com/

4.5 Fitting Data to Models: Correlation and Regression Analysis for Food Quality Prediction

Ivanka Zheleva, Todor Todorov, Hüseyin Bozkurt, and Fahrettin Göğüş

4.5.1 *Objectives and Learning Outcomes*

The main objective of this chapter is to present information about curve fitting, equations of approximating curves, the method of least squares, regression and fundamentals of correlation theory. The least square lines, explained and unexplained variation, coefficient of correlation are also considered.

After completion of this chapter the reader should be able to fit data to statistical models.

4.5.2 *Introduction*

A model is *a relationship between variables.* We fit a model to data from a sample almost always to make a statement about the model in the population. So we will deal with models here, under the heading of generalizing to a population. Thus a model is another way of summarizing data using effect statistics.

The Method of Least Squares

To avoid individual judgment in constructing lines or other approximating curves to fit sets of data, it is necessary to agree on a definition of a "best fitting lines", "best fitting formula", etc.

To motivate a possible definition, consider Fig. 4.29 in which the data points are given by (X_1,Y_1), (X_2,Y_2), . . .,(X_n,Y_n). For a given value of X, say X_1, there will be a difference between the value Y_1 and the corresponding value as determined from the curve C. As indicated in the figure we denote this difference by D_1, which is sometimes referred to as a *deviation*, *error* or *residual* and may be positive, negative or zero. Similarly, corresponding to the values X_2, . . ., X_n we obtain the deviations D_2,. . . D_n.

A measure of the "*goodness of fit*" of the curve C to the given data is provided by the quantity $D_1^2 + D_2^2 + \ldots + D_n^2$. If this is small the fit is good, if it is large the fit is bad. We therefore make the following.

Definition Of all curves approximating a given set of data points, the curve having the property that $D_1^2 + D_2^2 + \ldots + D_n^2$ is a minimum called *best fitting curve.*

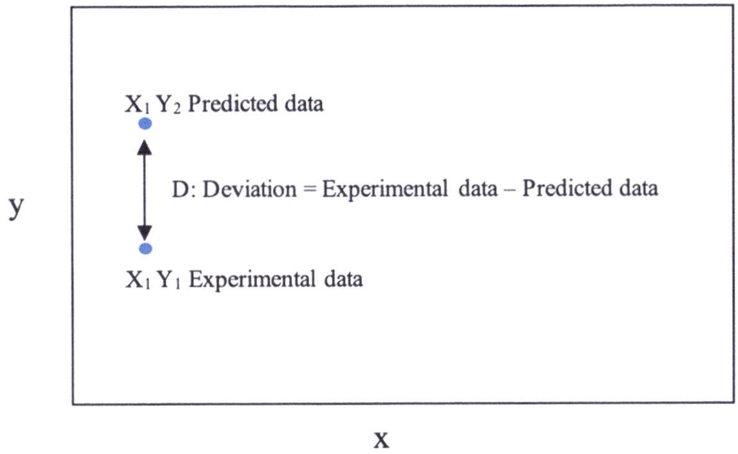

Fig. 4.29 Representation of experimental data, peredicted data and deviation.

A curve having this property is said to fit the data in the *least square sense and* is called *least square curve.*

Thus a line having this property is called a "*least square line*", a parabola with this property is called a "*least square parabola*", etc.

It is customary to employ the above definition when X is the independent horizontal instead of vertical deviations, which amounts interchange of the X and Y axes. These two definitions in general lead to different least square curves. Unless otherwise specified we shall consider Y as the dependent and X as the independent variable.

It is possible to define another least square curve by considering perpendicular distances from each of the data points to the curve instead of either vertical or horizontal distances. However, this is not used too often.

The *least square line* approximating the set of points $(X_1, Y_1), (X_2, Y_2), \ldots, (X_n, Y_n)$ has the equation

$$Y = a_0 + a_1 X \tag{4.1}$$

Where the constants a_0 and a_1 are determined by solving simultaneously the equations

$$\sum Y = a_0 n + a_1 \sum X \tag{4.2}$$

$$\sum XY = a_0 \sum X + a_1 \sum X^2 \tag{4.3}$$

Note that this is not derivation of the normal equations but simply a means for remembering them.

Note also that in Eq. 4.2 and 4.3 we have used the short notation ∑X, ∑XY, etc., in place of $\sum_{i=1}^{n} x_i$ and $\sum_{i=1}^{n} x_i y_i$, etc.

The labour involved in finding a least square line can sometimes be shortened by transforming the data so that $x = X - -X$ and $y = Y - -Y$. The equation of the least square line can then be written

$$y = \left(\frac{\sum xy}{\sum x^2}\right)x \quad or \quad y = \left(\frac{\sum xY}{\sum x^2}\right)x \tag{4.4}$$

In particular if X is such that $\sum X = 0$, *i.e.* $-X = 0$, this becomes

$$Y = \_Y + \left(\frac{\sum XY}{X^2}\right)X \tag{4.5}$$

From these equations it is evident that the least square line passes through the point $(-X, -Y)$, called the "*centroid*" or "*center of gravity*" of the data.

If the variable X is taken as dependent instead of independent variable, we write Eq. (4.1) as $X = b_0 + b_1 Y$. Then the above results hold if X and Y are interchanged and a_0 and a_1 are replaced by b_0 and b_1 respectively. The resulting least square line is in general not the same as that obtained above

The least square parabola approximating the set of points (X_1, Y_1), (X_2, Y_2), ... (X_n, Y_n) has the equation

$$Y = a_0 + a_1 X + a_2 X^2 \tag{4.6}$$

where the constants a_0, a_1 and a_2 are determined by solving simultaneously the equations

$$\begin{aligned} \sum Y &= a_0 N + a_1 \sum X + a_2 \sum X^2 \\ \sum XY &= a_0 \sum X + a_1 \sum X^2 + a_2 \sum X^3 \\ \sum X^2 Y &= a_0 \sum X^2 + a_1 \sum X^3 + a_2 \sum X^4 \end{aligned} \tag{4.7}$$

called the "*normal equations for the least square parabola* (4.6).

The Eq. (4.7) are easily remembered by observing that they can be obtained formally by multiplying Eq. (4.7) by *1*, *X*, and X^2 respectively and summing on both sides of the resulting equations. This technique can be extended to obtain normal equations for least square cubic curves, least square quadratic curves and in general any of the least square curves (Table 4.17).

As in the case of the least square line, simplifications of Eq. (4.7) occur if X is chosen so that $\sum X = 0$. Simplifications also occurs by choosing the new variables

$$x = X - \overline{X},\ y = Y - \overline{Y}$$

In Table 4.17 are listed several equations of approximating curves which are often used in practice:

Correlation and Regression

Often, on the basis of sample data, we wish to estimate the value of a variable *Y* corresponding to a given value of a variable *X*. This can be accomplished by estimating the value of *Y* from a least square curve which fits the sample data. The resulting curve is called a "*regression curve*" of *Y* on *X*, since *Y* is estimated from *X*.

If we desired to estimate the value of *X* from a given value of *Y* we would use a regression curve of *X* on *Y*, which amounts to interchanging the variables in the scatter diagram so that *X* is the dependent variable and Y is the independent variable. This is equivalent to replacing vertical deviations in the definition of least square curve by horizontal deviations.

In general the regression line or curve of *Y* and *X* is no the same as the regression line or curve of *X* on *Y*.

If the independent variable *X* is time, the data show the values of *Y* at various times. Data arranged according to time are called "*time series*". The regression line or curve of *Y* on *X* in this case is often called a "*trend line*" or "*trend curve*" and is often used for purposes of estimation, prediction or forecasting.

Problems involving more than two variables can be treated in manner analogous to that for two variables. For example, there may be a relationship between the three variables *X*, *Y* and *Z* which can be described by equation

$$Z = a_0 + a_1X + a_2Y \tag{4.8}$$

which is called *a linear equation in the variables X, Y and Z.*

In a three dimensional rectangular co-ordinate system this equation represents a plane and the actual sample points (X_1, Y_1, Z_1), (X_2, Y_2, Z_2),. . .,(X_n, Y_n, Z_n) may "scatter" not too far from this plane which we can call an "*approximating plane*".

By extension of the method of least squares, we can speak of a "least square plane" approximating the data. If we are estimating *Z* from given values of *X* and *Y*, this would be called a "regression plane" of *Z* on *X* and *Y*. The normal equations corresponding to the least plane (Eq. 4.8) are given by

$$\begin{aligned} \sum Z &= a_0N + a_1\sum X + a_2\sum Y \\ \sum XY &= a_0\sum X + a_1\sum X^2 + a_2\sum XY \\ \sum YZ &= a_0\sum Y + a_1\sum XY + a_2\sum Y^2 \end{aligned} \tag{4.9}$$

and can be remembering as obtained from (Eq. 4.8) by multiplying by 1, X and Y successively and then summing.

More complicated equations than (Eq. 4.8) can also be considered. These represent "*regression surfaces*". If the number of variables exceeds three, geometric intuition is lost since we then require four, five,. . ., dimensional spaces.

Problems involving estimation of a variable from two or more variables are called problems of "*multiple regression*".

In this Chapter we consider the closely related problem of *correlation*, or the degree of relationship between variables, which seek to determine *how well* a linear or the other equation describes or explains the relationship between variables.

If all values of the variables satisfy an equation exactly we say that the variables are *perfectly correlated* or that is *perfect correlation* between them. Thus the circumferences $\boldsymbol{C}$ and radii $\boldsymbol{r}$ of all circles are perfectly correlated since $\boldsymbol{C = 2\pi r}$. If two dice are tossed simultaneously 100 times there is no relationship between corresponding points on each die (unless the dice are loaded), i.e. they are *uncorrelated*.

When only two variables are involved we speak of *simple correlation* and *simple regression*. When more than two variables are involved we speak of a *multiple correlation* and *a multiple regression*. In this chapter, only simple correlation is considered.

If X and Y denote the two variables under consideration, a *scatter diagram* shows the location points *(X, Y)* on a rectangular co-ordinate system. If all points in this scatter diagram seem to lie near a line (Fig. 4.30), the correlation is called *linear*. In such cases, a linear equation is appropriate for purposes of regression or estimation.

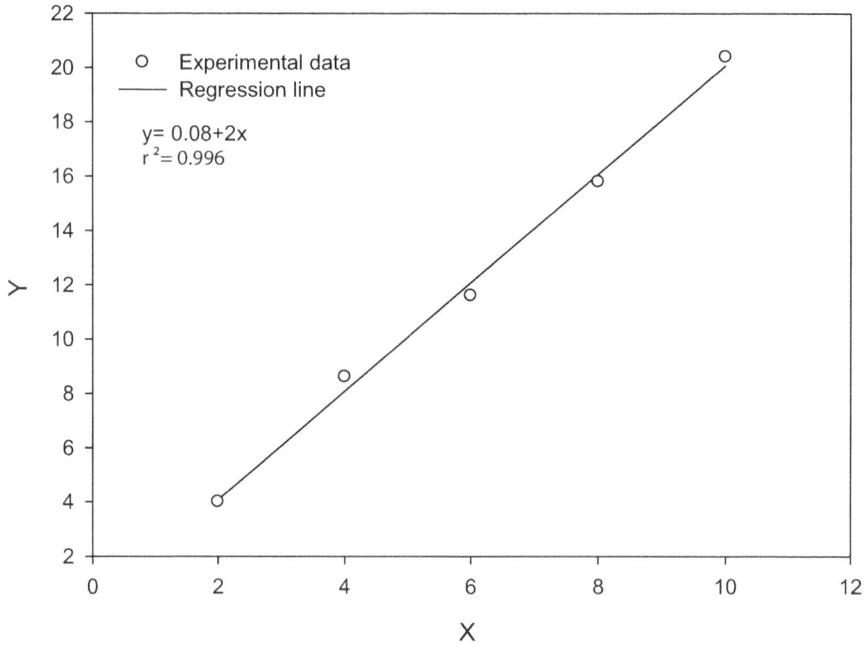

Fig. 4.30 Positive linear relation between Y and X

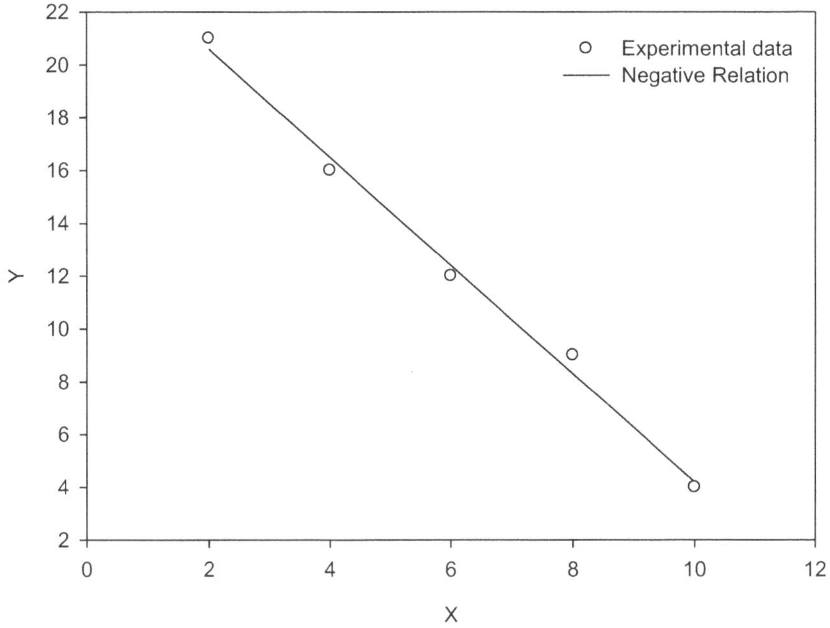

Fig. 4.31 Negative relationship between experimental data

If ϒ tends to increase as X increases, as in Fig. 4.30, the correlation is called *positive* or *direct correlation.* If ϒ tends to decrease as X increases, as in Fig. 4.31, the correlation is called *negative* or *inverse correlation.*

If all points seem to lie near some curve, the correlation is called *non-linear* and a non-linear equation is appropriate for regression or estimation. It is clear that non-linear correlation can be sometimes positive and sometimes negative.

If there is no relationship indicated between the variables, as in Fig. 4.32, we say that there is *no correlation* between them, i.e. they are *uncorrelated.*

We can determine in a *qualitative* manner how well a given line or curve describes the relationship between variables by direct observation of the scatter diagram itself. For example, it is seen that a straight line is far more helpful in describing the relation between X and ϒ for the data or in Fig. 4.30 than for the data of Fig. 4.31 due to the fact that there is less scattering about the line of Fig. 4.30.

If we are to deal with the problem of scattering of sample data about lines or curves in a *quantitative* manner it will be necessary for us to devise *measures of correlation*.

The Least Square Regression Lines

We first consider the problem of how well a straight line explains the relationship between two variables. To do this we shall need the equations for least square regression lines obtained earlier. As we have seen, the least square regression line

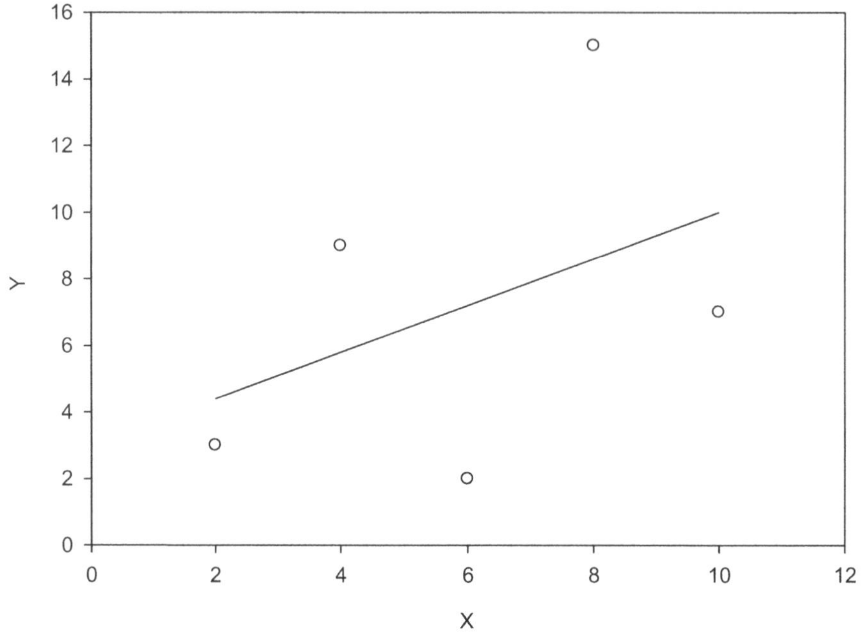

Fig. 4.32 No correlation between experiment data

of Υ on X is (Eq. 4.1) where a_0 and a_1 are obtained from the normal equations (Eq. 4.2) which yield

$$a_0 = \frac{(\Sigma \Upsilon)(\Sigma X^2) - (\Sigma X)(\Sigma X\Upsilon)}{N\Sigma X^2 - (\Sigma X)^2}$$
$$a_1 = \frac{N\Sigma X\Upsilon - (\Sigma X)(\Sigma \Upsilon)}{N\Sigma X^2 - (\Sigma X)^2} \tag{4.10}$$

Similarly, the regression line of X on Υ is given by

$$X = b_0 + b_1 \Upsilon \tag{4.11}$$

where b_0 and b_1 are obtained from the normal equations

$$\Sigma X = b_0 N + b_1 \Sigma \Upsilon$$
$$\Sigma X\Upsilon = b_0 \Sigma X + b_1 \Sigma \Upsilon^2 \tag{4.12}$$

which yield

$$b_0 = \frac{(\Sigma X)(\Sigma Y^2) - (\Sigma Y)(\Sigma XY)}{N\Sigma Y^2 - (\Sigma Y)^2}$$
$$b_1 = \frac{N\Sigma XY - (\Sigma X)(\Sigma Y)}{N\Sigma Y^2 - (\Sigma Y)^2} \tag{4.13}$$

The Eqs. 4.1 and 4.7 can also be written respectively as

$$y = \left(\frac{\Sigma xy}{\Sigma x^2}\right) \text{x and x} = \left(\frac{\Sigma xy}{\Sigma y^2}\right) \text{y} \tag{4.14}$$

where $x = X - \overline{X}$ and $y = Y - \overline{Y}$

The regression equations are identical if and only if all points of the scatter diagram lie on a line. In such case there is *perfect linear correlation* between X and Y.

Standard Error of Estimate

If we let Y_{est} represent the value of Y for given values of X as estimated from (Eq. 4.1), a measure of the scatter about the regression line of Y on X is supplied by the quantity

$$s_{y.x} = \sqrt{\frac{\Sigma(Y - Y_{est.})^2}{N}} \tag{4.15}$$

which is called the *standard error of estimate of* Y *on* X.

If the regression line (Eq. 4.1) is used, an analogous standard error of estimate of X on Y is defined by

$$s_{x.y} = \sqrt{\frac{\Sigma(X - X_{est.})^2}{N}} \tag{4.16}$$

In general, $s_{y.x} \neq s_{x.y}$

Equation (4.16) can be written

$$s_{x.y}{}^2 = \frac{\Sigma Y^2 - a_0\Sigma Y - a_1\Sigma XY}{N} \tag{4.17}$$

which may be more suitable for computation. A similar expression exists for (Eq. 4.15).

The standard error of estimate has properties analogous to those of the standard deviation. For example, if we construct lines parallel to the regression line of Y on X at respective vertical distances $s_{y.x}$, $2s_{y.x}$ and $3s_{y.x}$ from it, we should find, if N is large enough, that there would be included between these lines about 68%, 95% and 99,7% of the sample points.

Just as a modified standard deviation given by $\hat{s} = \sqrt{\frac{N}{N-1}}s$ was found useful for small samples, so a modified standard error of estimate given by $\hat{s}_{y.x} = \sqrt{\frac{N}{N-2}}s_{y.x}$ is useful. For this reason some statisticians prefer to define (Eq. 4.15) or (Eq. 4.16) with N − 2 replacing N in the denominator.

The *total variation of* Y is defined as $\Sigma(Y - \overline{Y})^2$, i.e. the sum of the squares of the deviations of the values of Y from the mean $\overline{Y}$. This can be written

$$\Sigma(Y - \overline{Y})^2 = \Sigma(Y - Y_{est.})^2 + \Sigma(Y_{est.} - \overline{Y})^2 \tag{4.18}$$

The first term on the right of (Equation 4.18) is called the *unexplained variation* while the second term is called the *explained variation,* so called because the deviations $Y_{est.} - \overline{Y}$ have a definite pattern while the deviations $Y - Y_{est.}$ behave in a random or unpredictable manner. Similar results hold for the variable X.

Coefficient of Correlation

The ratio of the explained variation to the total variation is called the *coefficient of determination.* If there is zero explained variation, i.e. the total variation is all unexplained, this ratio is zero. If there is zero unexplained variation, i.e. the total variation is all explained, the ratio is one. In other cases the ratio lies between zero and one. Since the ratio is always non-negative, we denote it by r^2. The quantity r, called the *coefficient of correlation,* is given by

$$r = \pm\sqrt{\frac{\text{explained variation}}{\text{total variation}}} \pm \sqrt{\frac{\Sigma(Y_{est.} - \overline{Y})^2}{\Sigma(Y - \overline{Y})^2}} \tag{4.19}$$

and varies between −1 and +1. The signs ± are used for positive correlation and negative linear correlation respectively. Note that r is a dimensionless quantity, i.e. it does not depend on the units employed.

By using the last formulae and the fact that standard deviation of Y is

$$s_y = \sqrt{\frac{\Sigma(Y - \overline{Y})^2}{N}} \tag{4.20}$$

we find that (Eq. 4.19) can be written, disregarding the sign, as

$$r = \sqrt{1 - \frac{s_{xy}^2}{s_y^2}} \quad \text{or} \quad s_{y.x} = S_Y\sqrt{1 - r^2} \tag{4.21}$$

Similar equations exist when X and Y are interchanged.

For the case of linear correlation the quantity r is the same regardless of whether X or Υ is considered the independent variable. Thus r is a very good measure of the linear correlation between two variables.

The definitions (Eq. 4.19) or (Eq. 4.21) for the correlation coefficient are quite general and can be used for non-linear relationship as well as linear, the only differences being that $\Upsilon_{est.}$ is computed from a non-linear regression equation in place of a linear regression equation and the signs ± are omitted. In such case equation (Eq. 4.19) defining the standard error of estimate is perfectly general. Equation 4.21, however, which applies to linear regression only, must be modified. If, for example, the estimating equation is

$$\Upsilon = a_0 + a_1 X + a_2 X^2 + \ldots. + a_n - X^{n-1} \quad (4.22)$$

Then Eq. 4.15 is replaced by

$$s_{y.x}{}^2 = \frac{\Sigma \Upsilon^2 - a_0 \Sigma \Upsilon - a_1 \Sigma X\Upsilon - \ldots - a_{n-1} \Sigma X^{n-1} \Upsilon}{N} \quad (4.23)$$

In such case the *modified standard of estimate* is $\hat{s}_{y.x} = \sqrt{\frac{N}{N-n}} s_{y.x}$, where the quantity $N - n$ is called the number of *degrees of freedom*.

It must be emphasized that the value of r computed on any case measures the degree of the relationship relative to the type of equation which is actually assumed. Thus if a linear equation is assumed Eq. 4.19 or Eq. 4.21 yields a value of r near zero, it means that there is almost *no linear correlation* between the variables. However, it does not mean that there is no correlation at all, since there may actually be a high *non-linear correlation* between the variables. In other words the correlation coefficient measures the goodness of fit of the equation actually assumed to the data. Unless otherwise specified, the term correlation coefficient is used to mean *linear* correlation coefficient.

It should also be pointed out that a high correlation coefficient (i.e. near 1 or -1) does not necessarily indicate a direct dependence of the variables. Thus there may be a high correlation between the number of book published each year and the number of football games played each year. Such examples are sometimes referred to as *nonsense* or *spurious correlations*.

Product-Moment Formula for the Linear Correlation Coefficient

If a linear relationship between two variables is assumed, Eq. 4.19 becomes

$$r = \frac{\Sigma xy}{\sqrt{(\Sigma x^2)(\Sigma y^2)}} \quad (4.24)$$

where $x = X - \overline{X}$ and $y = \Upsilon - \overline{\Upsilon}$. This formula, which automatically gives the proper sign of r, is called the *product-moment formula* and clearly shows the symmetry between X and Υ.

If we write

$$s_{xy} = \frac{\Sigma xy}{\text{N}}, s_{x^2} = \sqrt{\frac{\Sigma x^2}{\text{N}}}, \ldots s_{y^2} = \sqrt{\frac{\Sigma y^2}{\text{N}}} \tag{4.25}$$

then s_x and s_y will be recognized as the standard deviations of the variables X and Y respectively, while s_{x^2} and s_{y^2} are their variances. The new quantity s_{xy} is called the *covariance* of X and Y. In terms of symbols of Eq. 4.24 can be written

$$r = \frac{s_{xy}}{s_x s_y} \tag{4.26}$$

Note that r is not only independent of the choice of units of X and Ybut is also independent of choice of origin.

Equation 4.24 can be written in the equivalent form

$$r = \frac{\text{N}\Sigma \text{X}Y - (\Sigma \text{X})(\Sigma Y)}{\sqrt{\left[\text{N}\Sigma \text{X}^2 - (\Sigma \text{X})^2\right].\left[\text{N}\Sigma \text{X}^2 - (\Sigma Y)^2\right]}} \tag{4.27}$$

which is often used in computing r.

For data grouped as in a *bivariate table* or *bivariate frequency distribution*, it is convenient to use *coding method*. In such case Eq. 4.27 can be written

$$r = \frac{N\Sigma f u_x u_y - (\Sigma f_x u_x)(\Sigma f_y u_y)}{\sqrt{\left[N\Sigma f_x u_x^2 - (\Sigma f_x u_x)^2\right].\left[N\Sigma f_y u_y^2 - (\Sigma f_y u_u)^2\right]}} \tag{4.28}$$

For convenience in calculation using this formula, a *correlation table* is used. For grouped data, Eq. 4.25 can be written

$$s_{xy} = c_x c_y \left[\frac{\Sigma f u_x u_y}{\text{N}} - \left(\frac{\Sigma f_x u_x}{\text{N}}\right)\left(\frac{\Sigma f_y u_y}{\text{N}}\right)\right] \tag{4.29}$$

$$s_x = c_x \sqrt{\frac{\Sigma f_x u_x^2}{\text{N}} - \left(\frac{\Sigma f_x u_x}{\text{N}}\right)} \tag{4.30}$$

$$s_y = c_y \sqrt{\frac{\Sigma f_y u_y^2}{\text{N}} - \frac{\Sigma f_y u_y}{\text{N}}} \tag{4.31}$$

where c_x and c_y are the class interval widths (assumed constant) corresponding to the variables X and Yrespectively.

Instead of using precise values of the variables, or when such precision is unavailable, the data may be ranked in order of size, importance, etc., using the numbers *1, 2, …,* N. If two variables X and Y are ranked in such manner the *coefficient of rank correlation* is given by

$$r_{rank} = 1 - \frac{6\Sigma D^2}{N(N^2 - 1)} \tag{4.32}$$

where D = differences between ranks of corresponding values of X and Y; N = number of pairs of values (X, Y) in the data.

The Eq. 4.32 is called *Spearman's formula for rank correlation.*

Sampling Theory of Correlation

The N pairs of values (X, Y) of two variables can be thought of as a sample from a population of all possible such pairs. Since two variables are involved this is called a *bivariate population*, which we assume is a *bivariate normal distribution.*

We can think of a theoretical population coefficient of correlation denoted by p, which is estimated by the sample correlation coefficient r. Tests of significance or hypotheses concerning various values of p require knowledge of the sampling distribution of r. For $p = 0$ this distribution is symmetric and a statistic involving Student's distribution can be used. For $p \neq 0$ the distribution is skewed. In such case a transformation due to *Fisher* produces a statistic which is approximately normally distributed. The following test summarizes the procedures involved.

(i) **Test of Hypothesis** $\rho = 0$

Here we use the fact that the statistic

$$t = \frac{r\sqrt{N-2}}{\sqrt{1-r^2}} \tag{4.33}$$

has Student's distribution with $v = N - 2$ degrees of freedom.

(ii) **Test of Hypothesis** $\rho = \rho_0 \neq 0$

Here we use the fact that the statistic

$$Z = \frac{1}{2} \log_e\left(\frac{1+r}{1-r}\right) = 1.1513 \log_{10}\left(\frac{1+r}{1-r}\right) \tag{4.34}$$

where $e = 2 \cdot 71828$, is approximately normally distributed with mean and standard deviation given by

$$\mu_z = \frac{1}{2} \log_e\left(\frac{1+\rho_0}{1-\rho_0}\right) = 1 \cdot 1513 \log_{10}\left(\frac{1+\rho_0}{1-\rho_0}\right), \sigma_z = \frac{1}{\sqrt{N-3}} \tag{4.35}$$

Significance of a Difference Between Correlation Coefficients
To determine whether two correlation coefficients r_1 and r_2, drawn from samples of size N_1 and N_2 respectively, differ significantly from each other, we compute Z_1 and Z_2 corresponding to r_1 and r_2 using (Eq. 4.34). We then use the fact the test statistic

$$z = \frac{Z_1 - Z_2 - \mu_{z_1 - z_2}}{\sigma_{z_1 - z_2}} \tag{4.36}$$

where $\mu_{z_1 - z_2} = \mu_{z_1} - \mu_{z_2}$ and $\sigma_{z_1 - z_2} = \sqrt{\sigma^2_{z_1} + \sigma^2_{z_2}} = \sqrt{\frac{1}{N_1 - 3} + \frac{1}{N_2 - 3}}$ is normally distributed.

Analysis of Variance and Regression
Analysis of variance, commonly referred to as ANOVA, is a statistical technique that seeks to determine whether differences in the values of a variable can be explained by categorization of the observation. This technique was developed for application to experimental data. Most of the results can be obtained through regression, and we approach ANOVA from this point of view.

The traditional framework for analysis of variance gives rise to a special format for organizing some of the results of the regression estimation and hypothesis testing. This format, which is known as an ANOVA table, is shown generically for the results of a multiple regression in Table 4.18. Many computer programs print such a table for each estimated regression, replacing the formulas and symbols given here by their calculated values.

The coefficient of determination, R^2, can be computed as the ratio of the sum of squares in the first row to that in the third; equivalently, it is 1 minus the ratio of the second to the third:

$$R^2 = \frac{\sum \left(\widehat{Y}_i - \overline{Y}\right)^2}{\sum \left(Y_i - \overline{Y}\right)^2} = 1 - \frac{\sum \left(Y_i - \widehat{Y}_i\right)^2}{\sum \left(Y_i - \overline{Y}\right)^2} \tag{4.37}$$

The coefficient of determination, R^2, can be computed as the ratio of the sum of squares in the first row to that in the third; equivalently, it si 1 minus the ratio of the second to the third:

Table 4.18 ANOVA Table for Regression

Source of variation	Sum of squares	Degrees of freedom	Mean square	F ratio
Explained (by regression)	$\sum \left(\widehat{Y}_i - \overline{Y}\right)^2$	k	$\frac{\sum \left(\widehat{Y}_i - \overline{Y}\right)^2}{k}$	$\frac{\sum \left(\widehat{Y}_i - \overline{Y}\right)^2 / k}{\sum \left(Y_i - \widehat{Y}_i\right)^2 / (n-k-1)}$
Unexplained (due error)	$\sum \left(Y_i - \widehat{Y}_i\right)^2$	n − k − 1	$\frac{\sum \left(Y_i - \widehat{Y}_i\right)^2}{n-k-1}$	
Total	$\sum \left(Y_i - \overline{Y}\right)^2$	n − k		

$$R^2 = \frac{\sum\left(\widehat{Y}_i - \overline{Y}\right)^2}{\sum\left(Y_i - \overline{Y}\right)^2} = 1 - \frac{\sum\left(Y_i - \widehat{Y}_i\right)^2}{\sum\left(Y_i - \overline{Y}\right)^2} \tag{4.38}$$

The only item in the "*F*" ratio column is the ratio of the mean square to the second, and it is sometimes called the variance ratio. SSR_U is the same as the unexplained sum of squares here. SSR_R results from a regression on a constant, because all β_j are restricted to be zero. The resulting *SSR* is identical to the total sum of squares in the ANOVA table (Table 4.18). Hence $SSR_R - SSR_U$ is equal to the explained sum of squares in the ANOVA table, and thus the *F* ratios are identical.

$$F = \frac{(SSR_R - SSR_U)/k}{SSR_U/(n-k-1)} = \frac{R^2/k}{(1-R^2)/(n-k-1)} \tag{4.39}$$

4.5.3 Use of Statistic Software

For Linear Regression: plot the y versus x by using SigmaPlot 11.0. Select "regression" icon on the statistics menu. Then, "linear item" is selected and dependent (y) and independent (x) columns are defined. After that the regression is done by clicking "finish" icon. From regression wizard, linear line equation and its r^2 are obtained. The linear regression can also be done by using Excel, SPSS and other statistics package programmes. For Excel; first, graph was plotted, and then double click to "data" and select "add regression line" item. Then select "linear line".

For non-linear regression: plot the y versus x by using SigmaPlot 11.0. Select "nonlinear regression" icon on the statistics menu. Then, select "Regression Wizards" and select "model". And then, dependent (y) and independent (x) columns are defined. After that the nonlinear regression is done by clicking "finish" icon. From report its r^2 and other statistical results could be obtained.

4.5.4 Finding the Relationship Between Thiobarbituric Acid Reactive Substance of Meat and Time

Rancidity (2 thiobarbituric acid reactive substance; TBARS) value of meat was followed during the storage. Following data were obtained. Find the relationship between TBARS value and time. Determine r^2 and standard error of estimate.

Time(h)	0	20	40	60	80	100	120	140	160
TBARS (mg/kg)	0.21	0.29	0.34	0.42	0.39	0.46	0.62	0.66	0.72

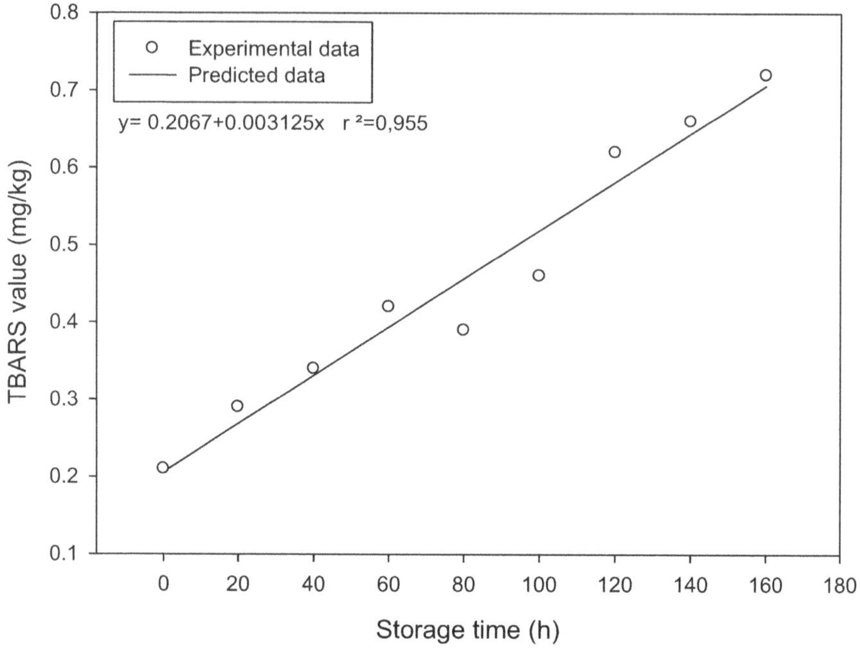

Fig. 4.33 Change in TBARS value during the storage

Plot the experimental data TBARS (y) versus time (x) by using SigmaPlot or other suitable programs. When graph is examined, it can be withdrawn that relation seem to be a linear (Fig. 4.33). Then apply linear regression given in Eq. 4.10 and r^2 given in Eq. 4.27.

X	Y	XY	X^2	Y^2
0	0.21	0	0	0.0441
20	0.29	5.8	400	0.0841
40	0.34	13.6	1600	0.1156
60	0.42	25.2	3600	0.1764
80	0.39	31.2	6400	0.1521
100	0.46	46	10,000	0.2116
120	0.62	74.4	14,400	0.3844
140	0.66	92.4	19,600	0.4356
160	0.72	115.2	25,600	0.5184
$\sum X = 720$	$\sum Y = 4.11$	$\sum XY = 403.8$	$\sum X^2 = 81600$	$\sum Y^2 = 2.1223$

$$a_0 = \frac{(4.11)(81600) - (720)(403.8)}{9x81600 - (720)^2} = 0.2067$$

$$a_1 = \frac{9x403.8 - (720)(4.11)}{9x81600 - (720)^2} = 0.003125$$

$$r^2 = \frac{9x403.8 - (720)(4.11)}{\sqrt{\left[9x81600 - (720)^2\right].\left[9x2.1223 - (4.11)^2\right]}} = 0.955$$

$$s_{x.y} = \sqrt{\frac{\Sigma Y^2 - a_0\Sigma Y - a_1\Sigma XY}{N-2}} = \sqrt{\frac{2.1223 - 0.2067^*4.11 - 0.003125^*403.8}{9-2}}$$
$$= 0.039$$

Report of SigmaPlot for This Linear Regression Analysis:
Equation: Polynomial; Linear. f = y0 + a * x

Analysis of Variance:

R	**Rsqr**	**Adj Rsqr**	**Standard Error of Estimate**
0.9773	0.9551	0.9487	0.039
	Coefficient		**Std. Error**
y0	0.2067	0.0244	8.4725
a	0.0031	0.0003	12.1988

	DF	**SS**	**MS**	**F**	**P**
Regression	1	0.2344	0.2344	148.8095	<0.0001
Residual	7	0.0110	0.0016		
Total	8	0.2454	0.0307		

4.5.5 Predicting the Drying Time of Macaroni as a Function of Temperature

Following data was obtained for the drying of macaroni. Predict the drying time of macaroni as a function of temperature.

Temperature (°C)	50	60	70	80
Total drying time (min)	180	150	130	105

Plot the total drying time versus temperature by using SigmaPlot 11.0 or other suitable programs. When graph is examined, it is observed that it seems to be a linear but negative relationship exist (Fig. 4.34). Then apply linear regression given in Eq. 4.10 and r^2 given in Eq. 4.27 as in Fig. 4.31.

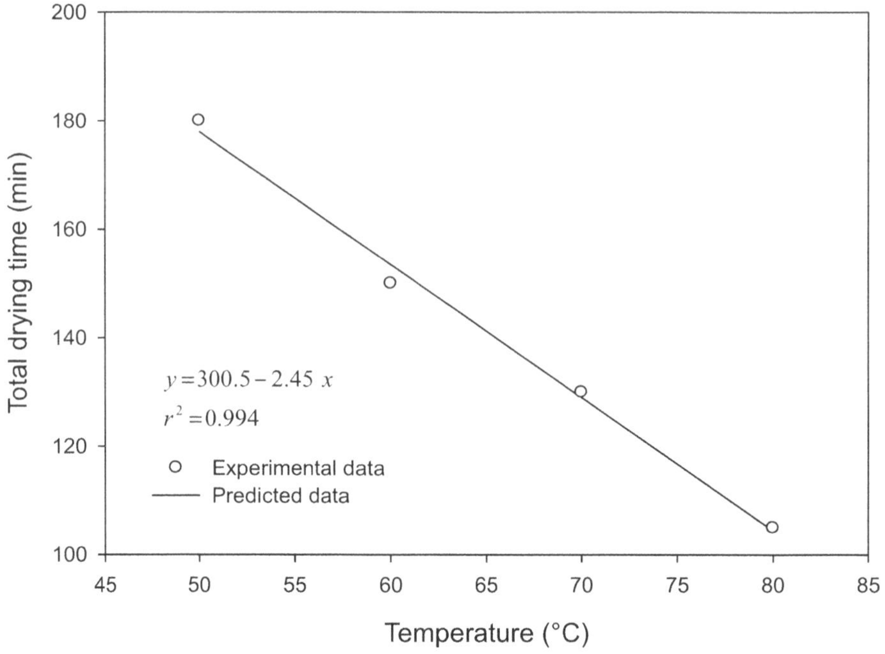

Fig. 4.34 Change of total drying time as a function of time.

Report of SigmaPlot for this linear regression analysis:
Equation: Polynomial; Linear f = y0 + a * x

Analysis of Variance:

R	Rsqr	Adj Rsqr	Standard Error of Estimate
0.9971	0.9942	0.9913	2.9580
	Coefficient	**Std. Error**	
y0	300.5000	8.7250	
a	−2.4500	0.1323	

	DF	SS	MS	F	P
Regression	1	3001.2500	3001.2500	343.0000	0.0029
Residual	2	17.5000	8.7500		
Total	3	3018.7500	1006.2500		

4.5.6 *Predicting the Growth of Salmonella in Chicken Meat as a Function of Time*

Growth of Salmonella (log CFU/g) in chicken meat during the storage was investigated. Following data were obtained. Predict the growth of Salmonella in chicken meat as a function of time.

Time(h)	0	2	4	6	8	10	12	14	16	18	20
Salmonella (log CFU/g)	0.30	0.30	0.47	0.70	0.96	1.23	1.54	1.70	1.79	1.79	1.80

Plot the number of *Salmonella* (log CFU/g) versus time by using SigmaPlot 11.0 or other suitable programs. When graph is examined, it is observed that it seems to be a sigmoidal (Fig. 4.35). For sigmoidal graphs, we can apply numerous models such as (Gompertz, Logistic, Richards, Stannard, Schnute and etc. We can use SigmaPlot, SPSS, Mathematica, or other suitable program (Table 4.19).

For that purpose, Logistic 4 parameter model (Table 4.18) was used (Fig. 4.36).

Report of SigmaPlot for this nonlinear regression analysis:

Nonlinear Regression, Equation: Sigmoidal; Logistic, 4 Parameter

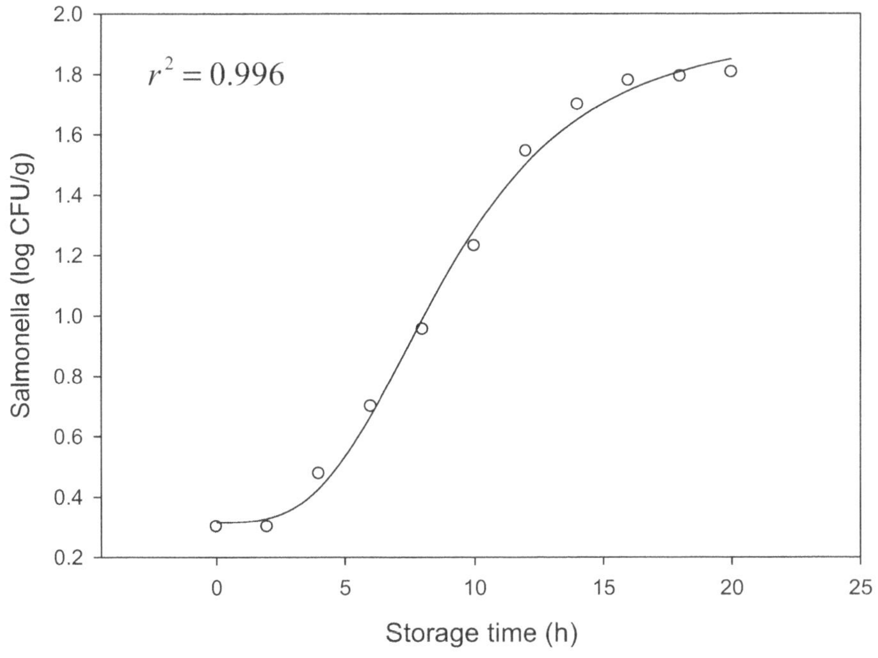

Fig. 4.35 Prediction of *Salmonella* growth in chicken meat during the storage.

Table 4.19 Some commonly used equations and curves

N	Equation	Curve
1	$Y = a_0 + a_1X$	Straight line
2	$Y = a_0 + a_1X + a_2X^2$	Parabola or Quadratic curve
3	$Y = a_0 + a_1X + a_2X^2 + a_3X^3$	Cubic curve
4	$Y = a_0 + a_1X + a_2X^2 + a_3X^3 + a_4X^4$	Quartic curve
5	$Y = a_0 + a_1X + a_2X^2 + a_3X^3 + \ldots + a_nX^n$	n-th degree curve
6	$Y = \frac{1}{a_0 + a_1X}$ or $\frac{1}{Y} = a_0 + a_1X$	Hyperbola
7	$Y = a.\ b^x$ or $\log Y = \log a + (\log b).$ $X = a_0 + a_1X$	Exponential curve
8	$Y = a.\ b^x + g$	Modified exponential curve
9	$Y = a.\ X^b$ or $\log Y = \log a + b.\ \log .\ X$	Geometric curve
10	$Y = a.\ X^b + g$	Modified geometric curve
11	$Y = \frac{a}{(bx)^c + 1}$ or $Y = a + \frac{b}{(cx)^d + 1}$	Logistic curves (Logistic 3 parameter and Logistic 4 parameter)
12	$Y = a_0 + a_1(\log X) + a_2(\log X)^2$	
13	$Y = a^* \exp\{- \exp[b - c^*x)]\}$	Gompertz
14	$Y = a^*\{1 + b^* \exp[c^*(d - x)]\}^{-1/b}$	Richards
15	$Y = a^*\{1 + \exp[-\frac{b + c^*x}{d}]\}^{-d}$	Stannard
16	$Y = \left\{y_1^a + (y_2^a - y_1^a) * \frac{1 - \exp[-b^*(x - c)]}{1 - \exp[-b^*(d - c)]}\right\}^{1/a}$	Schnute

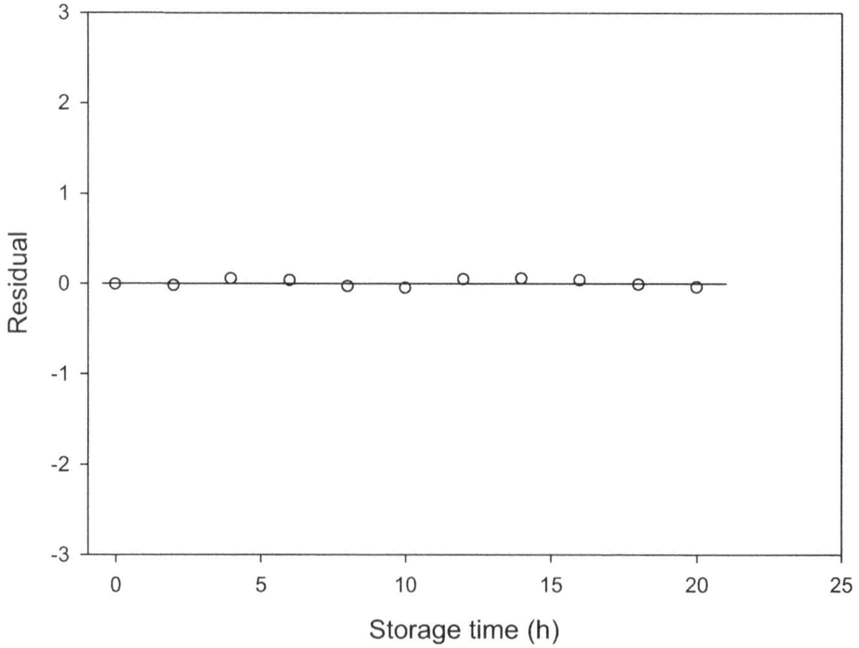

Fig. 4.36 Residual plot for Prediction of Salmonella growth in chicken meat during the storage

$$Y = a + \frac{b}{(cx)^d + 1}$$

R	Rsqr	Adj Rsqr	Standard Error of Estimate
0.9979	0.9959	0.9941	0.0476

	Coefficient	**Std. Error**
a	0.3155	0.0337
b	1.6503	0.0896
c	8.9558	0.3566
d	−3.2297	0.3801

Analysis of Variance:

	DF	**SS**	**MS**	**F**	**P**
Regression	3	3.8207	1.2736	562.3866	<0.0001
Residual	7	0.0159	0.0023		
Total	10	3.8365	0.3837		

4.5.7 *Predicting the Growth of Lactic Acid Bacteria in Turkish Dry Fermented Sausage During Storage*

Lactic acid bacteria (LAB) count (log CFU/g) in Turkish dry fermented sausage during the storage was investigated. Following data were obtained. Model the given data as a function of time.

Time(h)	0	5	10	15	22	26	33	42
LAB (log CFU/g)	5.54	5.73	4.85	3.63	2.99	2.59	2.19	2.00

Plot the LAB count (log CFU/g) versus time by using SigmaPlot 11.0 or other suitable programs. It is observed that the plot seems to be a sigmoidal (Fig. 4.37).

Let's use Gompertz model:

Nonlinear Regression
Equation: Sigmoidal; Gompertz, 4 Parameter
f = a + b * exp(−exp(−(x−c)/d))

R	**Rsqr**	**Adj Rsqr**	**Standard Error of Estimate**
0.9877	0.9755	0.9572	0.3102

	Coefficient	**Std. Error**
a	2.1081	0.2428
b	4.5852	1.2902
c	16.2021	3.4604
d	−10.5832	4.6806

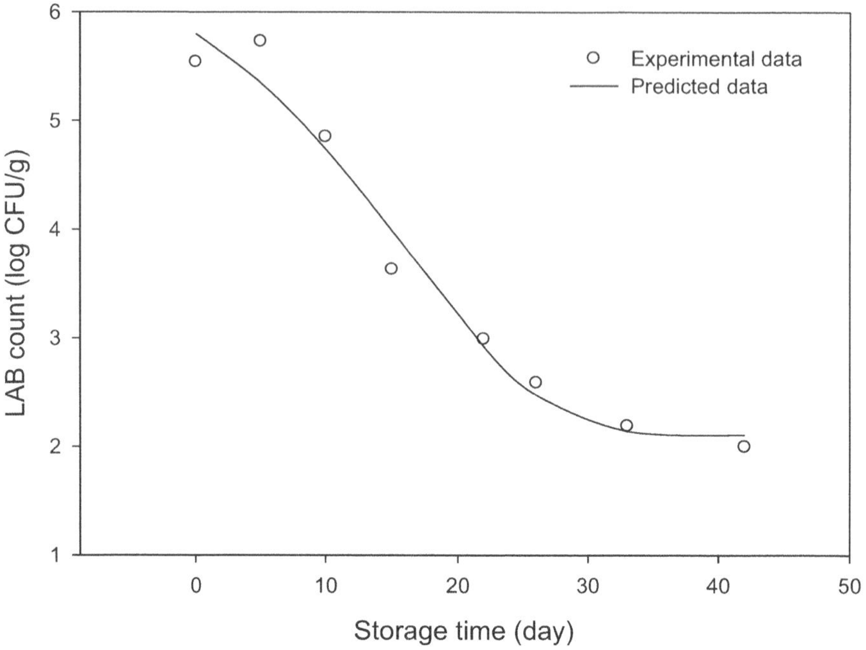

Fig. 4.37 LAB count in Turkish dry fermented sausage

Analysis of Variance:

	DF	SS	MS	F	P
Regression	3	15.3545	5.1182	53.1864	0.0011
Residual	4	0.3849	0.0962		
Total	7	15.7394	2.2485		

Appendix (Tables 4.20 and 4.21)

Table 4.20 Coefficients for control charts

Sample	Coefficients for control limits											Coefficients for central line			
size = n	A	A2	A3	B3	B4	B5	B6	D1	D2	D3	D4	C4	1/C4	d2	1/d2
2	2.121	1.880	2.659	0.000	2.605	0.000	2.605	0.000	3.686	0.000	3.267	0.7979	1.2533	1.128	0.8865
3	1.732	1.023	1.954	0.000	2.275	0.000	2.275	0.000	4.358	0.000	2.574	0.8862	1.1284	1.693	0.5907
4	1.500	0.729	1.628	0.000	2.038	0.000	2.088	0.000	4.598	0.000	2.282	0.9213	1.0854	2.059	0.4857
5	1.342	0.577	1.427	0.000	1.964	0.000	1.964	0.000	4.918	0.000	2.114	0.9400	1.0638	2.325	0.4299
6	1.225	0.483	1.287	0.030	1.970	0.029	1.874	0.000	5.078	0.000	2.004	0.9515	1.0510	2.534	0.3946
7	1.134	0.419	1.182	0.118	1.882	0.113	1.806	0.204	5.204	0.076	1.924	0.9594	1.0423	2.704	0.3698
8	1.061	0.373	1.099	0.185	1.815	0.179	1.751	0.388	5.306	0.136	1.864	0.9650	1.0363	2.847	0.3512
9	1.000	0.337	1.032	0.239	1.761	0.232	1.707	0.547	5.393	0.184	1.816	0.9693	1.0317	2.970	0.3367
10	0.949	0.308	0.975	0.284	1.716	0.276	1.669	0.687	5.469	0.223	1.777	0.9727	1.0281	3.078	0.3249
11	0.905	0.285	0.927	0.321	1.679	0.313	1.637	0.811	5.535	0.256	1.744	0.9754	1.0252	3.173	0.3152
12	0.866	0.266	0.886	0.354	1.646	0.346	1.610	0.922	5.594	0.283	1.717	0.9776	1.0229	3.258	0.3069
13	0.832	0.249	0.850	0.382	1.618	0.374	1.585	1.025	5.647	0.307	1.693	0.9794	1.0210	3.336	0.2998
14	0.802	0.235	0.817	0.406	1.594	0.399	1.563	1.118	5.696	0.328	1.672	0.9810	1.0194	3.407	0.2935
15	0.775	0.223	0.789	0.428	1.572	0.421	1.544	1.203	5.741	0.347	1.653	0.9823	1.0180	3.472	0.2880
16	0.750	0.212	0.763	0.448	1.552	0.440	1.526	1.282	5.782	0.363	1.637	0.9835	1.0168	3.532	0.2831
17	0.728	0.203	0.739	0.466	1.534	0.458	1.511	1.356	5.820	0.378	1.622	0.9845	1.0157	3.588	0.2787
18	0.707	0.194	0.718	0.482	1.518	0.475	1.496	1.424	5.856	0.391	1.608	0.9854	1.0148	3.640	0.2747
19	0.688	0.187	0.698	0.497	1.503	0.490	1.483	1.487	5.891	0.403	1.597	0.9862	1.0140	3.689	0.2711
20	0.671	0.180	0.680	0.510	1.490	0.504	1.470	1.549	5.921	0.415	1.585	0.9869	1.0133	3.735	0.2677
21	0.655	0.173	0.663	0.523	1.477	0.516	1.459	1.605	5.951	0.425	1.575	0.9876	1.0126	3.778	0.2647
22	0.640	0.167	0.647	0.534	1.466	0.528	1.448	1.659	5.979	0.434	1.566	0.9882	1.0119	3.819	0.2618
23	0.626	0.162	0.633	0.545	1.455	0.539	1.438	1.710	6.006	0.443	1.557	0.9887	1.0114	3.858	0.2592
24	0.612	0.157	0.619	0.555	1.445	0.549	1.429	1.759	6.031	0.451	1.548	0.9892	1.0109	3.895	0.2567
25	0.600	0.153	0.606	0.565	1.435	0.559	1.420	1.806	6.056	0.459	1.541	0.9896	1.0105	3.931	0.2544

SR ISO 8258+C1

Table 4.21 Gaussian Laplace table

z	0	0.01	0.02	0.03	0.04	0.05	0.06	0.07	0.08	0.09
0	0	0.004	0.008	0.012	0.016	0.0199	0.0239	0.0279	0.0319	0.0359
0.1	0.0398	0.0438	0.0478	0.0517	0.0557	0.0596	0.0636	0.0675	0.0714	0.0754
0.2	0.0793	0.0832	0.0871	0.091	0.0948	0.0987	0.1026	0.1064	0.1103	0.1141
0.3	0.1179	0.1217	0.1255	0.1293	0.1331	0.1368	0.1406	0.1443	0.148	0.1517
0.4	0.1555	0.1591	0.1628	0.1664	0.17	0.1736	0.1772	0.1808	0.1844	0.1879
0.5	0.1915	0.195	0.1985	0.2019	0.2054	0.2088	0.2123	0.2157	0.219	0.2224
0.6	0.2258	0.2291	0.2324	0.2357	0.2389	0.2422	0.2454	0.2486	0.2518	0.2549
0.7	0.258	0.2612	0.2642	0.2673	0.2704	0.2734	0.2764	0.2794	0.2823	0.2852
0.8	0.2881	0.291	0.2939	0.2967	0.2996	0.3023	0.3051	0.3049	0.3106	0.3133
0.9	0.3159	0.3186	0.3212	0.3238	0.3264	0.3289	0.3315	0.334	0.3365	0.3389
1	0.3413	0.3438	0.3461	0.3485	0.3508	0.3531	0.3554	0.3577	0.3599	0.3621
1.1	0.3643	0.3665	0.3686	0.3708	0.3729	0.3749	0.377	0.379	0.381	0.383
1.2	0.3849	0.3869	0.3888	0.3907	0.3925	0.3944	0.3962	0.398	0.3997	0.4015
1.3	0.4032	0.4049	0.4066	0.4082	0.4099	0.4115	0.4131	0.4147	0.4162	0.4177
1.4	0.4192	0.4207	0.4222	0.4236	0.4251	0.4265	0.4279	0.4292	0.4306	0.4319
1.5	0.4332	0.4345	0.4357	0.437	0.4382	0.4394	0.4406	0.4418	0.443	0.4441
1.6	0.4452	0.4463	0.4474	0.4485	0.4495	0.4505	0.4515	0.4525	0.4535	0.4545
1.7	0.4554	0.4564	0.4573	0.4582	0.4591	0.4599	0.4608	0.4616	0.4625	0.4633
1.8	0.4641	0.4649	0.4656	0.4664	0.4671	0.4678	0.4686	0.4693	0.47	0.4706
1.9	0.4713	0.4719	0.4726	0.4732	0.4738	0.4744	0.475	0.4756	0.4762	0.4767
2	0.4772	0.4778	0.4783	0.4788	0.4793	0.4794	0.4803	0.4808	0.4812	0.4817
2.1	0.4821	0.4826	0.483	0.4834	0.4838	0.4842	0.4846	0.485	0.4854	0.4857
2.2	0.4861	0.4865	0.4868	0.4871	0.4875	0.4878	0.4881	0.4884	0.4887	0.489
2.3	0.4893	0.4896	0.4898	0.4901	0.4904	0.4906	0.4909	0.4911	0.4913	0.4916
2.4	0.4918	0.492	0.4922	0.4925	0.4927	0.4929	0.4931	0.4932	0.4934	0.4936
2.5	0.4938	0.494	0.4941	0.4943	0.4945	0.4946	0.4948	0.4949	0.4951	0.4952

2.6	0.4953	0.4955	0.4956	0.4957	0.4959	0.496	0.4961	0.4962	0.4963	0.4964
2.7	0.4965	0.4966	0.4967	0.4968	0.4969	0.497	0.4971	0.4972	0.4973	0.4974
2.8	0.4974	0.4975	0.4976	0.4977	0.4977	0.4978	0.4979	0.498	0.498	0.4981
2.9	0.4981	0.4982	0.4983	0.4983	0.4984	0.4984	0.4985	0.4985	0.4986	0.4986
3	0.4986	0.4987	0.4987	0.4988	0.4988	0.4989	0.4989	0.4989	0.4989	0.499
3.1	0.499									
3.2	0.4993									
3.3	0.4995									
3.4	0.4997									
3.5	0.49977									
4	0.499968									
4.5	0.499997									
5	0.4999997									

References

Betteley GN, Mettrick E, Sweeney, Wilson D (1994) Using statistics in industry. Prentice Hall, New York, p 361

Bozkurt H (2002) Effects of some storage conditions and additives on quality and stability of Sucuk (Turkish dry-fermented sausage). PhD thesis, Gaziantep University, Gaziantep Turke

Chan LK, Cheng SW, Spring FA (1988) A new measure of process capability: C_{pm}. J Qual Technol 30(3):162–175

Crawshaw J, Chambers J (1990) A concise course in A-level statistics, 2nd edn. Stanley Thornes Ltd., London, 677 p

Feingenbaum AV (1983) Total quality control, engineering and management, 3rd edn. McGraw-Hill, New York

Grant EL, Leavenworth RS (1988) Statistical quality control, 6th edn. McGraw-Hill Book Company, New York, pp 115–361

Hubbard M (2003) Statistical quality control for the food industry, 3rd edn. Kluwer Academic, New York, pp 49–71

Juran JM (1988) Quality control handbook. McGraw-Hill, New York

Keuen Tae Cho (2003) Multicriteria decision methods: an attempt to evaluate and unify. Math Comput Model 37(2003):1099–1119

Luceno A (1996) A process capability index with reliable confidence intervals. Comm Stat Simul 25:235–245

Montgomery DM, Runger GC (2003) Applied statistics and probability for engineers, 3rd edn. John Wiley PH, New York

Montgomery DC, Keats BJ, Runger GC, Messina WS (1994) Integrating statistical process control and engineering process control. J Qual Technol 26(2):79–87

Pearn WL, Kotz S, Johnson NL (1992) Distributional and inferential properties of process capability indices. J Qual Technol 24:216–231

Petrescu E, Vodă VG (2002a) Process control charts. Theory and case studies. Ed. Economica, Bucureşti, pp 79–159

Petrescu E, Vodă VG (2002b) Process control charts. Theory and case studies. Ed. Economica, Bucureşti, pp 80–110

Pyzdek T (2003) Quality engineering handbook, 2nd edn. Keller PA (ed). Marcel Dekker, Inc., New York

Sanders Donald H, Franklin Murph A, Eng RJ (1980) Statistics. A fresh approach, 2rd edn. McGraw-Hill Book Co., New York, 428 pp

SigmaPlot 11.1 (2008) Exact graphs and data analysis, SYSTAT Software, Inc. www.systat.com

SPSS, 16.0 (2007) Family for windows, user manuals, SPSS Inc. www.spss.com

SR EN ISO 8258+ C1. Shewhart control charts

Vasconcellos JA (2003) Statistical tools and strategies for process control improvement. In: Quality Assurance for the Food Industry: A practical approach. CRC Press, eBook ISBN 9780429210006

Vod VG (2001) Process quality control—an open issue. Trbuna Calitatii IV(7–8), 23–26

Vodă VG (1998) ISO- Statistical procedures. Trbuna Calitatii I(10, 67–68), 6–10

Woodroofe M. Probability with applications. McGraw-Hill Kogakusha, Ltd., Tokyo, 372 p.

Chapter 5
Statistic for Sensory Science

Eugenios Katsanidis and Peter Ho

5.1 Preference and Analytical Discriminative Difference Tests in Sensory Science

Eugenios Katsanidis

5.1.1 Objectives and Learning Outcomes

The objective of this chapter is to discuss how to identify the cases of sensory analysis where the use of parametric tests is not appropriate and to present the analogous nonparametric tests. The most often used nonparametric tests will be considered in regard to when and how these tests should be applied. Three case studies where nonparametric tests are used to analyse data from sensory tests are also presented to better illustrate the application of these tests.

E. Katsanidis
Department of Food Science & Technology, Faculty of Agriculture, Aristotle University of Thessaloniki, Thessaloniki, Greece
e-mail: ekatsani@agro.auth.gr

P. Ho (✉)
School of Food Science and Nutrition, University of Leeds, Leeds, UK
e-mail: p.ho@leeds.ac.uk

G. Schleining et al. (eds.), *Statistics in Food and Biotechnology*, Integrating Food Science and Engineering Knowledge Into the Food Chain 16,
https://doi.org/10.1007/978-3-031-51568-2_5

5.1.2 Introduction

In sensory evaluation of food, human senses are used as instruments for the description of food properties and for the detection and identification of perceived differences in certain attributes (i.e. texture, flavor, color, appearance, liking) of food products.

Sensory tests can be divided into affective and analytical tests. Affective tests are used to evaluate preference and/or acceptance of products. The affective tests do not require any training of the panelists since they are aiming at measuring how well a product is liked or not. Analytical tests are used for laboratory evaluation of products in terms of differences or similarities and for the identification and quantification of sensory characteristics. Analytical tests can be further divided into discriminative and descriptive tests. Discriminative tests are employed when the researcher is looking for differences among similar products or for determining sensitivity thresholds. Descriptive tests are used to identify and quantify sensory characteristics. Discriminative tests require at least some training of the panelists and some descriptive tests may require extensive training and highly specialized panelists (Sensory Evaluation Division of the IFT 1981).

In this chapter, statistical analysis of preference tests and analytical-discriminative difference tests will mostly be discussed. Tests that fall into the first category include the paired preference test and ranking tests. In the second category belong tests like the paired comparison, the duo-trio, the triangle and the ranking test.

In the paired preference test, two samples are presented, usually simultaneously and the panelists are asked to express their preference (product A vs. product B). The ranking preference test is an extension of the paired preference test; several samples are presented simultaneously, and the panelists are asked to rank the products according to their preference.

For difference testing, the paired-comparison test is used to compare two samples. The comparison can either be simple or it can be directional. In the simple comparison the panelists are asked whether there is a difference between the two samples. In the directional comparison the panelists are asked to choose the sample that has the greater amount of the specified characteristic. Usually, the directional comparison is chosen because it is considered to be a very sensitive test.

In the duo-trio test three samples are presented to each panelist (two alike and one different). One of the samples is identified as the control and the panelists are asked to select the sample that is the same with the control. The probability of a panelist selecting the correct sample by chance is one-half (50%).

In the triangle test, three samples are again presented to each panelist (two alike and one different), but this time no control sample is identified. The panelists are asked to identify the odd sample. The probability of a panelist selecting the correct sample by chance is one-third (33.3%).

In the ranking test, several samples can be compared. Samples are presented to the panelists simultaneously and are ranked according to the intensity of a specified characteristic. Usually, no ties are allowed (Sensory Evaluation Division of the IFT 1981).

Sensory tests aiming at identifying preference or examining whether there is a perceptible difference (i.e. between two or more competitive or similar products) generate data that in most cases are ranks or of the type "product A" vs. "product B". Such data cannot be analyzed by typical parametric tests (t-test, ANOVA) because they do not follow a normal distribution but rather have a binomial distribution. So, data from forced choice tests such as the paired comparison, triangle or duo-trio tests must be analyzed by nonparametric procedures.

Several sensory tests produce results that can be analyzed by parametric tests such as the Student's t-test and ANOVA, discussed in other chapters of this book. Requirements for the use of parametric tests are that data have to be numerical, on an interval or ratio scale and be normally distributed.

Many sensory tests, however, produce results that are not numerical (i.e. outcomes like "product A" vs. "product B") or even if they are numerical they are on an ordinal scale (i.e. ranks: 1st, 2nd, 3rd, etc.). Numerical data on an ordinal scale are not normally distributed and should not be analyzed by parametric tests. There are also cases where data may seem to be on an interval scale, but in reality, they are not. A sensory test using a rating scale may fall into this category. It has been proven that sensory panelists tend to avoid the edges of a scale and use more the middle. For example, when panelists are asked to rate a product for saltiness or for sweetness on a 9-point scale, they will move from 4 to 5 more easily than from 8 to 9. In this case, the psychological distance from 4 to 5 and from 8 to 9 is not the same and the data, even though they appear to be on an interval scale, in reality are on an ordinal scale where 5 is bigger than 4 and 9 is bigger than 8, but their distances are not the same. Many analytical-discriminative and rating sensory tests fall into this category and require the application of nonparametric tests for the statistical analysis of the results.

Parametric tests should be used when possible because they make use of more information and tend to be more discriminating or powerful than nonparametric tests (O'Mahony 1986a, b). However, nonparametric tests are more robust and less likely to lead to erroneous conclusions (Lawless and Haymann 1998a, b). For this reason, it has been suggested that nonparametric tests could even be used as a check on conclusions drawn from parametric tests.

Depending on the sensory analysis data, different statistical tests may be utilized to identify differences.

If the sensory test produces only two outcomes, the binomial distribution can be used. Tests that can produce only two outcomes include forced-choice tests such as the paired-comparison test, the duo-trio test and any other kind of test in which the outcomes can only be in pairs (i.e., "yes – no", "1 – 2" or "product A – product B").

The Mann Whitney U-Test is analogous to the t-test for two independent samples, but it can be used when the data are on an ordinal scale or the normality assumption cannot be satisfied.

If there are two or more possible outcomes from a sensory test, one approach could be to compare the relative frequencies of the different responses by using the chi-square (χ^2) statistic.

The Friedman test is nonparametric analogue of the ANOVA test and it can be used for cases with more than two samples.

5.1.3 The Binomial Test

The binomial is one of the simpler statistical tests used to determine whether there are more cases of one possible outcome compared to the other possible outcome. It is very applicable in forced choice difference tests such as the paired-comparison, the duo-trio and the triangle tests. The binomial test gets its name from the binomial expansion $(p + q)^n$ used to calculate the probability of each possible outcome of n events; p denotes the probability of the one outcome and $q = 1 - p$ denotes the probability of the other outcome.

For small experiments with few observations (N < 25), Tables with critical values for the estimation of statistical significance are used. Tables for $p = 1/2$ and $p = 1/3$ are provided in many statistical books. These Tables cover the cases of the paired-comparison test ($p = 1/2$), the duo-trio test ($p = 1/2$) and the triangle test ($p = 1/3$). In the appendix of this chapter, Table 5.9 gives the values for both probabilities (1/2 & 1/3). N denotes the sample size and X denotes the minimum number of correct answers required to reject the null hypothesis (H_o) and establish a significant difference at $\alpha = 0.05$.

When the sample size is large (N > 25) the binomial distribution approximates the normal distribution, and the z-value can be used instead of Tables. The z-value can be calculated from equation (5.1), where X is the number of correct answers in N tests with a chance probability of p.

$$\mathbb{Z} = \frac{X - Np - 0.5}{\sqrt{Npq}} \tag{5.1}$$

For one-tailed tests and $a = 0.05$, the critical z-value is 1.645. For two-tailed tests and $\alpha = 0.05$, the critical z-value is 1.96. Critical z-values for other significance levels can be found in many statistical books.

Using the appropriate z-value and p-value (depending on the sensory test), the minimum number of correct answers (X) required to establish a difference can be calculated. For this calculation, X should always be rounded to the next integer since it refers to people (panellists).

Special attention should be paid to the decision whether to use a one-tailed or two-tailed test. In a paired-comparison preference test a two-tailed test needs to be applied. In a duo-trio test or in a triangle test a one-tailed test should be used.

Case Study 1: The Binomial Test Applied to the Results of a Triangle Test to Determine Sensorial Differences by Substituting Shortenings with a Low-Caloric Content Fat Replacer

A company that produces very tasty but high-caloric content cookies is reformulating the cookie recipe by substituting half of the shortening with a low-caloric content fat replacer. Because of the great success of the specific cookies, the marketing department requires that there should be no detectable differences between the old "regular" cookies and the new, low-calorie cookies by the consumers. Since a change in the shortening may cause changes in the texture, flavor, mouthfeel and even color of the cookies, but it is not known which—if any—change will take place, an analytical—discriminative test that will measure whether there is an overall detectable difference between the regular and the low-fat cookies must be selected.

In the triangle test each panelist is presented with a questionnaire and three samples. Two of the samples are the same and one is different. All samples are coded with random numbers and presented to the panelists at the same time. The panelists are asked to identify the odd sample. Ideally, half of the panelists should receive two samples of the regular and one of the new, low-fat cookie and the other half panelists should receive one sample of the regular and two of the new cookie, so that there is no bias against the one or the other sample.

The null hypothesis is that there are no perceptible differences in the two cookies and the probability of a panelist correctly selecting the odd sample is that of chance (1/3), so the null hypothesis can be stated as H_o: $P = 1/3$. The alternative hypothesis is that there is a perceptible difference between the two cookies and the panelists can tell them apart more often than by chance only, or H_a: $P > 1/3$. It is clear from the alternative hypothesis that an one-tailed test is appropriate.

Eighty persons (all consumers of the cookie brand) were selected to participate in this sensory evaluation of the cookies. Of the 80 panelists, 32 correctly identified the odd sample. Can we claim that there are no perceptible differences between the regular and the low-fat cookie? If the answers were given by chance only, then we would expect $80 \cdot (1/3) = 26.66$ or 27 panelists correctly selecting the odd sample by chance. In our case 32 panelists selected the odd sample. Is 32 much bigger than 27 (the chance occurrence), so that we can reject the null hypothesis?

We have a large number of observations ($N = 80$), so the distribution of the data can be considered to approximate the normal distribution; therefore, we can use Eq. (5.1) for the estimation of the z-value.

$$\mathbb{Z} = \frac{32 - 80 \cdot \frac{1}{3} - 0.5}{\sqrt{80 \cdot \frac{1}{3} \cdot \frac{2}{3}}} = 1.15$$

We see that the z-value of 1.15 is much lower than the critical value of 1.645 in order to establish significance at $\alpha = 0.05$. In this case we can say that there were not many more panelists correctly identifying the odd sample than what we would expect by chance only and that the panelists could not tell a difference between the regular and the low-fat cookie.

5.1.4 The Chi-Square Test

The chi-square or chi-squared (χ^2) test is a useful statistic used to compare the frequency of occurrence of two or more events. The chi-square test takes advantage of a different distribution, named the chi-square distribution. The chi-square statistic is computed and compared with the largest value that could occur on the null hypothesis (available from Tables in various statistical books). If the chi-square statistic is larger than the critical value, the null hypothesis is rejected (O'Mahony 1986a, b). The value of the chi-square statistic reflects the magnitude of the discrepancies between observed and expected values. When the differences are big, the chi-square statistic is large and the null hypothesis can be rejected (Devore and Peck 1986). The chi-square statistic is mostly used for independent observations and it is not considered appropriate for related observations by some writers (Lawless and Haymann 1998a, b). One advantage of the chi-square test is that the data can be classified along one dimension (i.e. preference) or two dimensions at the same time (i.e. preference X gender).

The chi-square statistic can be calculated by Eq. (5.2):

$$\chi^2 = \sum \left[\frac{(observed - expected)^2}{expected} \right] \tag{5.2}$$

The distribution of the statistic χ^2 is chi-square with $(r-1)(c-1)$ degrees of freedom, where r represents the number of rows in the two-way table and c represents the number of columns.

Case Study 2: The Chi-Square Test Applied to the Results of a Preference Test to Analyse Whether the Appeal of a Sugar-Free Drink Is the Same to Both Men and Women

A company wants to launch a sugar-free drink but there are questions whether the appeal of this drink is the same to both men and women. In order to investigate this, a test is conducted in a major shopping mall. The new sugar-free drink is offered to both men and women and they are asked to mark their approval by expressing whether (i) they like the product, (ii) they are neutral or (iii) they dislike the product. The results are summarized in Table 5.1.

The numbers in the cells of Table 5.1 represent the observed values. The expected values need to be calculated so that the χ^2 statistic can be calculated. If there was no difference in the liking of the products between men and women, one would expect roughly similar numbers of men and women liking or disliking the product, if the sensory test had used the same number of men and women. However, because there

Table 5.1 Responses of women and men on the preference of the new drink

	Women	Men	Total
Like	42	33	75
Neutral	20	27	47
Dislike	18	25	43
Total	80	85	165

Table 5.2 Calculation of expected values for the data of Table 5.1

	Women	**Men**	**Total**
Like	$(75 \cdot 80)/165 = 36.36$	$(75 \cdot 85)/165 = 38.64$	75
Neutral	$(47 \cdot 80)/165 = 22.79$	$(47 \cdot 85)/165 = 24.21$	47
Dislike	$(43 \cdot 80)/165 = 20.85$	$(43 \cdot 85)/165 = 22.15$	43
Total	80	85	165

are more men than women in this test, the expected frequencies need to be weighted for the proportion of men vs. women. Also, they need to be adjusted according to the fact that the panelists tend to like the product more than dislike it. One way to calculate the expected frequencies, taking into account all the above considerations, is to use the geometric mean for each cell of Table 5.2 as follows:

$$\text{Expected value} = (\text{row total} \times \text{column total})/(\text{grand total})$$

Using the above formula, we can calculate the expected values for our data as in Table 5.2

Now, the χ^2 statistic can be calculated according to Eq. (5.2):

$$\chi^2 = \left[\frac{(42 - 36.36)^2}{36.36} + \frac{(20 - 22.79)^2}{22.79} + \ldots + \frac{(25 - 22.15)^2}{22.15}\right] = 3.11$$

The degrees of freedom (df) are: $(3-1)(2-1) = 2$. The χ^2 critical value for 2 df and $\alpha = 0.05$ is 5.99. A two-tailed test was used because there was no assumption that men liked the sugar-free drink more or less than women.

We see that the χ^2 statistic that we calculated is smaller than the critical value, so we can claim that there is no difference in the preference of the sugar-free drink between men and women.

5.1.5 *The Friedman Test*

The Friedman test is a two-factor, multiple-sample difference test, applicable to related samples (repeated-measures). It is the nonparametric analogue of the two-sample, repeated-measures ANOVA design. The Friedman test, just like the ANOVA test is always two-tailed (O'Mahony 1986a, b) and it is preferred over the

Kramer test (Meilgaard et al. 2007). One disadvantage of the Friedman test it that it cannot be used to study interaction effects (Miller and Miller 2000).

In sensory evaluation the Friedman test is often used to analyze ranking data. The ranking data are tabulated so that scores for each different sample are presented in columns and the panelists are placed in different rows. The rank sums for each sample are calculated (column totals) and they are used to calculate the appropriate statistic.

The Friedman test statistic is based on the χ^2 distribution and it can be calculated by the formula shown in Eq. (5.3).

$$\chi_r{}^2 = \frac{12}{N(K)(K+1)} \sum_{i=1}^{K} R^2 - 3N(K+1) \tag{5.3}$$

where K is the number of samples (number of columns)
N is the number of panelists (number of rows)
R is the rank totals for each sample (column)

The degrees of freedom are $df = (\mathrm{K} - 1)$.

If there are no differences in the rankings of the different samples, the $\Sigma \mathrm{R}^2$ value will not be very large. If there are significant differences in the rankings, some samples will tend to have rather large rank sums, which when squared, produce a much larger $\Sigma \mathrm{R}^2$ value and, consequently, a large χ_r^2 statistic.

If the Friedman test shows that there are significant differences among the samples tested, a multiple comparison procedure will be required to identify specific differences.

The "least significant ranked difference" or $\mathrm{LSD}_{\mathrm{rank}}$ is analogous to the Fisher's LSD for ANOVA, and it can be calculated from equation (5.4).

$$LSD_{rank} = z_{\alpha/2} \cdot \sqrt{NK(K+1)/6} = t_{\alpha/2,\infty} \cdot \sqrt{NK(K+1)/6} \tag{5.4}$$

with $z_{a/2}$ or $t_{a/2,\,\infty}$—critical values derived from the appropriate statistical tables for the desired level of significance (always two-tailed).

An other, more conservative, multiple comparison test that can be used is the Tukey's HSD test for ranks. Equation (5.5) shows the formula for the calculation of the minimum statistically significant difference (Meilgaard et al. 2007).

$$HSD_{rank} = q_{a,K,\infty} \sqrt{NK(K+1)/12} \tag{5.5}$$

The $\mathrm{q}_{\alpha,\mathrm{K},\infty}$ critical values can be found from the studentized range statistic Tables available in many statistical books.

Table 5.3 Ranking of the three ice-cream flavors by children

Panelist	Bubblegum (A)	Cool mint (B)	Mixed berries (C)
1	1	2	3
2	1	3	2
3	2	1	3
4	1	3	2
5	1	3	2
6	1	2	3
7	3	2	1
8	1	3	2
9	2	1	3
10	1	2	3
11	1	3	2
12	1	2	3
13	2	1	3
14	3	2	1
15	1	3	2
16	1	2	3
17	1	3	2
18	2	1	3
19	1	3	2
20	1	2	3
Rank sum (R)	28	44	48

Case Study 3: The Friedman Test Applied to the Results of a Preference Test to Analyse Which of 3 Flavors of Ice-Creams Are Preferred

A company wants to extend their line of ice-cream flavors by creating a "kid's flavor" that will be liked by children. The research department has come up with three possible new flavors: "bubblegum", "cool mint" and "mixed berries" and needs to know which of the three flavors is preferred by children. In order to answer the question, a preference test is set up with 20 children ages 7–11 years old. Every child is presented with the three ice-cream flavors and is asked to rank them in terms of overall liking.

The null hypothesis is that there are no differences in the preference of the three ice-cream flavors (H_o: A = B = C) and the alternative (H_a) is that there is at least one flavor that is preferred more or less than the others (two-tailed).

The results are summarized in Table 5.3 with "1" representing the best liked flavor and "3" the least liked flavor. The rank sums (R) are also shown for each ice-cream flavor.

Using Eq. (5.3), we can calculate the statistic

$$\chi_r^{\ 2} = \frac{12}{20 \cdot 3 \cdot 4}\left(28^2 + 44^2 + 48^2\right) - 3 \cdot 20 \cdot 4 = 11.2$$

From the χ^2 distribution Table, for (K − 1) = 2 degrees of freedom and $\alpha = 0.05$ (two-tailed), we can see that the critical value is 5.99. Our statistic is much larger than the critical value and we can conclude that there is at least one flavor of ice-cream with different liking.

In order to know which flavor is liked better, we need to further analyze the results with a multiple-comparison test. The Tukey's HSD_{rank} will be used because it is more conservative and less likely to exaggerate any differences. The $q_{\alpha,K,\infty}$ critical values for $\alpha = 0.05$ and K = 3 is $q_{0.05,\ 3,\ \infty} = 3.32$. From Eq. (5.5) we can calculate the minimum significant difference to be:

$$HSD_{rank} = 3.32\sqrt{20 \cdot 3(3+1)/12} = 14.85$$

This means that for any two rank sums to be different, they must have a difference of at least 14.85. Comparing the rank sums, we can see that the "bubblegum" flavor is ranked higher than the other two flavors. There are no significant differences between the "cool mint" and "mixed berries" flavors. The results can be represented as follows, with the similar values underlined:

Bubblegum (A) <u>Cool mint (B) Mixed berries (C)</u>

5.2 Examples of Using R in Sensory Science

Peter Ho

5.2.1 Objectives and Learning Outcomes

The main objective of this chapter is to demonstrate how R can be used to conduct simple statistical tests, which is often present as built-in functions of the base program, to analyse typical sensory data from overall difference tests, attribute difference tests and preference tests. A complete R script is provided for each example which readers can use as a template for analysing their own data.

5.2.2 Introduction

R (R Development core team 2023) is a freely available software program that can be downloaded at http://www.r-project.org. It is available for Linux, Mac OSX and the Windows operating systems. R consists of the basic program which comes with a number of pre-installed libraries (also known as packages) of statistical functions. Additional packages can be found on the website (there are currently over 3000 packages available) which can be installed directly from within the program.

R Basics

The basic structure of R consists of a primary working window, known as the *workspace*. On starting an R session, the workspace will appear with some basic instructions (Fig. 5.1), followed by the command prompt. i.e., the > symbol.

The basic working principle is to directly type a line of R script or paste the R script that you have constructed using an external text editor or the in-built script editor after the command prompt. Typing enter would result in the execution of the R script. However, if the R script is incomplete the command prompt on the subsequent line will be replaced with a + symbol. This implies that there is something missing in the R script and only after the missing part of the script is added will the command prompt appear in the subsequent line. It is useful in some cases when creating an R script to provide text comments describing what the line of R script is to be used for. The # symbol is used at the beginning of the comment line. The comment line could also be used after the R script as demonstrated below.

Take for example the case where we might want to conduct a simple addition between two values and have the result of the calculation stored for later use. This can be done in R by using an assignment symbol <- which assigns the result of any R expression or function to a named R object. For example, the value `6+8` can be

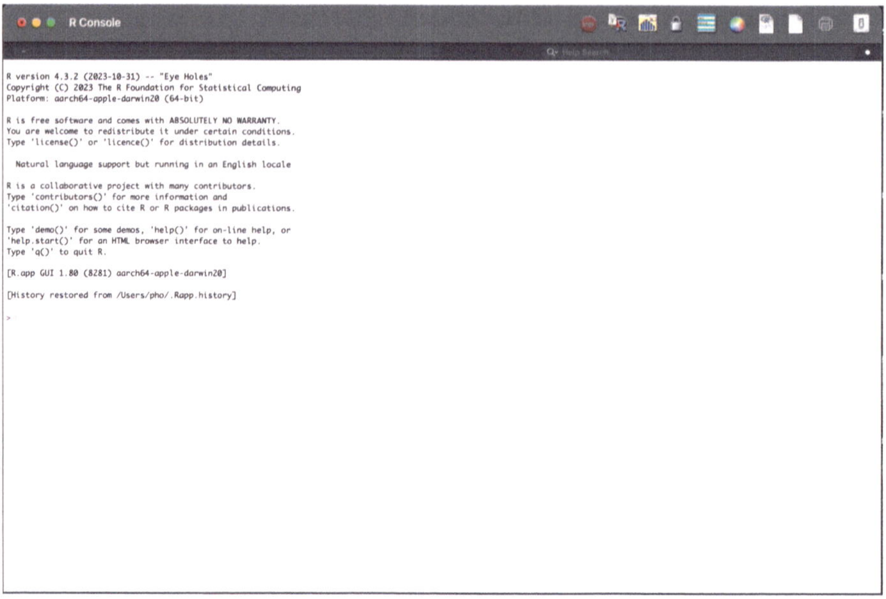

Fig. 5.1 The workspace window from R 4.3.2. under Mac OSX

determined and the results stored in an R object `a` without printing the result. separated by a comma.

```
a <- 6+8 #Add 6 to 8 and store result in a
```

If we want to know the values stored in `a` we have to type `a` after the command prompt. Values stored in R objects can be single values, vectors, data frames, matrices and values of results from a statistical analysis. Storing the results of an analysis can often be useful as it could be used for further analysis or for plotting data.

Data Entry

A vector which is a string of data values, similar to a column of data in a spreadsheet, can be created and stored as an R object using the concatenate function `c()`. For example, we can create a vector `b` by with four values, each separated by a comma.

```
b <- c(5,7,2,8,6)
```

Vectors can be used not only for storing data from a sensory analysis experiment but is often used to indicate the a factor of sensory assessors and a factor indicating the samples being examined. When we have more than one vector and we want to put them together, we can create a data frame.

Let's create two data vectors `b` and `d` each containing 6 values and a factor with two levels called `method`, using the function `factor()`. The R script uses the `c()` within `factor()`. Using a function over another function is a common method when we want to reduce the number of lines in a completed R script.

```
b <- c(5,7,2,8,6,3)
d <- c(9,11,25,22,16,19)
method <- factor(c(1,1,1,2,2,2))
mydata <- data.frame(b,d,method)
```

The last line of the R script then uses the function `data.frame()` to join the three vectors together and the result is stored in an R object `mydata`. We have created our first R script. Typing `mydata` after the command prompt will give us a data frame of three columns.

```
> mydata
  b  d method
1 5  9      1
2 7 11      1
3 2 25      1
4 8 22      2
5 6 16      2
6 3 19      2
```

R scripts such as the ones used for data entry will be used throughout the rest of this chapter together with functions from various packages to conduct a series of statistical tests for analysing some typical sensory analysis data. A typical function in R consists of a series of arguments (or parameters) that must be included within the brackets of a function. A help page for each function can be found by typing `help.start()` after the command prompt, which opens a web browser window. The help files can be found by clicking on the *Packages* link.

5.2.3 Problems to solve

The number of samples of interest that is to be compared during the sensory test (two or more than two samples), the type of measurement scale (nominal, ordinal or interval) and the sample presentation order (balanced complete or balanced incomplete block experimental design) will determine the choice of the statistical procedure that should be used to analyse the sensory data. In this chapter we shall only consider the case when all samples of interest are presented to each sensory assessor. i.e., a balanced complete block experimental design is used for the presentation order. The problems are presented in order of the complexity of the statistical testing, from comparing two samples to more than two samples. We shall be examining the following sensory tests:

1. Triangle test;
2. Rank difference test;
3. Paired preference test;
4. Attribute difference test;
5. Hedonic preference test.

5.2.4 How to Use the Binomial Test Function of R to Determine Sensorial Differences of a New Formulation of a Lemonade

A food company was interested in changing the artificial sweetener used in their lemonade drink to reduce their production costs. The NPD department decided to conduct an overall difference test using a triangle test to determine if there was a detectable difference between the new formulation (lemonade B) and the original formulation (lemonade A) at a 5% level of significance. A panel of 30 trained assessors were each presented with a single set of 3 coded samples. Each assessor was required to taste each sample and identify the odd sample. They were informed that two of the samples were identical and one was different.

Triangle and Duo-Trio tests are used to determine if overall differences between two samples exists. In a triangle test, assessors are presented with a single or multiple sets of 3 coded samples. Each set comprises of two identical samples and one different sample. Data from these tests can be analysed using the binomial proportion test, using the `binom.test()` function in R.

Table 5.4. shows the results of the triangle test using 30 trained assessors. Counting the number of "1"s in the column *Response* we find that there were 17 out of 30 assessors that correctly identified the different sample in the 3 sample set. A one-sided hypothesis test can be conducted comparing this result with the hypothesized probability of success for the Triangle test, $p_0 = \frac{1}{3}$, as there is a one in three chance of selecting the different sample. We can test the one-sided alternative hypothesis $H_1 : p > p_0$ using the `alternative = "greater"` argument at a 5% level of significance with the argument `conf.level = 0.95`.

The test can be conducted by typing the following R script into the R workspace.

```
binom.test(17, 30, p = 0.333, alternative ="greater",
      conf.level = 0.95) #R script for binomial test
```

The results of the exact binomial test given below indicates that the panel of 30 trained assessors were able to detect a difference between the two samples of lemonade at a p-value of 0.007142. Thus a statistical significant difference between the two formulations was detected at a p-value < 0.01.

Table 5.4 Results from the Triangle test for lemonade samples

Assessor	Response[a]	Assessor	Response
1	1	16	1
2	1	17	0
3	0	18	1
4	1	19	1
5	0	20	0
6	1	21	1
7	1	22	0
8	1	23	0
9	0	24	1
10	1	25	0
11	1	26	0
12	1	27	1
13	0	28	0
14	0	29	1
15	1	30	0

[a]1: correctly identified; 0: incorrectly identified

```
> binom.test(17, 30, p = 0.333, alternative
+ ="greater", conf.level = 0.95)

  Exact binomial test

data: 17 and 30
number of successes = 17, number of trials = 30,
p-value = 0.007142
alternative hypothesis: true probability of success
is greater than 0.333
95 percent confidence interval:
 0.4016289 1.0000000
sample estimates:
probability of success
        0.5666667
```

5.2.5 How to Use the Friedman Test Function of R to Determine Differences in Hardness of 5 Different Formulations of Cheddar Cheese

A local dairy working on the development of a new mild cheddar cheese is interested in comparing the hardness of 5 different formulations that have previously been assessed to have similar flavour properties. A ranking test was conducted by the sensory analyst to determine if there were any differences in hardness, at a 5% level of significance, with 20 assessors each receiving all 5 samples in a random presentation order based on a balanced complete block design. Assessors were required to rank samples from the hardest sample (ranked 1) to the softest (ranked 5).

Attribute rank difference tests can be conducted when there is an interest in comparing more than two samples and the objective is only determine the order of intensity of the attribute of interest. The ranking test can be used also in a preference test when only the order of preference is of interest. A complete randomised block design is used as when between 3 and 5 samples are presented to each assessor in a single session.

A two-sided hypothesis test can be carried out to determine if there were any difference in the hardness of the cheese samples at a 5% level of significance. The `friedman.test()` function can be used to conduct a Friedman two-way ANOVA to analyse data from an attribute rank difference test when all samples of interest are presented to each assessor in complete randomised block design.

The null hypothesis (*Ho*:) compares the equality of the rank sums of each sample with the alternative hypothesis (*H1*:) of the at least one comparison between the rank sums be unequal. This can be represented, for comparing samples A to E, as

$$H_0 : \sum R_A = \sum R_B = \sum R_C = \sum R_D = \sum R_E$$

$$H_1 : not\ all \sum R_i$$

The test can be conducted by typing the following R script into the R workspace, whereby a matrix is created from the data in Table 5.5 using the `matrix()` function. The data is added to the matrix column-wise using the argument `byrow=FALSE` with labels for each column added using the `dimnames` argument. The Friedman 2-way ANOVA test can then be simply determined by calling the `friedman.test()` function.

Table 5.5 Results from the ranking test for Cheddar cheese samples

Assessor	A	B	C	D	E	Assessor	A	B	C	D	E
1	2	4	5	3	1	11	4	5	2	3	1
2	4	5	3	1	2	12	2	4	3	5	1
3	1	4	5	3	2	13	5	3	4	2	1
4	1	2	5	3	4	14	3	5	2	4	1
5	1	5	2	3	4	15	1	2	5	4	3
6	2	3	4	5	1	16	3	4	5	1	2
7	4	5	3	1	2	17	2	3	4	5	1
8	2	3	5	4	1	18	1	4	2	5	3
9	1	3	4	5	2	19	4	5	3	2	1
10	1	2	5	3	4	20	2	3	5	4	1

```
#R script for Friedman 2-way ANOVA
hardness<-          matrix
(c(2,4,1,1,1,2,4,2,1,1,4,2,5,3,1,3,2,1,4,2,
     4,5,4,2,5,3,5,3,3,2,5,4,3,5,2,4,3,4,5,3,
     5,3,5,5,2,4,3,5,4,5,2,3,4,2,5,5,4,2,3,5,
     3,1,3,3,3,5,1,4,5,3,3,5,2,4,4,1,5,5,2,4,
     1,2,2,4,4,1,2,1,2,4,1,1,1,1,3,2,1,3,1,1),
nrow = 20,byrow = FALSE,
dimnames = list(1 : 20, c("A", "B", "C","D","E")))
friedman.test(hardness)
```

The results of conducting the Friedman 2-way ANOVA test indicates that there are differences in the hardness between the 5 cheese samples at a 0.001% level of significance, as the p-value was 0.0001073.

```
> friedman.test(hardness)

  Friedman rank sum test

data: Hardness
Friedman chi-squared = 23.36, df = 4, p-value = 0.0001073
```

The results of the Friedman 2-way ANOVA test indicates that at least one comparison between rank sums was likely to be significantly different. In order to determine which samples had rank sums that were significantly different, the absolute difference between each rank sum (e.g., $\sum R_A - \sum R_B$) can be compared against the Tukey's honestly significant difference (HSD) critical distance for ranks.

```
##Calculate difference between rank sums
##and critical distance for HSDRanks
Ranktotals <-
 c(abs(colSums(hardness)[[1]]-  colSums(hardness)[[2]]),
  abs(colSums(hardness)[[1]]-colSums(hardness)[[3]]),
  abs(colSums(hardness)[[1]]-colSums(hardness)[[4]]),
  abs(colSums(hardness)[[1]]-colSums(hardness)[[5]]),
  abs(colSums(hardness)[[2]]-colSums(hardness)[[3]]),
  abs(colSums(hardness)[[2]]-colSums(hardness)[[4]]),
  abs(colSums(hardness)[[2]]-colSums(hardness)[[5]]),
  abs(colSums(hardness)[[3]]-colSums(hardness)[[4]]),
  abs(colSums(hardness)[[3]]-colSums(hardness)[[5]]),
  abs(colSums(hardness)[[4]]-colSums(hardness)[[5]]),
qtukey(0.95,5,999)*sqrt((20*5*(5+1))/12))
## Add sample labels
Comparisons <- c("A-B","A-C","A-D","A-E","B-C", "B-D",
"B-E","C-D","C-E","D-E","critical distance for HSDRanks")
```

```
## Create table of results
HSDRank.comparisons <-
data.frame(Ranktotals, row.names=Comparisons)
```

In the R script above, the absolute differences between each rank sums are calculated using the `abs()` function and the rank sums are calculated using the `colSums()` function. For example, `colSums(hardness)[[1]]` gives us the rank sums for column 1 of the data matrix *hardness*, which is for sample A (i.e., 1 denoting sample A, 2 sample B and so on). We can construct a table of all the absolute differences between all sample rank sums, together with the Tukey's HSD critical difference for ranks using the `qtukey()` function, based on formula (5.5) in Sect. 5.1.

Tukey's HSD critical difference by ranks for a 5% level of significance was calculated to be 27.32765. Any rank comparison greater than the critical distance can be considered significantly different.

```
> HSDRank.comparisons
                          Ranktotals
A-B                         28.00000
A-C                         30.00000
A-D                         20.00000
A-E                          8.00000
B-C                          2.00000
B-D                          8.00000
B-E                         36.00000
C-D                         10.00000
C-E                         38.00000
D-E                         28.00000
critical distance for HSDRanks  27.32765
```

The table shows that the comparison between samples A–B, A–C, B–E, C–E and D–E were statistically significant. The results shows that Samples E (rank total of 38) is significantly harder than sample B (rank total of 74), C (rank total of 76) and D (rank total of 66), whereas sample A (rank total of 46) is significantly different from only samples B and C but not sample D.

5.2.6 How to Use the the t.test Function of R to Determine the Preference between Two Soup Formulations

A new soup product is being developed in the product development department in a local food company. Two different products with different ingredients that provide a salty flavour is compared to determine if there is an overall preference for one of the

Table 5.6 Results from the paired preference test for soup samples

	Samples			Samples			Samples			Samples	
No.	A	B	No.	A	B	No.	A	B	No.	A	B
1	6	6	16	8	7	31	6	8	46	6	7
2	8	7	17	8	6	32	5	7	47	6	8
3	5	6	18	8	6	33	9	7	48	7	6
4	7	8	19	8	7	34	8	7	49	7	7
5	5	6	20	7	5	35	8	6	50	4	5
6	8	8	21	7	4	36	7	6	51	8	8
7	7	8	22	8	7	37	6	5	52	6	2
8	8	7	23	7	8	38	6	7	53	7	5
9	8	6	24	5	6	39	5	6	54	4	5
10	3	5	25	8	7	40	8	7	55	3	6
11	3	6	26	6	7	41	4	7	56	2	5
12	5	8	27	5	7	42	7	8	57	7	4
13	4	5	28	8	7	43	8	7	58	4	5
14	6	7	29	6	4	44	6	7	59	7	6
15	6	7	30	4	6	45	5	7	60	4	7

two product formulations. A paired preference test was used to determine if a panel of 60 untrained assessors had a preference for either soup A or soup B. at a 5% level of significance. The sensory analyst was also interested to know the difference in the degree of acceptability between the two soups, so a 9-point hedonic scale was used. The test was conducted with 60 assessors. Sixty plates with soup "A" and 60 plates with soup "B" are coded with unique 3-digit random numbers. For 30 assessors, soup "A" was presented in session one and soup "B" in session two, for the 30 others soup "B" was presented in session one and soup "A" in session two (Table 5.6).

A paired preference test is often used when comparing two samples and there is an interest in determining which sample between the two are preferred and the difference in the degree of acceptability. The 9-point hedonic scale produces ordinal scaled data that is usually considered to be equivalent to interval scaled data, as it is customary to use a large number of untrained assessors (e.g., 60 assessors) for a preference test. This allows the analyst to use the usual parametric statistical test of a paired t-test. Preference test are not normally conducted with low number of untrained assessors, however, if this is done or if there is some doubt as to whether the assumptions of a parametric test is violated, then a nonparametric statistical test should be used instead. The Wilcoxon Matched-Pairs Signed Ranks test (Sheskin 2000) is a nonparametric statistical test which is equivalent to the paired t-test.

A two-sided hypothesis paired t-test can be conducted to test if there was a difference in acceptability between the two soups using the `t.test()` function with the `paired=TRUE` argument, indicating that a paired t-test is used instead of an independent t-test. The other arguments that must be included in the function is `alternative="two.sided"` to indicate that a two-side paired t-test should be conducted with a `conf.level=0.95` for a test at a 5% level of significance. The null and alternative hypotheses for this test for comparing two samples, sample A and sample B are

$$H_0 : \mu_A - \mu_B = 0$$
$$H_1 : \mu_A - \mu_B \neq 0$$

The R script for this test consists of first constructing a vector for data from sample and applying the `t.test()` function to the two vectors.

```
#Create a vector for sample A data
sample.A <-
c(6,8,5,7,5,8,7,8,8,3,3,5,4,6,6,8,8,8,8,7,
 7,8,7,5,8,6,5,8,6,4,6,5,9,8,8,7,6,6,5,8,
 4,7,8,6,5,6,6,7,7,4,8,6,7,4,3,2,7,4,7,4)
#Create a vector for sample B data
sample.B <-
c(6,7,6,8,6,8,8,7,6,5,6,8,5,7,7,7,6,6,7,5,
4,7,8,6,7,7,7,7,4,6,8,7,7,7,6,6,5,7,6,9,7,
8,7,7,7,7,8,6,7,5,8,2,5,5,6,5,4,5,6,7)
###Conduct the two-sided paired t-test
t.test(sample.A,sample.B,paired =TRUE,
alternative="two.sided",conf.level= 0.95)
```

The results of the paired t-test with the panel of 60 untrained assessors indicates that there was no preference for either of the two soups, as the p-value = 0.3755 was larger than the 5% level of significance (i.e., $\alpha = 0.05$).

```
> t.test(sample.A,sample.B,paired =TRUE,
alternative="two.sided",conf.level= 0.95)

  Paired t-test

data: sample.A and sample.B
t = -0.8929, df = 59, p-value = 0.3755
alternative hypothesis: true difference in means is
not equal to 0
95 percent confidence interval:
 -0.6481939 0.2481939
sample estimates:
mean of the differences
             -0.2
```

Table 5.7 Results from attribute difference test for chorizo sausage samples

	Sample A			Sample B			Sample C		
Assessor	1	2	3	1	2	3	1	2	3
1	2	2	1	3	4	5	5	4	3
2	2	1	1	2	2	1	3	4	3
3	1	2	1	2	3	2	5	4	4
4	3	3	3	4	5	6	5	6	4
5	2	4	3	4	3	4	3	5	6
6	2	4	1	3	2	4	4	6	5
7	1	1	1	2	2	3	3	5	3
8	6	4	3	4	6	3	4	6	3
9	2	2	2	3	3	5	4	6	5
10	1	4	3	2	4	3	5	4	5
11	3	4	2	1	3	4	6	5	3
12	2	2	1	3	2	1	3	4	6

5.2.7 *How to Use the ANOVA Function of R with Sample Replications to Compare the Spiciness of Three Different Formulations of Chorizo Sausages*

A product development team in a meat company was interested in developing a new spicy chorizo sausage. An attribute difference test was conduct to compare three different products (A,B and C), each made with a different spice blend mix, to determine any differences in spiciness at a 1% level of significance. The sensory test was conducted with 12 trained assessors each receiving all three samples in a random presentation order in three separate sessions. Assessors rated each sample separately using a 9-category labelled scale for spiciness from Not detectable (rating of 1) until Extremely intense (rating of 9) (Table 5.7).

Attribute difference tests with a rating scale, such as the labelled category rating scale) used in this example, can be used to compare differences in the intensity of an attribute between more than two samples. A complete randomised block design should be used for the presentation order of samples whenever we are comparing between 3 and 5 samples, or when the samples contain any intense flavors which could affect the judgement of the trained assessors. In this case, a single-factor within subjects analysis of variance (ANOVA) with sample replications is often used when analysing data from an attribute difference test with the trained panel. Sample replication is important in an attribute difference test and should almost always be used.

The `aov()` function can be used to conduct an analysis of variance. The ANOVA model for a single-factor within-subjects ANOVA with sample replications makes it possible to fit the complete ANOVA model that includes the main factor effects and the factor interaction effect. This can be represented as

$$X_{ijk} = \mu + T_i + B_j + (TB)_{ij} + \varepsilon_{ijk}$$

where

X_{ijk} is the value of all response for the ith sample for each jth assessor
μ is the overall mean
i is the number of samples ranging for 1,2,....t
j is the number of assessors ranging for 1,2,....b
T_i is the Sample effect (the Single factor) for ith sample
B_j is the Assessor effect (or Block factor) for jth assessors
$(TB)_{ij}$ is the interaction between Samples and Assessor effect
ε_{ijk} is the random error or residual

The null hypothesis of the attribute difference test was to compare the mean values of spiciness of each sample is defined as

$$H_0 : \mu_A = \mu_B = \mu_C = 0$$
$$H_1 : at\ least\ one\ \mu_i - \mu_j \neq 0$$

A two-sided hypothesis test can be carried out to determine if there were any difference in the spiciness of the chorizo samples from Table 9.3.4 at a 1% level of significance. A data frame is used instead of a matrix that consists of 3 columns, a Sample effect column with levels indicating the sample used, an Assessor effect column indicating the assessor and a Spiciness column for the assessor response to intensity of spiciness.

Examining Table 9.3.4 we see that a vector of assessors 1 to 12 should be repeated 9 times in a sequence. The `rep()` function for a sequence of 12 assessors is repeated `times=9` within the `c()` function, before apply the `factor()` function to indicate that *assessors* is the block factor. A sample vector is also created but this time each sample level ("A", "B" or "C") must be repeated 36 times before apply the `factor()` function to indicate that *samples* is the sample effect factor. A third vector, *spiciness*, for the assessors responses should also be created. We can then create the data frame, panel, using the `data.frame()` function. The R script for the data frame was

```
#Create a factor for assessor
assessors <-factor(c(rep(c(1:12),times=9)))
#Create a factor for samples
samples <-factor(c(rep(c("A"),times=36),
rep(c("B"),times=36),rep(c("C"),times=36)))
#Create a response vector for spiciness
spiciness<-c(2,2,1,3,2,2,1,6,2,1,3,2,
        2,1,2,3,4,4,1,4,2,4,4,2,
        1,1,1,3,3,1,1,3,2,3,2,1,
        3,2,2,4,4,3,2,4,3,2,1,3,
        4,2,3,5,3,2,2,6,3,4,3,2,
        5,1,2,6,4,4,3,3,5,3,4,1,
```

```
      5,3,5,5,3,4,3,4,4,5,6,3,
      4,4,4,6,5,6,5,6,6,4,5,4,
      3,3,4,4,6,5,3,3,5,5,3,6)
#Create a data frame, panel, for all three vectors
panel<-data.frame(samples,assessors,spiciness)
```

After the data frame has been created using the R script, we can fit the ANOVA model using the aov() function, saving the results of the ANOVA test in an R object called *panel.aov*.

```
panel.aov<- aov(spiciness~samples+assessors+
samples:assessors,panel)
```

The ANOVA model consists of the response *spiciness* on the left side of the equation, which is separated by the ~ symbol. This is followed on the right by the 3 factor terms in the full ANOVA model for the single-factor within subjects ANOVA, consisting of the main effect *samples*, the block effect *assessors*, and the two-way interaction effect *samples:assessors*. An interaction term in an ANOVA model can only be estimated when there are sample replications, as is the case in our example.

The aov() function in R uses the so-called Type II sums of squares. There are no differences in the results from an ANOVA test using either Type II or Type III sums of squares when a balanced complete block design is used. However, if any of the residuals from fitting the ANOVA model are found to be outliers (see regression diagnostics below on how to identify an outlier), they would have to be removed, as the least squares regression fitted line would be shifted towards the outlier and thus resulting in a bias fit (Neter et al. 1985). In that case, those outliers would have to be removed and this would result in an unbalanced data, which should then be analysed using Type II sums of squares (Langsrud 2003). The use of Type II sums of squares will be considered again in the next example.

The typical ANOVA table can be produced by calling the *panel.aov* object with the summary() function. The results shown in the ANOVA table indicate that both the sample effect and the assessor effect were highly significant at a 0.0001% level of significance. The sample and assessor interaction terms was not significant.

```
> summary(panel.aov)
           Df Sum Sq Mean Sq F value  Pr(>F)
samples        2 83.39  41.69 39.157 3.11e-12 ***
assessors     11 50.78   4.62  4.335 5.84e-05 ***
samples:assessors 22 22.83   1.04  0.975   0.505
Residuals     72 76.67   1.06
---
Signif.codes: 0 '***' 0.001 '**' 0.01 '*' 0.05 '.' 0.1 ' ' 1
```

It is important when conducting any ANOVA test to always check for the validity of model assumptions. The most common method is to conduct an analysis of the residuals after fitting the ANOVA model (Dean and Voss 1999; Montgomery 1997). Both Dean and Voss (1999) and Neter et al. (1985) recommend conducting an analysis of residuals to check the: (i) form of the model; (ii) presence of outliers; (iii) independence of error terms; (iv) error terms do not have constant variance; (v) error terms are normally distributed. In addition, Neter et al. (1985) suggest to also examine if any important variables have been omitted from the regression model.

Regression diagnostics, as it is sometimes called, can be conducted in R by applying the `plot()` function to the *panel.aov* object. This produces 4 graphs: residuals versus fitted values (top left); the Q-Q normal distribution plot of standardized residuals (top right); the scale-location plot of the square root of the absolute value of the standardized residuals (bottom left); and the Constant leverage plot of the standardized residuals versus each level of the sample factor (bottom right).

Figure 5.2 shows the 4 graphs of residuals for the results of fitting the full ANOVA model. The plot of residuals versus fitted values (top left) and the Scale-Location plot (bottom left) from Fig. 5.2 can be used to determine the linearity of the ANOVA regression function. The Scale-Location plot makes it easier to detect any dispersion trend effects, as it reduces the skewness of the distribution (Dalgaard 2002). The random pattern of residuals from the plot of residuals versus fitted values indicates that there was no real problem with nonlinearity, which can be identified when there is a curvature in form of the residuals, or any other systematic trend

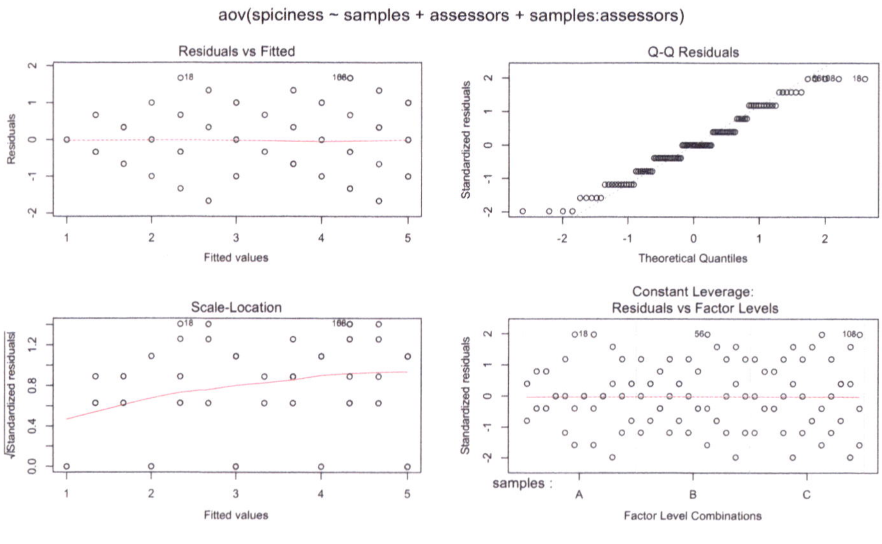

Fig. 5.2 Graphical analysis of residuals for checking ANOVA model assumptions from the attribute difference test for spiciness of chorizo sausages

(Neter et al. 1985). If the plot showed a "funnel" effect, this would have indicated that the error terms did not have constant variance. Montgomery (1997) attribute a violation of constance variance due to the error or background noise being a constant percentage of the size of the observation, non-normality of data or a skewed distribution. The Constant Leverage plot (bottom right), which plots the standardized residuals for each level of a factor, is used to determine if the error terms have constant variance. The plots shows the same extent of residual scatter for each sample (A, B and C), thus suggesting that the error variances were essentially constant.

The assumption of normality can be easily examined by the normal Quantile-Quantile (Q-Q) plot (top right) from Fig. 5.2. The normality of residuals is confirmed as the plot shows that the majority of the residuals fit along a straight line. Some degree of deviation or tailing off of a small percentage of the residuals can be normally expected and should not be cause for alarm. However, any other behaviour, such as heavy tailing, an "S" shape or a shift in the slope of the residuals would indicate a violation of the normality assumption.

The validity of model assumptions are can also be assessed by examining the number of outliers. The series of plots in Fig. 5.2 indicate a number of residuals that might be possible outliers (i.e., those that have been numbered in the plots). Dean and Voss (1999) suggests that model assumptions are held as long as 99.7% of the standardized residuals are within −3 to 3, 95% between −2 to 2 and 68% within −1 to 1. This seems to be the case with our example.

The independence of error terms can be assessed by plotting the standardized residuals versus the order in which data was collected to determine if error variances showed any increasing or decreasing trends (Neter et al. 1985). This would be useful to examine if sample presentation order affects the manner in which samples are evaluated by the assessors. The use of more than one session for analysing all samples and replications, such as having each assessor evaluate the required number of samples divided over two or three days, may also result in a session or day effect. If a session effect is present, then it is important for it to be included in the ANOVA model so that its significance can be estimated. For the single factor within subjects ANOVA model with replications to be valid, the session effect should be insignificant.

Now that we have checked the validity of model assumptions and found that there were no significant violations, the next step would be to examine the results from the ANOVA table. Factor effects plots are often a useful indicator to examine the way each assessor rates every sample and can be plotted before or after conducting an ANOVA using the `interaction.plot()` function in R. In our example, we can use the following R script, with the three vectors we created previously for the *samples* and *assessors* factors and the response *spiciness*.

```
interaction.plot(samples, assessors, spiciness,
  fixed=TRUE,col=0:11)
```

The plot in Fig. 5.3. gives us a rough estimate of the significance of the factors and can also be used to examine the effect of factor interactions. This plot shows

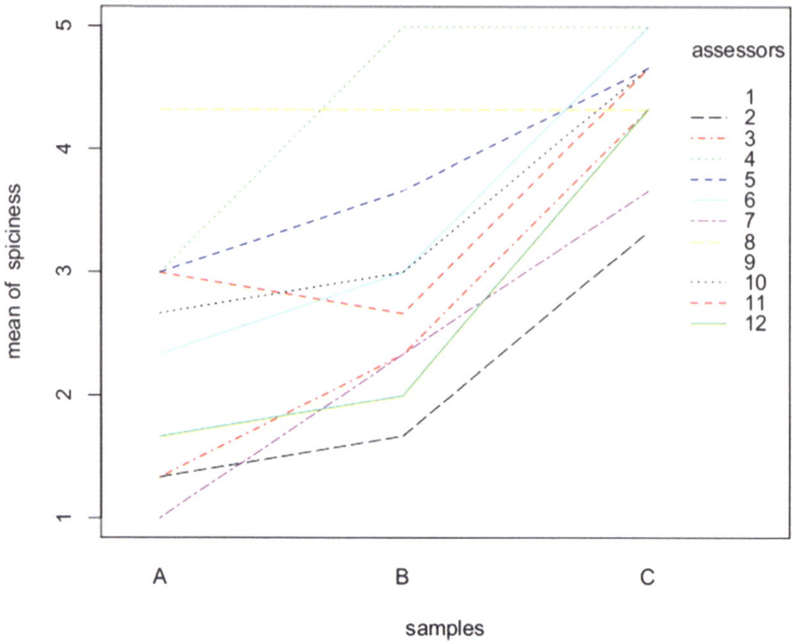

Fig. 5.3 Factor effects plot of samples versus spiciness for each assessor from the attribute difference test for spiciness of chorizo sausages

very clearly the significance of the assessor effect. Assessors 2, 7 and 12 rated all samples significantly lower than assessors 5, 6 and 9, even though they all rated the samples in the same order of increasing preference (i.e., sample A < B < C). If some of the assessors had rated the samples in a different order of preference, resulting in a lines crossing each other, this would be an indication of an interaction effect.

If we change the order by plotting the assessors instead of samples on the x axis of the factor effects plot, using the following R script, we can examine how samples are evaluated across assessors.

```
interaction.plot(assessors,samples,spiciness,
  fixed=TRUE,col=2:4)
```

Figure 5.4. gives us a clear picture of the sample effect and the order of preference for each assessors. Both factor effects plots should be plotted together to examine the significant of each factor in the ANOVA model.

The next step in our analysis would now be to determine which samples were significantly different from each other. Figure 5.4 gives us an indication that sample A was likely to be significantly different from sample C, although it is more difficult to determine if sample A was significantly different from sample B. The usual way to determine this is to conduct a multiple comparison test to determine if there is a significant different between the mean values of spiciness for each sample. There are

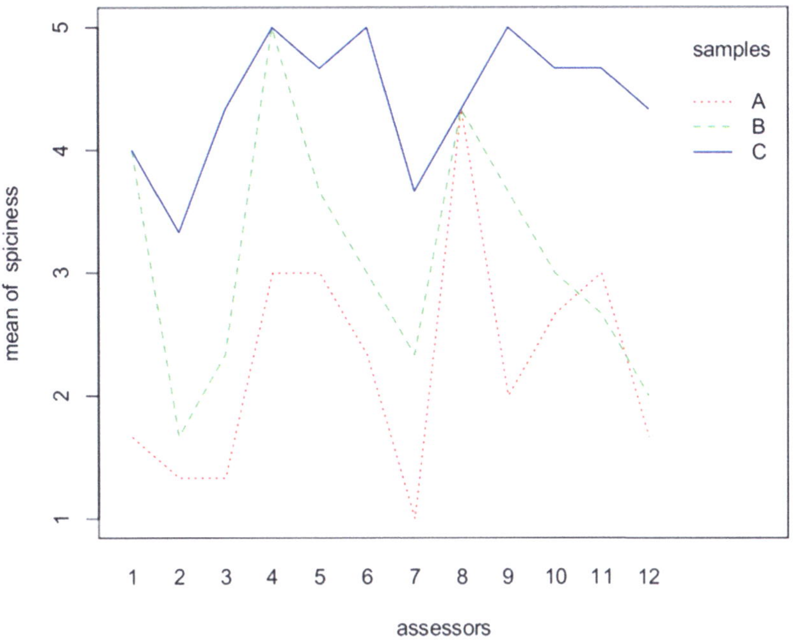

Fig. 5.4 Factor effects plot of assessors versus spiciness for each sample from the attribute difference test for spiciness of chorizo sausages

a number of different methods available, including a method to determine the difference between a control samples and other samples (Neter et al. 1985). Tukey's honestly significant difference (HSD) which was used for comparing differences in rank totals can also be used to determine differences between mean values of interval scaled data from an attribute difference test. The test hypothesis determines if the mean difference between any two samples is significantly higher than the Tukey's HSD critical distance. For any two samples (e.g., Sample A and Sample B), the `TukeyHSD()` function can be used to test the null hypothesis against the alternative hypothesis of using the *panel.aov* object at a 1% level of significance using a `conf.level=0.99` for all comparisons between Samples A, Sample B and Sample C.

```
#Calculate Tukey's HSD for the samples factor
TukeyHSD(panel.aov, "samples", conf.level=0.99)
```

The R script produces a table showing the difference in the means (`diff`) for each sample comparison, the lower (`lwr`) and upper (`upr`) levels of the error bar and the p-value (`p adj`) of the test significance. The results of the HSD test below shows clearly that there are significant differences between all three samples. These comparison can be visualized by typing the above R script within the `plot()` function to produce a graphical comparison of the results presented in the table as shown in Fig. 5.5.

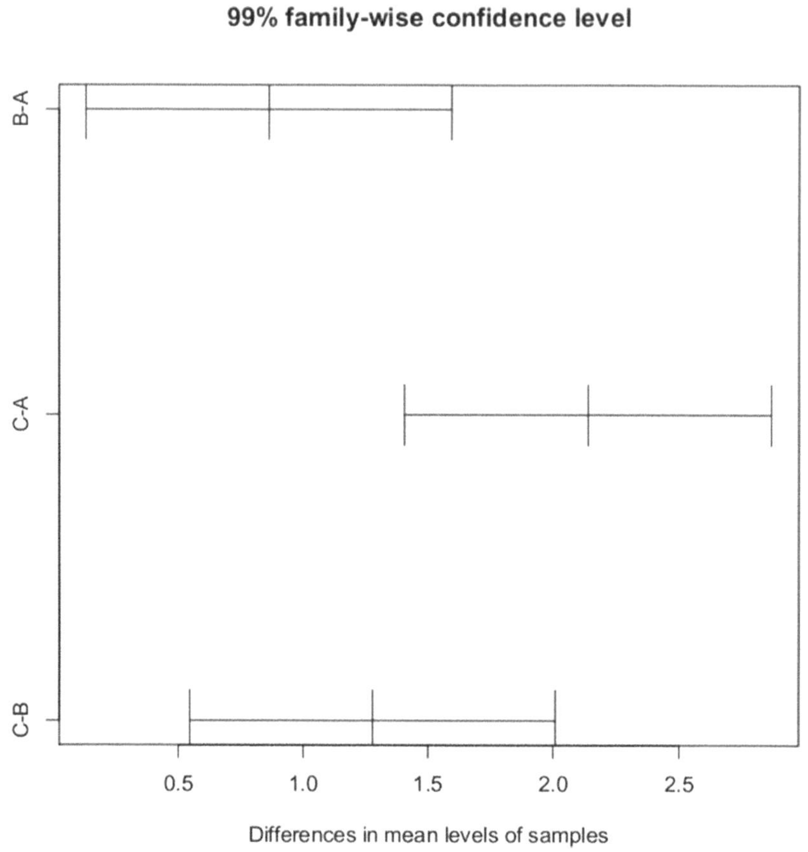

Fig. 5.5 Tukey's HSD plots comparing mean values of each sample (A, B and C) from the attribute difference test for spiciness of chorizo sausages

```
> TukeyHSD(panel.aov, "samples", conf.level=0.99)
 Tukey multiple comparisons of means
  99% family-wise confidence level

Fit: aov(formula = spiciness ~ samples + assessors + samples:assessors,
data = panel)

$samples
    diff     lwr     upr    p adj
B-A 0.8611111 0.1294117 1.592811 0.0020140
C-A 2.1388889 1.4071895 2.870588 0.0000000
C-B 1.2777778 0.5460783 2.009477 0.0000043
```

The graph shows that non of the error bars in each comparison crosses the mean difference of zero, thus indicating all comparisons are significantly different at a 1% level of significance. Sample C is spicier than both Sample B and Sample A.

5.2.8 *How to Use the ANOVA Function of R Without Sample Replications for Benchmarking the Overall Acceptability of Three Similar Products of Chorizo Sausages*

The product development team from the meat company that has been developing the new spicy chorizo sausage in example 4.2.7. decided to conduct a product benchmarking test for overall acceptability by comparing their least spicy (D) and most spicy (E) product with similar products on the market. 3 different commercial products (A to C) from different manufacturers were compared with the prototype samples (D & E) for overall acceptability using a 9-point hedonic scale to determine if any product differences exists at a 5% level of significance. 60 untrained assessors each received all 5 samples (one sample presented at a time) in a random presentation order based on a balanced complete block design and were instructed to taste each product sample and rate the overall acceptability.

The 9-point hedonic scale when used for comparing two samples in the paired preference test for two soup formulations in Sect. 4.2.6 can also be used in a preference with at least three different product samples. It is worth noting at this point that the number of assessors used in a preference test should be carefully considered when designing the sensory test. Just as in the case of the paired preference test, the statistical test that is to be chosen and the validity of the test chosen needs to be considered and this is essentially affected by the number of assessors and characteristics of the selected assessors.

There are two choices for a statistical methods for preference tests with three or more samples. i.e. the parametric analysis of variance (ANOVA) and the non-parametric versions of the ANOVA method. We have so far come across the Friedman two-way ANOVA in Sect. 4.2.5, which was used with rank data from an attribute rank test. We have also come across the single-factor ANOVA with sample replications in Sect. 4.2.7. Both of these methods can be used with the 9-point hedonic scale, with some modification to the raw data or the choice of the ANOVA model.

The statistical test most commonly used with the 9-point hedonic scale is the parametric single-factor ANOVA test. There is an important difference between the model used for a preference test compared to the one used in Sect. 4.2.7. In a preference test each sample is evaluated only once by each assessor. Therefore, there are no sample replications. Under this design restriction, it is not possible to use the ANOVA model with the interaction term as stated in Sect. 4.2.7. Instead, the appropriate ANOVA model is in fact the same ANOVA model described previously, but with the interaction term.

The aov() function is also used in this case to conduct the analysis of variance test. The ANOVA model for a single-factor within-subjects ANOVA without sample replications includes only the main factor effects. The model can be represented as

$$X_{ijk} = \mu + T_i + B_j + \varepsilon_{ijk}$$

where

X_{ijk} is the value of all response for the *i*th sample for each *j*th assessor
μ is the overall mean
i is the number of samples ranging for 1,2,....*t*
j is the number of assessors ranging for 1,2,....*b*
T_i is the Sample effect (the Single factor) for *i*th sample
B_j is the Assessor effect (or Block factor) for *j*th assessors
ε_{ijk} is the random error or residual

The null hypothesis of the hedonic preference test was to compare the mean values of overall acceptability of each sample is defined as

$$H_0 : \mu_A = \mu_B = \mu_C = 0$$
$$H_1 : at\,least\,one\,\mu_i - \mu_j \neq 0$$

A two-sided test can be carried out to determine if there were any difference in the degree of acceptability of the chorizo samples from Table 5.8 Results from preference test of chorizo sausage samples. at a 5% level of significance. As in the previous example, a data frame has to be created that consists of 3 columns, a Sample effect column with levels indicating the sample used (i.e. samples A to E), an Assessor effect column indicating the assessor (i.e. 1 to 60) and an Acceptability column for the assessor response to sample acceptability.

Examining Table 5.8 Results from preference test of chorizo sausage sampleswe see that a vector of assessors 1 to 60 should be repeated 5 times in a sequence. The rep() function for a sequence of 60 assessors is repeated times=5 within the c() function, before apply the factor() function to indicate that *assessors* is the block factor. A sample vector is also created but this time each sample level ("A", "B", "C", "D" and "E") must be repeated 60 times before apply the factor() function to indicate that *samples* is the sample effect factor. A third vector, *acceptability*, for the assessors responses should also be created. We can then create the data frame, panel, using the data.frame() function. The R script for the data frame will then be

```
#Create a factor for assessor
assessors <-factor(c(rep(c(1:60),times=5)))
#Create a factor for samples
samples <-factor(c(rep(c("A"),times=60),
rep(c("B"),times=60),
```

Table 5.8 Results from preference test of chorizo sausage samples

	Samples						Samples						Samples				
Assessor	A	B	C	D	E	Assessor	A	B	C	D	E	Assessor	A	B	C	D	E
1	7	7	9	3	8	21	6	4	9	5	7	41	7	3	9	5	5
2	8	6	6	5	5	22	8	6	5	7	9	42	7	6	9	3	7
3	6	7	9	5	8	23	7	7	4	2	5	43	9	5	6	5	8
4	8	5	6	7	6	24	7	6	4	3	9	44	8	3	9	3	8
5	7	7	7	2	8	25	9	5	4	5	6	45	7	2	6	2	5
6	7	6	8	3	6	26	9	3	4	3	7	46	9	7	7	3	7
7	8	5	8	5	9	27	7	2	7	2	5	47	8	9	6	5	5
8	8	3	4	3	7	28	7	7	8	3	7	48	6	4	9	5	9
9	7	2	9	4	5	29	8	5	6	5	5	49	9	6	8	7	5
10	9	7	9	3	7	30	6	4	9	5	6	50	7	1	4	5	8
11	8	9	6	5	5	31	9	5	9	7	4	51	7	6	8	3	6
12	6	4	9	5	9	32	7	7	4	2	8	52	6	5	6	5	6
13	9	5	7	7	4	33	7	6	4	3	7	53	9	3	4	3	5
14	7	5	9	2	9	34	8	5	4	5	5	54	7	2	8	6	5
15	7	6	7	3	4	35	6	3	4	3	7	55	7	7	9	3	7
16	8	5	6	5	5	36	7	2	7	2	5	56	8	9	6	5	5
17	7	3	4	3	9	37	8	7	7	3	7	57	6	4	9	5	6
18	7	2	8	6	5	38	8	9	6	5	5	58	9	6	3	7	4
19	6	7	8	3	7	39	6	4	9	5	5	59	7	8	9	4	9
20	8	9	6	5	5	40	7	3	6	7	6	60	7	6	8	3	7

```
rep(c("C"),times=60),
rep(c("D"),times=60),
rep(c("E"),times=60)))
#Create a response vector for acceptability
# Step 1. Create vector of data for samples A to E
Sample.A <- c(7,8,6,8,7,7,8,8,7,9,8,6,9,7,7,8,7,7,6,8,
6,8,7,7,9,9,7,7,8,6,9,7,7,8,6,7,8,8,6,7,
7,7,9,8,7,9,8,6,9,7,7,6,9,7,7,8,6,9,7,7)
Sample.B <- c(7,6,7,5,7,6,5,3,2,7,9,4,5,5,6,5,3,2,7,9,
4,6,7,6,5,3,2,7,5,4,5,7,6,5,3,2,7,9,4,3,
3,6,5,3,2,7,9,4,6,1,6,5,3,2,7,9,4,6,8,6)
Sample.C <- c(9,6,9,6,7,8,8,4,9,9,6,9,7,9,7,6,4,8,8,6,
9,5,4,4,4,4,7,8,6,9,9,4,4,4,4,7,7,6,9,6,
9,9,6,9,6,7,6,9,8,4,8,6,4,8,9,6,9,3,9,8)
Sample.D <- c(3,5,5,7,2,3,5,3,4,3,5,5,7,2,3,5,3,2,3,5,
5,7,2,3,5,3,2,3,5,5,7,2,3,5,3,2,3,5,5,7,
5,3,5,3,2,3,5,5,7,5,3,5,3,6,3,5,5,7,4,3)
Sample.E <- c(8,5,8,6,8,6,9,7,5,7,5,9,4,9,4,5,9,5,7,5,
7,9,5,9,6,7,5,7,5,6,4,8,7,5,7,5,7,5,5,6,
5,7,8,8,5,7,5,9,5,8,6,6,5,5,7,5,6,4,9,7)
#Step 2. Create single vector for all sample data
acceptability <-         c(Sample.A,Sample.B,Sample.C,Sample.D,
Sample.E)
#Create a data frame, panel, for all three vectors
panel<-data.frame(samples,assessors,acceptability)
```

There is a slight difference in the way the acceptability response vector was created in this example compared to the way it was done in Sect. 4.2.7. A data vector for each sample is first created and then we used the `c()` function to join all sample data into a single vector.

Since the single-factor ANOVA model that is to be used with this data does not included sample replications, the assumption made with this model is that the effect of the interaction term is insignificant. Tukey's one dergree of freedom test for additivity can be used to test this assumption using the `tukey.add.test()` function in the ***asbio*** library. The R script to conduct this test for the hedonic preference data in Table 9.3.5. is

```
#Conduct Tukey's one degree of additivity test
library(asbio)
tukey.add.test(panel$acceptability,panel$samples,
panel$assessors)
```

The results of this test., shown below, indicates that the assumption is valid as the p-value of 0.06 cannot be rejected at a 5% level of significance.

```
>tukey.add.test(panel$acceptability,panel$samples,
panel$assessors)

   Tukey's one df test for additivity

data: panel$samples and panel$assessors on panel$acceptability
F = 3.566, num.df = 1, denom.df = 235, p-value = 0.06021
```

The ANOVA model without the interaction term can then be fitted using the `aov ()` function, and saving the results of the ANOVA test in a R object called *prefereence.aov*.

```
#Conduct ANOVA with aov()
preference.aov<-aov(acceptability~samples+assessors,panel)
#Print out the ANOVA table
summary(preference.aov)
```

The ANOVA model consists of the response *acceptability* on the left side of the equation, which is separated by the ~ symbol. This is followed on the right by the 2 factor terms in the ANOVA model for the single-factor within subjects ANOVA, consisting of the main effect *samples and* the block effect *assessors*.

The typical ANOVA table can be produced by calling the *preference.aov* object with the `summary()` function. The results indicate that both the sample effect and the assessor effect were highly significant at a 0.0001% level of significance. We can examine these results after we have conducted an analysis of the residuals.

```
> summary(panel.aov)
       Df  Sum Sq Mean Sq F value Pr(>F)
samples    4 411.4 102.84  37.602  <2e-16 ***
assessors   59 154.2   2.61   0.955  0.571
Residuals 236 645.4   2.73
```

A graphical analysis of residuals should be conducted and the reader should refer to Sect. 4.2.7. for an explanation of the different plots presented in Fig. 5.6. The plots used to assess the assumptions of normality of residuals (Q-Q plot), the absence of significant number of outliers (both the Q-Q plot and the Constant leverage plot), the independence of error terms (Constant leverage plot), all indicate that the ANOVA model chosen was appropriate for analysing the data found Table 5.8.

Tukey's honestly significant difference (HSD) test, for multiple comparisons between all sample means, can now be conducted to determine which samples were significantly different. For a 5% level of significance, we can determine the confidence intervals for each comparison using the *preference.aov* object within the `TukeyHSD()` function, identifying `"samples"` as the factor to examine and using a confidence level of `conf.level=0.95`.

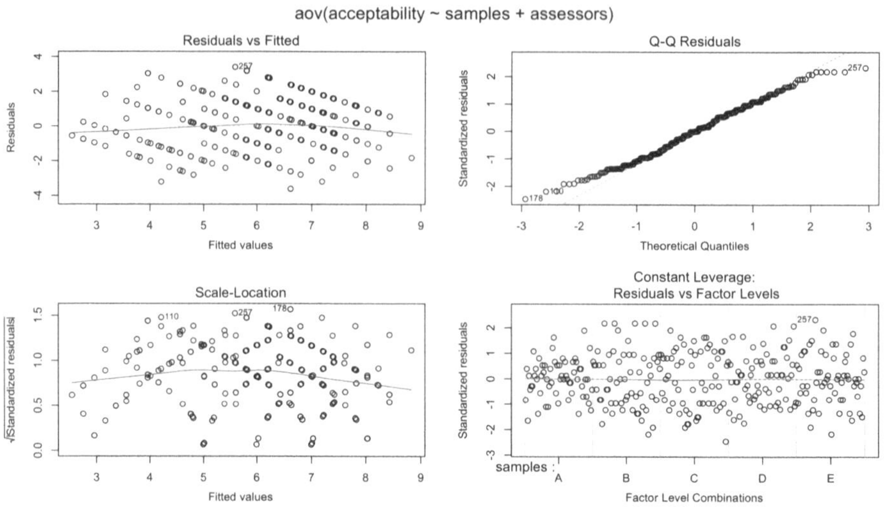

Fig. 5.6 Graphical analysis of residuals for checking ANOVA model assumptions from the hedonic preference test for chorizo sausages

```
#Calculate Tukey's HSD for the samples factor
TukeyHSD(preference.aov, "samples", conf.level=0.95)
```

The results of the HSD test below shows clearly that there are significant differences between B-A, D-A, E-A, C-B, D-B, E-B, D-C and E-D. These comparison, like the previous example, can be better visualized by typing the above R scrip within the `plot()` function to produce a graphical comparison of the results presented in the table. Sample E, which was spicier than the other prototype (i.e. Sample D) had the same level of acceptability as sample C. Both samples were considered to be "liked slightly". However, the most preferred sample was Sample A with a mean value of 7.4. i.e., between the categories "like moderately" (value of 7) and "like very much" (value of 8). It is important to note that Sample D was rated at a mean value of 4.2, suggesting that it was 'Dislike slightly". The panel preferred the spicier sausages with the companies prototype, sample E performing well against its competitors. However, the result would still indicate that improvements need to be made if the company wants to compete with the most preferred sample, that of Sample A (Fig. 5.7).

```
> TukeyHSD(preference.aov, "samples", conf.level=0.95)
  Tukey multiple comparisons of means
    95% family-wise confidence level

Fit: aov(formula = acceptability ~ samples + assessors,  data = panel)
```

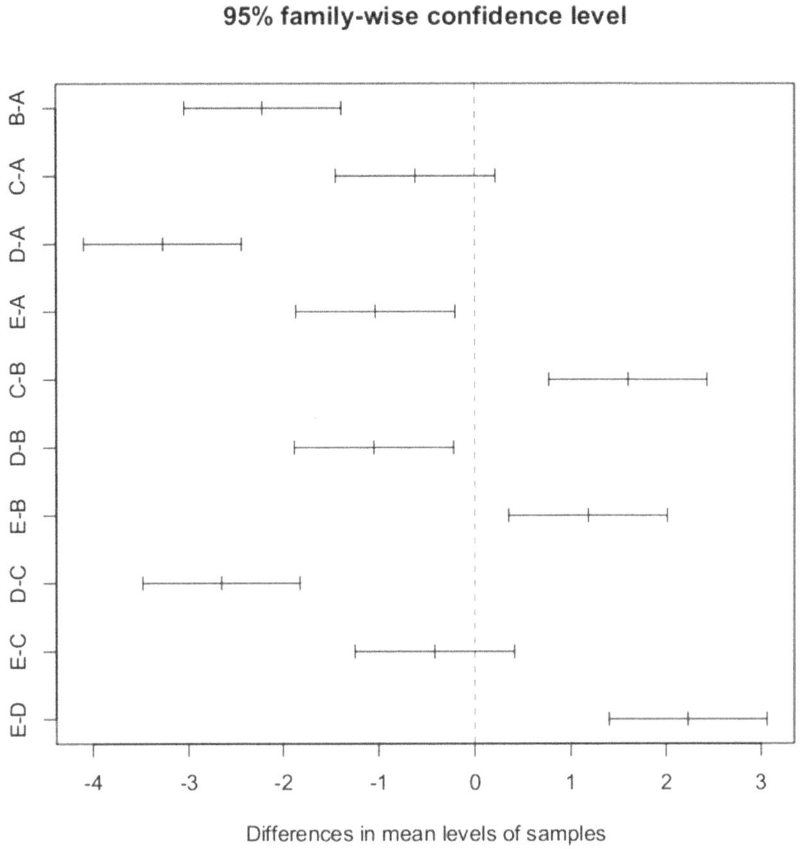

Fig. 5.7 Tukey's HSD plots comparing mean values of each sample (A, B, C, D and E) from the hedonic preference test for overall acceptability of chorizo sausages

```
$samples
      diff       lwr       upr     p adj
B-A -2.2166667 -3.0466814 -1.3866519 0.0000000
C-A -0.6166667 -1.4466814  0.2133481 0.2491619
D-A -3.2666667 -4.0966814 -2.4366519 0.0000000
E-A -1.0333333 -1.8633481 -0.2033186 0.0065118
C-B  1.6000000  0.7699853  2.4300147 0.0000026
D-B -1.0500000 -1.8800147 -0.2199853 0.0054010
E-B  1.1833333  0.3533186  2.0133481 0.0010937
D-C -2.6500000 -3.4800147 -1.8199853 0.0000000
E-C -0.4166667 -1.2466814  0.4133481 0.6411052
E-D  2.2333333  1.4033186  3.0633481 0.0000000
```

Appendix (Table 5.9)

Table 5.9 Minimum number of agreeing judgments[a] (X) required to establish significant difference for the Paired-Comparison and the Triangle Test at $\alpha = 0.05$

	Paired-Comparison Test		Triangle Test	
Number of trials (N)	One-tailed	Two-tailed	One-tailed	Two-tailed
7	7	7	5	6
8	7	8	6	6
9	8	8	6	7
10	9	9	7	7
11	9	10	7	8
12	10	10	8	8
13	10	11	8	9
14	11	12	9	9
15	12	12	9	10
16	12	13	9	10
17	13	14	10	10
18	13	14	10	11
19	14	15	11	11
20	15	15	11	12
21	15	16	12	12
22	16	17	12	13
23	16	17	12	13
24	17	18	13	14
25	18	18	13	14
30	21	21	15	16
40	26	27	19	20
50	32	33	23	24
100	59	61	42	44

[a]Values are derived from the formula $X = \left(z \cdot \sqrt{N} + 0.5N + 1\right)/2$ for the paired-comparison test and $X = z \cdot 0.47\sqrt{N} + 0.33N + 0.5$ for the triangle test, rounded up to the next integer. z-values are 1.645 for the one-tailed and 1.96 for the two-tailed test, for $\alpha = 0.05$

References

Dalgaard P (2002) Introductory statistics with R. Springer, New York

Dean A, Voss D (1999) Design and analysis of experiments. Springer, New York

Devore J, Peck R (1986) Probability and probability distributions. In: Devore J, Peck R (eds) Statistics: the exploration and analysis of data. West Publishing Company, St. Paul, pp 167–229

Langsrud Ø (2003) ANOVA for unbalanced data: use Type II instead of Type III sums of squares. Stat Comput 13:163–167

Lawless HT, Haymann H (1998a) Acceptance and preference testing. In: Lawless HT, Haymann H (eds) Sensory evaluation of food: principles and practices. Chapman & Hall, New York, pp 430–479

Lawless HT, Haymann H (1998b) Nonparametric and binomial-based statistical methods. In: Lawless HT, Haymann H (eds) Sensory evaluation of food: principles and practices. Chapman & Hall, New York, pp 679–700

Meilgaard MC, Civille GV, Carr BT (2007) Basic statistical methods. In: Meilgaard MC, Civille GV, Carr BT (eds) Sensory evaluation techniques. CRC Press, Boca Raton, pp 313–355

Miller J, Miller J (2000) Non-parametric and robust methods. In: Miller J, Miller J (eds) Statistics and chemometrics for analytical chemistry, 4th edn. Pearson Education Limited, Essex, pp 151–180

Montgomery DC (1997) Design and analysis of experiments, 5th edn. Wiley, New York

Neter J, Wasserman W, Kutner MH (1985) Applied linear statistical models, 2nd edn. Richard D. Irwin, Homewood

O'Mahony M (1986a) Additional nonparametric tests. In: O'Mahony M (ed) Sensory evaluation of food—statistical methods and procedures. Marcel Dekker, Inc, New York, pp 303–370

O'Mahony M (1986b) The binomial test: applications in sensory difference and preference testing. In: O'Mahony M (ed) Sensory evaluation of food—statistical methods and procedures. Marcel Dekker, Inc, New York, pp 57–90

R Development core Team (2023) R: a language and environment for statistical computing. R Foundation for Statistical Computing, Vienna. ISBN 3-900051-07-0. URL http://www.R-project.org/

Sheskin DJ (2000) Handbook of parametric and nonparametric statistical procedures, 2nd edn. Chapman & Hall, Boca Raton

The sensory evaluation division of the IFT (1981) Sensory evaluation guide for testing food and beverage products. Food Technol 35(11):50–59

Chapter 6
Statistic for Analytical Methods

Maciej Wojtczak, Ulrike Zitz, Semih Otles, Hakan Savas Sazak, Ebru Pelvan Evcil, Stefano Martin, and Sara Briguglio

Abbreviations

$(M + H)^+$	Protonated Molecule
$(M\text{-}H)^-$	Deprotonated Molecule
API	Atmospheric pressure ionization
EIC	Extract ion chromatogram
ESI	Electrospray ionization
LC-MS/MS	Liquid chromatography tandem mass spectrometer
LOQ	Limit of quantitation
QQQ	Triple quadrupole
TIC	Total ion chromatogram

M. Wojtczak (✉)
Lodz University of Technology, Łódź, Poland
e-mail: maciej.wojtczak@p.lodz.pl

U. Zitz
Department of Food Science and Technology, Institute of Food Sciences, University of Natural Resources and Life Sciences, Vienna, Austria
e-mail: ulrike.zitz@boku.ac.at

S. Otles · H. S. Sazak
Ege University, Food Engineering Department, Izmir, Turkey
e-mail: semih.otles@ege.edu.tr

E. P. Evcil
Ministry of Agriculture and Forestry, Izmir Food Control Laboratory Directorate, Izmir, Turkey

S. Martin · S. Briguglio
Analysis laboratory, Regional environmental agency of Friuli Venezia Giulia (Italy), Udine, Italy

G. Schleining et al. (eds.), *Statistics in Food and Biotechnology*, Integrating Food Science and Engineering Knowledge Into the Food Chain 16,
https://doi.org/10.1007/978-3-031-51568-2_6

6.1 Accuracy, Bias, Precision for Laboratory Measurements: A Case Study in Iron and Copper Measurements

Maciej Wojtczak

6.1.1 *Objectives and Learning Outcomes*

The objective of this chapter is to explain how the accuracy of the analytical method can be determined. The presented case study shows the procedure of determining trueness and repeatability of the analytical method using certified reference material (CRM). The most typical and frequently used procedure of checking the proper use of the analytical method is presented.

6.1.2 *Introduction*

Test result is always only an approximation of the true value which by its nature is an unknown value. Every test result is connected with some uncertainty. Uncertainty of measurement is a parameter associated with the results of a measurement that characterizes the dispersion of the values that could be attributed to the measurand (ISO/IEC Guide 43-1 1997). Uncertainty of measurement depends on many parameters which can be grouped in four categories:

- laboratory assistant—personnel (skill level, experience),
- test method—specific technical procedure for performing a test (characterization of method, selectivity, accuracy),
- laboratory equipment—apparatus, materials and reagents used for measurement (calibration, traceability, grade of reagents),
- environmental conditions and accommodation—measurement conditions (temperature, humidity, cleanness).

The extent to which the factors from the above categories contribute to the total uncertainty of measurement differs considerably between type of tests and between laboratories. The analyst should take into account these factors in choosing the method as well as during all analytical processes. To characterize the impact of all the factors on test result we must validate the method in our laboratory (ISO/IEC 17025 2001). It means that we must determine characteristic parameters of the method in our laboratory conditions and compare them with values from norms or literature. One of the main parameters which characterize uncertainty of analytical method is **accuracy** (trueness and precision) of measurement. Accuracy allows us to evaluate total impact of measurements conditions in our laboratory on test results and to show to what extent our results are close to the true value (ISO 5725-1-6 2002).

6.1.2.1 Accuracy

Accuracy—closeness of agreement between a test result and the accepted reference value which represents the true value (ISO 5725-1 2002). Accuracy concerns accidental errors as well as systematic errors (bias component). We can express the measurement result by the equation:

$$y_i = \tau + \overbrace{\Delta + \delta}^{\text{accuracy}} = \mu + \delta \tag{6.1}$$

where:

y_i: measurement result,
τ: true value (or accepted reference value)
Δ: systematic errors
δ: random errors,
μ: limiting mean (excepted value).

From Eq. (6.1) we see that the accuracy of measurement is a sum of two components: systematic—**trueness** and random—**precision**. Accuracy of single measurement result depends on trueness of method as well as on measurement precision (EURACHEM Guide 1998) (Fig. 6.1).

6.1.2.2 Trueness

Trueness is a closeness of agreement between the average value obtained from large series of test results and the true value (ISO 5725-1 2002). Because true value is usually unknown, in practice we determine the trueness by comparison of mean value with an accepted reference value. The accepted reference value is an estimation of true value with acceptable level of uncertainty. They can be defined as:

- a theoretical or established value, based on scientific principles,
- an assigned or certified value, based on collaborative experimental work,
- or, when the above are not available, the mean value of specified population of measurements (EURACHEM Guide 1998).

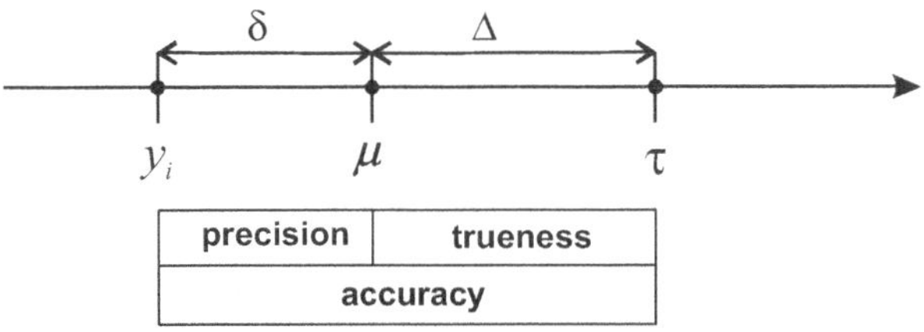

Fig. 6.1 Accuracy as a combination of precision and trueness

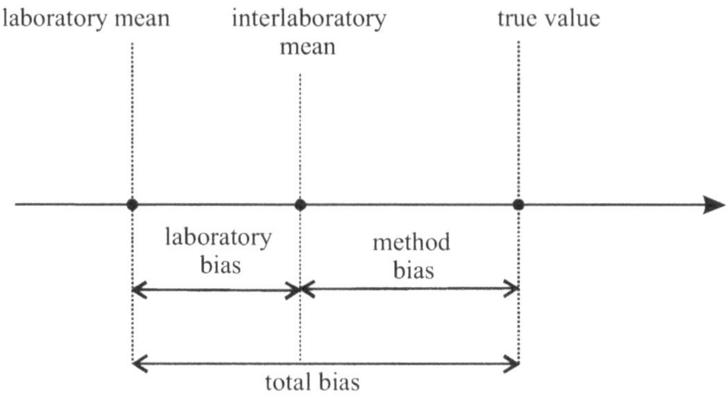

Fig. 6.2 Types of bias

Reference materials (RM) or certified reference materials (CRM) are usually used for the determination of the trueness value. To find the measurement trueness using a reference material we must determine the mean and standard deviation of a series of replicate tests and compare them with the certificated value of the reference material. Choosing the reference material we have to look for the reference material with the matrix closely similar to the sample of interest. Since finding ideally matched matrix CRM is sometimes difficult we may determine the trueness using:

- typical materials spiked with pure analyte,
- well-characterized materials checked in-house.

Trueness is usually expressed as the bias. **Bias** is the difference between the expectation of the test results and accepted reference value. Bias is the total systematic error as contrasted with random error. There may be one or more systematic error components contributing to the bias. A larger systematic difference from the accepted reference value is reflected by large bias value(ISO 5725-1 2002).

Two main components of bias are shown in Fig. 6.2—laboratory bias and method bias. The method bias arises from systematic errors inherent to the method, and it is independent of laboratory conditions. In Fig. 6.2 the biases have the same direction, but of course they may be opposite, however, the total bias is always the sum of these two components. In single laboratory tests we can only estimate the combined bias. The method bias may be determined in interlaboratory tests, and it is usually given with the standard method.

Acceptability of trueness should be decided by means of statistical significance tests, taking into account the precision of the method, the uncertainty of reference material values and the accuracy required by the end user of analytical results.

The most typical procedure of determining trueness is the analysis of certifited reference material sample (or reference material sample) by means of the examined analytical method. It is necessary to obtain test results (preferably repeated 10 times) of reference material and to compare the obtained mean value with reference value.

In order to check whether the obtained mean value does not differ in a statistically significant degree from the reference value we may use the Student's t-test. The null hypothesis in this test assumes that the obtained mean value does not differ from the reference value.

$$H_0: \ \bar{y} = \mu \tag{6.2}$$

In order to verify H_0 we must calculate experimental t according to the equation:

$$t_{\text{exp}} = |\mu - \bar{y}| \frac{\sqrt{n}}{s}, \tag{6.3}$$

where:

μ: reference value,

$\bar{y}$: mean value; $\bar{y} = \frac{\sum_{i=1}^{n} y_i}{n}$,
n: number of results,
s: standard deviation of results set.

The calculated t_{exp} value is compared with the critical value t_{crit} corresponding to the given degree of freedom (n-1) and the confidence level chosen (typically 0.05). The critical t values are given in Table 6.1 in appendix. If $t_{exp} > t_{crit}$ then H_0 is rejected, if not, H_0 is retained. When H_0 is rejected it means that there is a significant difference between our mean value and the assigned value, which means that our method is biased.

This simple statistic test does not take into account the uncertainty of reference value. If we know the accuracy of reference value (as in the case of CRM) it is better to check the trueness of the method in the following way.

Table 6.1 The obtained results of iron and cooper content in certified reference material—*Soya Bean Flour* **(INCT-SBF-4)**

No of replication	Cu mg/kg	Fe mg/kg
1	14.3	93.2
2	12.9	82.7
3	14.6	94.6
4	14.8	93.5
5	12.5	95.9
6	12.6	95.1
7	13.2	93.4
8	14.6	94.5
9	14.1	95.5
10	14.2	94.9
Certified value	**14.30 ± 0.46**	**90.8 ± 4.0**

First we should calculate the critical difference (CD) by the equation:

$$\mathrm{CD} = 2\sqrt{u_{\bar{y}}^2 + u_{ref}^2} \tag{6.4}$$

where:

u_y: uncertainty of mean value,
u_{ref}: uncertainty of reference value.

The difference between the mean value and the reference value $|\bar{y} - \mu|$ is compared with the value of critical difference. If $|\bar{y} - \mu| < 2\sqrt{u_{\bar{y}}^2 + u_{ref}^2}$ then the mean value is not significantly different from the reference value; if not, the test result obtained by the checked method is biased.

6.1.2.3 Precision

Precision is a closeness of agreement between independent test results obtained under prescribed (stipulated) conditions. Taking into consideration the prescribed conditions we distinguish two kinds of precision: **repeatability** and **reproducibility**.

Precision depends only on the distribution of random errors and does not relate to the true value. The measure of precision is usually expressed as a standard deviation of the test results (Fig. 6.3).

Repeatability is the precision obtained in repeatability conditions, that is, conditions in which independent test results of the same test items are obtained by means of the same methods in the same laboratory, by the same analyst with the same equipment in short time intervals.

Repeatability is usually determined as standard deviation of data set obtained in repeatability conditions. It is advisable to perform at least 10 independent measurements as a basis to calculate repeatability standard deviation (s_r) from the formula:

$$s_r = \sqrt{\frac{\sum_{i=1}^{n} (y_i - \bar{y})^2}{n-1}}. \tag{6.5}$$

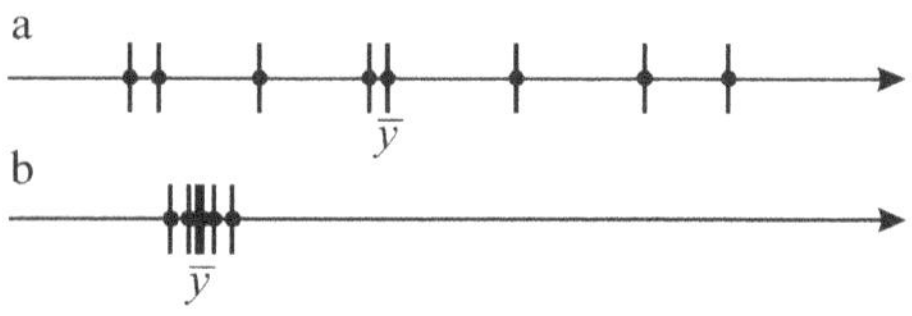

Fig. 6.3 Test results obtained by non-precise method (**a**) and precise method (**b**)

We can also express repeatability as a relative standard deviation (RSD) using the formula:

$$\mathrm{RSD}_r = \frac{s_r}{\bar{y}} \cdot 100\% \tag{6.6}$$

In practice it is convenient to calculate the repeatability limit (r). With 95% probability we may say that the repeatability limit should not be exceeded by the absolute difference between two test results obtained in the repeatability conditions.

$$r = 2.8 \times s_r \tag{6.7}$$

Reproducibility is the precision obtained in reproducibility conditions, that is, conditions in which test results of the same test items are obtained by means of the same methods in different laboratories by different analysts with different equipment's.

Reproducibility is determined as standard deviation of test results obtained by different laboratories, and is usually calculated from results of interlaboratory tests. The value of reproducibility is always higher than the value of repeatability because it depends upon a greater number of changing conditions.

6.1.3 Case Study: Accuracy of Iron and Copper Measurements by Means of Atomic Absorption Spectrometry in Flours

The basis of drawing conclusions in solving various problems are measurements. A measurement is a quantitative determination of an investigated feature by the experimental method. To assure correct conclusions on the basis of the results of measurement it is important to ascertain that the results are reliable. Reliability of a result is first of all manifested in its accuracy. Thus, beginning laboratory work by means of a given method we should first estimate the accuracy of the method. In order to perform it we should determine the trueness and repeatability of the method.

Our aim was to check the accuracy of iron and copper measurements by means of atomic absorption spectrometry in flours. For this purpose we used certified reference material—*Soya Bean Flour* (**INCT-SBF-4**), which has the following certified values for the above elements: **Cu—14.30 mg/kg ± 0.46 mg/kg; Fe—90.8 mg/kg ± 4.0 mg/kg**. The content of Cu and Fe was determined by means of atomic absorption spectrometry with standard method after microwave mineralization. Ten independent samples of 0.5 g of **INCT-SBF-4** were taken to analysis. The obtained results are shown in Table 6.1.

6.1.3.1 Detection of Outliers

First we should check our data set given in Table 6.1 for outliers. Outlier is defined as an observation considerably differing from the majority of the rest of the results. The rejection of suspect observations must be based on statistic tests like for instance, Dixon's Q-test or Grubbs' test. The Dixon's Q-test is the simplest test which allows us to check if one result from a small set of replicate results (from 3 to 10) can be rejected.

We must arrange the results in ascending order $y_1 < y_2 < \ldots < y_n$. For the content of copper we have:

$$12.5; 12.6; 12.9; 13.2; 14.1; 14.2; 14.3; 14.6; 14.6; 14.8;$$

and for the content of iron we have:

$$93.2; 82.7; 94.6; 93.5; 95.9; 95.1; 93.4; 94.5; 95.5; 94.9.$$

Now we should calculate experimental Q-value (Q_{exp}) for the lowest result (y_1) and for the highest result (y_n) from the equation:

$$Q_{exp}(y_1) = \frac{y_2 - y_1}{y_n - y_1}, \quad Q_{exp}(y_n) = \frac{y_n - y_{n-1}}{y_n - y_1} \tag{6.8}$$

The calculated Q_{exp} values are compared to the critical Q-value (Q_{crit}). The critical Q-value depends on the probability and on the number of the replications (n). The Q_{crit} values for 99%, 95% and 90% probability and for n from 3 to 10 are given in Table 6.34 in appendices. If $Q_{exp} > Q_{crit}$, then the suspect value can be characterized as an outlier and it can be rejected; if not, the suspect value must be retained and used in all subsequent calculations.

In our case we will use statistical tests with confidence level 95%, which is most typical for chemical analysis. Then in our example:

$$Q_{crit(10,\ 95\%)} = 0.466. \tag{6.9}$$

The calculated Q_{exp} values from formula (6.6) for data in Table 6.1 are:

$$\text{for Cu}: Q_{exp}(y_1) = 0.043, \quad Q_{exp}(y_n) = 0.087 \tag{6.10}$$

$$\text{for Fe}: Q_{exp}(y_1) = 0.795, \quad Q_{exp}(y_n) = 0.030 \tag{6.11}$$

Only one Q_{exp} value is higher than $Q_{crit(10,\ 95\%)}$ which means that we should reject result 88.7 mg/kg from our data set. If we reject this test result, we must check again iron results for the new lowest result—99.2 mg/kg. We should do it in the same way as given above:

$$Q_{\text{exp}}(y_2) = 0.074 < Q_{\text{crit}(9,95\%)} = 0.437 \tag{6.12}$$

Concluding, we found only one outlier value in our data set (88.7 mg Fe/kg) and thus for subsequent calculations we will use all 10 results for copper and only 9 results for iron.

6.1.3.2 Calculation of Precision

In our case we can express precision of the method only as repeatability because our results were obtained in one laboratory by one analyst in short time interval. From formula (6.5) we can calculate repeatability standard deviation (s_r):

$$\text{for Cu–}s_r = \sqrt{\frac{7.08}{9}} = 0.89 \quad \text{mg/kg}, \tag{6.13}$$

$$\text{for Fe–}s_r = \sqrt{\frac{7.39}{8}} = 0.96 \quad \text{mg/kg}. \tag{6.14}$$

Now we can calculate repeatability limits for both elements:

$$\text{for Cu–}r = 2.8 \cdot 0.89 = 2.5 \quad \text{mg/kg}, \tag{6.15}$$

$$\text{for Fe–}r = 2.8 \cdot 0.96 = 2.7 \quad \text{mg/kg}. \tag{6.16}$$

We found that our method has similar precision of determining the content of both iron and copper in flours. The calculated repeatability limits mean that two results obtained by this method in our laboratory can differ no more than 2.5 for Cu and 2.7 for Fe.

6.1.3.3 Checking the Method Bias

To check the method bias we must compare the mean value calculated from test results with the certified value of reference material. We can do it in two ways, as explained in Sect. 2.5.1.2.

Method 1

In this method we must calculate t_{exp} value and compare it with the critical value t_{crit} corresponding to the given degree of freedom (n-1) and the confidence level (typically 0.05). If $t_{exp} > t_{crit}$ then there is a significant difference between our mean value and the reference value, which means that our method is biased.

Let us count the t_{exp} for the obtained results for Cu and Fe, and take appropriate values of t_{crit} for 9 degrees of freedom for Cu, 8 degrees of freedom for Fe and for 95% probability from Table 6.35. We have:

	Cu	Fe
Certified value	**14.30 ± 0.46**	**90.8 ± 4.0**
Our mean value	**13.8**	**94.5**
s_r	0.89	0.96
t_{exp}	1.86	11.59
t_{crit}	2.26	2.31
$H_0: \ \overline{y}=\mu$	**Retain**	**Reject**

Since in the case of copper $t_{exp} = 1.86 < t_{crit} = 2.26$ H_o must be accepted, which means that the trueness of this method is good. A different situation takes place in the case of iron in which $t_{exp} = 11.59 > t_{crit} = 2.31$ so that H_o must be rejected. Thus there is a significant difference between our mean value and the certified value. This means that the trueness of the method of determining the content of iron is not acceptable.

Method 1 of checking the trueness based on Student's t-test does not take into account the uncertainty of reference value. If we use the certified material to check the trueness of the method we know the accuracy of the certified value and then it is better to check the method trueness by Method 2.

Method 2

In Method 2 we should compare the difference between the obtained mean value and reference value with critical difference (CD) calculated from formula (6.4), which takes into account the uncertainty of reference value. In used certified material INCT-SBF-4 the uncertainty is given as the expanded uncertainty which was obtained by multiplying u_c by a coverage factor $k = 2$. For this reason for the calculation of CD we will take the expanded uncertainty divided by 2. In our case standard deviation of the mean value will be taken as uncertainty of the mean value calculated from repeatability standard deviation by the formula:

$$s_{mean} = \frac{s_r}{\sqrt{n}}. \tag{6.17}$$

Let us calculate now CD for both elements:

	Cu	Fe
The expanded uncertainty of certified value (U)	**0.46**	**4.0**
The uncertainty of certified value $u_{ref} = U/2$	**0.23**	**2.0**
Reproducibility s_r	0.89	0.96
The uncertainty of mean value $u_{\overline{y}} = s_{mean} = s_r/\sqrt{n}$	0.28	0.32
$CD = 2\sqrt{u_{\overline{y}}^2 + u_{ref}^2}$	0.73	4.1
$\lvert\overline{y} - \mu\rvert$	**0.52**	**3.7**

We found that the critical difference for Cu amounts to 0.73 mg/kg while the difference between the determined mean content of Cu and its certified value is 0.52. Since the difference $\lvert\overline{y} - \mu\rvert$ is lower than CD we can state that the trueness of determining the content of Cu is acceptable. In the case of iron CD amounts to 4.1 mg/kg and is also higher than the absolute difference between the obtained mean content of Fe and the certified value—3.7 mg/kg. Using this method of checking the trueness we state that also the method of determining the content of iron has acceptable trueness. Earlier, on the basis of Student's t-test, we found significant bias of iron determining but, as was stated above, Student's t-test does not take into account the uncertainty of reference value. The uncertainty of certified Fe content is relatively high (4.0 mg/kg) and it significantly influences the evaluation of the method trueness, because the interval in which the true value is situated is considerably larger than in the case of the copper content.

Concluding, we found that our method of determining the content of iron and copper by means of atomic absorption spectrometry has acceptable precision and trueness.

6.2 Validation of Method to Determine Phosphorus Content in White Wine

Ulrike Zitz

6.2.1 Objectives and Learning Outcomes

The procedure of the characterization of an indirect quantitative analytical method is described exemplarily. This is followed by the verification of the obtained parameters to be able to decide, if the defined requirements can be achieved. It should be mentioned again that every validation procedure depends closely on the fields of application, so that for this reason the needs of the given application do not constitute a rigid process. Therefore, the proposed procedure should be used as an orientation, when dealing with method validation studies, if comparable. By covering the main steps of the statistical evaluation of this specific intended use, this example should help applicants to become familiar with the procedure.

6.2.1.1 Introduction

Method validation is one of the most important tasks, which enables laboratories to demonstrate both competence and integrity when analysing samples.

The appropriateness of analytical methods has to be proved before starting valid measurements. Especially when implementing newly developed methods, the fundamental validation parameters have to be evaluated by extensive characterization, focusing on the requirements which have to be met. Furthermore, during routine application, the defined parameters have to be continuously re-confirmed by a partial validation process, so that suitability for use is to be assured as well as documented at any time.

Validation is defined by ISO 17025 as: "Validation is the confirmation by examination and the provision of objective evidence that the particular requirements for a specific intended use are fulfilled." Based on the requirements which have to be defined at the beginning, the characteristic parameters must be determined. The whole validation process also includes the verification that the requirements can be fulfilled to be qualified and finished with a statement of validity. If the requirements cannot be met, the process has to be improved. Changes to the procedure of a validated method always require the assessment of influences on the results. If appropriate, a new validation should be carried out to assure their fitness for purpose. In some of the cases the use of alternative methods, wherever applicable, or the

Fig. 6.4 Validation as a comparison of requirements to process parameters

change of the quality criteria should be aimed at. In doing so, the agreement with the customer is of capital importance (see also Fig. 6.4).

Numerous attempts have been made by several experts to provide a standard procedure or something like a roadmap to be able to correctly carry out a validation procedure (e.g.: Funk et al. 1995, 2005; Garfield et al. 2000; Kromidas 2000; Prichard 1995; Thompson et al. 2002; Wegscheider 1996). Profound knowledge is necessary to decide how to conduct a validation process as well as to balance costs, risks and technical possibilities. Validation should be always done based on the problem. Therefore, this process should be as extensive as necessary to meet the needs of the given application, or field of application (Burgess 2000).

For quantitative methods the working range is determined by examining samples with different analyt concentrations. The concentration range, for which acceptable accuracy (trueness and precision) (for definitions see ISO 5725-1) can be achieved, has to be determined. According to Taylor the development of methods being validated should be done in two steps: The first step is to assure acceptable precision (repeatability) so that the method is "under statistical control" followed by the proof of trueness (cited in Wegscheider 1996).

6.2.1.2 Problem to Solve

The presented experiment is based on the 4-Phase model of analytical quality assurance (Funk et al. 1995, 2005). The aim of the study is to characterize, as well as to control, an analytical process within the expected working range by using a calibration experiment. In phase 1, the quality characteristics of a new analytical process are determined and, if necessary, the process is improved and documented in a reproducible manner. The provisional standard operating procedure (SOP) is defined. Phase 2 contains preparative quality control activities to make the process operational for routine analysis. Finally, phase 3 and 4 represent internal and external laboratory quality assurance measures of routine analysis. However, the last two phases will not be part of the example outlined below.

6.2.2 A Case Study: Validation of Method to Determine Phoshoros Content in White Wine

For quantitative analytical methods to be used routinely, the following procedure is proposed (Fig. 6.5):

6.2.2.1 Defining the Scope/Requirements of the Method to Be Validated

The targeting method is intended to be used in the food analysis to determine the total phosphorus content (both inorganic as well as organic) in white wine samples,

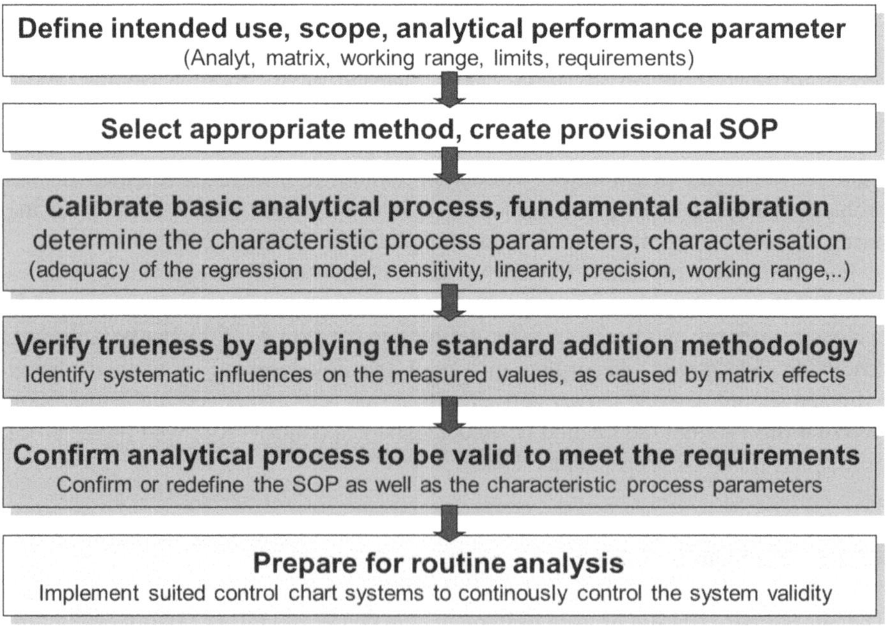

Fig. 6.5 Validation procedure for a quantitative method using a calibration function—chronology

expressed as gram phosphate per liter wine. In fact, the described method "Photometric method to determine phosphate in wine" (Klosterneuburg 30, 1980) does not constitute a standard method. Therefore, this method has to be characterised and proven to be adequate and reliable for the intended use and a procedure has to be developed to control the state of validity of the method before being applied routinely.

For this the following requirements have to be met:

According to the results of the analysis of typical white wine samples (Bandion and Valenta 1980; Bandion and Arangnitz-Rebolledo 2000) the concentration range should cover standard phosphate concentrations (expressed as P_2O_5) of dry white wine samples of 50–450 mg/L. For this the working range was set from 0.25 to 2.25 mg/L. Linearity of the calibration curve has also be proven within this area.

A minimum of relative analytical precision of 20% is claimed by the customer—the relative analytical precision especially at the lower limit of the range has to be verified.

The uncertainty of the results should not exceed 20 mg/L.

6.2.2.2 Planning of the Validation Experiment

The basic validation has to be representative of routine conditions (e.g., reagents, apparatus, persons, processing, etc.). Additionally, all quality-influencing factors have to be under control before starting the experiment. In any case of modification, a recalibration is required, when influences on the results are expected. The measurement system is only valid, if the "state of control" is verified and documented e.g. by using reference materials in combination with control charts.

At first, the fundamental analytical process is investigated only by analysing different concentrations of the analyt in pure solvent. For the fundamental calibration, calibration standards (S) are used being representative with respect to the targeted analyt. This means, the chemical structure and properties of the chosen substance has to be equal, or as similar as possible, to the native analyt. Furthermore, the concentration has to be accurately known to be able to validate the procedure. The concentration range of the standards represents the working range of the method, the concentration has to be defined in SI-units. An equidistant distribution of the chosen standard concentration steps is recommended. To avoid any systematic effect, the standards have to be prepared independently and measured in randomized order. Replication of the experiments is necessary to determine the random error. According to ISO 11095 a minimum of the calibration standards of $N \geq 3$, and of repeated measurements of $K \geq 2$ is proposed.

Recovery studies should be conducted by using the standard addition methodology. This should only be done when taking into account the following requirements: Typical matrix-materials should be chosen, whereas constant volumes have to be spiked with successively increasing amounts of the analyt (A). The analyt (surrogate) has to be added in the same state as present in the native sample. The concentration range should not exceed the working range of the fundamental calibration. Any change of the matrix caused by volume contraction should be avoided by adding only high concentrated standards and dealing with constant end-volumes. Calibration standards may also be used for spiking, if highly concentrated. For comparability reasons the spiked sample preparations have to be sufficiently homogenous.

To meet all requirements it is recommended to prepare a study plan including the whole working schema, which reflects the preparation and the measurement in the given randomized order. This is also to be used for the documentation of the calibration experiment.

6.2.2.2.1 Create the Experimental Design to Be Used for This Experiment

For the preparation of the calibration standards, the following stock-solutions containing KH_2PO_4 (plain salt) dissolved in 0.5 Mol/L HCl have been prepared in randomized order. Within the defined working range 0.25–2.25 mg/L expressed as P_2O_5 it was decided to select five standard concentrations in equidistant order for both, the fundamental calibration and the recovery experiment. S specifies standards to be used for the fundamental calibration, e.g. S0,25, with the P_2O_5 concentration of 0.25 mg/L, and A specifies standards to be added to the matrix at the recovery experiment (standard addition).

Fundamental Calibration $N = 5$, $K = 2$ (S0,25 – S0,75 – S1,25 – S1,75 – S2,25) each of 50 μL + 50 μL HCl per tube; In terms of variance homogeneity testing $K = 6$ was defined for S0,25 and S2,25 respectively; Blank measurements were only done to qualify the used reagents. Based on the fact that both, the sample measurement as well as the calibration process, include the blank value, a blank correction was not found to be necessary.

Recovery Experiment $N = 5$, $K = 2$ (A0,25 – A0,50 – A0,75 – A1,25), each of 50 μL + 50 μL wine matrix per tube; Additionally, the matrix was measured 4 times with no standard added: 50 μL matrix +50 μL HCl per tube. A defined white wine sample with low Phosphate concentration was chosen due to the lack of suitable matrix reference materials or CRM's not containing phosphate.

Sample Measurements The white wine sample is analysed for 4 times, each of 50 μL sample + 50 μL HCl.

Control Chart System 2 defined standards, RM1 and RM2 (standard or wine within the working range preferably located near the lower and the upper border line) were analysed for at least 10 days, each of them with 50 μL standard +50 μL HCl.

Except those measurements needed for the control chart system, the whole experiment has to be performed in randomized order on the same day.

6.2.2.2.2 Evaluation of all Influencing Factors

Before starting the experiment all influencing factors have to be validated for the defined scope. For the outstanding measurement process this means:

- To prove the quality of the relevant volume of the piston-operated pipette (Eppendorf Research 20–200 μL). According to ISO 8655-6 (former DIN 12650-6) the tolerance limit of the nominal volume of 50 μL is specified as 49.2 μL $\leq \mu \leq$ 50.8 μL.
- To perform a functional test of the photometer (Shimadzu U-1800) by using a set of suited grey-filters (NIST SRM 930, 4 filters with the nominal values expressed as extinctions: 0.200 ± 0.002; 0.600 ± 0.004; 1.220 ± 0.008; 1.7600 ± 0.016 at 540 nm).
- To adjust the base-line to zero before starting the calibration and sample measurements at 720 nm and
- To assure that the person who is carrying out the analysis has been well trained and is competent.

Pipette-Check to assure that the targeted volume of 50 μL is located within the tolerance limits according to ISO 8655-6 (DIN 12650-6).

For this the targeting volume was weighted for 20 times (± 0.5 mg) after draining the piston-operated pipette adjusted to 50 μL and filled with distilled water ($T = 22$ ° C), the mean of the weights was converted to the corresponding volume by using the relevant value of the water density at 22 °C ($\rho_{H2O(22°C)} = 0.9978$ g/L). For a mean of the weight of 0.0503 g a mean of the volume of 0.0502 mL was obtained with the standard deviation of the volume $s = 0.000249$ mL and the variation coefficient $CV = 0.50\%$. The variation coefficient is quite low which indicates that the analyst is well trained and qualified for using piston-operated pipettes. Furthermore, the relevant volume of $\mu = 50$ μL is assured since the calculated mean 50.2 μL ± 2*0.249 μL does not exceed the given confidence limit (50 μL ± 0.8 μL).
Photometer check to assure that the measurement of absorbance is being within the limits certified by NIST.

All filters measurements were within the given limits so the accuracy of the photometer is proven within the entire range of signals.

6.2.2.2.3 Starting the Measurement Process

In Table 6.2 an overview of the raw-data is given.

The statistical evaluation is performed by using STATGRAPHICS Centurion XVI (Statgraphics Centurion XVI user manual 2010). For the characteristic process parameters additional calculations were made by using an EXCEL-Macro. It should be considered that the Excel-Macro has to be validated before starting the calculations, e.g. by using a pocket-calculator.

Table 6.2 Overview on the measurements

Fundamental CALIBRATION				Recovery studies by STANDARDADDITION				QCC / Sample / BLANC		
No	CODE	CONC	EXT	No	CODE	STDadd	EXT	No	CODE	EXT
28	S0,25	0,248	0,068	1	A0,25	0,248	0,213	6	RM1	0,066
37	S0,25	0,248	0,065	29	A0,25	0,248	0,217	25	RM1	0,064
17	S0,75	0,742	0,167	2	A0,5	0,496	0,264	5	RM2	0,479
22	S0,75	0,742	0,166	38	A0,5	0,496	0,261	19	RM2	0,472
30	S1,25	1,239	0,271	4	A0,75	0,742	0,312	33	Sample	0,158
34	S1,25	1,239	0,272	24	A0,75	0,742	0,312	27	Sample	0,159
12	S1,75	1,733	0,372	23	A1,0	0,991	0,366	7	Sample	0,157
20	S1,75	1,733	0,375	36	A1,0	0,991	0,37	26	Sample	0,162
8	S2,25	2,227	0,471	18	A1,25	1,239	0,413	14	Blanc	0,016
10	S2,25	2,227	0,48	32	A1,25	1,239	0,419	15	Blanc	0,015

Variancehomogeneity				Variancehomogeneity			
No	CODE	CONC	EXT	No	CODE	CONC	EXT
28	S0,25	0,248	0,068	8	S2,25	2,227	0,471
37	S0,25	0,248	0,065	10	S2,25	2,227	0,48
11	S0,25	0,248	0,065	13	S2,25	2,227	0,475
21	S0,25	0,248	0,061	16	S2,25	2,227	0,478
3	S0,25	0,248	0,066	31	S2,25	2,227	0,478
9	S0,25	0,248	0,065	35	S2,25	2,227	0,481

Abreviations used:

CONC	Concentration of standards
EXT	Extinction, signal *y* obtained
STDadd	Standard concentration added
RM	Referencematerial RM1 and RM2

For this example the wine sample material was used as matrix material too.

6.2.2.3 Characterization of the Fundamental Calibration Process

6.2.2.3.1 Determination of the Calibration Function and Process Data

The most appropriate way to be aware of the relationship of concentration to signal is to illustrate the data graphically, e.g. by an *X-Y*-plot of the raw data. At first, the linearity has to be checked by the naked eye whether all measured values are being located on a straight line. Then the processed data can be described by a first degree calibration function of the type: $y = b_0 + b_1.x$. By reasons of simplicity this model should be preferred, if appropriate. Otherwise a more demanding model has to be used (for this see Funk et al. 1995, 2005).

This is followed by the estimation of the parameters of the linear calibration function by assuming a constant standard error of estimation in accordance to ISO 11095: 6.2. Using a linear calibration model, based on the last square method, the following assumptions have to be made:

- Only *y*, the measured signals, possesses variations, and *x*, the predefined standard concentrations are known without any error.
- The measured values are independent of each other and Gaussian's distributed.
- There exists a linear relationship between concentration and measured signals.
- The precision is constant within the working range so that homogeneity of variances is given.

Regression model:

$$y = b_0 + b_1.x + \varepsilon_i \tag{6.18}$$

Where

ε_i: is the random error which is independent and normally distributed with μ and σ^2
x: is the given concentration (CONC), exemplarily expressed as (mg/L)
y: is the measured value or signal (EXT), expressed as extinction (Ext)

Intercept b_0 and slope b_1, when $n = N^*K$:

$$b_0 = \overline{y} - b_1{}^*\overline{x} = \frac{\sum y_i}{n} - b_1{}^* \frac{\sum x_i}{n} [Ext] \tag{6.19}$$

$$b_1 = \frac{\sum[(x_i - \overline{x}) * (y_i - \overline{y})]}{\sum (x_i - \overline{x})^2} [\mathrm{Ext}/(mg/L)] \tag{6.20}$$

If the calibration function can be sufficiently described by a first degree linear model following a straight line, then the **sensitivity** is expressed by the slope b_1.

The standard error of estimation or residual standard deviation s_{res} (also as s_y):

$$s_{res} = \sqrt{\frac{\sum (y_i - \widehat{y}_i)^2}{n-2}} [Ext] \tag{6.21}$$

6.2.2.3.1.1 *Regression Analysis of the Fundamental Calibration*

Simple Regression: EXT Versus CONC (raw data Tab. 6.3; regression analysis Tab. 6.4; regression line Fig. 6.6; lack-of-fit test Tab. 6.5)

6.2.2.3.2 Verification of the Fundamental Calibration

The adequacy of the used calibration model has to be checked by verifying the linearity as well as the precision.

The residual analysis is a useful feature to discover trends, irregularities and/ or deviations from linearity as well as from constant residual standard deviation. A

Table 6.3 Raw-data of the fundamental calibration

No	CODE	CONC	EXT	Predicted	Residuals	s residuals	Leverages
28	S0,25	0.248	0.068	0.0656991	0.00230087	1.03383	0.3
37	S0,25	0.248	0.065	0.0656991	−0.000699135	−0.294398	0.3
17	S0,75	0.742	0.167	0.168.013	−0.00101317	−0.388.966	0.150182
22	S0,75	0.742	0.166	0.168.013	−0.00201317	−0.798744	0.150182
30	S1,25	1.239	0.271	0.270949	5.14639E-05	0.0189952	0.1
34	S1,25	1.239	0.272	0.270949	0.00105146	0.392326	0.1
12	S1,75	1.733	0.372	0.373263	−0.00126257	−0.4876	0.150061
20	S1,75	1.733	0.375	0.373263	0.00173743	0.681414	0.150061
8	S2,25	2.227	0.471	0.475577	−0.0045766	−2.77534	0.299757
10	S2,25	2.227	0.48	0.475577	0.0044234	2.59027	0.299757

Table 6.4 Regression analysis of the fundamental calibration

Regression Analysis - Linear model: Y = a + b*X

Dependent variable: EXT
Independent variable: CONC

Parameter	Estimate	Standard Error	T Statistic	P-Value
Intercept	0,014335	0,00171637	8,35192	0,0000
Slope	0,207113	0,00120704	171,588	0,0000

Analysis of Variance

Source	Sum of Squares	Df	Mean Square	F-Ratio	P-Value
Model	0,210127	1	0,210127	29442,53	0,0000
Residual	0,0000570948	8	0,00000713685		
Total (Corr.)	0,210184	9			

Correlation Coefficient = 0,999864
R-squared = 99,9728 percent
Standard Error of Est. = 0,00267149

The equation of the fitted model is **EXT = 0,014335 + 0,207113*CONC**

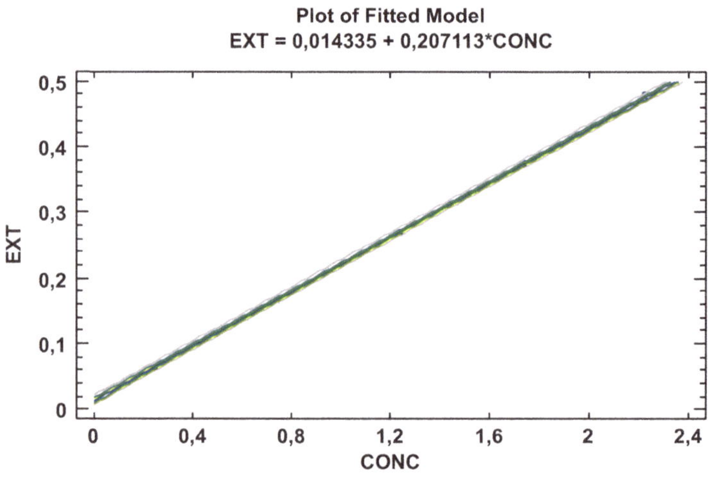

Fig. 6.6 Regression line of the fundamental calibration

Table 6.5 Lack of Fit test of the fundamental calibration

```
Analysis of Variance with Lack-of-Fit
-----------------------------------------------------------------------------
Source               Sum of Squares     Df  Mean Square     F-Ratio      P-Value
-----------------------------------------------------------------------------
Model                      0,210127      1      0,210127    29442,53       0,0000
Residual               0,0000570948     80,00000713685
-----------------------------------------------------------------------------
   Lack-of-Fit        0,00000659482     30,00000219827        0,22        0,8803
   Pure Error             0,0000505      5     0,0000101
-----------------------------------------------------------------------------
Total (Corr.)              0,210184      9

Since the P-value for lack-of-fit in the ANOVA table is greater to 0.10, the model
appears to be adequate for the observed data.
```

graphic representation of the residuals, defined as the vertical distances of the observations from the regression curve ($d_i = y_i – \hat{y}_i$) in dependency on the concentration (x_i), allows for a quick and comfortable check of the quality gstatus of the obtained data within the whole working range. If the residuals are randomly distributed, around the zero-line, the model seems to be adequate to describe the data. Furthermore, homogeneity of the variances can be assumed and trends may not exist. In any case of abnormalities, e.g. the data distribution is curved instead of linear, a selective statistical analysis should be applied to clarify and confirm the effect occurred. Some typical patterns of the residual analysis are shown at Funk et al. (1995, 2005).

For the verification of the **linearity,** e.g. the lack of fit test and/or the Mandel's fitting test can be used.

The lack of fit test is performed according to ISO 11095: 6.5.2 by comparing the variability of the current residuals with the variability of individual observations (each with N replications on K levels) obtained from the independent variable x. If the p-value in the ANOVA table is greater or equal to 0.10 (90%), the model appears to be adequate to describe the data. For better reasoning the pattern of the residuals can be used. If the decision is not clear the Mandel's fitting test (Funk et al. 1995, 2005) could be used additionally to secure the adequacy of the model.

The verification of the **precision** within the targeted working range is necessary to be aware of the quality status of the obtained data. Therefore an outlier check and a check of the variance homogeneity within the working range should be performed. Furthermore, the lower range limit should be secured by determining the relative analytical precision.

Given that outlier can seriously interfere with the quality parameters determined, it is essential to keep data material free of outlier. According to Funk et al., a special F-test can be used comparing the corresponding residual standard deviations ($s_{res} = s_y$), with and without the outlier-data pair, to check for significant differences:

$$PW = \frac{((N.K)_{A1} - 2) * s^2_{y_{A1}} - ((N.K)_{A2} - 2) * s^2_{y_{A2}}}{s^2_{y_{A2}}} \quad (6.22)$$

If $PW < F(f_1 = 1, f_2 = N.K_{A2}-2, P = 99\%)$: The suspected data pair is retained in the data collective as it is not considered to be an outlier.

Alternatively, a student's *t*-test can be applied, where the prediction interval of the regression line without the outlier data pair is used to verify, if the suspected value is positioned within this interval or not (for details see Funk et al. 1995, 2005). If an outlier is proven statistically, it is absolutely essential to review the whole process to be able to identify and eliminate the error source and to repeat the fundamental calibration in due course.

To be able to apply a linear regression model constant variances within the working range are required. For the verification of the homogeneity of variances, both standard concentrations of the predefined lower range limit (x_1) as well as of the higher range limit (x_N) are separately determined for at least 3 times (according to ISO 8466-1: $N_i \geq 3$) – Funk et al. propose to operate a tenfold approach. For both data sets the variances were calculated and compared with an *F*-test as follows:

$$s_i^2 = \frac{\sum (y_{ij} - \bar{y}_i)}{N_i - 1} \quad (6.23)$$

$$PW = \frac{s_N^2}{s_1^2} \quad (6.24)$$

If $PW > F(f_1 = N_{x1} - 1; f_2 = N_{xN} - 1; P = 99\%)$, the variances differ significantly. It is recommended to reduce the working range and to repeat the procedure. By using suited statistic software, weighted regression or multiple curve fitting may be applied alternatively.

Securing the lower limit of the concentration range by determining the relative analytical precision at the lowest concentration level is necessary, whenever an analyt is quantified. The limit of decision (x_p) has to be calculated according to DIN 32645 and it has to be proven, if the lower limit of the concentration range (x_1) is higher then the limit of decision. Only when this requirement ($x_1 > x_p$) is fulfilled, the entire working range will become statistically sound. Otherwise, the calibration has to be done again after starting the working range above x_p.

Proceed as in the following:

Calculate the critical signal value y_p, when $n = N*K$:

$$y_p = b_0 + s_{res}.t.\sqrt{\frac{1}{n} + 1 + \frac{\bar{x}^2}{\sum (x_i - \bar{x})^2}}[Ext] \quad (6.25)$$

Calculate the limit of decision x_p:

$$x_p = 2.s_{x0}.t\sqrt{\frac{1}{n} + 1 + \frac{(y_p + \overline{y})^2}{b_1^2 . \sum (x_i - \overline{x})^2}}[mg/L] \tag{6.26}$$

If a minimum precision of the obtained results of analysis is required by the customer, the relative analytical precision at the lower concentration range VB_{rel} (x_1) has to be verified as follows:

$$VB(x_1) = s_{x0}.t.\sqrt{\frac{1}{n} + 1 + \frac{(x_1 - \overline{x})^2}{\sum (x_i - \overline{x})^2}}[\mathrm{mg/L}] \tag{6.27}$$

$$VB_{rel}(x_1) = \frac{VB(x_1)}{x_1}.100[\%] \tag{6.28}$$

If the calculated $VB_{rel}(x_1)$ is higher than the given value, one is asked to improve the analytical precision or to reduce the working range by moving the lowest standard to a higher concentration level.

6.2.2.3.2.1 Evaluation of Further Characteristic Process Parameters

The process standard deviation s_{x0} quantifies the absolute precision of the results expressed as concentration:

$$s_{x_0} = \frac{s_{res}}{b_1}[mg/L] \tag{6.29}$$

The process variation coefficient also known as relative process standard deviation CV_{x0} is obtained by:

$$CV_{x_0} = \frac{S_{x_0}}{\overline{x}}^* 100[\%] \tag{6.30}$$

6.2.2.3.2.2 Verification of the Results of the Validation Experiment

Linearity Check

Regarding the *X/Y*-plot as well as the plot of the fitted model all data points showed excellent linearity. According to the lack-of-fit test the model appeared to be adequate to describe the observed data. The equation of the fundamental calibration was defined by:

$$\boldsymbol{EXT = 0.014335 + 0.207113^{*}CONC}$$

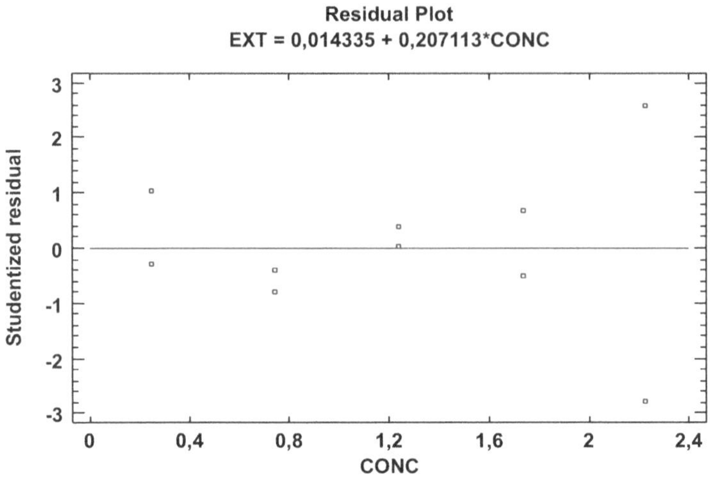

Fig. 6.7 Residual plot of the data of the fundamental calibration

The value of the slope b_1 = 0.2071 [Ext.mg^{-1}.L] expressed the sensitivity of the method. Furthermore, a standard error of estimation was obtained with $s_{res} = 0.0027$ [Ext].

Precision Check

The fact that the residual analysis showed only a random variation of the data within the entire working range there were not any indications of anomalies or additional effects (Fig. 6.7).

For checking the variance homogeneity the variance of the upper concentration range to the lower concentration range obtained from 6 single measurements respectively were compared by using the F-test (6.24):

$$PW = 1.3.10^{-5}/5.2.10^{-6} = 2.571 \quad F(5;5;\ 0.99) = 10.97$$

As $PW < F$ (5;5; 0.99), variance homogeneity could be assumed with $P = 99\%$ for the entire working range.

Based on the fact that there were 2 unusual residuals exceeding the value 2 but none of them exceeding the value 3, the existence of outlier was not evident since this has been confirmed by the outlier check.

For the assurance of the precision of the lower range limit the following parameters were obtained by using Eqs. 6.25–6.28 and summary statistics (Table 6.7):

Table 6.6 Unusual residuals and influential point of the fundamental calibration

Unusual Residuals

Row	X	Y	Predicted Y	Residual	Studentized Residual
9	2,227	0,471	0,475577	-0,0045766	-2,78
10	2,227	0,48	0,475577	0,0044234	2,59

Influential Points

Row	X	Y	Predicted Y	Studentized Residual	Leverage

Average leverage of single data point = 0,2

Table 6.7 Summary statistics of the fundamental calibration

	CONC (x)	EXT (y)
Count (n – N*K)	10	10
Average (mean x,y)	1,2378	0,2707
Variance (s^2)	0,544281	0,0233538
Standard deviation	0,737754	0,152819
Minimum	0,248	0,065
Maximum	2,227	0,48
Range	1,979	0,415
Stnd. skewness	-0,00196863	0,00421665
Stnd. kurtosis	-0,862708	-0,872298

The following variables show standardized skewness values which were all inside the expected range.
The following variables show standardized kurtosis values with none of them outside the expected range.

$$y_p = \mathbf{0.0217}\,[\mathbf{Ext}] \qquad \textbf{range of the signals}: \mathbf{0.061\text{–}0.481}\,[\mathbf{Ext}]$$

$$x_p = \mathbf{0.0703}\,[\mathbf{mg/L}] \qquad \textbf{working range}: \mathbf{0.25\text{–}0.025}\,[\mathbf{mg/L}]$$

$$\boldsymbol{VB_{(x1)}} = \mathbf{0.034}[\mathbf{mg/L}]$$

$$\boldsymbol{VB_{rel(x1)}} = \mathbf{13.67}\%$$

Given that the calculated decision limit of 0.07 mg/L was considerably lower than the lower working range limit of 0.25 mg/L, the determined precision parameters could be considered as reliable within the entire working range. Furthermore, the minimum of the relative analytical precision required by the customer could be assured because the value of 13.7% was clearly below the given limit of 20%.

The following characteristic process parameters derived from the fundamental calibration were evaluated for the defined working range:

- the process standard deviation (6.29) s_{x0} = **0.013 [mg/L]**
- the process variation coefficient (6.30) CV_{x0} = **1.042%**

6.2.2.4 Verification of Trueness: Identification of Systematic Influences Caused by Matrix Influences and/or Individual Process Steps

6.2.2.4.1 Determination of the Recovery Function and Process Data by Using the Standard Addition Methodology

Constant systematic as well as proportional systematic errors have been defined by several authors (Funk et al. 1995, 2005; Kromidas 2000).

For the evaluation of systematic effects caused by the matrix, recovery experiments are used based on standard addition methodology (Funk et al. 1995, 2005; DIN 32633). Based on the fundamental calibration a recovery function is established taking into account the predefined working range.

6.2.2.4.1.1 Establishment of the Recovery Function

The analytical results (x_f) were calculated by using the found signal values (y_f) as well as the parameters b_0 and b_1 originating from the fundamental calibration.

$$x_f = \frac{(y_f - b_0)}{b_1}[mg/L] \tag{6.31}$$

To obtain the recovery function the added standard concentrations (x_c), which are known, are plotted against the found concentrations (x_f).

$$x_f = b_{0,f} + b_{1,f}{}^{*}x_c \tag{6.32}$$

For our experiment the following definitions applied: $x_f = STD_{found}$ and $x_c = STD_{add}$, therefore, the following equation was obtained for the recovery function:

$$STD_{found} = b_{0,f}(\pm\ s_{b0}) + b_{1,f}(\pm\ s_{b1}) * STD_{add} \tag{6.33}$$

As no matrix blank was available, matrix-material containing a defined amount of the analyt (x_{Matrix}) was used. Even though, their analyt concentrations should be as low as possible, it was necessary to equalize the total phosphate concentration ($CONC_{found}$) to obtain the recovered part of the added analyt (surrogate recovery). Hence, x_{Matrix} had to be pre-examined with sufficient accuracy ($m \geq 4$) firstly to be subtracted by using Eq. 6.34.

$$\left(STD_{found} = CONC_{found} - \overline{x}_{Matrix}\right) \tag{6.34}$$

It should be mentioned that in such cases only proportional systematic errors can be verified with statistical relevance. The evidence of constant systematic errors caused by the measurement system can be excluded, if intercept b_0 of the fundamental calibration do not differ significantly from zero.

6.2.2.4.1.2 Regression Analysis of the Recovery Function

Simple Regression: STD_{found} Versus STD_{add} raw data (Tab. 6.8); regression analysis (Tab. 6.9); regression line (Fig. 6.8); lack-of-fit test (Tab. 6.10).

By applying Eq. 6.34 with a matrix mean value of 0.6985 (Table 6.13), the *STDfound* values were calculated. Then the differences *STDdiff* were calculated by subtracting *STDadd* from *STDfound* (Tables 6.8).

6.2.2.4.1.3 Verification of the Recovery Function

Linearity Check
Again all data show good linearity, which has been affirmed by a further lack-of-fit test. The equation of the recovery function was defined as:

$$\boldsymbol{STD_{found} = 0.0165265 + 0.989263 * STD_{add}}$$

This was obtained after subtraction of the actual P_2O_5-content of the wine sample used as matrix material. Due to the fact that the given concentrations were drawn to the found concentrations the standard error of estimation was expressed in concentration units, $s_{yf}\,(= s_{res}) = 0.0147$ mg/L (Fig. 6.8).

Precision Check
By considering the residual plot no discrepancies or effects could be identified (Fig. 6.9). Neither unusual residual nor influential points were found, so that absence of outlier data could be assumed (Tab. 6.11).

6.2.2.4.2 Assessment and Verification of Systematic Influences

Again linearity as well as precision has to be proven over the entire range. Therefore, the same procedure applies as shown at the fundamental calibration.

According to the proposed procedure the verification of systematic effects requires homogeneity of precision with respect to both calibration functions (fundamental calibration and recovery function).

Hence, the corresponding precision parameters of both calibration functions are compared simply by applying an F-test. Within this context it should be mentioned to always operate with the same units when calculating the PW-value, e.g. PW = concentration units per concentration units as described by Eq. 6.35.

Table 6.8 Raw data of the recovery function

No	CODE	STDadd	EXT	CONCfound	STDfound	STDdiff	Abreviation used	
1	A0,25	0.248	0.213	0.95921067	0.26071067	0.01271067	STDadd	Standard concentration added to matrix aliquot
29	A0,25	0.248	0.217	0.9785238	0.2800238	0.0320238		
2	A0,50	0.496	0.264	1.20545306	0.50695306	0.01095306	EXT	Extinction, signal y obtained
38	A0,50	0.496	0.261	1.19096822	0.49246822	–0.00353178	CONCfound	Concentration x calculated by using Eq. 6.31, contains both, standard and matrix component
4	A0,75	0.742	0.312	1.4372106	0.7387106	–0.0032894		
24	A0,75	0.742	0.312	1.4372106	0.7387106	–0.0032894		
23	A1,00	0.991	0.366	1.69793784	0.99943784	0.00843784	STDfound	Standard concentration found at the recovery experimen
36	A1,00	0.991	0.37	1.71725097	1.01875097	0.02775097		
18	A1,25	1.239	0.413	1.9248671	1.2263671	−0.0126329	STDdiff	Difference of standard concentrations, (STDfound–STDadd)
32	A1,25	1.239	0.419	1.95383679	1.25533679	0.01633679		

Table 6.9 Regression analysis of the recovery function

```
Regression Analysis - Linear model: Y = a + b*X
-----------------------------------------------------------------------------
Dependent variable: STDfound
Independent variable: STDadd
-----------------------------------------------------------------------------
                               Standard           T
Parameter        Estimate        Error         Statistic         P-Value
-----------------------------------------------------------------------------
Intercept       0,0165265     0,0109291         1,51216          0,1689
Slope            0,989263     0,0133019         74,3702          0,0000
-----------------------------------------------------------------------------


                           Analysis of Variance
-----------------------------------------------------------------------------
Source              Sum of Squares     Df  Mean Square     F-Ratio      P-Value
-----------------------------------------------------------------------------
Model                       1,2009      1       1,2009     5530,93       0,0000
Residual                  0,001737      8  0,000217125
-----------------------------------------------------------------------------
Total (Corr.)              1,20264      9

Correlation Coefficient = 0,999278
R-squared = 99,8556 percent
Standard Error of Est. = 0,0147352

The equation of the fitted model is   STDfound = 0,0165265 + 0,989263*STDadd
```

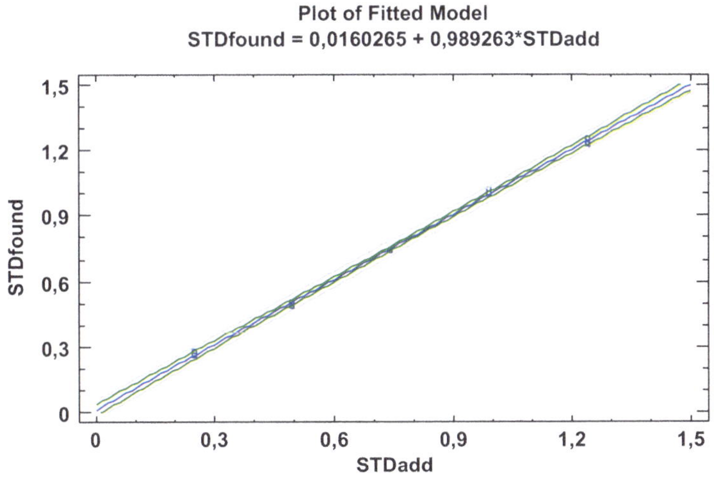

Fig. 6.8 Regression line of the recovery function

Table 6.10 Lack of Fit test of the regression of the recovery function

Analysis of Variance with Lack-of-Fit

Source	Sum of Squares	Df	Mean Square	F-Ratio	P-Value
Model	1,2009	1	1,2009	5530,93	0,0000
Residual	0,001737	8	0,000217125		
Lack-of-Fit	0,000839474	3	0,000279825	1,56	0,3094
Pure Error	0,000897524	5	0,000179505		
Total (Corr.)	1,20264	9			

Since the P-value for lack-of-fit in the ANOVA table is greater or equal to 0.10, the model appears to be adequate for the observed data.

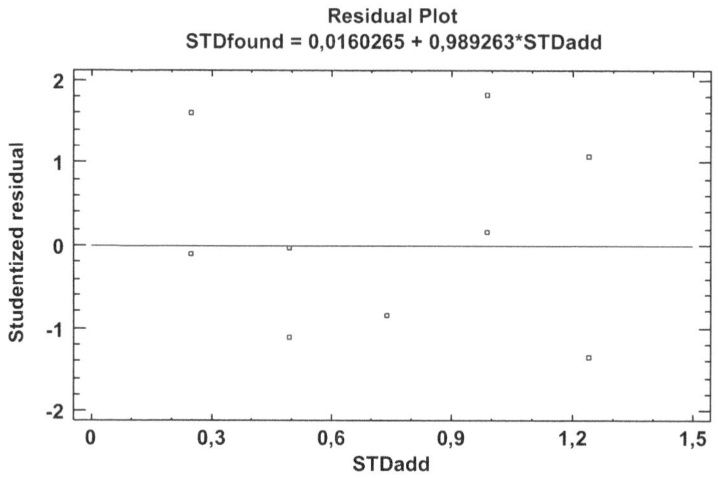

Fig. 6.9 Residual plot of the recovery function

$$PW = \left(\frac{s_{yf}}{s_{xoc}}\right)^2 \tag{6.35}$$

If $PW > F(f_f = (N.\ K)_f - 2; f_c = (N.\ K)_c - 2; P = 99\%)$ the precision of both calibration functions differs significantly. In such cases a reliable statement and the evidence of systematic effects would not be possible. It is recommended to identify the sources of imprecision and to repeat the recovery experiment. If homogeneity of the variances can be assumed, the parameters of the recovery function can be used for verification. By means of the confidence intervals, of the intercept b_{0f} (Eq. 6.36)

Table 6.11 Unusual residuals and influential points of the recovery function

Unusual Residuals

Row	X	Y	Predicted Y	Residual	Studentized Residual

Influential Points

Row	X	Y	Predicted Y	Studentized Residual	Leverage

Average leverage of single data point = 0,2

and of the slope b_{1f} (Eq. 6.37), the presence of influencing systematic effects can be checked.

$$b_{0f} \pm s_{b0}{}^{*} t_{(P=95\%, f=Nf.Kf\text{-}2)} \tag{6.36}$$

$$b_{1f} \pm s_{b1}{}^{*}\ t_{(P=95\%, f=Nf.Kf\text{-}2)} \tag{6.37}$$

If the confidence interval of the intercept does not include the value $b_{0f} = 0$, a constant systematic error is present with 95% probability—and furthermore, if the confidence interval of the slope does not include the value $b_{1f} = 1$, the presence of a proportional systematic error is proven. To be able to visualise these effects over the entire concentration range a graphical presentation can be chosen by plotting the differences of the concentrations (STD_{found}–STD_{add}) to the given standard concentrations.

6.2.2.4.2.1 Variability Check of Both Calibration Functions

Firstly, the process standard deviation of the fundamental calibration (s_{x0c}) and the residual standard deviation of the recovery function (s_{yf}) were compared by the use of Eq. 6.35:

$$PW = 0.0147^2/0.0133^2 = 1.105 \quad F(8; 8;\ 0.99) = 6.03$$

Since the *PW*-value was below the *F*-value (8;8; 0.99), there was no significant difference between the variances of the two calibration functions, so that the verification of additional systematic effects could be done.

Table 6.12 Confidence intervals of the parameters of the recovery function

95,0% confidence intervals for coefficient estimates

Parameter	Estimate	Standard Error	Lower Limit	Upper Limit
CONSTANT	0,0165265	0,0109291	-0,00867602	0,0417291
STDadd	0,989263	0,0133019	0,958589	1,01994

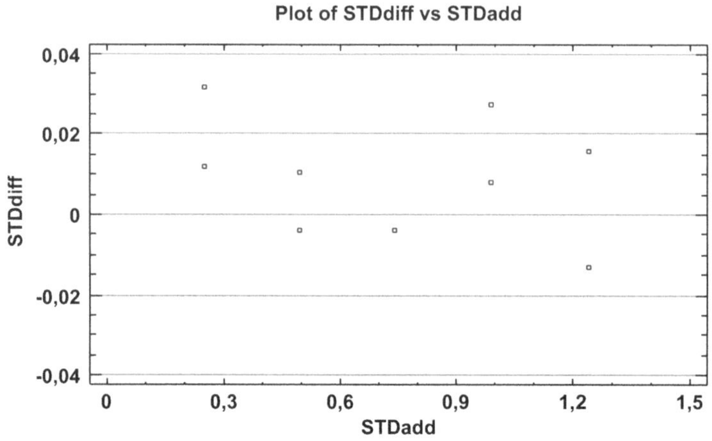

Fig. 6.10 *X-Y* Plot of the *d*-values (STD_{diff}) against the added standard concentrations (STD_{add})

6.2.2.4.2.2 *Verification of Systematic Influences*

The evidence of additional effects was evaluated by the confidence intervals of the intercept and the slope of the recovery function (Eqs. 6.36 and 6.37, respectively):

Since the confidence interval of the intercept *(CONSTANT)* included 0, no constant systematic error could be detected and furthermore, since the confidence interval of the slope (STD_{add}) included 1, no proportional systematic error could be identified (Table 6.12).

Despite a slight shift of the *d*-values was evident, the *X-Y* plot (Fig. 6.10) showed a nearly randomized pattern of the obtained data points originating from the recovery experiment.

6.2.2.4.3 Defining the Characteristic Validation Parameters of the Method

In case of having identified systematic effects, the fundamental calibration function is not valid to determine the sample analyt concentration. In such cases it is recommended to apply the standard addition method according to DIN 32633 to determine the sample analyt concentration. The absence of any constant systematic error has to be checked. Alternatively, if only proportional systematic effects have been verified, the slope of the recovery function may be used for correction of results based on the fundamental calibration function. The recovery rate (*RR*) can also be given with the results, which is determined by $RR = b_{1f}{}^{*}100$ [%]. However, each correction of the result automatically leads to an extension of the uncertainty budget, hence this alternative is not really recommended (Wegscheider 2000).

Since the absence of additional systematic effects could be proven the characteristic validation parameters of the fundamental calibration function were considered to be valid.

6.2.2.5 Determination of the Results and Their Uncertainty

6.2.2.5.1 Determination of the Results Derived by Applying the Analysis Function

The result of the analysis is obtained by using the inverse function of the calibration function solved for x, also called the analysis function. When m sample measurements are conducted, the mean value of all signals is used to calculate the analyt concentration of the sample:

$$x_{\mathrm{Pr}} = \frac{(\bar{y}_{\mathrm{Pr}} - b_0)}{b_1} \tag{6.38}$$

For the white wine sample "Müller Thurgau 2004, A-2352" the following result could be derived from the arithmetic mean of the four single measurements (Table 6.13) by using Eq. 6.38:

The total amount of phosphate of the measured sample was obtained as 141.14 [mg/L] after multiplication of the mean concentration by the dilution factor of 202.

6.2.2.5.2 Defining the Confidence Interval of the Results Obtained from the Sample Measurements

Several procedures are proposed to define the confidence interval of sample measurements (e.g. ISO 11095, DIN 38402). According to the ISO 11095 (see ISO 11095 chapter 7.2.1 Model with constant residual standard deviation) the calculation

Table 6.13 Results from the sample measurements

CODE	EXT	CONC	Amount
Sample	0.158	0.694	140.16
Sample	0.159	0.699	141.14
Sample	0.157	0.689	139.19
Sample	0.162	0.713	144.06
Mean Y(W)	0.159	0.699	141.14

can be made as demonstrated below, if model adequacy with constant residual standard deviation is given:

The estimation of the confidence interval within periods of proven validity of the calibration function is based on the pooled variance s^2_{cal}. For this the reference standards were measured to determine the differences of the concentrations d_j, known minus found concentration. The type and the number of the reference standards as well as the frequency of the measurements within a given period have to be defined. At least 2 standards (i and m) should be chosen, the index j characterizes the measurement points within a period of measurement times J (see Eq. 6.39).

$$S_{cal} = \sqrt{\frac{\sum_{j=1}^{J}\left(d_{ij}^2 + d_{mj}^2\right)}{2J}} \tag{6.39}$$

The confidence interval is then calculated as:

$$x_{Pr} \pm s_{cal}{}^{*} t_{(1-\alpha/2),(2J)} \text{ with } \alpha = 0.05 \tag{6.40}$$

The confidence interval of the sample measurements of the validated method was calculated by using the results of the reference standards RM1 and RM2 measured on the day the calibration experiment was performed (Table 6.2). Firstly, the d-values ($STD_{found} - STD_{added}$) were examined, and then inserted in Eq. 6.39 to derive the pooled standard deviation s_{cal}. Again, for the result the dilution factor of “202” had to be considered.

$$s_{cal} = 2.2044 \quad \text{with } t\,(0.05;4) = 2.776$$

Equation 6.40 then gave the confidence interval of:

$$VB_{(xquerPr)} = \pm\ s_{cal} \cdot t_{(0.05;\,2J)} = \pm\ 6.119\ [\text{mg/L}]$$

The complete result obtained with the white wine analysis was rounded to two significant figures and was then expressed as follows:

For the analysis of the white wine sample “Müller Thurgau 2004, A-2352”, a total amount of phosphate of **141.1 ± 6.1 mg/L**, calculated as P_2O_5, was obtained.

6.2.2.5.3 Determination of the Uncertainty of the Measurement

According to the International Vocabulary of Basic and General Terms in Metrology (VIM) the uncertainty of measurement is defined as “A parameter associated with the result of a measurement that characterises the dispersion of the values that could reasonably be attributed to the measurand” (ISO/IEC Guide 99:2007).

The calculation of the uncertainty as illustrated in the example presented includes the following contributions:

1. **the calibration process** (calibration standards, calibration, calculation of the results by using the analysis function)
2. **the sample analysis process** (sample, sample preparation, measurement)

Therefore, the following assumptions have to be made:

- To use the calibration parameter derived from the fundamental calibration, the absence of additional systematic effects has to be proven. For the significance of the recovery function homogeneity of the variability is required with respect to both calibration functions.
- The method has to be in "state of control": This can be easily demonstrated by maintaining a suitable quality control chart system.
- The treatment of the samples during measurement has to be equal to the calibration procedure (used pipettes, dilution scheme, reagents, apparatus, person, ...)

ad 1, uncertainty of the calibration process:

- uncertainty of the calibration standards and preparation of the standards

To avoid any error propagation due to dependencies while preparing and diluting of the standards, all calibration standards were made independently and in randomized order. Thus, the uncertainty that originated from the calibration standards was considered to be negligible.

- uncertainty contribution caused by the analysis function of the calibration when used for the calculation of the results

$$u(\widehat{x}_i)_{Cal} = \frac{s_{res}}{b_1}\sqrt{\left(\frac{1}{m}+\frac{1}{n}+\frac{(\widehat{x}_i-\overline{x})^2}{\Sigma(x_i-\overline{x})^2}\right)} \quad (6.41)$$

According to Eq. 6.41 the estimated uncertainty contribution caused by the use of the calibration model was $u_{cal} = 0.00825$ mg/L.

ad 2, uncertainty of the sample analysis process:

- variabilities caused by the sample: The sample material has to be sufficiently homogenous. This item was considered as negligible, because homogeneity can be easily achieved after gentle mixing of liquids such as white wine samples.
- variabilities derived from the sample preparation: This item is not applicable, because the white wine samples can be directly measured. Additional sample preparation steps are not necessary, e.g. cleaning, concentration or extraction. Nevertheless, it should be mentioned that in some cases we observed some problems associated with pipetting, which could lead to disproportional

variabilities. To avoid any underestimation of the uncertainty budget, it was decided to estimate the variability of the sample measurement separately by using repeatability measures.

- uncertainty of the sample analysis process: Eq. 6.41 does not take into account extraordinary uncertainties caused by the sample measurement or preparation. To solve this problem two approaches are presented in the following to obtain the combined standard uncertainty u_c.

ad 2.1 comprehensive approach

The actual standard deviation of the signals obtained from a sufficient number of independent sample measurements is determined and converted to concentration units by using b_I. Then, Eq. 6.41 is extended with the sample repeatability term to obtain the following comprehensive approach (Rösslein 1999):

$$u_c(\widehat{x}_i) = \frac{1}{b_1}^* \sqrt{\left(\frac{\sum_{j\mathrm{Pr}=1}^{m} \left(y_{i\,\mathrm{Pr}} - \overline{y}_{\mathrm{Pr}}\right)^2}{(m-1) * m}\right) + s^2{}_{res}{}^* \left(\frac{1}{n} + \frac{(\widehat{x}_i - \overline{x})^2}{\sum (x_i - \overline{x})^2}\right)} \tag{6.42}$$

According to Eq. 6.42 a combined uncertainty was obtained (for a mean value of m replicate determinations with $m = 4$) as:

$$u_c(xi) = 0.02786 \text{ mg/L}$$

Based on the bias observed at small sample sizes in the standard deviation a correction factor of $f = 1.08402$ has also to be applied according to Rösslein et al. (2007).

ad 2.2 alternative approach

The two uncertainty components, u_{cal} and u_{Pr}, are determined separately to be combined by Eq. (6.43):

$$u_c(\widehat{x}_i) = \sqrt{\left(u^2{}_{(\mathrm{Pr})} + u^2{}_{(Cal)}\right)} \tag{6.43}$$

Based on the fact that the problem mainly occurred when taking aliquots of the sample, the repeatability of pipetting 50 μL of white wine was used to estimate the contribution of u_{Pr}. For this, the targeted volume was weighted 20 times (± 50 mg) after draining the piston-operated pipette adjusted to 50 μL and filled with white wine. A variation coefficient (relative standard deviation) of the weights *CV %* = 5.292% was obtained, which was converted into phosphate concentration units by multiplication with the calculated sample mean of 0.699 mg (phosphate in white wine). This resulted in a *CV%* = 3.699% which corresponded to the standard uncertainty of $u_{Pr} = 0.02008$ mg/L ($= CV\%*f/(100*\sqrt{4})$ for a mean value of $m = 4$ replicate sample measurements and a correction factor of $f = 1.08402$). By the use of Eq. 6.43 the following combined uncertainty value was obtained alternatively:

$$u_c(xi) = 0.02171 \text{ mg/L}$$

When combining estimates, care has to be taken to avoid double counting. It should be mentioned that the uncertainty contribution of the calibration model u_{Cal} according to Eq. 6.41 also accounts for sample variations of the whole sample measurement process with the term ($1/m$). However, this term was considered to be negligible, when it was compared to the amount of the extraordinary sample variation u_{Pr} caused by pipetting of white wine only.

Finally, the expanded uncertainty U is calculated by multiplying u_c by an appropriate coverage-factor k, typically in the range of 2 to 3. With this, an interval is provided, where a large fraction of distribution of values attributed to the result may be expected to be included. For methods to be used in the routine normally a factor of $k = 2$ (with $P = 95\%$) is proposed to be sufficient (ISO-Guide "GUM", EURACHEM/ CITAC-Guide) but in any case k should be selected on the basis of confidence as required for the interval.

$$U(\widehat{x}_i) = u(\widehat{x}_i) * 2 \tag{6.44}$$

Equation 6.44 then gave the extended uncertainty $U(x_i)$ by using a coverage factor of $k = 2$. Furthermore the dilution factor of "202" was used to express the result as mg/L phosphate in wine. The fact that both measurement processes (calibration as well as sample measurements) were done by applying the same dilution schema by using the same equipment (pipettes, dispensing unit, etc.), the uncertainty contribution of the dilution factor was considered as being neutralized.

The complete result of the analysis was then expressed as the arithmetic mean of the measured samples and their relevant uncertainty budget as follows:

Comprehensive approach: **141.1 mg/L wine with $U(c_{P2O5}) = 11.3$ mg/L ($k = 2$)**
Alternative approach: **141.1 mg/L wine with $U(c_{P2O5}) = 8.8$ mg/L ($k = 2$)**

Comparing the uncertainty contribution caused by the calibration model $U(c_{P2O5}) = 3.3$ mg/L ($k = 2$) with those uncertainty budgets also taking into account extraordinary variations due to the sample preparation, it was evident that sample contributions should not be ignored. Nevertheless, the uncertainty budget of 11.3 mg/L (Comprehensive approach) could be improved with respect to the estimate for the sample measurement variability by enhancing the extent of independent sample measurements. It is likely that this would reduce the extended uncertainty to a value of about 9 mg/L or even lower.

6.2.2.6 Development of a Control Chart System to Be Used in Routine

For routine usage of a method the validity of the calibration curve should be checked. Especially when the determined calibration function is to be used for an extended time period, it is recommended to implement a control chart system. By continuously monitoring additional sources of undesired variations, trends can be quickly detected to decide, if there is a necessity for recalibration of the system. According to ISO 11095 the control chart system is based on the variability of the valid calibration function. Given that constant residual standard deviation is proven the upper (U_d) and the lower control limit (L_d) is calculated as:

$$U_d = +(s_{res}/b_1).\ t_{(1-x\xi/2),(NK-2)} \tag{6.45}$$

$$L_d = -(s_{res}/b_1).t_{(1-\xi/2),(NK-2)} \tag{6.46}$$

ξ = Significance level of each of the reference standards to ensure the global significance level of α where approximately it can be said $\xi \sim \alpha/m$.

For the control chart system the differences of the measured reference standards ($d = x_f - x_c$) are used to be plotted to the measurement time. The decision on the state of validity of the measurement system is made by periodically checking the position of the d-values. If they are located within the control limits L_d and U_d, the parameters of the calibration function are still valid and applicable. After a d-value exceeds one of the limits, the measurement system is "out of control" and has to be re-calibrated.

Initially both limits of the QCC have been estimated on the basis of the parameters of the fundamental calibration process by using Eqs. 6.45 and 6.46:

$$\boldsymbol{U_d} = 0.0027/0.2071^*2.306^*202 = +\mathbf{6.01}\ [\mathbf{mg/L}]$$

$$\boldsymbol{U_l} = 0.0027/0.2071^*2.306^*202 = -\mathbf{6.01}\ [\mathbf{mg/L}]$$

The raw data (listed in Table 6.14) were obtained from 20 measurement points over a time period of 10 days by measuring two different RM alternatively. Each of the standards was examined on two measurement points a day to be able to detect possible time influences or trends.

The d-values (P_2O_5 found—P_2O_5 given) of the measurements are then plotted on the quality control chart (QCC) shown in Fig. 6.11:

Considering the pattern of d-values obtained from the single measurements all differences (P_2O_5 found—P_2O_5 given) are more or less randomly distributed within the calculated QCC limits. Additionally, no trend between two consecutive single measurements could be identified so that the absence of time effects can be assumed. The fact that some of the d-values are relatively close to the QCC limits indicates that the limits could be too restricted. Therefore, an extension of the QCC limits should be envisaged on the basis of a larger period of measurements, e.g. the results of the initial phase taken from 20 days, to re-define the QCC limits. The larger extent of the uncertainty budget that has been calculated for the entire process, including the

Table 6.14 Raw data and calculations of the developing phase from the control chart system

No.	Extinction	CONC calc	P2O5 found	P2O5 given	d-value	d square
1	0.067	0.25428148	51.3648588	50.066	1.2988588	1.68703417
2	0.066	0.2494532	50.3895458	50.066	0.3235458	0.10468189
3	0.479	2.24353372	453.193812	450.176	3.01781207	9.10718971
4	0.475	2.22422059	449.29256	450.176	−0.8834399	0.78046606
5	0.063	0.23496835	47.4636068	50.066	−2.60239318	6.77245025
6	0.064	0.23979663	48.4389198	50.066	−1.62708018	2.64738992
7	0.471	2.20490747	445.391308	450.176	−4.78469187	22.8932763
8	0.477	2.23387716	451.243186	450.176	1.06718609	1.13888614
9	0.065	0.24462491	49.4142328	50.066	−0.65176719	0.42480047
10	0.066	0.2494532	50.3895458	50.066	0.3235458	0.10468189
11	0.47	2.20007918	444.415995	450.176	−5.76000487	33.1776561
12	0.471	2.20490747	445.391308	450.176	−4.78469187	22.8932763
13	0.06	0.2204835	44.5376678	50.066	−5.52833216	30.5624564
14	0.063	0.23496835	47.4636068	50.066	−2.60239318	6.77245025
15	0.479	2.24353372	453.193812	450.176	3.01781207	9.10718971
16	0.481	2.25319029	455.144438	450.176	4.96843806	24.6853768
17	0.063	0.23496835	47.4636068	50.066	−2.60239318	6.77245025
18	0.064	0.23979663	48.4389198	50.066	−1.62708018	2.64738992
19	0.47	2.20007918	444.415995	450.176	−5.76000487	33.1776561
20	0.473	2.21456403	447.341934	450.176	−2.83406589	8.03192945

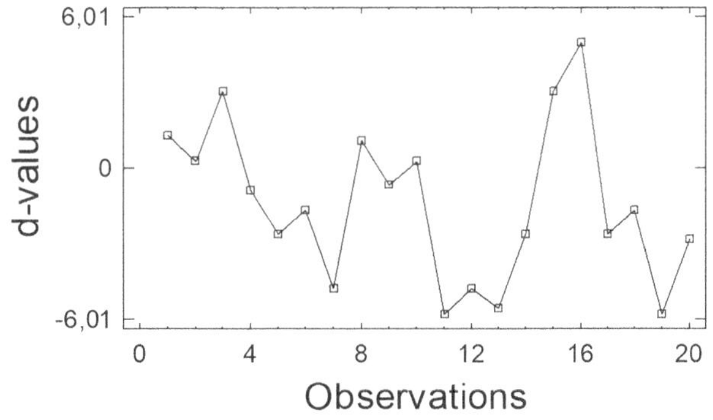

Fig. 6.11 QCC—Pattern of the *d*-values (measurements) characterizing the initial phase

calibration as well as the sample measurement, would also indicate this presumption and should be verified. If appropriate, it is recommended to operate with the uncertainty values for defining the QCC limits.

6.2.2.7 Discussion of the Results

The interpretation of the results was basically done by comparing the defined requirements to the achieved quality parameters:

For the routine analysis of the whole phosphate content of typical white wine samples the defined working area of **50–450 mg/L** could be secured. Within the range linearity could be proven to apply an first degree regression model of the type of $y = b_0 + b_1 . x$. Given that variance homogeneity could be assumed the precision of the method was sufficient for the intended use and clearly positioned below the required value of 20%. Additionally, the lower concentration limit of the working range was significantly higher than the decision limit. Hence, the working range was secured with adequate precision. The results of the recovery experience showed that neither constant nor proportional systematic effects could be detected. Additionally, there were no indications for the existence of any time dependencies of the method parameters. To be able to continuously control the state of validity of the determined calibration parameters, a QCC system was established. This was based on the differences of concentrations, found minus given, expressed as d-values. The following control chart limits were derived from the fundamental calibration process:

$$\boldsymbol{U_d = + \ 6.01\,[\mathrm{mg/L}]}$$

$$\boldsymbol{U_l = - \ 6.01\,[\mathrm{mg/L}]}$$

The system seemed to be under control and the calibration function did not need to be updated as of day 10. However, at some observations the distance to one of the limits was very close and for this higher attention is demanded. In case of any problems, it is recommended to apply the calculated uncertainty estimates for adjustment in order to improve the control chart system.

Nevertheless, within the monitoring period none of the periodically measured d-values exceeded the defined control limits. Hence, the system was declared to be in the state of validity. Within periods of validity the determined parameters could be applied and the results were derived by using the analysis function:

$$\mathbf{CONC = (EXT - 0.014335)/0.207113^{*}\ 202 \qquad [mg/L]}$$

The confidence interval and the uncertainty of the results were satisfactorily positioned below the required value of 20 mg/L.

$$\mathbf{141.1\ mg/L \pm 6.1\ mg/L\ P_2O_5\ in\ wine}$$

$$\mathbf{141.1\ mg/L\ P_2O_5\ in\ wine\ with\ \boldsymbol{U}\ (\boldsymbol{c_{P2O5}}) = 8.8\ mg/L\ (\boldsymbol{k} = 2)}$$

Based on the higher reliability of the sample variability contribution (20 repeated measurements were used to estimate the variability), it was decided to apply the alternative approach for the uncertainty estimation.

Taking into account the results of this validation experiment, it could be sufficiently proven that the characterized method was qualified to be used within the intended scope.

6.3 Optimization of LC-MS/MS Signal Response in Pesticides

Semih Otles, Hakan Savas Sazak, and Ebru Pelvan Evcil

6.3.1 *Objectives and Learning Outcomes*

The objective is to evaluate the parameters which directly affect the peak responses and process these parameters in the RSM. To use RSM statistical approach to explore how to select parameters that could efficiently optimize the objective functions.

6.3.2 *Introduction*

Response surface methodology (RSM) is a collection of statistical and mathematical techniques used to study the interactions of different variables. RSM is very useful for modelling and analyzing a complex process, allowing the construction of a model that contains a full description of the independent variables that are effective under optimal conditions. The main advantage of RSM is the small number of experimental trials needed to evaluate multiple parameters and their interactions, making the optimization process more efficient and cost-effective in terms of both manpower and resources. RSM has been used successfully in a variety of experimental processes (Yugui et al. 2008).

Instrument based analysis techniques produces large amount of data, to arrange, classify, optimize, and convert to meaningful ones, collaboration with statistical approaches is essential. RSM was used in optimization to increase sensitivity of analysis and decrease the limit of quantitation (LOQ) levels of pesticide. In this sense, liquid chromatography/tandem mass spectrometry (LC-MS/MS) with triple quad (QQQ) in multiple reaction monitoring (MRM) mode is the most widely used techniques for quantitation of pesticides in food (Takino and Tanaka 2008; Hiemstra and Kok 2007).

The aim of this study is to develop an analytical strategy for adapting any pesticide residue to LC-MS/MS with a QQQ mass analyzer, to investigate the effects of several factors on peak signals. Careful optimization is required to achieve maximum sensitivity and very low detection limits, which enable us to analyze very complex samples with a high degree of confidence. This includes adjustment of typical parameters like ionization voltage, temperature of the spraying/nebulizing gases, pressure, and flow rates (Asperger et al. 2001). The investigations were based on API-MS-MS analyses of both positive and negative ionization mode at systematically varied collision energy, fragmentor voltage, nebulizer pressure, gas flow rate and temperature.

6.3.3 Problem to Solve

The operation conditions of the chromatographic system are critical in residue analysis. Analysts must demonstrate that the obtained results are reliable and fit with purpose. By this way, different instrumental parameters that are affecting the signal responses are tested. RSM is used in data analysis to evaluate the effect of different instrumental parameters on the analytical responses and detection limits.

6.3.4 Procedure

RSM methodology was introduced by Box and Wilson (1951) to find the optimum values or levels of design variables to maximize or minimize a response variable. Several packages such as Matlab are used to produce several combinations for the values of the design variables. As a next step, for each level, several trials are conducted to produce values of the response variable. Afterwards, the maximized or minimized value of the response variable is explored by using regression analysis and RSM graphics. Regression analysis is used to explore the relationship between the exploratory variables and the response variable. The right regression model for the data is selected by using several techniques such as backward elimination, forward and stepwise selection, etc. depending on the situation. These techniques are accompanied by RSM graphics for visualization.

RSM will be conducted by using Matlab in this study. As we mentioned earlier, the first step of RSM is the production of several values or levels of design variables. For example, let us assume that there are three design variables X_1, $X_2$and $X_3$with the possible values-1,0,1. For this example, all the possible produced values are given in Fig. 6.12. Afterwards, several trials are done for the response variable for each level of these produced values and the optimum levels for these design variables are explored through regression analysis and RSM graphics.

Regression analysis is used to explore the relationship between the exploratory variables and the response variable (see Draper and Smith 1981; Neter et al. 1996 for more details). The multiple linear regression model is given as follows:

$$Y = \beta_0 + \beta_1 X_1 + \ldots + \beta_p X_p + \varepsilon$$

Y: response variable
$X_1,\ldots,X_p$: design variables

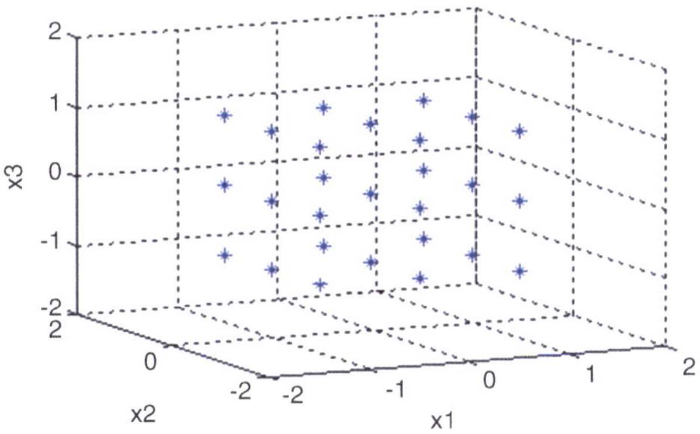

Fig. 6.12 The produced values of the design variables

β_0: intercept
$\beta_1, \ldots, \beta p$: slope parameters
ε: error term.

Several models are possible, but the simplest models are always preferred. If there is just one design variable X_1, the simplest model is the simple linear regression model given below:

$$Y = \beta_0 + \beta_1 X_1 + \varepsilon$$

When this model is not enough to explain the variability in the response variable, quadratic, cubic models, and models containing interaction terms are used. When there is one design variable X_1, the quadratic and cubic models are respectively given below:

$$Y = \beta_0 + \beta_1 X_1 + \beta_2 X_1^2 + \varepsilon$$

$$Y = \beta_0 + \beta_1 X_1 + \beta_2 X_1^2 + \beta_3 X_1^3 + \varepsilon$$

When there is more than one design variable, models containing interaction terms are also possible. For example, if we have two design variables X_1 and X_2, the quadratic model with an interaction term as followings:

$$Y = \beta_0 + \beta_1 X_1 + \beta_2 X_2 + \beta_3 X_1^2 + \beta_4 X_2^2 + \beta_5 X_1 X_2 + \varepsilon$$

In the model selection process, several techniques are utilized such as backward elimination, forward and stepwise selection, etc. There are also some criteria used in the comparison of several models. R^2 or the coefficient of determination shows the explained percentage of the variability in the response variable by the model constructed. The regression analysis is accompanied by RSM graphics which are very important tools in the optimization of the response variables.

In this study, the effect of several instrumental factors on signal of two different pesticide molecules under different polarization modes by LC-MS/MS was investigated. The study included (a) Preparation of pesticide standards and LC conditions, (b) method development by LC-MS/MS, (c) find optimum fragmentor value by using RSM, (d) preparing another LC-MS/MS methodology by using defined fragmentor, (e) find optimum collision energy value by using RSM through quadratic and cubic regression models.

6.3.5 *Case Study*

6.3.5.1 MS-QQQ Parameters

Mass spectrometer works by ionizing molecules and then sorting and identifying the ions according to their mass-to-charge (m/z) ratios. Two key components in this process are the ion source, which generates ions, and the mass analyzer, which sorts the ions. As an ionization technique, electrospray ionization (ESI) was used in this study.

- Fragmentor (F) is the voltage (V) to apply to the exit end of the capillary. The range is 0 to 400 V.
- Collision Energy (Ce) is the voltage to apply to the ions in the collision cell. The range is 0 to 250 V.
- Gas Temperature is the temperature of the nitrogen drying gas. The default value is 300 °C. The temperature required depends on the LC flow rate, the drying gas flow rate, the type of analysis (ESI or APCI), and the thermal stability of the sample (Table 6.15).

To optimize performance and sensitivity, the fragmentor voltage was selected to produce the highest signal intensity for the precursor ion. The collision energy was then adjusted to produce the highest intensity for the main fragment. From the total ion chromatogram (TIC), the corresponding $[M + H]^+$ for ESI+ and $[M - H]^-$ for ESI− were used to obtain the extracted ion chromatogram (EIC) for the different fragmentor voltages in steps of 10 V between 80 and 160 V. According to fragmentor energy value, the height of the peaks for each injection was recorded and fragmentor energy was optimized by using RSM. With this optimized fragmentor voltage, an additional injection experiment was performed to optimize collision energy. Each time segment of the analysis was operated with increasing collision energy in steps of 1 V between 0 and 20 V. The optimum collision voltages

Table 6.15 Source Parameters

Source	Minimum	Maximum
Drying gas temp	0 °C	350 °C
Drying gas flow	0 L/min	13 L/min
Nebulizer pressure	0 psi	60 psi

were defined as generating the highest signal for the main fragment ion by using quadratic and cubic regression models.

6.3.5.2 Data Acquisition

In this study RSM is used to predict the effect of nebulizer pressure, drying gas temperature and flow rate, fragmentor and collision energy on the peak height for both positive and negative ionization mode. A quadratic function is obtained as expected. To obtain this quadratic function to be investigated, design points are created for the fragmentor (80, 90,. . ., 160), drying gas temperature (200, 250, 300), drying gas flow (7, 9, 11, 13) and nebulizer pressure (15, 30, 45, 60). Then, the peak heights for all combinations of these values are obtained. As a final step, by using the optimum values of the variables mentioned above, the peak heights are obtained for 21 design points of collision energy (0, 1,..., 20).

6.3.5.3 Statistical Packages

Matlab and SPSS package programs are used in the study. First, Matlab RSMtool (rstool) is used to find initial results for the full quadratic model. Second, by using the backward elimination procedure of SPSS, the quadratic function with significant effects is constructed. Then, the response surface graphs are created to investigate the quadratic function obtained again by using Matlab. Finally, to find the optimum collision energy value for the optimum variable levels found by investigating the response surface graphs, quadratic and cubic regression models and their graphs are constructed by the help of Matlab and SPSS package programs.

6.3.5.4 Optimization of Fragmentor and Collision Energy in Positive Ionization Mode

LC-MS/MS with electrospray ionization (ESI) is technique most often employed for multiresidue pesticide analysis both negative and positive ionization modes. Identification of the precursor ion for Amitraz was performed in the full scan positive ionization mode that produces molecular ion forms for more polar and less volatile pesticides and it was observed as 294.2 m/z. For each fragmentor voltage (F), nebulizer pressure (N), gas flow rate (f) and gas temperature (T) values, the peak height of precursor ion is obtained (Fig. 6.13) and the relationship is investigated by using Matlab.

The Matlab rstool shows that the full quadratic model of peak height on the factors indicated here takes its maximum value 9109829.8821 with fragmentor = 109, drying gas temperature = 287, drying gas flow = 10.1 and nebulizer = 60 (Fig. 6.14).

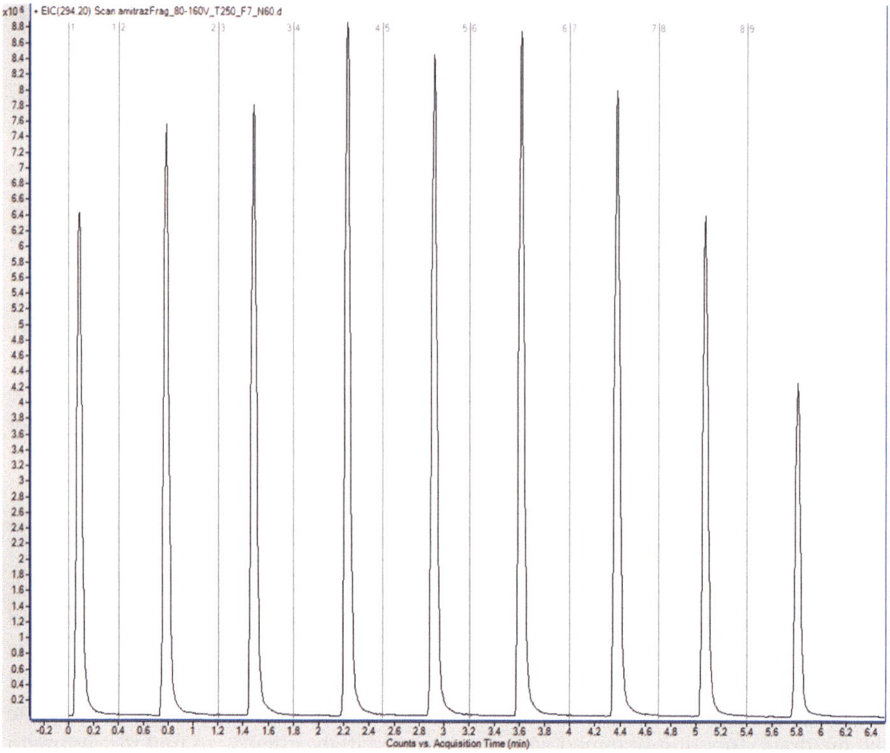

Fig. 6.13 Extract ion chromatogram for Amitraz (F:80–160 V, T: 250 °C, f:7 L/min, N:60 psi)

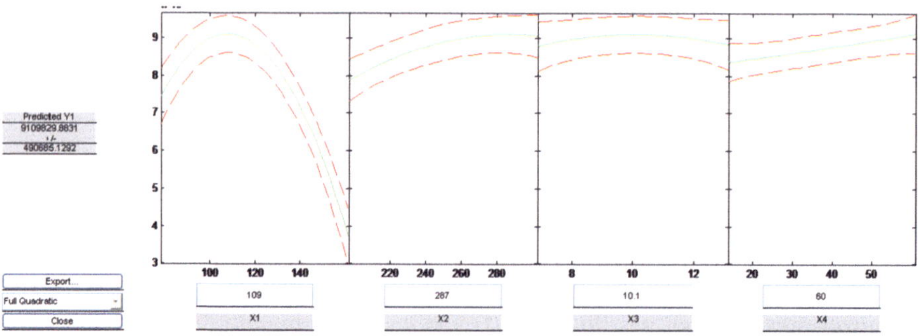

Fig. 6.14 Full quadratic model with optimum values

First, an outlier (the observation with high standard error) is deleted from the data and then using the backward elimination procedure in SPSS. The model summary is given in Table 6.16 The highest significant effect is noticed inmodel 4 with the coefficient of multiple determinations (R^2) is 0.821 or 82.1% of the variability in peak height can be explained by the final model.

Table 6.16 Model summary

Model	R	R square	Adjusted R square	Std. Error of the estimate
1	.907[a]	.823	.817	665741.66555
2	.907[b]	.823	.817	664988.33227
3	.907[c]	.823	.817	664444.70958
4	.906[d]	.821	.817	665689.91237

[a]Predictors: (Constant), nebpres^2, gastemp, gasflow, fr^2, flow_neb, fr_neb, temp_neb, fr_flow, fr_temp, temp_flow, nebpres, flow^2, fragmentor, temp^2
[b]Predictors: (Constant), gastemp, gasflow, fr^2, flow_neb, fr_neb, temp_neb, fr_flow, fr_temp, temp_flow, nebpres, flow^2, fragmentor, temp^2
[c]Predictors: (Constant), gastemp, gasflow, fr^2, flow_neb, fr_neb, temp_neb, fr_flow, fr_temp, temp_flow, flow^2, fragmentor, temp^2
[d]Predictors: (Constant), gastemp, gasflow, fr^2, fr_neb, temp_neb, fr_flow, fr_temp, temp_flow, flow^2, fragmentor, temp^2
[e]DependentVariable: height

Table 6.17 Final model

Model		Unstandardized coefficients		StandardizedCoefficients		
		B	Std. Error	Beta	t	Sig.
	(Constant)	−36757039.650	2450647.488		−14.999	.000
	fragmentor	509343.708	16424.555	8.452	31.011	.000
	gastemp	96179.524	14597.681	2.531	6.589	.000
	gasflow	776798.582	194942.519	1.119	3.985	.000
	fr_temp	−218.712	30.384	−1.150	−7.198	.000
	fr_flow	−3085.342	555.060	−.748	−5.559	.000
	fr_neb	−275.608	59.366	−.403	−4.642	.000
	temp_flow	653.990	350.849	.294	1.864	.063
	temp_neb	159.923	28.801	.465	5.553	.000
	fr^2	−1847.061	54.915	−7.393	−33.635	.000
	temp^2	−155.006	27.242	−2.043	−5.690	.000
	flow^2	−29464.222	8016.571	−.852	−3.675	.000

[a]DependentVariable: height

The final model in Table 6.17, the main effects fragmentor, drying gas temperature, drying gas flow, the interaction of fr-temp, fr-flow,fr-neb, temp-flow, temp-neb, and the squares of fragmentor, gas temperature, and flow exist. The coefficients of these effects are given in Table 6.19. When the magnitudes of the t values are greater, the corresponding factors have more significant effect on the peak height. The response surface plot for flow = 7, neb = 60 in top view is given in Fig. 6.15). Here, the maximum peak height is obtained at fragmentor = 114 and gas temperature = 248 with peak height = 6330300. It can be verified from the contour plot given in Fig 6.15b and the response surface plot of the side view is given in Fig 6.15c.

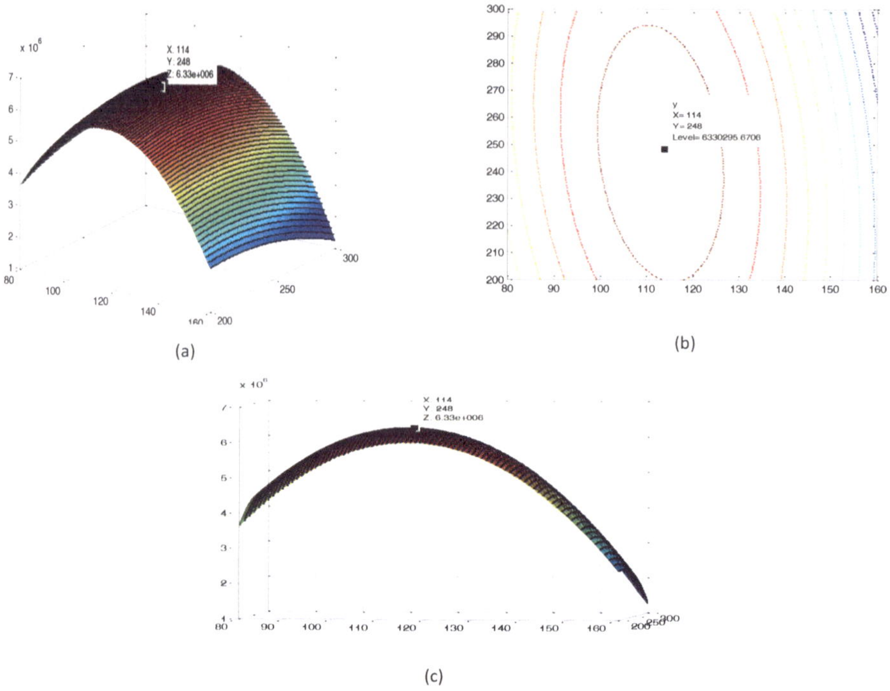

Fig. 6.15 Response surface plot of fragmentor and gas temperature with gas flow = 7 mL/min and nebulizer = 60 psi (X: fragmentor in Volt, Y: gas temperature in °C, Z & Level: peak height-Detector Signal (**a**) Response surface plot in top view (**b**) Contour plot (**c**) Response surface plot in side view

It is found in Fig. 6.17 that the maximum peak height is obtained at fragmentor = 114 and gas temperature = 248. By using these values of fragmentor and gas temperature, the optimum value of collision energy was investigated by increasing 1 V between 0 and 20 V. Due to the limitations of the software installed on the instrument, the gas temperature value 248 was rounded up to 250. It was found that Amitraz has two product ions as 253.2 (Y1) and 163.1 (Y2). Therefore, for both, separate regression models are constructed. The observed values are given in Table 6.18. The model summary is given in Table 6.19. Here, 95.8% of the variability in peak height can be explained by the cubic model.

Y1 has its maximum at x = 8 with Y1_fitted = 519180. It is easily observed in Fig. 6.16 that the fitted cubic model is appropriate for the regression points. The same type of analysis is done for Y2 but with a quadratic model. The summary and the details of the final model for Y2 are given in Table 6.20. By this quadratic model, 97.7% of the variability in peak height can be explained.

For the quadratic regression model, Y2 has its maximum at x = 12 with Y2_fitted = 8407700. It is easily seen in Fig. 6.17 that the fitted quadratic model is very appropriate for the regression points. Since we have to select only one value

Table 6.18 Peak height of the product ions for each collision energy voltage

Ce (X)	Peak height for 253.2 (Y1)	Peak height for 163.1 (Y2)
0	222920	2247269
1	262882	3087212
2	315469	3967448
3	384495	4771554
4	435665	5439340
5	467933	5838600
6	513818	6355860
7	529478	7345843
8	537758	8075568
9	508806	8532325
10	539151	8205156
11	519964	8432415
12	440670	8367963
13	415785	8768903
14	382844	7987223
15	370267	7934276
16	324890	7346709
17	290033	7227775
18	293986	7143936
19	253541	6605823
20	264500	6269693

Table 6.19 Model summary

Model	R	R square	Adjusted R square	Std. Error of the estimate
1	.979[a]	.958	.951	23761.320

Model		Unstandardized coefficients		Standardized coefficients	t	Sig.
		B	Std. Error	Beta		
1	(Constant)	180277.586	17526.175		10.286	.000
	x	99554.447	7778.805	5.775	12.798	.000
	x^2	−8700.633	916.962	−10.455	−9.489	.000
	x^3	193.955	30.104	4.512	6.443	.000

Note: x: collision energy, x^2: quadratic term, x^3: cubic term, y1: peak height 1
[a]Predictors: (Constant), x^3, x, x^2
[b]Dependent Variable: y1

for x but we do not a common value for x in the maximization of Y1 and Y2, we have to sacrifice some amount from Y1 and Y2 and find a compromise point between x = 8 (maximized for Y1) and x = 12 (maximized for Y2). Then, the ideal value would be x = 10 with Y1_fitted = 499710 (just 3.75% less than Y1_fitted = 519180 for X = 8) and Y2_fitted = 8190500 (2.58% less than Y2_fitted = 8407700 for X = 12). The total loss percentage is 6.33% which is the lowest value among all the choices.

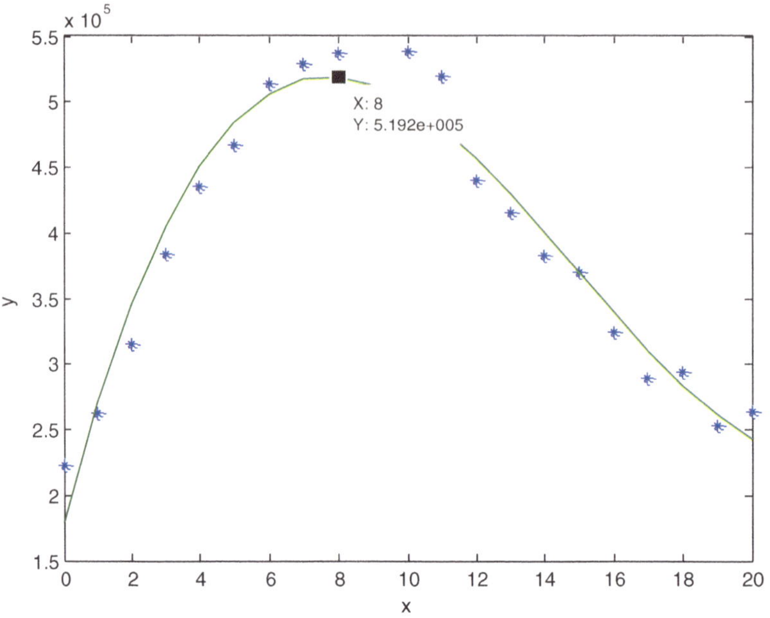

Fig. 6.16 Scatter plot of collision energy (x) and peak height 1 (Y) and the fitted cubic line

Table 6.20 Model summary

Model	R	R square	Adjusted R square	Std. Error of the estimate
1	.989[a]	.977	.975	294005.200

Model		Unstandardized coefficients		Standardized coefficients		
		B	Std. Error	Beta	t	Sig.
1	(Constant)	2110005.360	175493.248		12.023	.000
	x	1024285.722	40664.055	3.428	25.189	.000
	x^2	−41623.423	1962.974	−2.886	−21.204	.000

Note: x: collision energy, x^2: quadratic term, y2: peak height 2
[a]Predictors: (Constant), x^2, x
[b]Dependent Variable: y2

6.3.5.5 Optimization of Fragmentor and Collision Energy in Negative Ionization Mode

Identification of the precursor ion for 2,4-D acid was performed in the full scan negative ionization that produces molecular ion under negative ionization mode and it was observed as 219 m/z. For each fragmentor voltage (F), nebulizer pressure (N), gas flow rate (f) and gas temperature (T) values, the peak height of precursor ion is obtained (Fig. 6.18) and the relationship is investigated by using Matlab. For the negative ionization mode, the Matlab rstool shows that the full quadratic model of

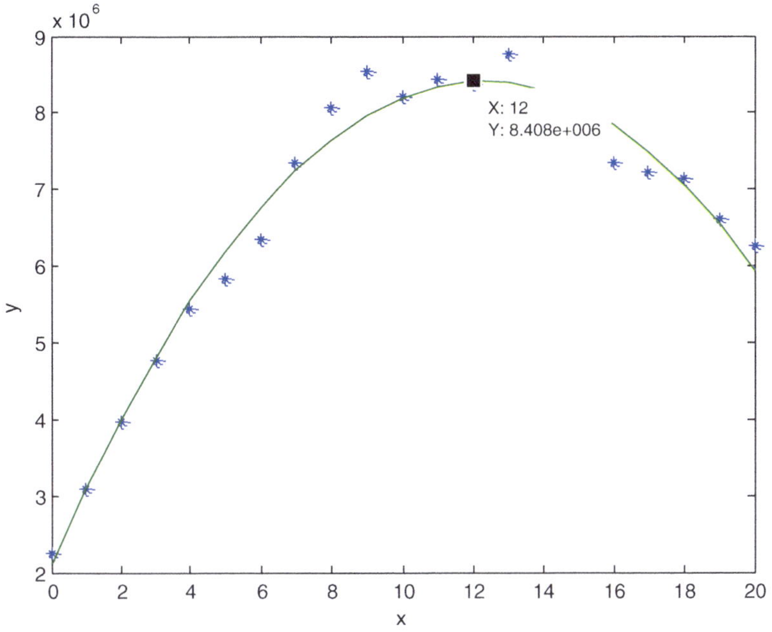

Fig. 6.17 Scatter plot of collision energy (x) and peak height 2 (y) and the fitted quadratic line

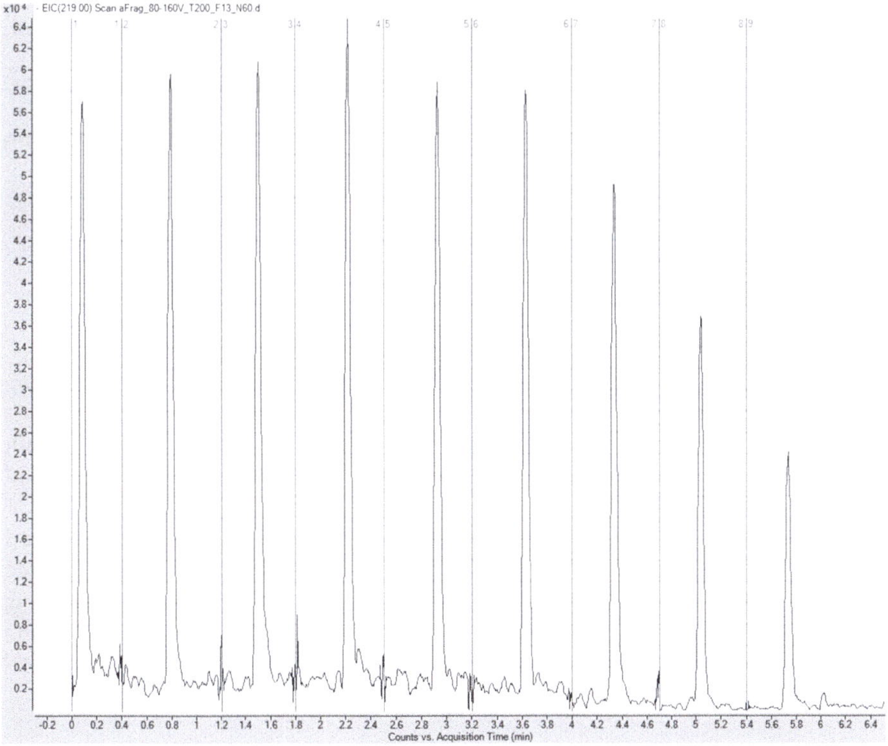

Fig. 6.18 Extract ion chromatogram for 2,4-D Acid (F:80–160 V, T: 200 °C, f:13 L/min, N:60 psi)

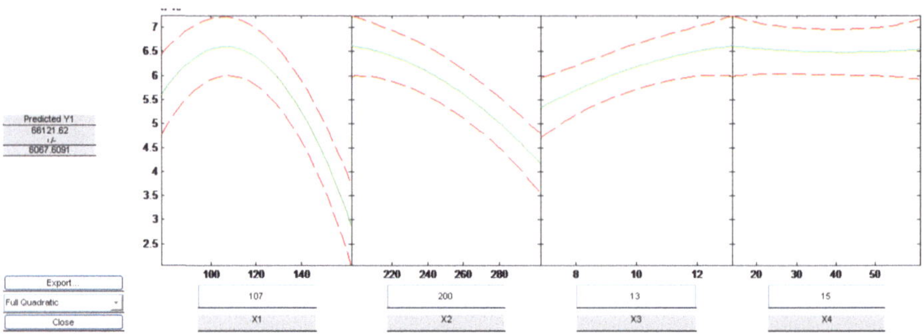

Fig. 6.19 Full quadratic model with optimum values

Table 6.21 Model summary

Model	R	R square	Adjusted R square	Std. error of the estimate
1	.941[a]	.885	.881	4725.74955
2	.940[b]	.885	.881	4720.96815
3	.940[c]	.884	.881	4716.47813
4	.940[d]	.884	.881	4719.40813
5	.940[e]	.883	.881	4724.91141
6	.940[f]	.883	.880	4732.25475

[a]Predictors: (Constant), nebpres^2, temp^2, fragmentor, flow^2, flow_neb, fr_neb, temp_neb, fr_flow, temp_flow, fr_temp, fr^2, nebpres, gasflow, gastemp
[b]Predictors: (Constant), temp^2, fragmentor, flow^2, flow_neb, fr_neb, temp_neb, fr_flow, temp_flow, fr_temp, fr^2, nebpres, gasflow, gastemp
[c]Predictors: (Constant), temp^2, fragmentor, flow^2, flow_neb, fr_neb, temp_neb, fr_flow, temp_flow, fr^2, nebpres, gasflow, gastemp
[d]Predictors: (Constant), temp^2, fragmentor, flow^2, flow_neb, fr_neb, temp_neb, fr_flow, temp_flow, fr^2, gasflow, gastemp
[e]Predictors: (Constant), temp^2, fragmentor, flow^2, flow_neb, fr_neb, fr_flow, temp_flow, fr^2, gasflow, gastemp
[f]Predictors: (Constant), temp^2, fragmentor, flow^2, flow_neb, fr_neb, temp_flow, fr^2, gasflow, gastemp
[g]Dependent Variable: height

peak height on the factors indicated here takes its maximum value 66121.62 with fragmentor = 107, drying gas temperature = 200, drying gas flow = 13 and nebulizer = 15 (Fig. 6.19).

First taking out seven outlier values from the data and applying the backward elimination procedure in SPSS, the final model is obtained in Table 6.21 (see Hoaglin et al. 1983 for details about detecting outliers). The model summary is given in Table 6.22. The model 6 is used as the final model in which 88.3% of the variability of the peak height can be explained by the quadratic model.

Table 6.22 Final model

Model		Unstandardized coefficients		Standardized coefficients	t	Sig.
		B	Std. error	Beta		
1	(Constant)	−223765.877	15904.273		−14.070	.000
	fragmentor	2666.911	95.511	5.047	27.923	.000
	gastemp	884.607	101.185	2.632	8.742	.000
	gasflow	8889.973	1349.612	1.445	6.587	.000
	fr_neb	−1.250	.370	−.210	−3.381	.001
	temp_flow	−28.042	2.557	−1.415	−10.967	.000
	flow_neb	12.322	4.427	.174	2.783	.006
	fr^2	−12.318	.392	−5.622	−31.418	.000
	$temp^2$	−1.454	.194	−2.169	−7.490	.000
	$flow^2$	−105.338	57.455	−.345	−1.833	.067

[a]Dependent Variable: height

In the final model, the main effects fragmentor, drying gas temperature, drying gas flow, the interaction fr-neb, temp-flow, temp-neb, flow-neb, and the squares of fragmentor, gas temperature and flow exist. The coefficients of these effects are given in Table 6.22. The response surface plot for flow = 13, neb = 60 in top view is given in Fig. 6.20a. The maximum peak height is obtained at fragmentor = 105 and gas temperature = 200 with peak height = 65810. It can be verified from the contour plot given in Fig. 6.20b and the response surface plot in side view given in Fig. 6.20c.

It is found in Fig. 6.20 that the maximum peak height is obtained at fragmentor = 105 and gas temperature = 200. By using these values of fragmentor and gas temperature, the optimum value of collision energy was investigated by increasing 1 V between 0 and 20 V. It was observed that 2,4-D acid has one product ions as 161. The observed values are given in Table 6.23. These data are best fitted by a cubic regression model. The summary table and the details of this cubic model are given in Table 6.24. The 77.1% of the variability can be explained by the model. The value is not as high as the previous models that were constructed because of the higher variability in the observations. This fact can be verified by examining the Fig. 6.20.

For the fitted cubic regression line, Y has the maximum at x = 7 with y_fitted = 26789. The scatter plot and the fitted cubic line for the given regression points are shown in Fig. 6.21.

In this study RSM technique was used to establish the optimum nebulizer pressure, gas flow rate and gas temperature values to obtain optimum fragmentor and collision energy. By using response surface plots, the optimum values were illustrated graphically. It was concluded that, optimal conditions for Amitraz are obtained at fragmentor = 114 V, gas temperature = 250 °C, gas flow rate = 7 l/min, nebulizer pressure = 60 psi and collision energy = 10 V, and for 2,4-D acid, the optimal levels are obtained at fragmentor = 105 V, gas temperature = 200 °C, gas

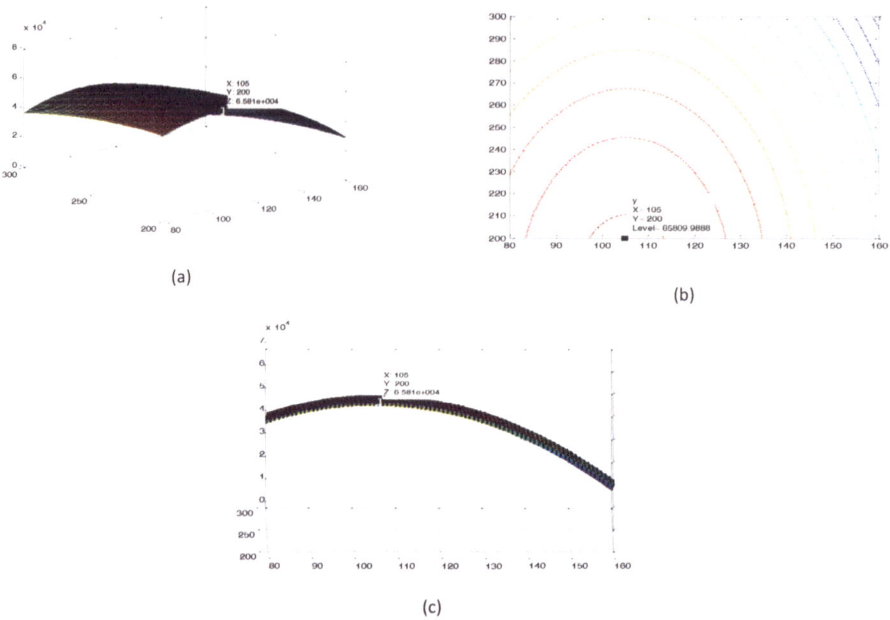

Fig. 6.20 Response surface plot of fragmentor and gas temperature with gas flow = 13 mL/min and nebulizer = 60 psi (X: fragmentor in Volt, Y: gas temperature in °C, Z & Level: peak height-Detector Signal (**a**) Response surface plot in top view (**b**) Contour plot (**c**) Response surface plot in side view

Table 6.23 Peak height of product ion for each collision energy voltage

Ce(X)	Peak height for 161 (Y)	Ce(X)	Peak height for 161 (Y)
0	14619	10	23736
1	20547	11	22169
2	20112	12	21111
3	16682	13	23399
4	23386	14	22243
5	25166	15	19882
6	26535	16	13142
7	31105	17	16206
8	32668	18	16658
9	25200	19	13249
		20	11316

flow rate = 13 L/min, nebulizer pressure = 60 psi, and collision energy = 10 V. The t values of the final model shows that the fragmentor and fr^2 effects the peak height of the precursor ions more than the others for Amitraz. This is also similar with 2,4-D acid. For 2,4-D acid, besides the fragmentor and fr^2, tempflow interaction also

Table 6.24 Model summary

Model	R	R square	Adjusted R square	Std. Error of the estimate
1	.878[a]	.771	.731	2945.027

Model		Unstandardized coefficients		Standardized coefficients		
		B	Std. Error	Beta	t	Sig.
1	(Constant)	14037.180	2172.230		6.462	.000
	x	3914.635	964.121	4.278	4.060	.001
	x^2	−351.577	113.650	−7.960	−3.093	.007
	x^3	7.513	3.731	3.293	2.013	.060

Note: x: collision energy, x^2: quadratic term, x^3: cubic term, y: peak height
[a]Predictors: (Constant), x^3, x, x^2
[b]Dependent variable: y

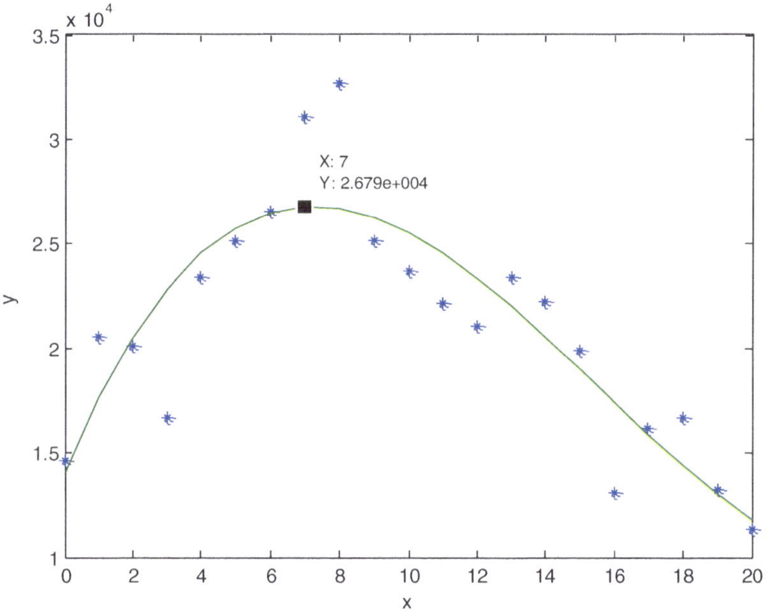

Fig. 6.21 Scatter Plot of Collision energy (x) and Peak height(y) and the fitted cubic line

significantly effects the peak heights of the precursor ion.This case study was a very good example of the efficient usage of the advanced statistical tools in food chemistry and the results will have essential impacts on the area.

6.4 Standardization of a Quantitative Microbiological Method Based on Colony Count Techniques

Ulrike Zitz

6.4.1 *Objectives and Learning Outcomes*

International collaborative studies are approved methodologies to define the capability of newly developed methods. However the applicability of such studies on microbiological scopes is difficult because of the complexity of biological systems as well as resulting problems. Whereas chemical methods are highly automated processes, the classical microbiological techniques exhibit considerable error components depending on the materials to be analysed, the various methodical influences on the result and especially the individual work being performed.

Due to the lack of experience within this area this study points out critical steps in detail when preparing, organizing and performing such international collaborative studies as well as visualises the importance of detailed planning and intensive support of such projects to achieve a successful outcome. Moreover, special efforts were put on the improvement of the reliability of the obtained precision parameters of the method by the application of robust methods to be used for the statistical data analysis.

6.4.2 *Introduction*

High quality of food products on a constant quality level is required by the customers. As regulated by legislation this should be continuously guaranteed and maintained by the producers as well as by official control laboratories. For detection and/or enumeration of microorganisms in food numerous microbiological methods have been developed and are extensively used in routine analyses. Particularly methods which are powerful and easy to apply as well as proven for a specific use are needed to be provided. To be able to compare results, harmonization of these methods is an essential requirement. Hence, organizations such as ISO and IDF highly promote horizontal standards and standardization of analytical methods for mutual recognition. Since the result depends considerably on the selected method, the definition of quality criteria of food according to the Codex Alimentarius is combined with a standardized method of analyses (e.g. International Standard) to be able to meet these requirements. Nevertheless, it should be mentioned that consensus is not easy to obtain and this usually leads to a rigid and slow evaluation of standards (Leclercq et al. 2000; Bulletin of IDF 395 2005).

Due to the lack in availability of suitable reference materials the capability of many microbiological methods has to be assessed by a standard substitute and/or inter-laboratory trial procedures (IDF 161A:1995). The fact that the components measured are only defined by the method itself a well-defined description of the method (SOP) has to be strictly followed. Only a world-wide application of the relevant International Standard enables comparability of the obtained results whereas all systematic deviations (bias) are defined to be set to Zero (Rösslein 1998; ISO/IEC Guide 98–3:2008).

The main purpose of method validation studies (or also known as method performance studies) is to establish the precision characteristics of the developed and optimized method when routinely applied. For this all collaborating laboratories have to be qualified to be competent and representative for laboratories which use the method in practice.

6.4.3 Problem to Solve

Bifidobacteria play an important role as probiotic microorganisms in fermented dairy products. Depending on the individual properties of the strains as well as on their density in the product and the survival conditions, several beneficial effects have been demonstrated for certain bifidobacteria in numerous studies. In this context, selective viable count and bacterial viability have been identified as key criteria for determining the quality of probiotic products. Usually, culture methods are used for routine examination of these products. As far as bifidobacterial enumeration is considered, several methodologies have been proposed during the last two decades, however, with limited applicability due to laborious procedures and/or lacking selectivity. On this account, a project team within the IDF "Lactic acid bacteria" group of experts has addressed this task and elaborated a procedure for bifidobacterial enumeration, which is easy to perform and of high selectivity (Zitz 2007, 2008; Zitz et al. 2007).

Furthermore a method validation study has to be organized and successfully conducted to be able to perform a sound statistical evaluation of the performance characteristics in practical application of the developed method. Finally this method has to be approved as ISO/IDF International Standard to be useful and valid to routinely enumerate bifidobacteria in milk products.

6.4.4 A Case Study: Standardization of a Quantitative Mikrobiological Method on Colony Count Techniques

6.4.4.1 Defining the Scope/Aim of the Method to Be Validated

Before starting with the development of the method the scope and the intended use of the method have to be defined preliminarily. This method should be routinely applied at quality control laboratories to enumerate commercial cultures of bifidobacteria to be present and viable in milk products (fermented as well as non-fermented) with comparable results at accepted precision. Concurrently, the growth of characteristic starter cultures (mostly lactic acid bacteria) usually used for the fermentation of fermented milk products has to be suppressed totally.

This project was done in the frame of a specific IDF working group (JAT—"lactic acid bacteria, E 104") which was established in particular for this purpose. Within this context the viable count of presumptive bifidobacteria expressed as CFU/g sample has been defined to be measured. Furthermore, the pour plate technique, a colony count technique at 37 °C which is easy and universal to apply, has been chosen. According to the FAO/WHO Codex Standard No. 243 labelled microorganisms (such as bifidobacteria in probiotics) have to be present in fermented milk products with a concentration of at least 10^6 CFU/g product total. This has to be valid up to the date of minimum durability to be able to guarantee the claimed health effects of the probiotic dairy product (Bulletin of the IDF 352, 2000; FAO/WHO Experts Report, 2001; Bulletin of the IDF 380, 2003; EC 1924/2006 "health claims"). Based on the literature several basic media as well as selective supplements were evaluated within some collaborative IDF-studies to be able to decide, whether this is suitable for the intended purpose or not. In parallel the developed method was characterized as well as optimized in many in-house studies at BOKU University, Vienna.

6.4.4.2 Development, Characterization and Optimization of the Method

6.4.4.2.1 Collaborative Trials

6.4.4.2.1.1 Qualification of Mupirocin as Selective Supplement

The accompanying microflora is a criterion of crucial relevance and therefore various attempts to develop a suitable cultural method for bifidobacterial enumeration have failed so far (Pacher and Kneifel 1996; IDF-"Bifido"-Guideline 1999). This microflora usually exists in high amounts in fermented milk products and also prefers similar growth requirements. It was not before the study of Rada and Koc (2000), that Mupirocin (MUP) was discovered as selective agent suppressing the growth of commonly used lactic acid bacteria. Since this antibiotic did not influence bifidobacterial growth the selectivity of MUP was confirmed subsequently in a series of tests carried out by eight laboratories of the IDF working group in 2003.

Experiences were collected on the growth properties of a representative number of bifidobacterial reference strains and isolates. In addition, the growth performance of the typical yogurt microflora (*S. thermophilus, L. delbrueckii* spp. *bulgaricus*) as well as of mesophilic cultures (e.g. *L. lactis*), and of *L. acidophilus, L. casei* and *L. rhamnosus* were examined. As a result of these trials, the selectivity of MUP as a supplement could be confirmed.

6.4.4.2.1.2 Qualification of TOS-Propionat Agar

Although most of the bifidobacterial strains grew on TPY (Scardovy 1986), parallel efforts were successfully made by the Japanese members of the action team to investigate the potential of TOS (transgalactosylated oligosaccharide mixture) propionate agar (Expert Group on Selective Enumeration of Bifidobacteria 2000). The ability of growth of eight strains of bifidobacteria to be stimulated by the oligosaccharide TOS has been described by Tanaka et al. (1983). A further collaborative IDF-trial was carried out examining three commercially available probiotic yogurt products by six laboratories used the TPY-MUP agar in comparison with the TOS-MUP agar. The trial was repeated on the day the examination was conducted. To be able to statistically assure the conclusions resulted of this media comparison, the study plan for the determination of the analyst performance developed by Weiss and Dahms (1993) and later described in ISO 14461-1 was firstly applied within this context. It could be shown that all participating laboratories were qualified in applying the pour plate technique on a comparable quality level. Both the Bland-Altman plot (Altman and Bland 1983; Bland and Altman 1996) shown in Fig. 6.22

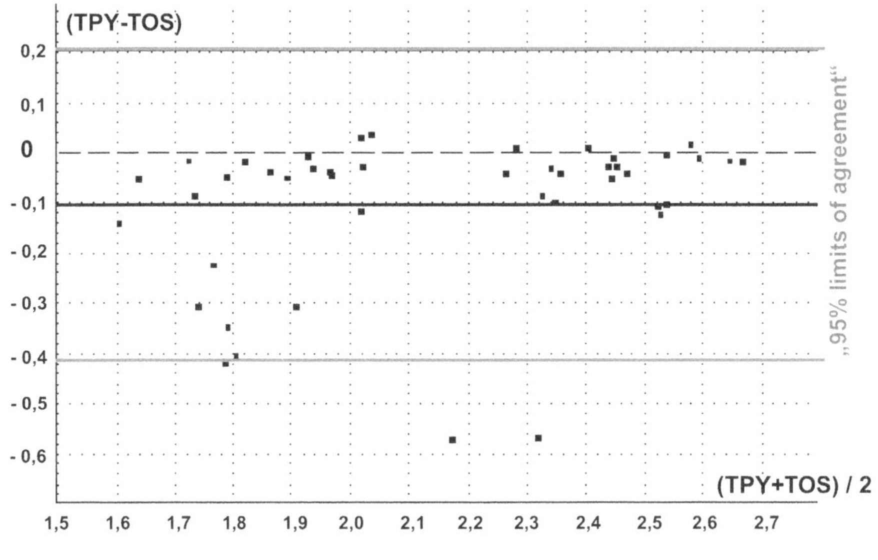

Fig. 6.22 Comparison of the TPY medium versus the TOS medium (Bland-Altman-plot)

as well as the scatter plot clearly have shown, that the results of the colony counts on TOS were significantly higher than those results obtained on TPY. After subtraction of the results of "TPY-TOS" 39 of 44 differences have been found negative which can be also expressed as 72–92% ($p = 95\%$) of growth on TPY-MUP in comparison to TOS-MUP. The Bland-Altman plot, where the difference of uniform results of two methods are plotted to their means, were found to be very appropriate to graphically illustrate systematic effects, provided that these methods possess comparable variability of the results.

6.4.4.2.2 In-House Studies

6.4.4.2.2.1 Influences of the Working Quality

Based on the proven selectivity of MUP as well as on the better qualification of TOS the developed method has been characterized and optimized on the basis of the TOS-MUP medium within the internal studies. The influence of the working quality has been shown to be of crucial relevance when applying this method. Therefore securing the analyst performance according to ISO 14461 is a prerequisite to be able to compare results. This biometric approach mainly consists of a two step quality system for standardized microbiological colony count procedures to evaluate the methodological variation on the basis of the Poisson distribution. The assessment of the analyst performance was done by applying two G^2-tests (Likelihood ratio test; for testing the homogeneity of the plating as well as for testing the overall homogeneity of all counts) combined with a hierarchical variance analysis to evaluate qualitatively as well as quantitatively methodological variabilities at the classical determination of viable counts. Besides of the interpretation of the occurred total variability, which should not exceed the value of 1, different methodological potential error sources can be recognized to be systematically influencing the working quality. Within this context this approach was also used to confirm the pour plate technique as being more precise and more robust than the surface spread technique. Hence it was decided to continue the method characterization studies with the pour plate technique.

6.4.4.2.2.2 Investigation of Additional Influencing Factors

For the investigation of the methodological influences on the obtained results the entire process of analysis has been systematically fragmented reflecting all potential effects as shown in Fig. 6.23 by using the Ishikawa diagram (cited at Zollodz 2001). Based on the "IDF-Bifido"-Guideline (1999) as well as on experience made during the development phase, consequentially, the most probable effects have been selected to be analysed by a ruggedness test of Youden conducted as 2^{5-2} study plan (Youden and Steiner 1975).

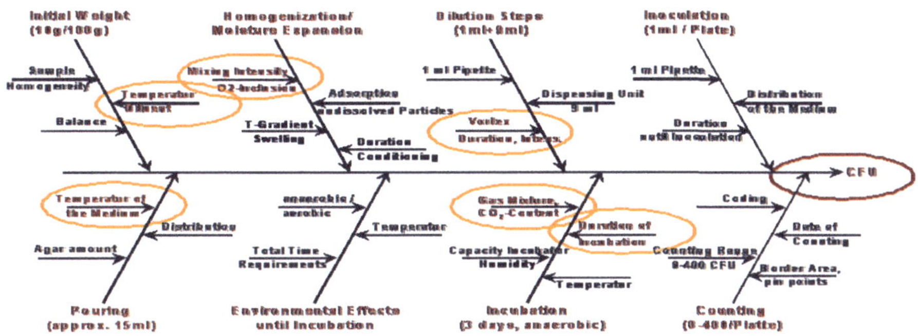

Fig. 6.23 Ishikawa-diagram to visualize potential methodological effects

The main effects, which have been chosen for the factorial design, are:

- Temperature depending influences as the temperature of the diluent when preparing the initial suspension or the temperature of the medium when pouring the plates
- The intensity of mixing which is a parameter for the O_2-incoorperation while homogenization
- The conditions of the anaerobic incubation of the plates

Consequentially, the following factor levels were set for the fractional design:

5 experimental factors (A-E)	Min.:−	Max.:+
A. Pouring temperature	47 °C	49 °C
B. Incubator or Jars	Incubator	Jars (non cont.)
C. Duration incubation	69 h	75 h
D. Temperature diluent	15 °C	25 °C
E. Mixing intensity	10x/2 s	20x/3 s
Response: CFU/g		

Hence the following plan-matrix (Table 6.25) was defined to later obtain the following results:

According to the factorial analysis non of the investigated factors shows a significant effect on the CFU-results. Under critical consideration this result could be distorted due to a higher error impact ($s_r = 0.092$ log CFU) which is typical for microbiological methods and have probably covered significant effects. Moreover this effect might be intensified by the working load of the method which has led to a limitation of the number of replicates. Finally the presence of influences mainly caused by temperature dependent effects could be confirmed by performing some selective trials made on the basis of paired comparisons. Therefore all methodological conditions relevant to be regulated have been clearly defined by the standard operation procedure.

Table 6.25 Plan-matrix for the 2^{5-2} study plan including also the results of the study

No.	A: T Pouring	B: Incub/Jars	C: h Incubation	D: T Diluent	E: Mixing	logCFU/mL
1	−1	−1	−1	1	−1	2.249877
2	1	−1	−1	−1	1	2.089279
3	−1	1	−1	−1	−1	2.280329
4	1	1	−1	1	1	2.190205
5a	1	1	−1	1	1	2.221476
6	−1	−1	1	1	1	2.220281
7	1	−1	1	−1	−1	2.222631
8a	1	−1	1	−1	−1	2.092453
9	−1	1	1	−1	1	2.150577
10	1	1	1	1	-1	2.205909

[a] Replication to evaluate the experimental error

6.4.4.2.2.3 Performance Characteristics of the Method

Based on the results of several in-house trials the performance characteristics of repeatability and within laboratory reproducibility could be defined for different probiotic yogurt-products (firm and liquid products) as well as for a probiotic infant milk formula available as powder product. Within the entire working range linearity could be proven by using the ISO 14461 approach. Based on the variation coefficient of the CFU's the best fitted counting range could be defined as from 200 to 400 CFU/ plate by considering the cultures investigated. According to Lightfoot and Maier (2002) an improvement can only be achieved with an increase of the particle counts, since low particle concentrations already lead to considerable random variations. Even when the colonies can clearly be separated a routine use requires an upper limitation of the plate counts. Furthermore mass effects have been described, whereas the growth of adjacent colonies were negatively influenced (Niemelä 1965).

The standard operation procedure (SOP) is then defined as simple, comprehensible and as clearly arranged as possible.

Finally the following conditions have been qualified to be best fitted for this field of application:

Determination of the viable count on the basis of the Koch's pour plate technique by using the elective TOS-propionate agar together with Li-Mupirocin to be used as selective supplement and anaerobic incubation at 37 °C for 72 h (see also Fig. 6.24).

6.4.5 Method Validation Study as Collaborative Study

After several phases of testing, evaluation and modification, the best qualified method was optimised as well as characterised before starting the validation study. For this all effects which are significantly influencing the results have been identified and a procedure has been defined to control these influences to be "under statistical

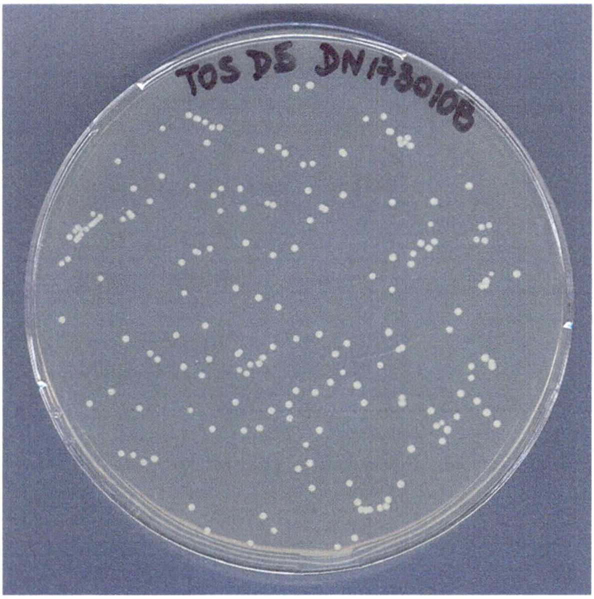

Fig. 6.24 Colonies of *B. animalis* subsp. *lactis* on TOS-MUP agar

control". Furthermore standardized quality assurance techniques for microbiological laboratories have been considered to be essential. Only when the analyst performance is qualified to be on a constant and well accepted level, valid and reliable results are ensured. By processing this final international collaborative study, the capability of the method in practical application was examined to approve this method as ISO/IDF Standard.

6.4.5.1 Planning of the Study

According to IDF 135B:1991 the following definition apply: "A collaborative study is an inter-laboratory study in which each laboratory uses the defined method of analysis to analyse identical portions of homogenous materials to assess the performance characteristics obtained for that method of analyses" (Fig. 6.25).

One of the most important steps during the planning phase was to create an applicable study design (Fig. 6.26) with realizable time frames.

The method validation study was structured into several parts. The recruitment and the qualification of each of the participants, the investigation of the probiotic powder samples as well as the yogurt-products and finally the statistical analysis of the data and the defining of the provisional standard draft were some of these steps. The most important part was certainly the preparatory phase of the validation study, which comprised the optimization of the method, the definition of the SOP, the selection of the samples to be prepared for delivery as well as the sequence planning of the study including both, the statistical study plan and the time schedule.

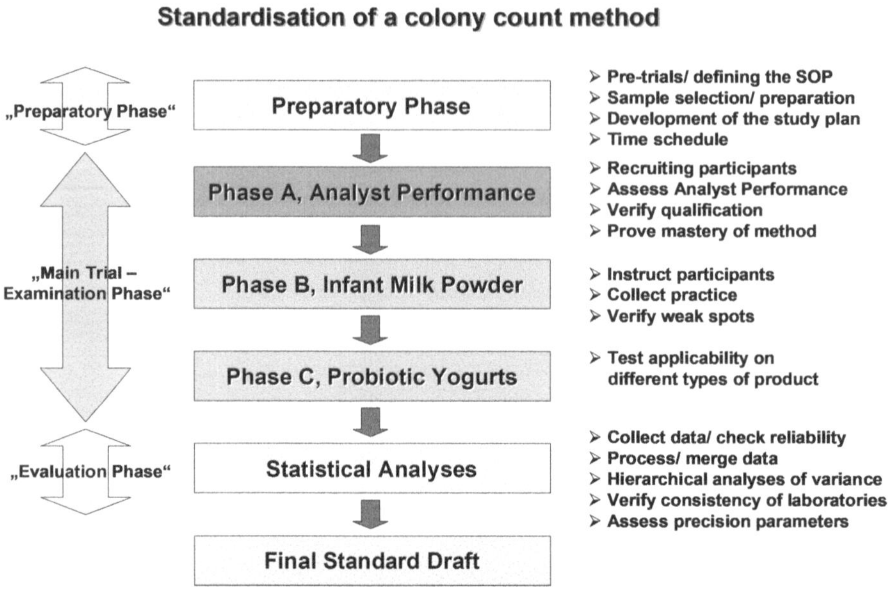

Fig. 6.25 Overview on the method validation study

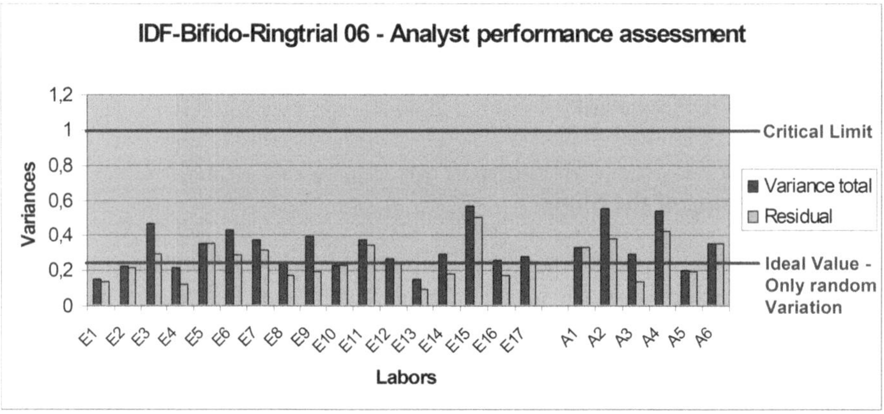

Fig. 6.26 Results of the Phase A of all participants according to ISO 14461/IDF 169

6.4.5.2 Selection of the Participants as well as Samples

A total of 21 laboratories have contributed to the collaborative trial to provide 23 valid data sets. The fact that each of these data sets was obtained by a different person performing the analysis independently. Hence in the following the total number of laboratories was defined as 23. Depending on technical reasons the investigation of the yogurt samples was conducted within two defined centres (Europe and Asia). Three different types of probiotic yogurt samples each selected to be representative for this scope as well as sufficiently homogenous and stable to assure comparability of the results were analysed. Additionally an infant milk formula containing bifidobacteria was analysed by all participants.

6.4.5.3 Qualification of the Laboratories

A special experimental design and appropriate statistical analysis (GLP-test) have been applied (ISO 14461), to be able to assess the analyst performance of the analysts when applying the pour plate technique. The objective is to quantify significant errors which are independent of natural random effects in colony count techniques. The attempt is also made not only to detect possible defects but also to suggest how to enhance the quality-level of the relevant analyst performance.

Furthermore, this test can be used to assess whether the analyst is well-trained in performing this technique, which proves the qualification of the involved laboratory personal.

Statistical analyses of the data from Phase A were performed by using a software programme developed by the Institute for Biometrics and Data Processing (http://web.fu-berlin.de/glp-analyse/).

Only when qualification of the participating laboratories could be proven, the respective qualified laboratory was selected. The results of the qualification of the participants of this study as shown in Fig. 6.26 were comparable on a good level.

The overall variability of the obtained results were positioned within the range of ≥0.25 up to 1 defined as acceptable. The ideal value of 0.25 is caused by the random error of the Poisson's distribution, the critical limit of 1 should not be exceeded because in that case it can be assumed that serious deviations or failures in conducting the pour plate method really exist. Moreover, it was found that even at results close to the critical limit of 1 failure in conducting the pour plate technique have been identified so that GLP-conformity could not be assumed. Therefore, this critical limit should be better reduced to a lower level.

Based on that all results were clearly below the critical value of 1, the qualification of all participants could be proven. Furthermore, they all were more or less on a comparable level so that it can be assumed that the working quality would not influence the analytical results of the study.

At two concrete examples improvement of the qualification of performance of the analysts involved could be demonstrated despite of the weak results initially obtained. Based on the hierarchical study plan error sources could be localized to be jointly investigated and at the end to be successfully improved. As shown in Fig. 6.27 individual operations which were incorrectly performed as well as some confusion in coding of the plates could be identified as sources of the exorbitant variations. As a consequence, more attention has to be drawn to the mixing procedure to be automated during inoculation to avoid unreasonable deviations in the future.

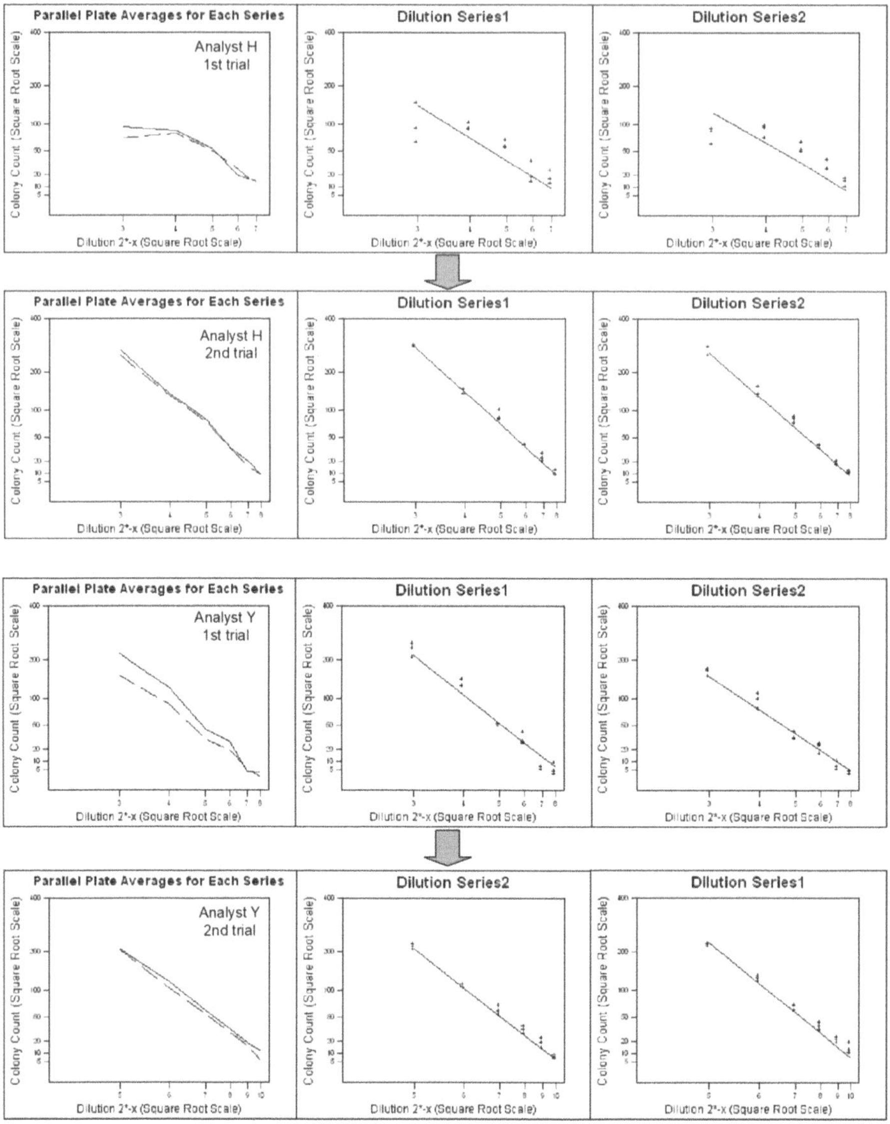

Fig. 6.27 Two examples for the successful improvement of the analyst performance

6.4.5.4 Preparation of the Samples

Identical sample units when the analysis is performed have to be guaranteed to reliably determine the precision parameter of the method (ISO 5725-1, ISO 5725-2, IDF 135B:1991). This is absolutely required to assure comparability of the data.

Furthermore, the food and target micro-organisms as well as the representative background micro flora or interfering components have to be capable of being homogenised and should remain stable both in transit and for the duration of the analyses (ISO 16140). Especially the transport of the yogurt samples has been considered as critical, because of their limited shelf life guaranteed under refrigerated conditions. Therefore, homogeneity and stability studies have been conducted prior to dispatching test materials from the organizing laboratory (BOKU University, Vienna). Additionally with a transport simulation study the influence of transport conditions on the CFU-results has been investigated and the best combination of the selected samples (size of units, total number of samples, ..) together with the most effective materials (dimension and material of the transport boxes, cooling units) have been chosen to protect the samples against variations on the transport.

Based on the findings the transport of the yogurt samples was done under defined and comparably cool conditions within each of the defined centres. Transport time was strictly limited to avoid considerable changes of the product quality, the maximum time contingent, where constant temperature conditions can be assured, were determined and reported in advance by using a data logger. Finally the two regional centres (Europe and Japan) performed Phase C in parallel with separated organization of each of the trials.

6.4.5.5 Study Design, Statistical Approach

The selection of the best suited study design of the experiment was done by balancing of the statistical power to the total working load to be able to find a useful compromise. Special attention has to be drawn on realizable dimensions of the work to be done by each of the laboratories within 1 day under conditions as used in routine. To make all occurring deviations visible and moreover to be able to assign certain amounts of individual contribution to certain working steps, a hierarchical structure was chosen (Fig. 6.28). This structure has been shown to be best suited, since abnormalities of some of the datasets occurred, a variance component analysis could been additionally made to detect possible error sources. Furthermore this structure was used to calculate the precision parameters (repeatability, r, and reproducibility, R) according to the well known equation "$s^2_R = s^2_{Lab} + s^2_r$" after reducing the hierarchical design to two levels (Laboratories and Analysis). This was achieved by defining the "first values" of the analysis corresponding to the relevant laboratories. Since the Wilcoxon-Matched-Pairs test had shown no differences to the "second", the "third" and the "fourth" values, the selection of the "first values" could be supported.

6.4.5.6 SOP's and Other Useful Instructions

Working documents and instructions have been regarded as critical when carrying out studies. To provide documents which are as comprehensively and as well

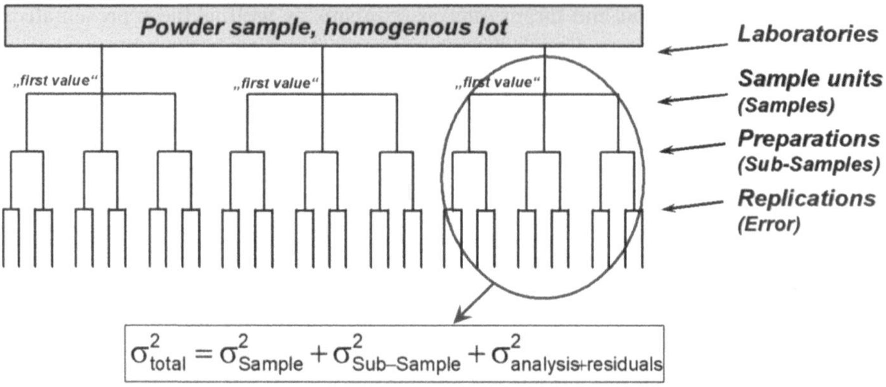

Fig. 6.28 Hierarchical study plan exemplarily shown for the analysis of the powder samples of Phase B—"first value" symbolizes the reduced dataset by using only the first result of analysis/sample unit

arranged as possible are the main target to avoid deviations from the defined procedure. Hence graphical illustrations as well as briefly listed and chronological instructions were applied by focusing good reproducibility (traceability) of the procedure even also from the point of view of persons being not so deeply involved in the developing process of the method. To obtain independent data coding of the plates from all dilution steps which are within the countable range according to a defined randomization plan have been done just before counting.

6.4.5.7 Procedure for the Data Assessment Suited for This Purpose

Another critical point is to apply suitable procedures to prepare the raw data for the statistical analysis as to select appropriate statistical procedures itself. For this, the following approach can be recommended:

1. Verification of the plausibility of the collected raw data followed by determining of the reliability of colony counts of parallel plates and subsequent dilution steps according to ISO 14461-2/IDF 169-2
2. Merging and structuring of the valid data material after calculation of the logarithms of the weighted arithmetical mean values of CFU's counted for all dilution steps within the countable range. For this the weighted arithmetical mean values were calculated according to Farmiloe et al. (1954) followed by the log10-transformation which was done for approximation to normality (IDF 161A:1995).
3. Graphical evaluation of all of the valid data reflecting the different variations possibly occur within and/or between all participating laboratories by using the graphical consistency technique (ISO 5725-2; Weiss 2005; Wilrich and Jarvis 2008; Amendment to ISO 16140). This can be combined with the analysis of

abnormalities within data sets to allocate possible error sources to be identified in close collaboration with the involved persons (Zitz 2007).

4. Robust statistical approaches (Wilrich and Jarvis 2008; Amendment to ISO 16140) based on the calculation of the Rousseouw's scale estimator Q_n should be preferably applied to be used for the data analyses. Thus the influence of outlier data can be avoided until a percentage rate of about 25% outlier total, whereas these outlier data may remain in the data material and also be used for the statistical analysis (Wilrich 2007). The simulation of the conditions of a routine usage should also be taken into account, e.g. by including the "first values" only.

Within this study preferably scatter diagrams and the graphical consistency technique performed in the classical approach (ISO 5725-2) as well as the robust approach (Wilrich and Jarvis 2008; Amendment to ISO 16140) were used as graphical illustration of the data to be compared with. This enables that data can be more clearly represented to be able to assess the obtained data rapidly.

6.4.5.8 Results of the Method Validation Study

6.4.5.8.1 Evaluation of the Results from Phase B

This phase of the ring-trial was performed by all 23 laboratories according to Fig. 6.28, where sample units originated from a defined and homogenous lot of an infant milk formula were analysed in parallel. Based on the good stability of this probiotic product the sample shipment did not need special requirements. The raw data in randomized order were decoded and verified according to the ISO Standard 14461-2/IDF 169-2. Then the reliability of the colony counts of parallel plates as well as of subsequent ten-fold dilution steps were assessed. Data exceeding the limits of agreement were excluded before the calculation of the logarithms of the weighted arithmetical mean values.

Firstly, the evaluation of the quality of the individual data sets were made by applying the graphical consistency technique to express the performance of the laboratories between and within each of the laboratories. The Mandel's k statistic shows the differences between the repeatability standard deviation of the laboratories which expresses the within consistency whereas the Mandel's h statistic demonstrates the differences between the mean of each laboratory compared with the overall mean in relation to the standard deviation between the laboratories also known as the between consistency (ISO 5725-2; Wilrich and Jarvis 2008, Amendment to ISO 16140). In Fig. 6.29 the variability of the mean values is shown with reference to all participating laboratories. The Mandel's h and k statistics indicates that only some laboratories produced some deviating results out of the tolerance limits. Possible sources of error were located by using hierarchical variance analyses of noticeable data sets (Table 6.26) to provide important details for the followed fact finding. Based on this method, non-conformities could be identified and subsequently discussed with the laboratories.

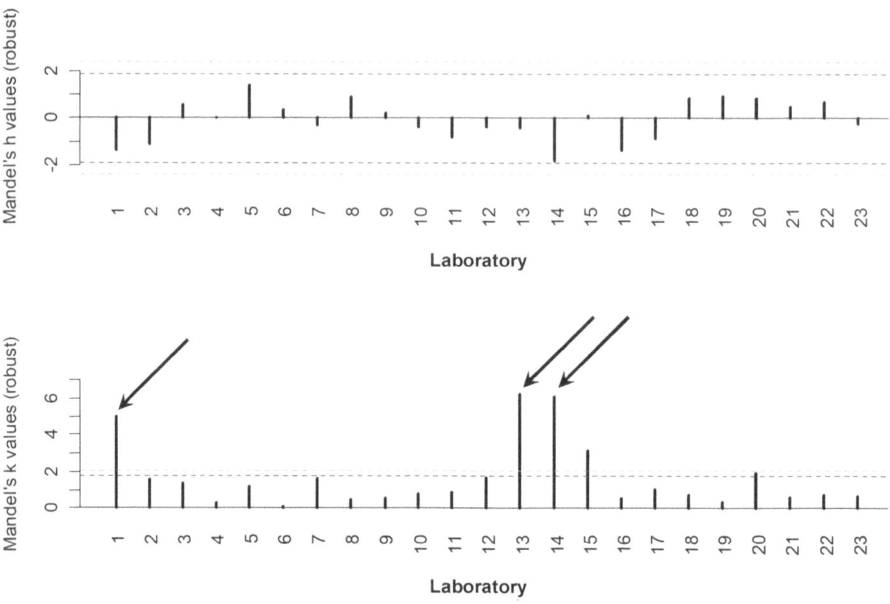

Fig. 6.29 Results of Phase B—Mandel-*h*- and Mandel-*k*-statistic according to robust analysis (Wilrich and Jarvis 2008)—higher variances are marked with errors

Table 6.26 Results of the variance component analysis of Phase B

	1		13		14		15		6	
Laboratory	Variance	%	Variance	%	Variance	%	Variance	%	Variance	%
Sample	0.038	66.2	0.071	93	0.069	92.6	0.018	88.1	0.000	0
Sub-sample	0.016	27.7	0.003	4.3	0.000	0	0.000	0	0.000	0
Error	0.004	6.1	0.002	2.7	0.006	7.4	0.002	11.9	0.002	100

By considering the pattern of Figs. 6.29 and 6.30 it is evident that the two laboratories (13 and 14) were differing from the others (expressed as striking Mandel's *k* values in Fig. 6.29 as well as higher standard deviations in Fig. 6.30). However, according to the Mandel's *h* values in Fig. 6.29 these two laboratories fitted well into the community that consistency between the laboratories can be considered. Marked deviations were also observed with the results from Lab 1.

In Table 6.26, the results of the variance component analysis of that laboratories with significant deviations are shown in comparison to Lab 6, where definitely no additional deviations occurred.

Variances due to the "sample" component were identified to be highest at laboratory 13. Since homogeneity was proven this was not caused by the sample material itself. The reason was more likely resulting from deviations in the sample

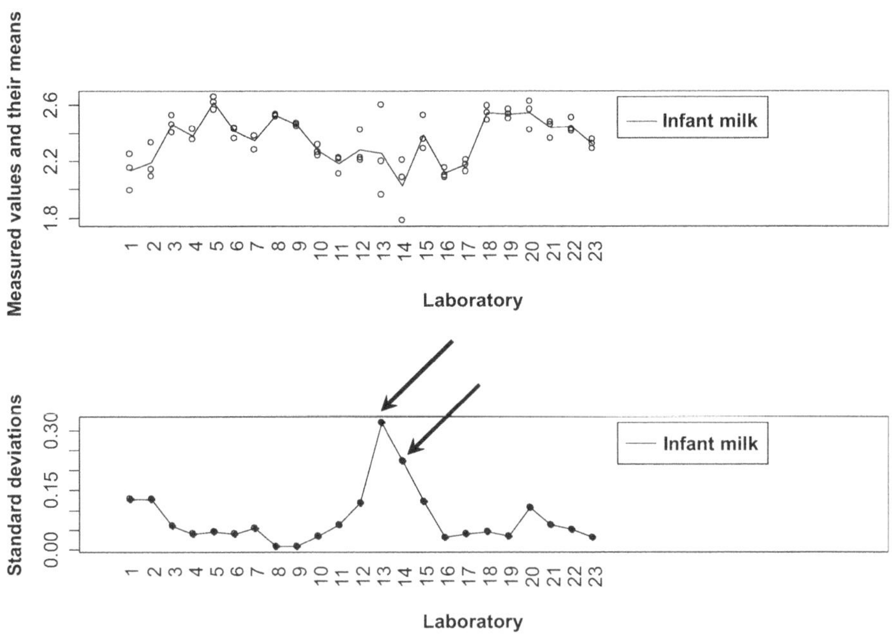

Fig. 6.30 Results and corresponding standard deviations of Phase B with reference to all laboratories—infant milk formula bifidobacterial counts [log CFU/g]—higher deviations are marked with arrows

treatment, e.g., during the preparation of sub-samples. In contrast, Laboratory 6 had variations within the parallel plate components only, which is also visualized by the calculated random error. The low "error-value" of 0.002 of both laboratories demonstrates that these laboratories are well-trained in applying this technique. This was in agreement with the results of the analyst performance assessment.

Based on these findings, the following reasons were identified in cooperation with the involved laboratories:

Lab 1: Different treatment of the two sub-samples had probably led to the marked deviation at Lab 1

Lab 13 and Lab 14: Deviating from the original protocol, the trials in Lab 13 were carried out on two subsequent days. Due to similarities of the results of Lab 13 and Lab 14 obviously, the same explanation might have been valid for laboratory 14 but could not been verified.

The results as measured values and their associated standard deviations are shown in Fig. 6.30, whereas also higher variabilities are clearly evident at Lab 13 as well as Lab 14. Using only the firstly measured values of the whole approach, routine conditions could have been reflected.

6.4.5.8.2 Evaluation of the Results from Phase C

Phase C was split into two sub-trials performed under analogous conditions at two centers analyzing different yogurt products of local significance. The evaluation of the raw data and the statistical analysis were carried out analog to phase B at the European centre. Again the graphical consistency technique was applied to evaluate the performance of the laboratories (Fig. 6.31), where almost all laboratories were within the given limits and exhibited comparable patterns of results. Findings of laboratories 2 and 23 at the Mandel's *h* statistic were probably caused by systematic deviations and not by additional within-variances.

Figure 6.32 shows the results of the yogurt samples investigated at the European (Laboratories 1–17) as well as the Japanese sub-trial (Laboratories 18–23) of Phase C. Some noticeable values could be sporadically identified at some of the dataset of certain products which were additionally examined by variance component analysis.

Again different treatment of the sub-sample could be identified causing higher deviations of individual results of yogurt Nr. 2 in Laboratory 1. At the Japanese sub-trial higher error components especially at Laboratory 21 and 22 could be attributed to inhomogeneities of the selected lot. Thus these results were excluded from the calculation of the precisions data.

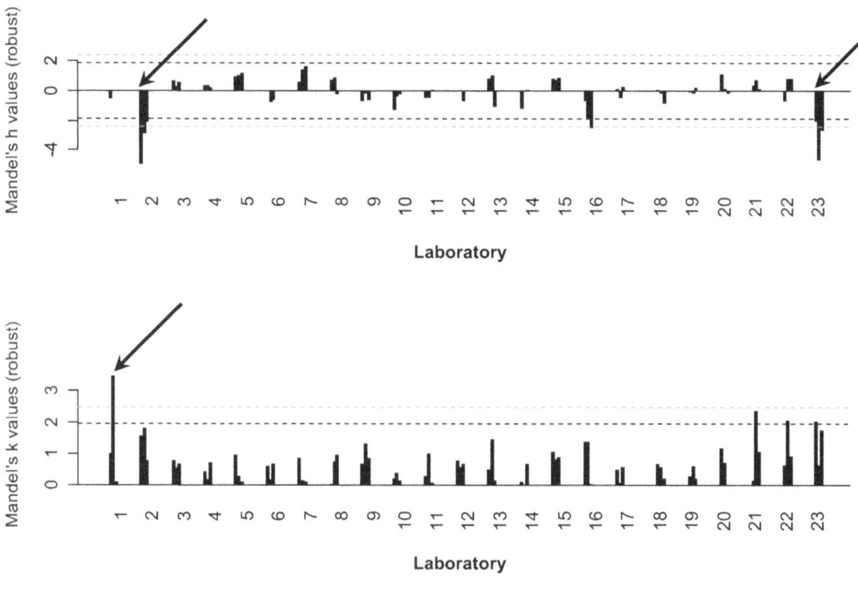

Fig. 6.31 Results of Phase C—Mandel-h- and Mandel-k-statistic according to robust analysis (Wilrich and Jarvis 2008)—higher variances are marked with errors

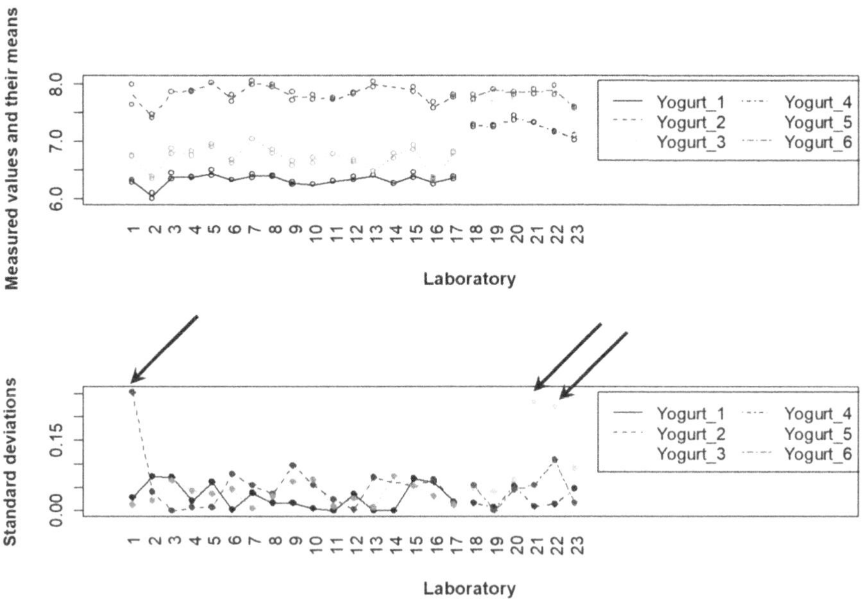

Fig. 6.32 Results and corresponding standard deviations of Phase C with reference to all laboratories—bifidobacterial counts [log CFU/g] of infant milk formula—higher deviations are marked with arrows

6.4.5.8.3 Precision Parameters

Taking into account the results from phases B and C, the standard deviations of the repeatability and of the reproducibility were calculated according to the classical (ISO 5725-2) as well as the robust analysis approach (Wilrich and Jarvis 2008). Figure 6.33 illustrates the obtained precision parameters (classical, robust) with respect to the corresponding mean values of the CFU's from the examined products expressed as repeatability as well as reproducibility. By comparing the classical with the robust approach the calculated values differ not so much, in the most of the cases the robust results are slightly higher than the classical results. The fact that the robust results correspond better to the reality the precision parameters were defined on the basis of the robust approach.

Based on similarities of some of the products (liquid or firm types of products and the solid product as powder type) the precision characteristics were merged to define precision data for these types of products. Therefore, the following parameters could be derived for the group of drink-yogurts (liquid products), the more creamy or jelly-like yogurts (firm products) and the spray-dried infant milk formula (powder products) listed in Tables 6.27 and 6.28:

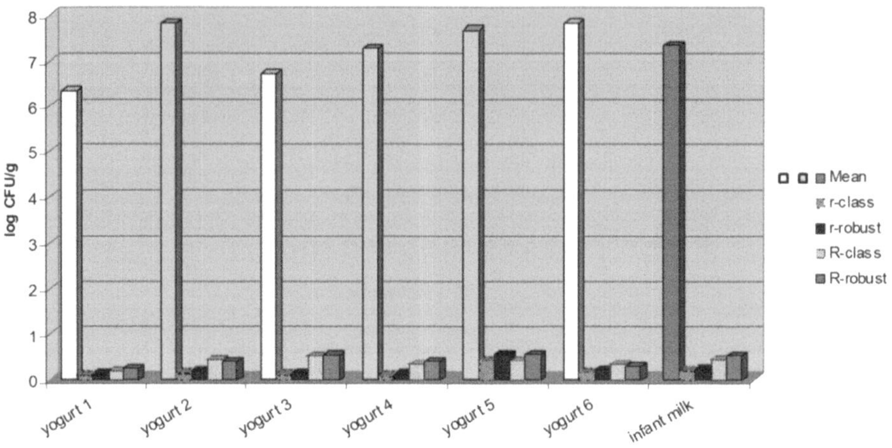

Fig. 6.33 Precision data (repeatability, reproducibility)—classical and robust—of the tested probiotic products with respect to the corresponding mean values of the colony counts [log CFU/g]—white and gray colums of the mean values represent the liquid and firm yogurt type, respectively, and dark gray columns represent the infant milk sample

Table 6.27 Precision data (robust) for defined product groups on the basis of the results firstly measured

Type	S_r collective	S_r collective	r (log CFU/g)	R (log CFU/g)
Liquid	0.042	0.139	0.12	0.39
Firm	0.071	0.144	0.20	0.40
Powder	0.079	0.189	0.22	0.53

Table 6.28 Precision data (robust) for defined product groups on the basis of all of the results measured

Type	S_r collective	S_R collective	r (log CFU/g)	R (log CFU/g)
Liquid	0.046	0.132	0.13	0.37
Firm	0.064	0.139	0.18	0.39
Powder	0.074	0.197	0.21	0.55

Moreover, it was tried to calculate precision data derived from all of the measured results also taking into account conditions of the routine usage. For this the mean value were built from the precision data obtained from all of the first results, the second results, the third and the fourth results (Zitz 2007).

These data can be considered as comparable to those primarily obtained in Table 6.27. The fact that a broader community is represented this data should be considered as more reliable and valid to be used to express the performance characteristics of this method. An overview on the results of the interlaboratory trial is also given in Table 6.36 (Appendices).

6.4.5.8.4 Preparation of the Preliminary Draft-Standard

With this international method validation study best applicability of the method has been proven and valuable precision parameters (reproducibility, reproducibility) have been obtained to be valid for the practical use. Therefore the experiences and the results of this approach have essentially contributed to prepare the Draft of the International Standard which had undergone several evaluation phases to be commented by international experts and was released in 2009 (ISO/FIS IDF/IDF 220 2009).

6.4.6 Summary and Conclusions

It could be shown that preparation as well as planning constitutes the most critical parts in conducting studies with this dimension and is crucial for a successful outcome. Especially the characterization as well as the optimization of the method has to be sufficient and well performed by the organizing laboratory prior to the validation study. The influence of each step of the procedure should have been carefully checked, and factors that may influence the final result have to be identified for the user of the method. Therefore a description of the method has to be prepared which is as comprehensive and clear as possible. All critical steps should have been pointed out and procedures have to be defined to avoid unintended influences to the results. Moreover the procedure should be consistent to the relevant Standards with reference to the related ISO Directives. Validation studies should only be conducted with methods under statistical control and sufficiently robust procedures.

For comparability reasons the qualification of the participants to be trained in applying the technique is another point of crucial relevance which has to be proven as well as the securing of a standardized and equal treatment of the sample-units to be investigated.

During the realization phase of the validation study best information of the participants is needed and close contact to get sufficient feed-back must be maintained. Coding and randomized counting of the plates to obtain independent data as well as the controlling of the critical steps, e.g. the sample shipment, is also required. As visualized in Fig. 6.34 the most of the working load is located in the preparation, the organization and the well timed material providing of the project. Project monitoring was one of the main topics of the organizing laboratory to achieve smooth collaboration during the entire project. The main phase including the real sample analysis (Phase B, Phase C) illustrated as cascade located in the centre of the schedule constitutes only a small part of the study.

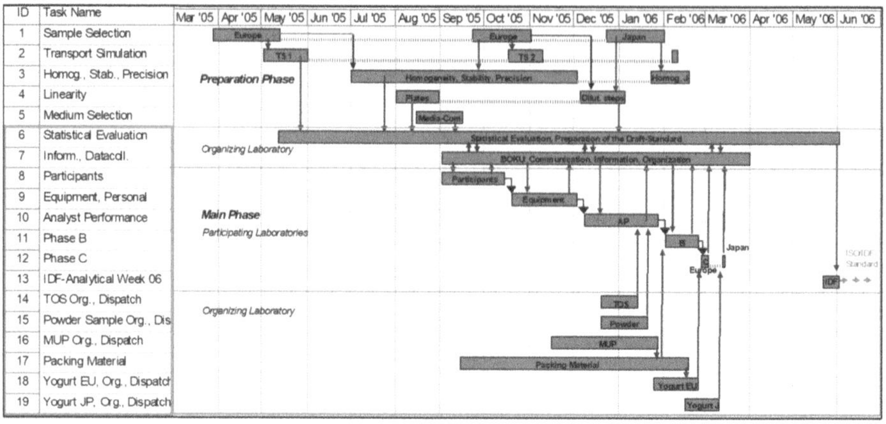

Fig. 6.34 Project schedule of the whole method validation study

An essential criterion to be successful during the project was the involvement and conviction of all concerned parties, especially to win each of them by focusing on their personal benefits. Therefore motivation is being considered as to be the most effective factor to improve the impact as well as the quality of the contributions. This includes to give feedback, which has to be well timed, to let each of them take part in the success and also to jointly find explanations with frankness and trust. On the collective basis all the intensive efforts finally have been turned to account with the high quality of the obtained data.

This was the reason that in this case the statistical evaluation based on robust methods was found to be not necessary, however these statistical approaches are extremely valuable to be usually used for applications especially in this field.

The International Standard was finally released in 2010 (IIDF 220/ IDF 220: 2010).

Finally it should be mentioned that this study was the first method validation study within this microbiological area, where a successful outcome with parameters on a high quality level could be achieved. This study especially points out critical steps during the preparation, the organization and the performance of international validation studies as well as demonstrates the importance of detailed planning and project monitoring. Hence, the presented study shall be used as a kind of roadmap within this field of application to be successfully applied in future.

6.5 Measurement Uncertainty of Chemical Test Results

Stefano Martin and Sara Briguglio

6.5.1 Introduction

The new ISO 17025: 2017, "General requirements for the competence of testing and calibration laboratories" reports that "The laboratory shall use appropriate methods and procedures for all laboratory activities and, where appropriate, for evaluation of the measurement uncertainty as well as statistical techniques for analysis of data."

The most likely or 'true' value may thus be considered as the measured value including a statement of uncertainty which characterizes the dispersion of possible measured values. As the measured value and its uncertainty component are at best only estimates, it follows that the true value is indeterminate. Uncertainty is caused by a combination of "errors" which create dispersion around the estimated value of the measurand, the smaller the dispersion, the smaller the uncertainty.

By definition, the term error (or measurement error) is the difference between the true value and the measured value.

Furthermore, the ISO 15189:2022, states in Sect. 7.3.4 that: "The measurement uncertainty of measured quantity values shall be evaluated and maintained for its intended use, where relevant. The measurement uncertainty shall be compared against performance specification and documented. The measurement uncertainty information shall be made available to laboratory users on request.

So, the uncertainty provides an interval within which the value of the measure and can be found with a stated probability. Practically, uncertainty tell us how good the result and it is used for instance in calibration or in verifying the results of a test.

The standard uncertainty may be thought of as equivalent to 'one standard deviation', but we may wish to have an overall uncertainty stated at another level of confidence, e.g. 95 percent. This re-scaling can be done using a *coverage factor, k*

.

A particular value of coverage factor gives a particular confidence level for the expanded uncertainty. Most commonly, we scale the overall uncertainty by using the coverage factor $k = 2$, to give a level of confidence of approximately 95 percent ($k = 2$ is correct if the combined standard uncertainty is normally distributed).

Some other coverage factors (for a normal distribution) are:

k: 1 for a confidence level of approximately 68 percent
k: 2 .58 for a confidence level of 99 percent
k: 3 for a confidence level of 99.7 percent

$$U = K\,U =$$

$$k = \text{coverage factor, usually } 2$$

In the next paragraphs will see how to calculate the "uncertainty of measurement" (MU) in a chemical testing laboratory.

6.5.1.1 Approaches to Estimate Measurement Uncertainty

There are a few different approaches to estimate measurement uncertainty (MU) but the most popular are the bottom-up or top-down calculations.

The bottom-up approach described in the ISO/IEC Guide of the Expression of Uncertainty in Measurement", commonly called GUM, combines all the uncertainties associated with all individual operations of an analytical procedure and then calculate the combined uncertainty that is multiplied by a coverage factor (95% confidence range) to obtain the expanded uncertainty.

For guidance for both approach and numerous worked examples, refer to the published the QUAM-guide (Quantifying Uncertainty in Analytical Measurement), which was revised in 2000 and 2012. This guide was written to help analytical chemists to use the approach described in GUM, including examples of analytical chemistry.

Measurement uncertainty comprises, in general, many components. Some of these may be evaluated by Type A evaluation from the probability distribution of the values from series of measurements and can be characterized by standard deviations. The other components, which may be evaluated by Type B evaluation, can also be characterized by standard deviations, evaluated from probability distributions based on experience or other information. Practically, Type B evaluation of a component of measurement uncertainty is determined by means other than a Type A evaluation.

6.5.2 Objectives and Learning Outcomes

After completion of this chapter the reader will be able to calculate the uncertainty associated to an analytical result obtained in a chemical analysis.

6.5.3 Problem to Solve

Estimate the uncertainty associated to an analytical result by using different methodologies in two practical examples: Estimation of the measurement uncertainty in the determination of ochratoxin A in cereals by the holistic criterion and Quassin in food by the metrological criterion.

6.5.4 Procedures

6.5.4.1 Ochratoxin in Cereals

To determine the amount of ochratoxin A present in cereals, the standardized method ISO 15141: 2018—"Cereals and cereal products—Determination of ochratoxin A—High performance liquid chromatographic method with immunoaffinity column cleanup and fluorescence detection", was used. The analytical procedure involves numerous steps: an extraction of the mycotoxin with a suitable organic solvent, a passage on an immunoaffinity column for SPE (solid phase extraction), an elution of the mycotoxin from the column with a suitable solvent and subsequent analysis in HPLC liquid chromatography with fluorescence detection.

However, the standardized method ISO 15141 reports information on the performance of the method (repeatability and reproducibility), obtained through proficiency tests organized over the years to evaluate the performance of the method, and consequently it is possible to estimate the measurement uncertainty with the criterion called "holistic" or top down or deconstructive approach.

This type of approach uses the results of the same measurement, performed in different laboratories. The variability of the results thus obtained can be considered random and the tests obtained can be considered independent of each other. This condition is the one that comes closest to the conditions of maximum independence of repeated measures.

Considering that, in conditions of maximum independence, the uncertainty coincides with the precision obtained by the various laboratories, the significant quantity to be used as a fundamental parameter for the evaluation of the uncertainty is the reproducibility of the method.

As stated before, the expanded uncertainty is estimated, applying a coverage factor to the combined standard uncertainty to provide a level of confidence of approximately 95%. In this manner, laboratories should be 95% confident that the reported result include the true value.

6.5.4.1.1 Calculation

A prerequisite for proceeding with the measurement uncertainty estimation with the holistic method is to verify whether the laboratory is able to respect the strict repeatability reported in the standardized method. If this check is positive, the laboratory is authorized to use the reproducibility standard deviation as the combined uncertainty (or standard uncertainty).

In the method under investigation, some reproducibility values are reported as a function of some types of matrices, but still concerning cereals, and as a function of the various proficiency tests organized over the years. Therefore, it is necessary, as a preliminary step, to carefully analyze the repeatability and reproducibility data reported on the method to verify if the statistical processing has always been carried out appropriately and in compliance with the reference standards of the ISO 5725 series. In Table 6.29 these data are reported.

Table 6.29 Data declared by the method

Level	Matrices	X_{mean} (ppb)	N°	σ_r	σ_R	γ
1	Wheat naturally contaminated	2.8	42	0.1	0.26	0.38
2	Wheat naturally contaminated	6.7	42	0.28	0.79	0.35
3	Corn naturally contaminated	2.4	39	0.09	0.29	0.31
4	Corn naturally contaminated	4.1	42	0.17	0.49	0.35

Where:

X_{mean}: average concentration of ochratoxin in ppb
N°: number of test performed
σ_r: the repeatability standard deviation
σ_R: the reproducibility standard deviation
γ: is the ratio σ_r/σ_R

From the data of the table, it is possible to deduce that the proficiency tests obtained good results since the factor γ (relationship between reproducibility and repeatability) remains in the range between 2 and 3 for all the matrices considered. In fact, if the ratio deviates greatly from this range, it is recommended to discard the specific datum.

By the holistic criterion, the value of the standard deviation of reproducibility can be attributed to the result of the compound under investigation. Furthermore, the expanded uncertainty, as previously stated, can be obtained by multiplying the standard uncertainty by the coverage factor of 2.

$$Uc = S_R$$

$$Eu = 2\ Uc$$

Where:

Uc: compound uncertainty or standard uncertainty
S_R: reproducibility type deviation
Eu: extended uncertainty

Subsequently, the remaining reproducibility values are used, the coverage factor equal to two is applied and a correlation is made of the extended uncertainty thus obtained as a function of the concentration which, in most cases, is linear but can also be quadratic; this is to obtain an uncertainty function extended throughout the measurement range of the method.

Some statistical tests, for example the analysis of variance, assume that variances are equal across groups or samples. Equal variances across samples are called homogeneity of variances. The Bartlett test can be used to verify that assumption.

$$Ho = {\sigma_1}^2 = {\sigma_2}^2 = \ldots = {\sigma_k}^2$$

$$Ha = {\sigma_i}^2 \neq {\sigma_j}^2 \quad \text{for at least one pair } (i, j).$$

Table 6.30 The Barlett test

Bartlett–test		
M	X^2	Outcome
63.0495	7.8147	Negative

In the specific case, the software OQlab (www.oqlab.com) performs the calculations and the output data are shown in the following Table 6.30.

So the Barlett tes confirms the homogenity of the variances and the software calculate the best relationship between the concentration (µg/Kg) and the extended uncertainty ($U_{extended}$).

In this case the software calculate three relationships:

Linear	Y = 0.2256 X
Weighted	Y = −0.0949+ 0.2484 X
Esponential	$Y = 0.1994^* X^{1.0951}$

On the basis of the residual standard deviation of the equations, the sofware recommend the use of the linear relationship: (Non sono sicuro di questa traduzione)

$$Y = Eu = 0.2256\ C$$

Where

Eu: is the expanded uncertainty
C: concentration

A graphical reprentation can be found below:

Nota: La figura può essere fatta in inglese?

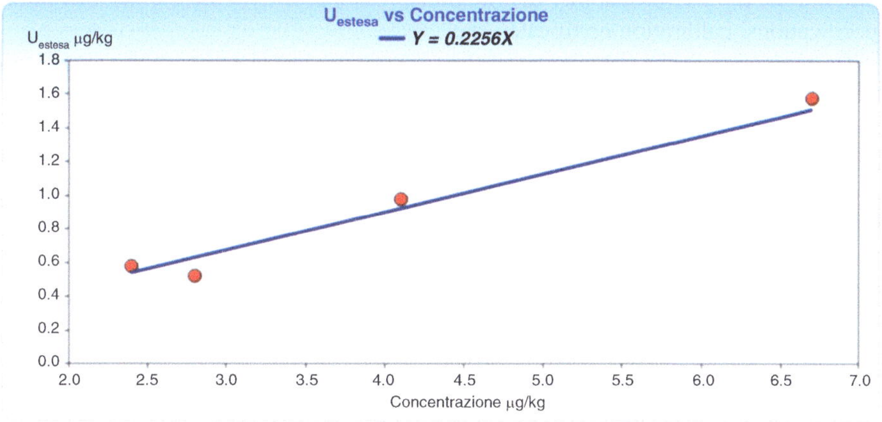

6.5.4.2 Second Example: Determination of Quassin in Food

Another example for the estimation of the measurement uncertainty concerns the determination of quassin (food flavoring component). In this example the metrological criterion (or bottom-up) has been used.

6.5.4.2.1 Introduction

The method in exam concerns the determination of quassin, a substance naturally present in flavorings and food ingredients with flavoring properties, as required by EC Reg. 1334/2008. This provision establishes specific maximum limits for each substance contained in both flavorings and finished foods. The principle of the method involves a thousand-fold dilution of the food (in this case a soft drink) and analysis by liquid chromatography (UPLC) coupled to a high-resolution mass spectrometer (Orbitrap generation). Considering that no standardized methods are available, the method was designed and developed in the laboratory. The measurement uncertainty was estimated using the metrological criterion called also bottom-up.

The methodology is the most rigorous because it considers all the various contributions to uncertainty, derived from the physical sciences, transferring the general principles from physical measurements to chemical measurements.

This approach is taken from the UNI CEI 70098-3 standard which provides for the distinction of the variability, during the execution of the analysis, two well-defined groups: category A variability and category B variability, which both contribute to the formation of uncertainty. The former is obtained through statistical analysis procedures of data obtained from repeated tests, while the latter are taken from known a priori probability distributions, reasonable hypotheses, manufacturer specifications, calibration certificates, etc.

To proceed with the calculation of the uncertainty associated with the analytical method, it is, therefore, necessary to preliminarily evaluate the complete sequence of operations which are necessary for the execution of the analysis and attribute the associated variability to each one. The approach involves dividing the entire process into distinct phases that will allow to calculate the final expression of the uncertainty. These steps include definition of the measurand, identification of the sources of uncertainty, distinction between category A and category B contributions, calculation of the value of the measurand, quantification of the components of uncertainty, including any covariance terms, calculation of sensitivity and uncertainty coefficients compound (or standard uncertainty), calculation of expanded uncertainty.

It is therefore necessary to define which are the variables that have an influence on the result (input quantities) and in what relationship they are with the measurand (output quantity). These relationships are usually derived from analytical methods or test or measurement procedures and in our case are represented in Fig. 6.35.

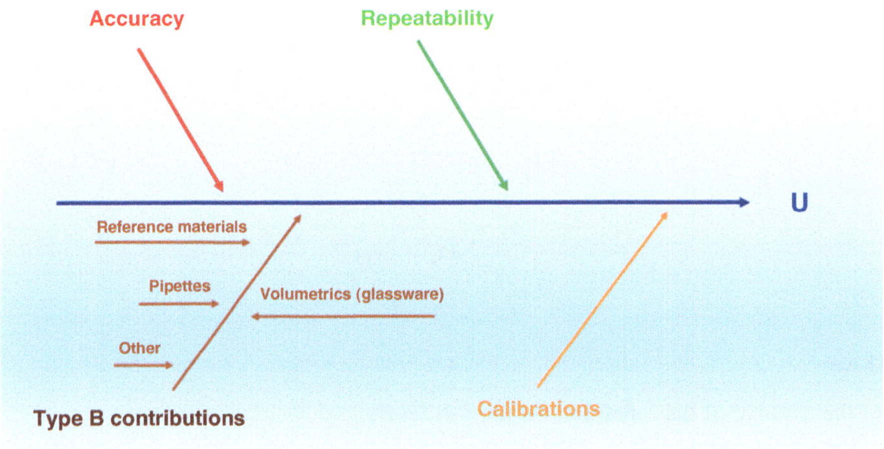

Fig. 6.35 Different contribution to the uncertainty calculation

Finally, the mathematical relationship for calculating the uncertainty (general law of error propagation) is applied, entering all the significant variances previously identified, each multiplied by the square of the respective sensitivity coefficient.

$$\dot{u}_s = \sqrt{\sum \dot{u}_i^2}$$

Where:

$\dot{u}_s$: is the compound or relative standard uncertainty.
$\dot{u}_i^2$: are the various contributions related to the uncertainty.

In the specific example under investigation:

$$\dot{u}_s = \sqrt{\dot{u}_{rip}{}^2 + \dot{u}_{tar}{}^2 + \dot{u}_{rif}{}^2 + \dot{u}_{rec}{}^2 + \dot{u}_{vol}{}^2}$$

Where:

$\dot{u}_{rip}{}^2$: relative uncertainty due to repeatability (type A);
$\dot{u}_{tar}{}^2$: relative uncertainty due to instrumental calibration (type B);
$\dot{u}_{rif}{}^2$: relative uncertainty of the reference material used for the construction of the calibration curve (type B);
$\dot{u}_{rec}{}^2$: relative uncertainty of the recovery contribution (type B);
$\dot{u}_{vol}{}^2$: relative uncertainty of glassware contribution and volume measurements (type B).

6.5.4.2.2 Repeatability Contribution

The repeatability contribution is estimated by calculating the standard deviation of the n replicates performed experimentally using the following formula:

$$\overline{x}_1 = \frac{\sum_{j=1}^{n} x_j}{n}$$

$$s_1^2 = \frac{\sum_{j=1}^{n} \left(x_j - \_x_1\right)^{\ 2}}{n-1}$$

Dove:

$\overline{x}_i$: the average of the repetitions at the i-th level;
$x_{j:}$ the single result of the n repetitions.
n: number of replicas performed to estimate repeatability.
$s_{i:}^2$ standard deviation at the i-th level.

From the repeatability study, carried out on six levels of concentration, we obtain the repeatability contribution (type A contribution) which is shown in the following table for the six levels studied (Table 6.31).

Forse è meglio inserire le formule che portono a calcolare la relative standard deviation e poi la expandedd uncertainty.

6.5.4.2.3 Contribution to Uncertainty of the Linear Regression for the Instrumental Calibration

When instrumental calibrations are made, studies are performed to determine a function that correlates the concentration with the instrumental signal. Normally linear regression is applied, which has been applied in this specific case. When this linearization is carried out there is a contribution of variability due to how the calibration line is constructed which includes how the data fit together. The algorithm for estimating this contribution is:

Table 6.31 Calculations of the expanded uncertainty at different level of concentration

Repeatability levels	Relative standard deviation %	Expanded Uncertainty %
0.101	6.97	585.60
0.103	2.74	578.02
1.746	2.85	40.45
2.051	9.36	40.70
4.313	4.27	28.12
4.623	6.43	29.36

Table 6.32 Residual standard deviations

Repeatability levels	Average (mg/Kg)	Expanded Uncertainty %
1	0.101	292.78
2	0.103	288.76
3	1.746	13.78
4	2.051	16.26
5	4.313	6.55
6	4.623	6.15

$$U_{tar} = Sy/x\sqrt{\frac{1}{N}+\frac{1}{n}+\frac{(x-\overline{x})^2}{\sum_i x_i^2-\left(\sum_i x_i\right)^2/n}}$$

Dove:

Sy/*x*: standard deviation of the residuals.
N: number of replicates used when making a real sample determination.
n: total number of replicates used for the construction of the calibration line.

From the calculations we calculate the variability shown in the table below for the 6 levels under investigation (Table 6.32):

Forse è meglio inserire le formule per il calcolo.

6.5.4.2.4 Contribution of Recovery to Uncertainty

The recovery of the analytical method also contributes to the measurement uncertainty. This is evaluated with the following formula:

$$R\%_i = \frac{\overline{x}_i}{\mu_i} \cdot 100$$

Where:

$R\%_i$: percentage recovery at the i-th level
$\overline{x}_i$: mean of the determinations at the i-th level;
μ_i: theoretical concentration of the i-th level.

From the calculations we calculate the variability for the 6 samples shown below:

Test (mg/kg)	1	2	3	4	5	6
	0.1063	0.1018	0.1041	0.0983	0.1044	0.1024
Average (mg/kg)	0.103					
Standard deviation	0.003					
Average recovery%	102.88					
Certified value mg/kg	0.100					

By using the approach reported in the NORDTEST document 537 the following relationship were used to calculate the U_{bias} (%) of the recovery:

$$bias_{i,rel} = \frac{x_i - x_{RM}}{x_{RM}}$$

$$RMS_{bias,rel} = \sqrt{\frac{\sum (bias_{i,rel})^2}{n}}$$

$$\mathbf{U_{bias}(\%) = 3.82}$$

Ho fatto I calcoli a mano ma non ho ottenuto 3.82. Sicuramente ho sbagliato. Controllare.

6.5.4.2.5 Volumetric Contributions to Uncertainty

When taking volume samples and dilutions, errors are made which can be obtained from the manufacturer's data. These variables must also be estimated and considered in the calculation of the overall uncertainty. For the estimation of the volumetric variability a rectangular distribution was considered. The software is able to calculate all the contributions of pipettes, precision syringe and glassware utilized in every single operation of the method. Foreseen by the method. The rectangular interval was used. Data are listed in the following Table 6.33.

6.5.4.2.6 Contribution of Reference Material

The uncertainty of the reference material is also a source of error and therefore of uncertainty. This is obtained from the certificate of the reference material. In our case the CRM quassin solution 100 ± 0.0175 μg/mL was used.

Table 6.33 Contribution of the glassware used in the method

Instruments	Operations	Contribution %
Pipette	Solvent addition work std. 10 μg/L	2.5 E-2
Pipette	Addition of solvent curve level 0.5 mg/kg	2.5 E-2
Pipette	Addition of solvent curve level 5 mg/kg	2.5 E-2
Pipette	Addition of solvent dahugter solution 1 mg/L	2.5 E-2
Pipette	Solvent addition work std. 0.1 μg/kg	2.5 E-2
Pipette	Solvent addition work std. 2.5 μg/kg	2.5 E-2
Pipette	Real sample taking	2.5 E-2
Pipette	Solvent addition work std. 100 μg/L	2.5 E-2
Pipette	Addition of solvent curve level 1 mg/kg	2.5 E-2
Pipette	Addition of solvent curve level 10 mg/kg	2.5 E-2
Precision syringe	Solvent addition work std. 10 μg/kg	2.5 E-2
Precision syringe	Sampling from work solution 100 μg/L for preparation of curve level 10 mg/kg	2.5 E-2
Precision syringe	Sampling from daughter solution for preparation work standard 100 μg/L	2.5 E-2
Precision syringe	Sampling from work solution 10 μg/L for preparation of curve level 0.1 mg/kg	2.5 E-2
Precision syringe	Sampling from work solution 100 μg/L for preparation of curve level 5 mg/kg	2.5 E-2
Precision syringe	Sampling from daughter solution for preparation work standard 10 μg/L	2.5 E-2
Precision syringe	Sampling from stock solution for preparation daughter 1 mg/L	2.5 E-2
Precision syringe	Sampling from work solution 100 μg/L for preparation of curve level 1 mg/kg	2.5 E-2
Precision syringe	Sampling from work solution 100 μg/L for preparation of curve level 2.5 mg/kg	2.5 E-2
Precision syringe	Sampling from work solution 10 μg/L for preparation of curve level 0.5 mg/kg	2.5 E-2
Glassware	Actual sample dilution	4.08 E-4

6.5.4.2.7 Calculation of the Compound Uncertainty or Standard Uncertainty

The compounded uncertainty, or standard uncertainty, is estimated by compounding all previously estimated relative contributions to the squared sub-square root. This operation is performed for the 6 estimated concentration levels.

The OQLab software provides the final result considering the various uncertainty components shown in the following table.

1	2	3	4	5	6	7	8	9
Repeatability level (mg/kg)	Repeatability U (%)	Replicates	Calibrat. U (%)	Accuracy U (%)	Replicates	Type B contribution U(%)	Standard U (%)	Degre of freedc
0.101	3.12	5	292.78	3.82	6	11.04	292.80	4
0.103	1.12	7	288.76	3.82	6		289.35	4
2.051	3.54	6	13.78	3.82	6		20.35	18
1.746	1.08	7	16.26	3.82	6		20.22	9
4.313	1.51	8	6.55	3.82	6		14.06	77
4.623	2.27	8	6.15	3.82	6		16.68	115

La tabella è ben fatta e completa ma ho tolto qualche colonna. Bisogna inserire, a mio avviso la formula per passare dalla standard U% alla extended uncertainty in mg/kg.

The table reports the following rows:

1. Levels of the tested concentration,
2. U% percentage repeatability contribution
3. Number of replicates carried out for the repeatability study
4. Calibration contribution due to the instrument calibration line
5. Contribution of the deviation or of the BIAS or of the recovery
6. Number of replicates used for the recovery estimate
7. B contributions for the volumetric and for the reference material uncertainties
8. Standard uncertainty
9. Number of degrees of freedom for the estimation of the factor k
10. Coverage factor used for the calculation of the extended uncertainty
11. Final extended uncertainty expressed in mg/kg

From the previous table at low aroma concentrations the significant contributions are the instrumental calibration and, as the concentration increases, the other contributions tend to rise.

6.5.4.2.8 Calculation of Expanded Uncertainty

Using the values of the combined uncertainties estimated for the 6 concentration levels and applying a coverage factor equal to two, a correlation of the expanded uncertainty as a function of the concentration is obtained which in most cases is linear but can also be quadratic; this allows to obtain an uncertainty function extended throughout the measurement range of the method.

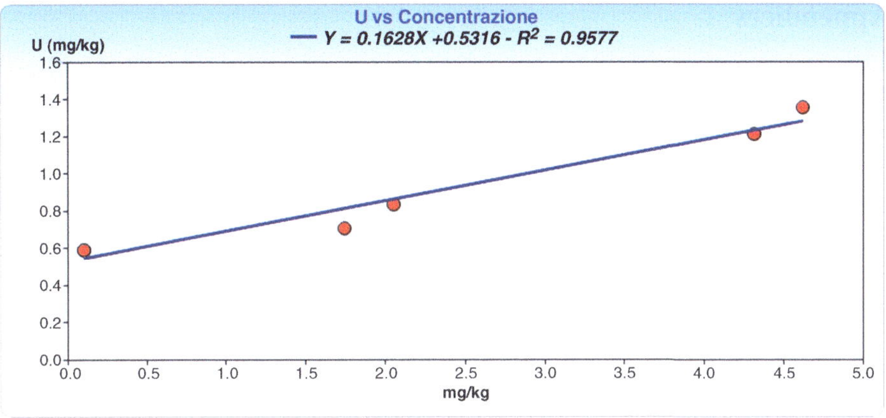

$$U_e = 0,1628C + 0,5316$$

From the graph we can say that we have obtained a function with a correlation coefficient close enough to one, therefore it is possible to use the function to represent the extended uncertainty of the analytical method over the entire measurement range.

6.5.5 Conclusions

Uncertainty of a measurement refers to the doubt, which exists for the result of any measurement within the laboratory . There are a number of factors which must be considered when calculating uncertainty, including the chosen method, bias, analytical errors and so on. If uncertainty is quantified it is no longer uncertainty but the confidence interval within which the results fall. Uncertainty should be assessed regularly, and attempts made to improve the value.

Appendices

Table 6.34 Dixont's Q-test. Critical values distribution for probability 90%, 95% and 99%

n	90%	95%	99%
3	0.886	0.941	0.988
4	0.679	0.765	0.889
5	0.557	0.642	0.780
6	0.482	0.560	0.698
7	0.434	0.507	0.637
8	0.399	0.468	0.590
9	0.370	0.437	0.555
10	0.349	0.412	0.527

Table 6.35 Student's t-test. Critical values distribution for confidence level $\alpha = 0.05$ and $\alpha = 0.01$ and for different degrees of freedom

df	$\alpha = 0.05$	$\alpha = 0.01$	df	$\alpha = 0.05$	$\alpha = 0.01$
1	12.71	63.66	11	2.20	3.11
2	4.30	9.92	12	2.18	3.05
3	3.18	5.84	13	2.16	3.01
4	2.78	4.60	14	2.14	2.98
5	2.57	4.03	15	2.13	2.95
6	2.45	3.71	20	2.09	2.85
7	2.36	3.50	40	2.02	2.70
8	2.31	3.36	60	2.00	2.66
9	2.26	3.25	100	1.98	2.63
10	2.23	3.17	∞	1.96	2.58

ASEW

Table 6.36 Overview on the results of the interlaboratory trial

	Samples						
	Internat.	Europe			Asia		
Parameter	K, L, N	A, B	C, D	E, F	A, B	C, D	E, F
Participants	23	17	17	17	6	6	6
Samples	3	2	2	2	2	2	2
Replicants	4	2	2	2	2	2	2
Examinations/ participant	12	4	4	4	4	4	4
Obtained results	276	68	64	68	24	24	24
Mean value (log 10 CFU/g)	7.344	6.337	7.813	6.723	7.262	7.674	7.824
Standard deviation (SD)	0.076	0.021	0.045	0.054	0.03	0.04	0.046

(continued)

Table 6.36 (continued)

Parameter	Samples						
	Internat.	Europe			Asia		
	K, L, N	A, B	C, D	E, F	A, B	C, D	E, F
Precision-parameter based on "first values"	**Infant Milk K, L, N**	**Yogurt 1 A, B**	**Yogurt 2 C, D**	**Yogurt 3 E, F**	**Yogurt 4 A, B**	**Yogurt 5 C, D**	**Yogurt 6 E, F**
Used datasets	23	17	16	17	6	6	6
Units/product examined once	3	2	2	2	2	2	2
Results used to estimate *r*, *R*	69	34	32	34	12	12	12
Repeatability *SD*, s_r robust[a]	0.079	0.041	0.065	0.044	0.042	0.194	0.076
Repeatability *r* robust[a]	0.221	0.115	0.182	0.123	0.118	0.543	0.213
Reproducibility *SD*, s_R robust	0.189	0.081	0.139	0.192	0.143	0.194	0.104
Reproducibility *R* robust	0.529	0.227	0.389	0.538	0.400	0.543	0.291
Precision-parameter based on "all values"	**Infant Milk K, L, N**	**Yogurt 1 A, B**	**Yogurt 2 C, D**	**Yogurt 3 E, F**	**Yogurt 4 A, B**	**Yogurt 5 C, D**	**Yogurt 6 E, F**
Used datasets	23	17	16	17	6	6	6
Units/product examined twice	3*2[b]	2	2	2	2	2	2
Results used to estimate *r*, *R*	276	68	64	68	24	24	24
Repeatability *SD*, s_r robust	0.074	0.039	0.057	0.064	0.033	0.150	0.072
Repeatability *r* robust	0.207	0.108	0.160	0.179	0.092	0.420	0.202
Reproducibility *SD*, s_R robust	0.197	0.070	0.143	0.180	0.120	0.154	0.117
Reproducibility *R* robust	0.552	0.195	0.401	0.505	0.336	0.432	0.327
Infant milk formalae	Commercial probiotic infant milk product containing *B.animalis* (BB12®)						
Yogurt 1	Comm. Europ. Probiotic yogurt product cont. *B.animalis* (BB12®), *L. acidophilus*. (LA5®), *S. thermophilus*						
Yogurt 2	Comm. Europ. Probiotic yogurt product cont. *B. animalis (Dig. essensis cultur)*; *L. bulgaricus*, *S. thermophilus*						
Yogurt 3	Comm. Europ. Probiotic yogurt product cont. *B.animalis*, *L.casei* (*belactiva 3*), *S. thermophilus*						
Yogurt 4	Comm. Asian probiotic yogurt product cont. *B.breve* (*SBR3212*), *L.casei* (*AST-8*), *S. thermophilus*						

(continued)

Table 6.36 (continued)

	Samples						
	Internat.	Europe			Asia		
Parameter	K, L, N	A, B	C, D	E, F	A, B	C, D	E, F
Yogurt 5	Comm. Asian probiotic yogurt product cont. *B.longum* (*SBT2928*), *L.gasseri*, *L. bulg.*, *S. thermophilus*						
Yogurt 6	Comm. Asian probiotic yogurt product cont. *B.animalis* (*BB12*®), *L.acidophilus*, *S. thermophilus*						

[a]Based on robust analyses of all "first values" as well as "all values" obtained within the ringtrial, none of the datasets had to be excluded. Furthermore thus the precision-parameter reflecting real conditions whilst usual application
[b]Two sub-samples were prepared per sample unit, each of them was examined twice

References

Altman DG, Bland JM (1983) Measurements in medicine: the analysis of method comparison studies. The Statistican 32:307–318

Asperger A, Efer J, Koal T, Engewald W (2001) On the signal response of various pesticides in electrospray and atmospheric pressure chemical ionization depending on the flow-rate of eluent applied in liquid chromatography-tandem mass spectrometry. J Chromatogr A 937:65–72

Bandion F, Arangnitz-Rebolledo E (2000) Analysen authentischer Weine des Jahrgangs 1999. Mitt Klosterneuburg 50(4–5):144–176

Bandion F, Valenta M (1980) Analysen authentischer Weine des Jahrgangs 1979. Mitt Klosterneuburg 30:120–121

Barna J, Grill F (1980) Ergebnisse von Untersuchungen über die Eignung einer modifizierten Molybdänblau-Methode zur Bestimmung des Phosphatgehaltes von Säften und Weinen. Mitt Klosterneuburg 30:117–119

Bland JM, Altman DG (1986) Statistical methods for assessing agreement between two methods of clinical measurements. Lancet 8(2):1986

Box GEP, Wilson KB (1951) On the experimental attainment of optimum conditions. J R Stat Soc Ser B 13(1):1–45

Burgess C (2000) Valid Analytical Methods & Procedures. Royal Society of Chemistry, UK

DIN 12650-6 Mechanische, physikalische und elektrische Laborgeräte—Volumenmessgeräte mit Hubkolben—Gravimetrische Prüfung der messtechnischen Zuverlässigkeit

DIN 32633 Verfahren der Standardaddition

DIN 32645 Nachweis- Erfassungs- und Bestimmungsgrenzen

DIN 38402 Deutsche Einheitsverfahren zur Wasser-, Abwasser- und Schlammuntersuchung

Draper NR, Smith H (1981) Applied regression analysis. Wiley, New York

EC 1924/2006 "Health-chlaims" Regulation of the European Parliament and of the Council of 20 December 2006 on nutrition and health claims made on foods

EURACHEM Guide (1998) The Fitness for Purpose of Analytical Methods. A Laboratory Guide to Method Validation and Related Topics, Firste Internet Version

EURACHEM/CITAG-Guide (2000) "Quantifying uncertainty in Analytical Measurements", second edition (ISBN 0 948926 15 5)

Expert Group on Selective Enumeration of Bifidobacteria. (2003) Method of Enumeration of Bifidobacteria in Fermented Milks and Fermented Milk Drinks. Japanese Association of Fermented Milks and Fermented Milk Drinks

FAO/WHO Experts'Report (2001) Health and nutritional properties of probiotics in food including powder milk with live lactic acid bacteria
FAO/WHO-CODEX Alimentarius: 2003, Codex Standard No. 243
Farmiloe FJ, Cornford SJ, Coppock JPM, Ingram M (1954) The survival of *Bacillus subtilis* spores in the baking of bread. J Sci Food Agric 5:292–304
Funk W, Dammann V, Donnevert G (1995) Quality Assurance in Analytical Chemistry. Wiley-VCH, Weinheim
Funk W, Dammann V, Donnevert G (2005) Qualitätssicherung in der Analytischen Chemie. Wiley-VCH, Weinheim
Garfield FM, Klesta E, Hirsch J (2000) Quality assurance principles for analytical laboratories. AOAC International, Gaithersburg
Hiemstra M, Kok A (2007) Comprehensive multi-residue method for the target analysis of pesticides in crops using liquid chromatography-tandem mass spectrometry. J Chromatogr A 1154:3–25
Hoaglin DC, Mosteller F, Tukey JW (1983) Understanding robust and exploratory data analysis. Wiley, New York
IDF Bulletin No. 352. (2000) Culture-Containing Dairy Products in Human Nutrition and Health
IDF Bulletin No. 380. (2003) Health Effects of Probiotics and Culture-Containing Dairy Products in Humans
IDF Bulletin No. 395. (2005) Method Standardization in the Third Millenium
IDF Int. Standard 135B: 1991 Milk and Milk Products—Precision characteristics of analytical methods—Outline of collaborative study procedure
IDF Int. Standard 161A: 1995 "Milk—Quantitative Determination of Bacteriological Quality—Guidance on Evaluation of Routine Methods"
IDF-"Bifido"-Guideline (1999) (IDF Group E104) "lactic acid bacteria and starters. Guideline for the enumeration of Bifidobacteria in fermented dairy products." Bull Int Dairy Fed 340:19–23
International Standard ISO 5725-1: 1994 (cor 1: 1998) Accuracy (trueness and precision) of measurement methods and results—Part 1- General principles and definitions
International Standard ISO 8466-1: 1990 Water quality—Calibration and evaluation of analytical methods and estimation of performance characteristics—Part 1: Statistical evaluation of the linear calibration function
International Standard ISO 8466-2: 2001 Water quality—Calibration and evaluation of analytical methods and estimation of performance characteristics—Part 2: Calibration strategy for non-linear second-order calibration functions
International Standard ISO 8655-6: 2002 Piston-operated volumetric apparatus—Part 6: Gravimetric methods for the determination of measurement error
International Standard ISO 11095: 1996 "Linear calibration using reference materials"
International Standard ISO 17025: 2005 (cor 1: 2006) General requirements for the competence of testing and calibration labors
International Standard ISO 5725-2: "Accuracy (trueness and precision) of measurement methods and results—Part 2: Basic method for the determination of repeatability and reproducibility of a standard measurement method"
International Standard ISO 14461-1/IDF 169-1 "Milk and Milk Products—Quality control in microbiological laboratories—Part 1: Analyst performance assessment for colony counts"
International Standard ISO 14461-2/IDF 169-2 "Milk and Milk Products—Quality control in microbiological laboratories—Part 2: Determination of the reliability of colony counts of parallel plates and subsequent dilution steps"
International Standard ISO 16140 "Microbiology of food and animal feeding stuffs—Protocol for the validation of alternative methods" and Amendments
International Standard ISO 5725-1: "Accuracy (trueness and precision) of measurement methods and results—Part 1- General principles and definitions"
International Standard ISO/FIS IDF/IDF 220 Standard Final-Draft-Version Mai 2009. "Milk products—Enumeration of presumptive bifidobacteria—Colony count technique at 37°C"

ISO 5725-1-6 (2002) Accuracy (trueness and precision) of measurement methods and results. Part s1–6
International Standard ISO IDF/IDF 220 Standard Version February 2010. "Milk products - Enumeration of presumptive bifidobacteria—Colony count technique at 37 °C"
ISO/IEC 17025 (2001) General requirements for the competence of testing and calibration laboratories
ISO/IEC Guide 43-1 (1997) Proficiency testing by interlaboratory comparisons—Part 1: Development and operation of proficiency testing schemes
ISO/IEC Guide 98-1: 2008 Uncertainty of measurement—Part 1: Introduction to the expression of uncertainty in measurement
ISO/IEC Guide 98-3:2008 Uncertainty of measurement—Part 3: Guide to the expression of uncertainty in measurement (GUM:1995)
ISO/IEC Guide 99: 2007 International vocabulary of metrology—Basic and general concepts and associated terms (VIM)
IUPAC (1997) Compendium of Chemical Terminology—The Gold Book. http://goldbook.iupac.org/
Kromidas S (2000) Handbuch Validierung in der Analytik. Wiley-VCH, Weinheim
Leclercq A, Lombard B, Mossel DA (2000) Standardization of food microbiological analyses methods. Science-des-Aliments 20(2):179–202
Lightfoot NF, Maier EA (2002) Mikrobiologische Analysen: Richtlinien zur Qualitätssicherung. Lebensmittel- und Wasseruntersuchungen. Springer-Verlag
Neter J, Kutner MH, Nachstheim CJ, Wasserman W (1996) Applied linear statistical models. McGraw-Hill, USA
Niemelä S (1965) Quantitative estimation of bacterial colonies on membrane filters. Ann Acad Sci Fenn Ser A VI Biologica:1–65
NORDTEST REPORT TR 537: Handbook for calculation of measurement uncertainty in environmental laboratories
Pacher B, Kneifel W (1996) Development of a culture medium for the detection and enumeration of Bifidobacteria in fermented Milk products. Int Dairy Journal 6:43–64
Polkowska-Mortenko H, Dybczyński R, Chajduk E, Dank B, Kulisa K, Samczyński Z, Sypuła M, Szopa Z (2006) Polish Reference Material Soya Bean Flour (INCT-SBF-4) for Inorganic Trace Analysis—Preparation and Certification. Institute of Nuclear Chemistry and Technology, Warszawa
Prichard E (1995) Quality in the analytical chemistry laboratory. Wiley, New York
Rada V, Koc J (2000) The use of mupirocin for selective enumeration of bifidobacteria in fermented milk products. Milchwissenschaft 55:65–67
Rösslein M (1999) Messunsicherheit in der Analytischen Chemie, Eine Einführung, 3.Auflage, EMPA St. Gallen
Rösslein M, Wolf M, Wampfler B, Wegscheider W (2007) A forgotten fact about the standard deviation. Accred Qual Assur 12:495–496
Scardovi V (1986) "Genus *Bifidobacterium*" Orla-Jensen 1924, 472AL. In: Sneath PHA, Mair NS, Sharpe ME, Holt JG (eds) Bergey's manual of systematic bacteriology, vol 2. The Williams & Wilkings Co., Baltimore, pp 1418–1434
Softwar: OQLAB S.r.l. 70132 Bari (Italy) website www.oqlab.com; email info.oqlab@gmail.com
Statgraphics Centurion XVI User Manual (2010) StatPoint Technologies, Inc., available at www.statgraphics.com
Takino M, Tanaka T (2008) Determination of 44 pesticides in Foodstuffs by LC/MS/MS Application Note, www.agilent.com/chem
Tanaka R, Takayama H, Morotomi M, Kuroshima T, Ueyama S, Matsumoto K, Kuroda A, Mutai M (1983) Effects of administration of TOS and *Bifidobacterium breve* 4006 on the human fecal Flora. Bifidobacterial Microflora 2:17–24
Thompson M, Ellison SLR, Wood R (2002) Harmonized guidelines for single-laboratory validation of methods of analysis (IUPAC technical Report). Pure Appl Chem 74(5):835–855

Wegscheider W (1996) Validation of analytical methods. In: Günzler H (ed) Accreditation and quality Assurance in Analytical Chemistry. Springer, Berlin, pp 135–158

Wegscheider W (2000) Grundlagen der Validierung. In: Wegscheider W (ed) QS-Leitfaden, Vorlesungsunterlagen, pp 1–49

Weiss H (2005) Harmonization—horizontal approach in collaborative studies. Bulletin of IDF 395

Weiss H, Dahms S (1993) Statistically based analyst performance assessment for microbiological analysis. In Analytical quality assurance and good laboratory practice in dairy laboratories, IDF Special Issue 9302, p 304–322

Wilrich P-T (2007) Robust estimates of the theoretical standard deviation to be used in interlaboratory precision experiments. Accred Qual Assur 12:231–240

Wilrich P-T, Jarvis B (2008) Proposal for an amendment of chapter 6.3 of ISO 16140—Interlaboratory Study

Youden WJ, Steiner EH (1975) Statistical manual of the AOAC. Association of Analytical Chemists

Yugui T, Yaoming W, Shilei Y, Lianbin Y (2008) Optimization of omethoate degradation conditions and a kinetics model. International Biodeterioration&Biodegradation 62:239–243

Zitz U (2007) Strategie der Standardisierung einer Analysenmethode zur Bestimmung der Bifidobakterienkeimzahl in Milchprodukten sowie deren qualitätsorientierte Beurteilung von Labormessdaten. Dissertation an der Universität für Bodenkultur, Institut für Lebensmittelqualitätssicherung

Zitz U (2008) Kulturelle Bestimmung der Bifidobakterienkeimzahl in Milchprodukten—*Ein Leitfaden zur Standardisierung quantitativer mikrobiologischer Methoden.* Monographie, SVH—Südwestdeutscher Verlag für Hochschulschriften AG & CoKG, 242 Seiten, ISBN 978-3-8381-0034-0

Zitz U, Kneifel W, Weiss H, Wilrich P-T (2007) Selective enumeration of bifidobacteria in dairy products: development of a standard method. Bulletin of the IDF 411:3–20

Zollondz H-D (2001) Lexikon Qualitätsmanagement. Handbuch des Modernen Managements auf der Basis des Qualitätsmanagements. R. Oldenburg Verlag München Wien

Index

G. Schleining et al. (eds.), *Statistics in Food and Biotechnology*, Integrating Food Science and Engineering Knowledge Into the Food Chain 16,
https://doi.org/10.1007/978-3-031-51568-2

SPRINGER NATURE

GPSR Compliance

The European Union's (EU) General Product Safety Regulation (GPSR) is a set of rules that requires consumer products to be safe and our obligations to ensure this.

If you have any concerns about our products, you can contact us on ProductSafety@springernature.com

In case Publisher is established outside the EU, the EU authorized representative is:

Springer Nature Customer Service Center GmbH
Europaplatz 3
69115 Heidelberg, Germany

The manufacturer's authorised representative in the EU is Springer
Nature Customer Service Centre GmbH, Europaplatz 3, 69115 Heidelberg,
Germany. If you have any concerns regarding our products, please
contact ProductSafety@springernature.com

Printed and bound by CPI Group (UK) Ltd, Croydon, CR0 4YY
01/10/2026
02225915-0015